Introduction to DIFFERENTIAL EQUATIONS

ODE, PDE, and Series

Introduction to DIFFERENTIAL EQUATIONS

ODE, PDE, and Series

RICHARD E. WILLIAMSON
Dartmouth College

PRENTICE-HALL, Englewood Cliffs, New Jersey 07632

Library of Congress Cataloging in Publication Data

WILLIAMSON, RICHARD E.
    Introduction to differential equations.

    Includes index.
    1. Differential equations I. Title.
QA371.W55 1986   515.3′5   85–6550
ISBN 0-13-480989-0

Printed in the United States of America

10 9 8 7 6 5 4 3 2 1

Editorial/production supervision and interior design: Paul Spencer
Cover design: Wanda Lubelska
Manufacturing buyer: John Hall

The cover illustration depicts the solution trajectory of the Lotka-Volterra
system

$$\dot{x} = x(3 - y)$$
$$\dot{y} = y(x + z - 3)$$
$$\dot{z} = z(2 - y),$$

for three species, with initial values $x(0) = .001$, $y(0) = 3$, $z(0) = 2$.

ISBN 0-13-480989-0

Prentice-Hall International (UK) Limited, *London*
Prentice-Hall of Australia Pty. Limited, *Sydney*
Prentice-Hall Canada Inc., *Toronto*
Prentice-Hall Hispanoamericana, S.A., *Mexico*
Prentice-Hall of India Private Limited, *New Dehli*
Prentice-Hall of Japan, Inc., *Tokyo*
Prentice-Hall of Southeast Asia Pte. Ltd., *Singapore*
Editora Prentice-Hall do Brasil, Ltda., *Rio de Janeiro*
Whitehall Books Limited, *Wellington, New Zealand*

Contents

vi   Contents

2 HIGHER-ORDER EQUATIONS   55

3 LAPLACE TRANSFORMS   109

4 FIRST-ORDER SYSTEMS   130

5   LINEAR SYSTEMS   188

Preface

We can trace the roots of most branches of mathematics to problems in geometry, in number theory, or in applied mathematics. For the study of differential equations, these roots lie primarily in geometry and in the application of mathematics to the physical world. Contact with these sources of inspiration has continued right up to the present and has been responsible for most of the subject's vitality ever since the time of Newton. This book depends very much on applied mathematics for motivation, but it is after all a mathematics text, so the fundamental mathematical ideas, including their geometric interpretation, appear before the applications. Encouraging a geometric point of view toward the subject is particularly important, because finding a neat solution formula is a relatively rare, though welcome, occurrence in practice, and it is very often the geometric properties of solutions that unify our view of them.

Students who use the book should have had the standard introduction to calculus in one dimension and to the elements of parametrized curves in two and three dimensions. It's necessary to know what a partial derivative is only for Chapters 7 and 8. The tradition of using an introduction to differential equations as an occasion for drill on integration technique has been somewhat slighted in this book by avoidance of exercises that are particularly demanding in that way; this is a personal preference, based on the idea that there are more important matters to emphasize in an introductory course. However, there are many routine exercises of the simpler sort. The essential ideas from linear algebra are included in the text, but the treatment is very much geared to differential equations, and the algebra topics are not treated in any greater generality than necessary.

The inclusion of numerical techniques is intended partly to take advantage of the availability of computers with graphics capability for displaying the geometry of solutions. Graphs of solutions are often difficult or impossible to draw using only the traditional methods of calculus, but simple numerical and graphical analysis can give very satisfying results, results that can be particularly satisfying in the case of non-linear equations. There is no attempt here to discuss ways of making highly accurate approximations; accuracy enough to produce good pictures has been the primary aim, though the results are usually better than they need to be for that purpose. Furthermore, numerical approximations appear along with each type of equation rather than being lumped into a single chapter. In this way a geometric feeling for the subject can be maintained throughout the book without restriction to examples that happen to be analytically simple. Nevertheless, the book is designed so that anyone who chooses to ignore the numerical work completely will still have a coherent text.

Many students come to a first course in differential equations having had some experience with linear equations of order one and with second-order constant coefficient equations. In spite of this, it is not feasible to skip that material altogether in an introductory differential equations course, so there is the question of what new things to do at this stage. One response is to take up more applications, which is done in this book. An additional one is to introduce Wronskians and order reduction for second-order linear equations. Wronskians appear later in this book, after students have had a brief look at linear independence in at least two slightly different contexts. What is done instead is to develop the Green's functions for the second-order constant coefficient equations. The resulting formulas are easy to apply and even easy to remember. Differential operator notation is very useful here, and its introduction at this point is a help later when it comes to using it to solve simple linear systems. Using operators fairly systematically allays the suspicion some students get that, while manipulating operators is convenient, there is something slightly lowbrow about it.

Chapter 5 on general techniques for constant coefficient linear systems begins with the essential matrix algebra. The method described for computing e^{tA} is quite efficient whether it is carried out by hand or by symbolic calculator, its main limitation being that it requires a prior computation of the eigenvalues of A. Additional results on matrix algebra are collected at the end of the chapter.

The Picard existence theorem appears very late in the book, though questions about existence and uniqueness are raised, and answered concretely, when they come up naturally at earlier points.

The chapter on partial differential equations is fairly traditional except for the use, due to Euler, of exponential solutions to motivate the classification of second-order linear operators. The chapter on infinite series will be viewed by many people as a sort of appendix, so it is placed at the end of the book to be used at an instructor's discretion; it is designed to be used either for review or as an introduction to the topic for students who have not studied it systematically. An introduction to power series solutions of differential equations is at the end of the series chapter.

The routine exercises are intended to be used for technical drill and to underline general principles. Other exercises are there to broaden a student's perspective. An occasional exercise is starred to indicate that it is more difficult than most of the others. All of Chapter 7 and Section 4 of Chapter 9 are significantly more theoretical than the rest of the book.

While I was preparing the manuscript I had very valuable help at Dartmouth from Joe Bonin, Peter Bushell, Bob Christy, Nancy French, Peter Holland, Paul Latiolais, and Randy Weis, and I am very grateful to them. My thanks go also to Bob Sickles of Prentice-Hall for his sensitive approach to the entire project and to Paul Spencer for his careful work in seeing the book through production. Most of the book has been used in a preliminary version at Dartmouth and I want to thank the students who used it; their helpful comments have worked their way into the book at many points.

Richard E. Williamson

Introduction to DIFFERENTIAL EQUATIONS

ODE, PDE, and Series

1

First-Order Differential Equations

1 SOLUTIONS

1A General Formulas

In many applications of mathematics it is possible to establish an equation relating an unknown function $y = y(x)$ to one or more of its derivatives. Simple examples are

$$y'(x) + y(x) = x \quad \text{and} \quad y''(x) - y(x) = 0,$$

or alternatively,

$$\frac{dy}{dx} + y = x \quad \text{and} \quad \frac{d^2y}{dx^2} - y = 0,$$

where the equation is to hold for x in some interval $a < x < b$, perhaps $-\infty < x < \infty$. Such an equation is called a **differential equation.** In this chapter we will deal primarily with **first-order** differential equations, that is, those in which there occur derivatives of order one at most. Of the two previous equations, only the first is of first order; the other has order 2.

Equations are often meant to be solved, and differential equations are no exception. A **solution** of a differential equation is a function $y = y(x)$, which, when substituted along with its derivatives into the differential equation, satisfies the equation for all x in some specified interval that we will call the **domain** of the solution. In practice, there are sometimes elementary formulas for solutions, although by no means always. It is important to bear in mind that, while a strictly algebraic equation like $x^2 + 3x + 2 = 0$ has numbers for solutions, a differential equation has functions for solutions. In the following examples we will simply verify that certain formulas provide solutions; derivation of the solutions will be discussed later.

EXAMPLE 1   The differential equation

$$y' + y = x$$

has the solution $y = x - 1$ for $-\infty < x < \infty$. The reason is that $y' = 1$, so

$$y' + y = 1 + (x - 1) = x.$$

More generally, the same differential equation has the solution

$$y = x - 1 + ce^{-x}$$

for each choice of the constant c, for then

$$y' + y = (1 - ce^{-x}) + (x - 1 + ce^{-x}) = x.$$

Thus there are actually infinitely many distinct solutions, and the solution $y = x - 1$ can be obtained from the general formula above by setting $c = 0$. Figure 1(a) shows the graphs of the solutions corresponding to $c = 0, c = 0.15$, and $c = 0.25$. Notice that these graphs converge together since the term ce^{-x}, by which they differ, tends to 0 as $x \to +\infty$.

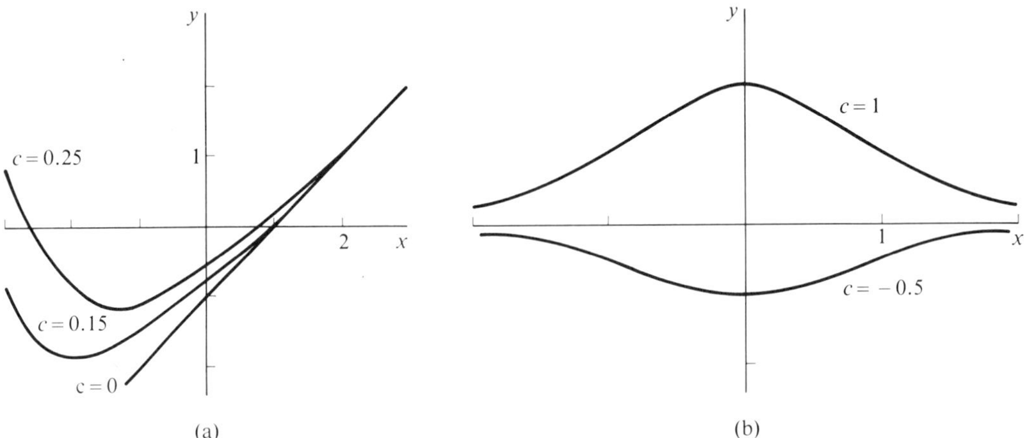

Figure 1

EXAMPLE 2   The differential equation

$$y' + xy = 0$$

has the solutions $y = ce^{-x^2/2}$, one for each value of c, since

$$y' + xy = (-cxe^{-x^2/2}) + x(ce^{-x^2/2}) = 0.$$

Figure 1(b) shows the graphs of the solutions corresponding to $c = -0.5$ and $c = 1$. For $c = 0$, we get $y = 0$ for $-\infty < x < \infty$, and the graph of this solution coincides with the x-axis.

EXAMPLE 3   Suppose $f(x)$ defines a continuous function on some interval $a \leq x \leq b$. The differential equation

$$y' = f(x)$$

then has the solution

$$y = \int_a^x f(t)\, dt + c$$

for each choice of the constant c. In this example, the role of the constant c distinguishes the various solutions from one another by making a constant vertical shift in the graph. Figure 2(a) shows some examples for the case in

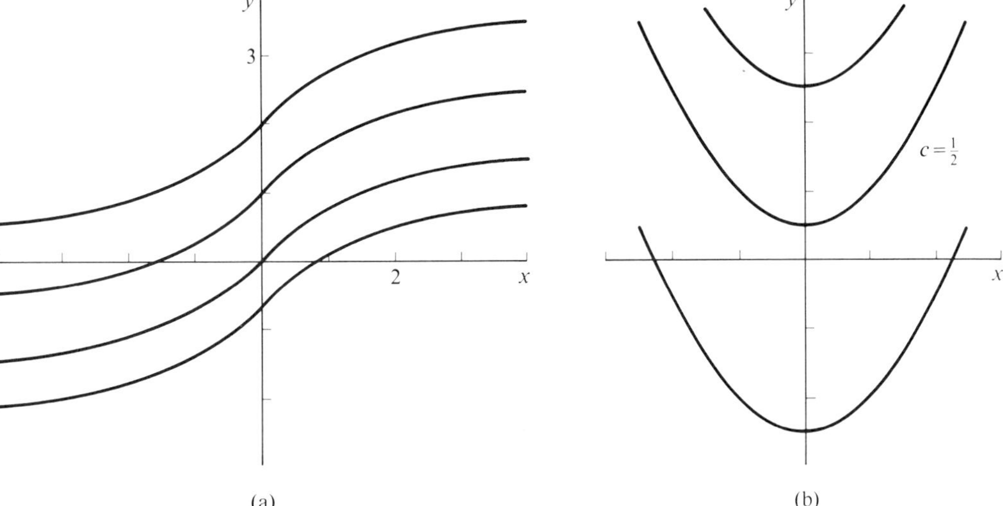

Figure 2

which $f(x) = 1/(1 + x^2)$. Choosing $a = 0$, we have

$$y = \int_0^x \frac{dt}{1 + t^2} + c$$

$$= \arctan x + c.$$

It might be that $f(x)$ was some complicated function that did not have an indefinite integral in terms of an elementary formula. We could still find an approximation to the values of a solution by numerical integration over successive intervals of the form $a \le t \le x$. Quite good results can be obtained this way using Simpson's rule, for example. While this may sound like a lot of work, recall that studying and computing a function like arctan x for the first time is also a lot of work. On the other hand, arctan x is "well known" and has many nice properties, so a solution that can be expressed in terms of it is often to be preferred over a strictly numerical computation.

1B Initial Values

A typical first-order differential equation has infinitely many solutions on some interval, and each of these solutions has a graph passing through infinitely many points. To distinguish one solution from another in a convenient way, we can pick a point x_0 in the common domain of several solutions. Then let $y_0 = y(x_0)$, where $y(x)$ is some particular solution, and label that solution with the pair (x_0, y_0). Geometrically, what we have proposed is to pick a point (x_0, y_0) on the given solution curve and then use that point to specify the solution. The choice (x_0, y_0) is called an **initial condition** and is usually expressed in the form $y(x_0) = y_0$. The question as to when such a condition specifies one and only one solution is taken up in Chapter 7; in the meantime we will concentrate on examples for which the choice of a point determines a unique solution curve containing the point.

EXAMPLE 4   The very simple differential equation

$$y' = x$$

has solutions

$$y = \int_0^x t \, dt + c$$

$$= \tfrac{1}{2}x^2 + c.$$

To pass through the point $(x_0, y_0) = (-1, 1)$, the solution must satisfy

$$1 = \tfrac{1}{2}(-1)^2 + c.$$

Then $c = \tfrac{1}{2}$, and the desired solution is $y = \tfrac{1}{2}x^2 + \tfrac{1}{2}$; this and two other solutions are shown in Fig. 2(b).

EXAMPLE 5   The differential equation

$$\frac{dy}{dt} = -ty$$

has solutions

$$y = ce^{-t^2/2}$$

To pass through the point $(t_0, y_0) = (0, 2)$, a solution must satisfy

$$2 = ce^0 = c.$$

Thus $c = 2$, and the particular solution selected by the condition $(t_0, y_0) = (0, 2)$ is $y = 2e^{-t^2/2}$.

EXAMPLE 6   When a drug is administered to a human body it has at time t a certain concentration $c(t)$ in the body's fluids. Careful measurements show that this concentration typically decreases, as time increases, according to an exponential law, $c(t) = c_0 e^{-at}$, a constant, $a > 0$, where c_0 is the concentration at $t = 0$: $c(0) = c_0$. Since $c'(t) = -ac_0 e^{-at} = -ac(t)$, it follows that $c(t)$ satisfies the first-order differential equation

$$\frac{dc}{dt} = -ac.$$

In words, the differential equation says that the rate at which the concentration decreases over time is proportional to the concentration at any time, the constant of proportionality being $a > 0$; the minus sign accounts for decreasing concentration as opposed to increasing. If another dose of the same size as the initial dose is administered at time $t_1 > 0$, the concentration in the body jumps to the sum of the two concentrations:

$$c_0 + c_0 e^{-at_1} = c_0(1 + e^{-at_1}).$$

After n equal doses, spaced at equal time intervals t_1, the concentration is expressible as a geometric series:

$$c(nt_1) = c_0(1 + e^{-at_1} + \cdots + e^{-(n-1)at_1}) = c_0 \frac{1 - e^{-nat_1}}{1 - e^{at_1}}.$$

Exercise 8 asks for a derivation of this last equation and some conclusions that can be drawn from it.

The previous example shows that it may be of some interest to start with a formula for a family of solutions to a differential equation and then find a differential equation that is satisfied by all solutions in the family. In principle, this is just a matter of differentiating the given solution formula with respect to the chosen independent variable and then algebraically eliminating the constant.

EXAMPLE 7   The formula $x + cy = 0$ describes a family of lines through the origin of the (x, y) plane, each line with a different slope $-1/c$. Differentiating with respect to x gives

$$1 + c\frac{dy}{dx} = 0.$$

But $c = -x/y$, so the differential equation is

$$1 + \left(\frac{-x}{y}\right)\frac{dy}{dx} = 0 \quad \text{or} \quad x\frac{dy}{dx} = y.$$

EXERCISES

1. For each of the following differential equations, check to see whether the given function is a solution on the specified interval.

   (a) $y' + y = 0$; $y = ce^{-x}$, $-\infty < x < \infty$, c const.

   (b) $y' = 2y$; $y = ce^{2x}$, $-\infty < x < \infty$, c const.

   (c) $\dfrac{dy}{dt} + 2ty = 0$; $y = ce^{-t^2}$, $-\infty < t < \infty$, b const.

   (d) $y' = y + x$; $y = ce^{-x} - x - 1$, $-\infty < x < \infty$, c const.

   (e) $x\dfrac{du}{dx} = 1$; $u = \ln(cx)$, $0 < x$, c const.

   (f) $\dfrac{dy}{dx} = 1 + y^2$; $y = \tan x$, $\dfrac{-\pi}{2} < x < \dfrac{\pi}{2}$; $y = \tan(x - c)$,

   $\dfrac{-\pi}{2} + c < x < \dfrac{\pi}{2} + c$, c const.

   (g) $\dfrac{dz}{dx} = 2\sqrt{|z|}$; $z = x|x|$, $-\infty < x < \infty$.

   (h) $\dfrac{dy}{dx} = 3x^2 y + 1$; $y = e^{x^3}\displaystyle\int_0^x e^{-t^3}\,dt + ce^{x^3}$, $-\infty < x < \infty$, c const.

   (i) $\dfrac{dy}{dx} = \sqrt{1 + x^3}$; $y = \displaystyle\int_0^x \sqrt{1 + t^3}\,dt$, $-1 < x < \infty$.

   (j) $\dfrac{dy}{dx} = 2x\sqrt{1 + x^6}$; $y = \displaystyle\int_0^{x^2} \sqrt{1 + t^3}\,dt$ $-\infty < x < \infty$.

   (k) $\dfrac{d^2 y}{dt^2} + 4y = 0$; $y = c_1 \cos 2t + c_2 \sin 2t$, $-\infty < t < \infty$, c_1, c_2 const.

   (l) $\dfrac{d^2 y}{dt^2} - 4y = 0$; $y = c_1 e^{2t} + c_2 e^{-2t}$, $-\infty < t < \infty$, c_1, c_2 const.

   (m) $\dfrac{dy}{dt} = \sqrt{y}$; $y = 0$, $-\infty < t < \infty$.

   (n) $\dfrac{dy}{dt} = \sqrt{y}$; $y = \begin{cases} 0, & t < 0 \\ \dfrac{t^2}{4}, & t \geq 0. \end{cases}$

2. (a) The differential equation

$$\left(\frac{dx}{dt}\right)^2 + x^2 = 0$$

has only one real-valued solution on a given interval $a < t < b$; explain why. What is the unique solution?

(b) The differential equation

$$\left(\frac{dx}{dt}\right)^2 + t^2 = 0$$

has no real-valued solution on a given interval $a < t < b$; explain why.

3. Each of the following formulas provides a family of solutions to some differential equation. In each case, find the value of the constant c or k for which the given initial condition is satisfied. Then sketch the graph of that solution.
   (a) $y = ce^{3x}$; $y(0) = 7$.
   (b) $y = ke^{-7x}$; $y(0) = 7$.
   (c) $x = \ln(ct)$; $x(1) = -2$.
   (d) $x = c \ln t$; $x(2) = -1$.
   (e) $y = k \cos x$; $y(\pi) = 2$.

4. For each of the solution formulas in Exercise 3, find a first-order differential equation satisfied by all members of the solution family.

5. By substitution into the differential equation, find all values of r such that the given formula provides a solution to the differential equation.
   (a) $x^2 y'' + 5xy' + 4y = 0$; $y = x^r$.
   (b) $y'' + 5y' + 4y = 0$; $y = e^{rx}$.

6. The definition of the exponential function is often introduced along with the **organic growth law.** According to this law, $P(t)$, the size of a population at time t, satisfies

$$\frac{dp}{dt} = kP, \qquad k \text{ const.}, \qquad k > 0.$$

This equation is satisfied by $P(t) = Ce^{kt}$, where C is a constant.
   (a) Show that $C = P(0)$.
   (b) If $P(0) = 2$ and $P(2) = 6$, find k.
   (c) Show that a population satisfying the organic growth law increases by the factor $R > 1$ in any time interval of length $t_1 = (\ln R)/k$.

7. The **radioactive decay equation** is

$$\frac{dQ}{dt} = -k, \qquad k \text{ const.}, \qquad k > 0.$$

This equation is satisfied by the quantity $Q(t)$ of a radioactive substance remaining after time t, starting with some initial amount $Q(0)$. Clearly, a solution has the form

$$Q(t) = Ce^{-kt},$$

where C is a constant.
   (a) Show that $C = Q(0)$.
   (b) If $Q(0) = 1$ and $Q(3) = 4$, find k.
   (c) Show that a radioactive substance satisfying the decay equation decreases by the factor r, $0 < r < 1$, in any time interval of length $t_1 = -(\ln r)/k$.

8. Read Example 6 of the text.
   (a) Show that after three equal doses, t_1 time units apart, the concentration is

$$c(3t_1) = c_0(1 + e^{-at_1} + e^{-2at_1}).$$

(b) Show inductively that after n equal doses, t_1 time units apart, the concentration is

$$c(nt_1) = c_0(1 + e^{-at_1} + \cdots + e^{-(n-1)at_1})$$

$$= c_0 \frac{1 - e^{-nat_1}}{1 - e^{-at_1}}.$$

(c) By considering what happens as $n \to \infty$ in part (b), show that the concentration satisfies

$$c(t) < \frac{c_0}{1 - e^{-at_1}}, \qquad \text{for } t \geq 0.$$

(d) How should t_1 be chosen so that the upper bound for $c(t)$ established in part (c) will be $2c_0$?

9. Let g (~ 32.2 feet per second per second) be the acceleration of gravity near the surface of the earth. Then a body falling with negligible air resistance has acceleration

$$\frac{dv}{dt} = g,$$

where $v = v(t)$ is the velocity of the body at time t. Derive the following relations:

(a) $v(t) = gt + v_0$, where v_0 is the velocity at time $t = 0$.

(b) $s = \frac{1}{2}gt^2 + v_0 t + s_0$, where $s(t)$ is the distance covered in time t by the body and $s_0 = s(0)$.

(c) $ds/dt = \sqrt{2gs}$, assuming $v_0 = 0$, $s_0 = 0$, and $s(t) \geq 0$.

10. Find a first order differential equation satisfied by $y = y(x)$ and such that

(a) the slope of the graph of $y(x)$ at (x, y) equals the distance from (x, y) to $(0, 0)$.

(b) the tangent line to the graph of $y(x)$ at (x, y) contains the point $(x + y, x + y)$.

(c) the rate of change of $y(x)$ with respect to x is proportional to x and inversely proportional to y.

2 DIRECTION FIELDS

Just as the graph of a function can be said to be a complete geometric representation of the function, so the direction field of a first-order differential equation, described next, represents the differential equation. This fundamental idea can be used to get information about solutions without going to the trouble of finding formulas for solutions.

2A Examples

A differential equation of the form

$$y' = F(x, y), \qquad a < x < b,$$

can be thought of as assigning a slope to the point (x, y) in the (x, y)-plane. Such an assignment is called a **direction field.** The reason is that, if $y(x)$ provides a solution to the differential equation, then

$$y'(x) = F(x, y(x)).$$

But since $y'(x)$ is the slope of the tangent to the graph of the solution at the point with coordinates $(x, y(x))$, the formula $F(x, y(x))$ is necessarily also equal to that same slope; the slope then specifies a direction along the solution curve $y = y(x)$. From one point of view, we can think of starting with the solution $y = y(x)$ and locating a short segment of tangent line at each point of the graph of $y = y(x)$; some of these are shown in Figure 3(a). A more useful alternative turns out to be to draw a segment of slope $F(x, y)$ at selected points (x, y). In this way, we get a sketch of the direction field associated with the differential equation $y' = F(x, y)$. If the sketch is skillfully made, it is often possible to get a good idea of what the graphs of solutions $y = y(x)$ should look like; the principle is that a solution graph should be tangent to the segment located at each point that the graph passes through.

Figure 3(b) shows such a sketch of the direction field, along with the graphs of four curves that are tangent to the segments in the direction field. The direction field is associated with the differential equation

$$y' = x - xy.$$

Thus the function F in the general discussion is $F(x, y) = x - xy$. In particular, $F(1, \frac{1}{2}) = \frac{1}{2}$, so that the segment through $(x, y) = (1, \frac{1}{2})$ has slope $\frac{1}{2}$. Notice particularly that there are apparently infinitely many possible curves that could be drawn tangent to the direction field. Of these infinitely many curves, Figure 3(b) shows just four; others could easily be drawn in.

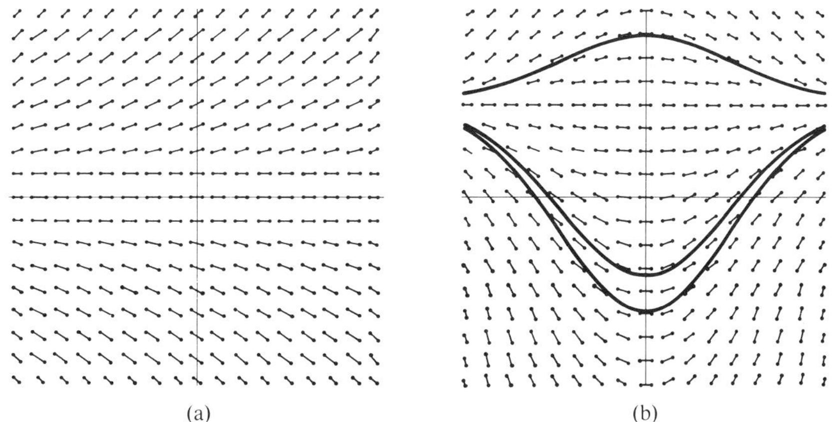

(a)   (b)

Figure 3

EXAMPLE 1   The differential equation

$$\frac{dy}{dx} = \frac{y^2}{x^2 + y^2}$$

defines a direction field for all (x, y) except $(0, 0)$. To sketch the field, we can first make a table of slopes. In this example, we will restrict ourselves to the first quadrant, because replacing x by $-x$ or y by $-y$ leaves the slope unchanged. We find

(x, y)	dy/dx	(x, y)	dy/dx
(0, 1)	1	(2, 3)	$\frac{9}{13}$
(0, 2)	1	(3, 0)	0
(0, 3)	1	(3, 1)	$\frac{1}{10}$
(1, 0)	0	(3, 2)	$\frac{4}{13}$
(1, 1)	$\frac{1}{2}$	(3, 3)	$\frac{1}{2}$
(1, 2)	$\frac{4}{5}$	(4, 0)	0
(1, 3)	$\frac{9}{10}$	(4, 1)	$\frac{1}{17}$
(2, 0)	0	(4, 2)	$\frac{1}{5}$
(2, 1)	$\frac{1}{5}$	(4, 3)	$\frac{9}{25}$
(2, 2)	$\frac{1}{2}$	(4, 4)	$\frac{1}{2}$

These points, plus a few more are plotted in Figure 4 along with segments having the assigned slopes. The pattern seems fairly evident. It is also worth noticing what happens as $x \to \infty$ and $y \to \infty$. If $x \to \infty$ and y is held fixed, then the slopes tend to zero. If $y \to \infty$ (x held fixed) then the slopes tend to 1. Two solution

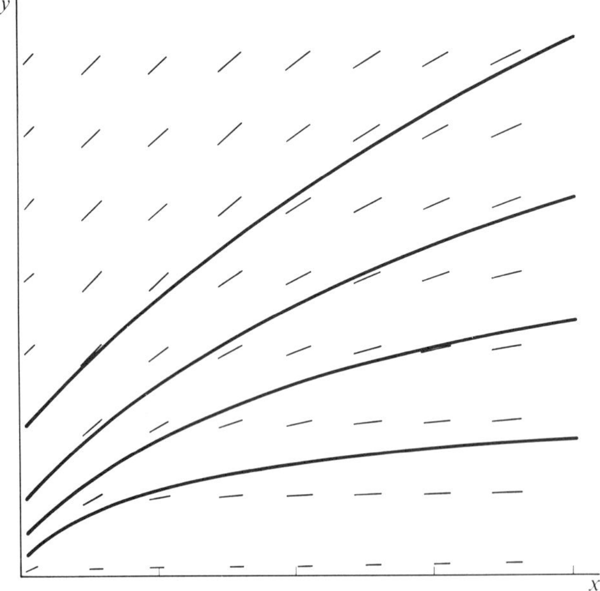

Figure 4

curves are drawn in, showing their tangency to the segments in the sketch of the direction field.

2B Isoclines

Sketching a direction field involves using a certain amount of judgment about how much information to include in the sketch. As with a caricature of a person's face, the trick is to include just enough information to allow the viewer to recognize what is being represented; too much detail can be confusing and too little can fail to show any recognizable pattern. One way to organize a sketch is to begin by sketching some **isoclines,** that is, curves along which the directions of the field are equal; this enables us simply to sketch parallel segments along each isocline, making sure to consider sufficiently many isoclines to make a good sketch of the field. An isocline is *not* in general the graph of a solution, although that can happen exceptionally.

EXAMPLE 2   The differential equation

$$y' = x^2 + y^2$$

cannot be solved using an elementary formula. Furthermore, listing the slopes associated with a rectangular array of points (x, y) requires computing many different slopes. Let's look for curves on which the field segments are parallel, that is, for isoclines. We consider, therefore, curves

$$x^2 + y^2 = c \qquad c > 0.$$

These curves are circles centered at $(0, 0)$ of radius $\sqrt{c}$. The direction field segment of a point of such a curve has slope c. Some of these are shown in Figure 5 (see p. 12), along with some sketches of solution curves tangent to the field.

2C Continuity

The associated direction field is a complete geometric representation of a differential equation $y' = F(x, y)$, but the sketch that we make of a direction field is very often incomplete for practical reasons. In any case the underlying geometric concept is fundamental to our understanding. For example, it seems reasonable to expect that if $F(x, y)$ varies discontinuously as x and y vary, the differential equation $y' = F(x, y)$ may not have any solutions at all. However, it can be shown that if $F(x, y)$ is continuous in some neighborhood of (x_0, y_0) then there is always a solution of $y' = F(x, y)$ satisfying $y_0 = y(x_0)$ and defined on some interval, perhaps very small, containing x_0. If in addition the partial derivative $F_y(x, y)$ of $F(x, y)$ with respect to y is continuous on a neighborhood of (x_0, y_0), then there is only one such solution. Exercise 3 shows that there can be more than one such solution if $F_y(x, y)$ fails to be continuous.

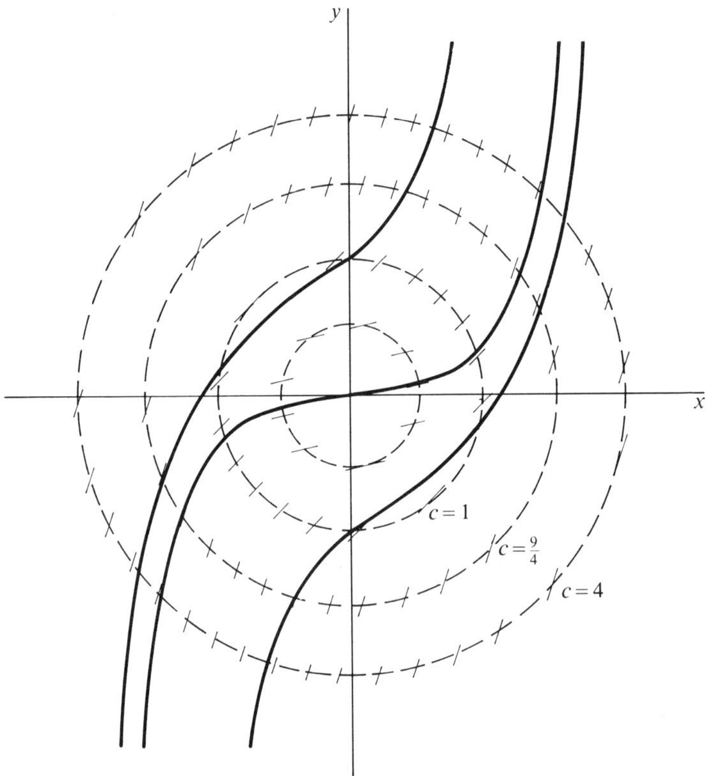

Figure 5

EXERCISES

1. For each of the following differential equations, sketch the associated direction field at points (x, y) with $-2 \le x \le 2$ and $-2 \le y \le 2$. Include enough points in your sketch so that a pattern becomes apparent. Then draw in a few solution curves.

   (a) $y' = x.$

   (b) $y' = y.$

   (c) $y' = -1.$

   (d) $y' = \begin{cases} 0, & x \le 0, \\ x, & x > 0 \end{cases}.$

   (e) $y' = xy.$

   (f) $y' = x + y.$

2. For each of the following differential equations, sketch the associated direction field at points of definition (x, y) with $-4 \le x \le 4$ and $-4 \le y \le 4$. These sketches can easily be done by first drawing some isoclines. After sketching the field, draw a few solution curves.

   (a) $y' = 2x^2 + y^2.$

   (b) $y' = x^2 + y.$

   (c) $y' = x - y$

   (d) $y' = \sqrt{x^2 + y^2}.$

   (e) $y' = \dfrac{x}{1 + y^2}.$

   (f) $y' = \sqrt{1 - x^2 - y^2}.$

3. Define

$$F(y) = \begin{cases} 0, & y \le 0, \\ \sqrt{y}, & y > 0, \end{cases}$$

and consider the differential equation

$$y' = F(y).$$

   (a) Sketch the direction field of the differential equation for $-2 \le x \le 2$, $-2 \le y \le 2$.
   (b) Verify that there are two distinct solution curves passing through each point $(x_0, 0)$.
   [*Hint:* Consider $y = (x - x_0)^2/4$ for $x > x_0$.]
4. (a) Show that the isoclines of the direction field for $y' = y/x$ are actually solution curves for the differential equation.
   (b) Show that no isocline of the direction field for $y' = x$ is a solution curve of the differential equation.
5. The differential equation $y' = \sqrt{1 + y^4}$ has a fairly simple direction field, but there is no elementary formula for its solution curves. Use the direction field to sketch solution curves $y = y(x)$ satisfying (a) $y(0) = 0$ and (b) $y(0) = 1$.
6. Write a computer graphics program to display a direction field, and apply it to the following equations:
   (a) $y' = x^2 + y^2$, $-2 \le x \le 2$, $-2 \le y \le 2$.
   (b) $y' = x^2 - y^2$, $-2 \le x \le 2$, $-2 \le y \le 2$.
   (c) $y' = e^{-x^2 - y^2}$, $-2 \le x \le 2$, $-2 \le y \le 2$.
7. Let $a > 0$ be constant in $y' = a(1 + y^2)$.
   (a) Verify that $y = \tan(ax)$ is a solution that is valid for $-\pi a/2 < x < \pi a/2$, but for no larger interval.
   (b) Sketch the direction field of $y' = a(1 + y^2)$ for $a = 1$, and use your sketch to account for the behavior of the solution $y = \tan x$.

3 INTEGRATION

3A Y' = F(X)

A differential equation of the simple form $y' = F(x)$ can in principle be solved on any interval $a \le x \le b$ over which F is continuous. We can write the general solution as

$$y(x) = \int F(x)dx + C \quad \text{or} \quad y(x) = \int_a^x F(u)du + C.$$

EXAMPLE 1   The differential equation $y' = x$ has general solution

$$y(x) = \int x \, dx + C = \tfrac{1}{2}x^2 + C,$$

valid for all real x. Similarly the differential equation $y' = \sin(1 + x^2)$ has general solution

$$y(x) = \int_0^x \sin(1 + u^2)\,du + C,$$

the main difference between this and the earlier example being that here we cannot find an explicit elementary formula for the solution. Nevertheless we can get quite a bit of information about the solution by using numerical integration to produce an accurate table of values. And the derivatives of $y(x)$ are readily available; since we know that $y'(x) = \sin(1 + x^2)$, it will follow that $y'' = 2x \cos(1 + x^2)$ and so on for higher order derivatives.

When the independent variable represents time we denote it by t rather than x, and we can interpret an equation of the form $y' = v(t)$ as a condition that the position $y(t)$ at time t of a one-dimensional motion has velocity $v(t)$. If $v(t)$ is differentiable then $v' = a(t)$ is called the acceleration of the motion at time t.

EXAMPLE 2   Near the surface of the earth, the acceleration due to gravity for a falling body has a nearly constant value $g = 32.2$ feet per second per second. Thus velocity satisfies the equation $v' = g$. Integration with respect to t gives $v = gt + C_1$. The constant C_1 is determined by observing the velocity v_0 at some time t_0. If $t_0 = 0$ we find

$$v(t) = gt + v_0.$$

Since position is related to velocity by $y' = v$, we can write a first order equation for $y(t)$ in the form

$$y' = gt + v_0.$$

Integration gives $y = gt^2/2 + v_0 t + C_2$. If $y(0) = y_0$, as measured down from some reference point above ground level, then the constant C_2 must by y_0, so

$$y(t) = \tfrac{1}{2}gt^2 + v_0 t + y_0.$$

If we were to measure up from the surface of the earth, the acceleration of gravity would act to decrease the velocity so we would replace g by $-g$ in the preceding derivation. Thus a projectile thrown straight up from the surface of the earth ($y_0 = 0$) with velocity $v_0 = 10000$ feet per second would have altitude in feet at t seconds given by

$$y(t) = -16.1t^2 + 10000t.$$

The maximum altitude $y_{\max}$ occurs when the velocity $v = y'$ is zero, so we find it from

$$v = -32.2t + 10000 = 0$$

to be $y_{\max} = 10000/32.2 = 310.56$ feet. The effect of air resistance has been assumed to be negligible here, but is taken into account in Section 4A of Chapter 2.

EXERCISES

1. Solve each differential equation, and find the particular solution that satisfies the associated initial condition.
   (a) $y' = x(1 - x)$, $y(0) = 1$.
   (b) $ds/dt = (t + 1)^2$, $s(1) = 2$.
   (c) $y' = x/(1 - x^2)$, $y(0) = 1$.
   (d) $du/dv = v^2 + 1$, $u(-1) = 1$.
   (e) $y'' = \sin x$, $y(0) = 1$, $y'(0) = 1$.
   (f) $y''' = 1$, $y(0) = y'(0) = y''(0) = 0$.

2. A projectile is fired up from ground level with an initial velocity of 5000 feet per second. What is the maximum altitude attained, and how long does it take to get there?

3. A weight is dropped from 5000 feet above ground. How long does it take to reach the ground, and with what final velocity does it hit?

4. Suppose the two objects described in Exercises 2 and 3 are released at the same time and are aimed directly at each other.
   (a) How long after release do they meet, and at what height above ground?
   (b) What initial velocity should the projectile be given so that the two objects meet 2500 feet above ground?

5. A block of wood is kicked across a smooth ice surface with initial velocity 20 feet per second. Friction with the ice produces a constant negative acceleration a. If the block stops sliding 4 seconds after the initial kick, what is the value of a? How far does the block slide?

6. Suppose a uniform horizontal beam sags by an amount $y = y(x)$ at a distance x measured from one end. For a fairly stiff beam with uniform loading, $y(x)$ typically satisfies a fourth order differential equation $y'''' = R$, where R is a constant depending on the load being carried and on the characteristics of the beam itself. If the ends of the beam are supported at $x = 0$ and at $x = L$, then $y(0) = y(L) = 0$. The extended beam also behaves as if its profile had an inflection point at each support, so that $y''(0) = y''(L) = 0$.
   (a) Show that the sag at point x is

$$\tfrac{1}{24}R(x^4 - 2Lx^3 + L^3x), \ 0 \le x \le L.$$

   (b) Where does the maximum amount of sag occur, and how much is it?
   (c) Could the value of R be computed for a particular beam by making appropriate measurements? Explain.

3B Variables Separable

A differential equation that can be written in the form

$$g(y)\frac{dy}{dx} = f(x)$$

can in principle always be solved by integration. By assuming that $y = y(x)$ is a solution of the equation, we can try to integrate both sides of the differential equation with respect to x, getting

$$\int g(y)\frac{dy}{dx}\,dx = \int f(x)\,dx. \tag{1}$$

Now suppose we can find indefinite integrals G and F such that $F' = f$ and $G' = g$. By the chain rule,

$$\frac{dG(y)}{dx} = G'(y)\frac{dy}{dx}$$

$$= g(y)\frac{dy}{dx},$$

so on the left side we have

$$\int g(y)\frac{dy}{dx}\,dx = \int \frac{d}{dx}G(y)\,dx$$

$$= G(y) + C_1.$$

The right side of the differential equation can be written

$$\int f(x)\,dx = F(x) + C_2.$$

Hence $G(y) + C_1 = F(x) + C_2$, or

$$G(y) = F(x) + c,$$

where $c = C_2 - C_1$. If we can now solve the equation for y in terms of x, we have a formula for solutions $y = y(x, c)$.

The process outlined is usually called **separation of variables** because it involves getting x's on one side of the equation and y's on the other. The chain rule allows us to cancel the dx's on the left side of equation (1). The resulting formal equation,

$$\int g(y)\,dy = \int f(x)\,dx,$$

still leads to the same solution: $G(y) = F(x) + c$, where $G' = g$ and $F' = f$. Equation (1) is sometimes written in the symmetric differential form

$$g(y)\,dy = f(x)\,dx,$$

which can be interpreted as either

$$g(y)\frac{dy}{dx} = f(x) \quad \text{or} \quad g(y) = f(x)\frac{dx}{dy}.$$

Unfortunately, these separated forms are not always attainable for a given differential equation.

EXAMPLE 1   In biological studies it is often observed that the size $P(t)$ of a growing bacteria population is proportional to the rate of change $(dP/dt)(t)$. Expressing this proportionality in the form

$$\frac{dP}{dt} = kP, \qquad k \text{ const.,} \tag{2}$$

gives us a first-order differential equation for P. Experience with the exponential function allows us to guess one solution. If we let

$$P(t) = Ce^{kt}, \qquad C \text{ const.,}$$

we see that

$$\frac{dP}{dt}(t) = kCe^{kt} = kP(t)$$

for all real numbers t; in other words, $P = Ce^{kt}$ is a solution of the differential equation. Since we guessed this solution, the possibility remains that there are others. Under the assumption that the population size $P(t)$ is never zero, we can proceed as follows. Divide equation (2) by P, getting

$$\frac{1}{P}\frac{dP}{dt} = k.$$

Now integrate both sides with respect to t:

$$\int \frac{1}{P}\frac{dP}{dt}\,dt = \int k\,dt.$$

The integral on the left is $\ln|P| + c_1$, and the integral on the right is $kt + c_2$. Hence we can lump the constants of integration together and write

$$\ln|P| = kt + c.$$

Since $\exp(\ln z) = z$, we can apply the exponential function to both sides to get

$$|P(t)| = e^{kt+c}$$
$$= e^{c}e^{kt} = Ce^{kt},$$

where C is a necessarily positive constant: $C = e^{c}$. But if $P(t)$ is to be continuous, and of course it must even be differentiable, then we must have either

$$P(t) = Ce^{kt} \quad \text{or} \quad P(t) = -Ce^{kt},$$

so the constant can be either positive or negative in theory. If $P(t)$ does in fact represent a population size, then we use the first formula with $C > 0$; thus $P(0) = C$, so $P(t) = P(0)e^{kt}$. Graphs of solutions satisfying $P(0) = 1$ are shown in Figure 6(a) for $k = 0.2$, 0.5, and 0.7.

EXAMPLE 2   The differential equation

$$\frac{dy}{dx} = x(1 + y^2)$$

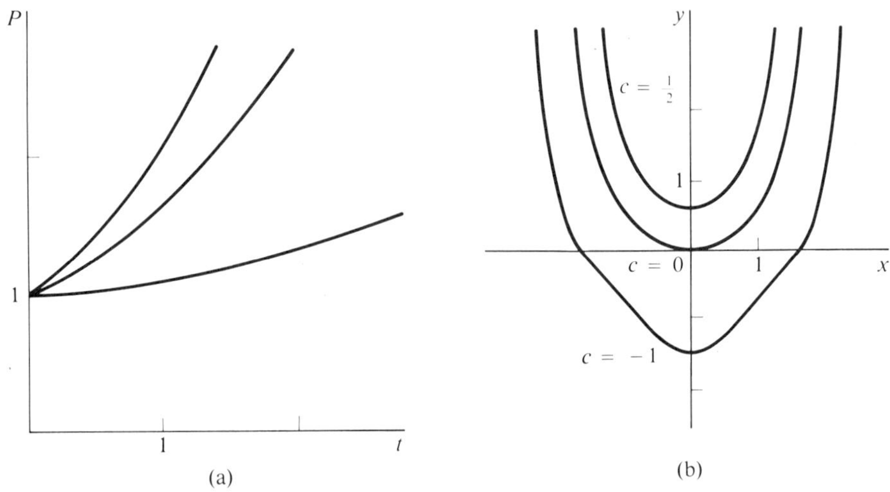

Figure 6

can be written in differential form as

$$\frac{dy}{1 + y^2} = x\,dx.$$

Integrating both sides gives

$$\arctan y = \tfrac{1}{2}x^2 + c.$$

Solving for y, we find

$$y = \tan\left(\tfrac{1}{2}x^2 + c\right).$$

The graphs of these solutions are shown in Figure 6(b), for $c = -1, 0$, and 0.5.

3C Transformations

When the variables do not separate in a first-order equation, it may still be possible to change variables in such a way that the new equation separates. After solving the new equation, then change variables back again.

EXAMPLE 3   Suppose we have an equation that can be written in the form

$$\frac{dy}{dx} = f(ax + by + c).$$

If we let $u = ax + by + c$, then

$$\frac{du}{dx} = a + b\frac{dy}{dx}, \text{ so } \qquad \frac{dy}{dx} = \frac{1}{b}\frac{du}{dx} - \frac{a}{b}.$$

The given equation becomes

$$\frac{1}{b}\frac{du}{dx} - \frac{a}{b} = f(u),$$

so the variables u and x can be separated. For instance, in

$$\frac{dy}{dx} = (x + y)^2$$

we let $u = x + y$, so $dy/dx = du/dx - 1$. Then

$$\frac{du}{dx} = u^2 + 1 \quad \text{or} \quad \frac{du}{1 + u^2} = dx.$$

Integrating, we find

$$\text{arctan } u = x + c,$$

so

$$\text{arctan}(x + y) = x + c, \quad \text{or} \quad y = -x + \tan(x + c).$$

EXAMPLE 4   Suppose we have an equation that can be written in the form

$$\frac{dy}{dx} = g\left(\frac{y}{x}\right).$$

If we let $u = y/x$, then

$$\frac{du}{dx} = \frac{x(dy/dx) - y}{x^2} = \frac{1}{x}\frac{dy}{dx} - \frac{y}{x^2}.$$

We solve for $\dfrac{dy}{dx}$ to get

$$\frac{dy}{dx} = x\frac{du}{dx} + \frac{y}{x} = x\frac{du}{dx} + u.$$

The given equation becomes

$$x\frac{du}{dx} = g(u) - u,$$

so the variables can be separated. For instance, instead of

$$x^2 \frac{dy}{dx} = x^2 + xy + y^2,$$

we can write

$$\frac{dy}{dx} = 1 + \left(\frac{y}{x}\right) + \left(\frac{y}{x}\right)^2.$$

The function $g(u)$ is evidently $g(u) = 1 + u + u^2$. Letting $u = y/x$, we transform the given equation to

$$x \frac{du}{dx} = (1 + u + u^2) - u = 1 + u^2.$$

In separated form, we find

$$\frac{du}{1 + u^2} = \frac{dx}{x}.$$

We can integrate directly to get

$$\arctan u = \ln |x| + c.$$

Then put $u = y/x$ back again:

$$\arctan \frac{y}{x} = \ln |x| + c.$$

To simplify the relation between y and x, we can set $c = \ln k$, $k > 0$. This gives

$$\arctan \frac{y}{x} = \ln |x| + \ln k$$

$$= \ln k|x|.$$

Then $y/x = \tan(\ln k|x|)$, so

$$y = x \tan(\ln k|x|).$$

Two remarks are in order. First, it is usually impossible to find a nice formula that expresses y in terms of x, or vice versa, as we did in this problem; we may have to make do with an implicit formula of the form $F(x, y, c) = 0$. Second, for a fixed value of $k \geq 0$ the solution to $x^2 y' = x^2 + xy + y^2$ fails to exist when $x = 0$ and for those values of x for which $\tan(\ln k|x|)$ fails to exist. Some solution curves are shown in Figure 7.

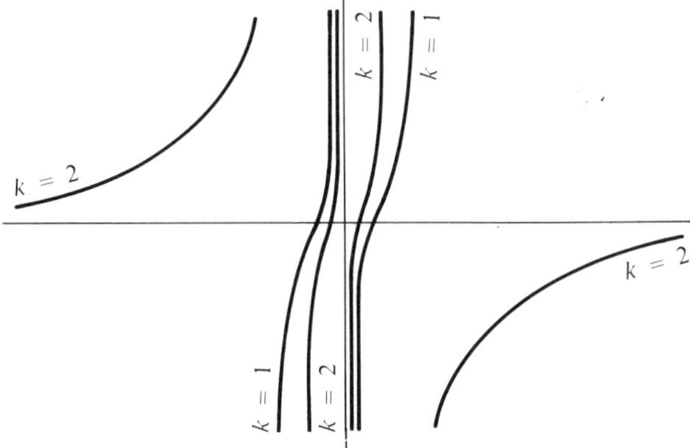

Figure 7

Equations of the type discussed in Example 4 are sometimes called **homogeneous** although the terminology conflicts with another use we have for that word.

EXERCISES

1. Solve each of the following equations by separating the variables. Then determine the constant of integration so that the given initial condition is satisfied.

   (a) $y \dfrac{dy}{dx} = x^2$; $y = 1$ when $x = 2$.

   (b) $y' = x^2(1 + y^2)$; $y = 0$ when $x = 0$.

   (c) $dx + e^x y^2 \, dy = 0$; $y = 1$ when $x = 0$.

   (d) $s \, dt + t \, ds = 0$; $s = 1$ when $t = 1$.

   (e) $\cos u \, du - \sin v \, dv = 0$; $u = \pi/2$ when $v = \pi/2$.

   (f) $y\sqrt{1 - x^2} \, dy = dx$; $y = \sqrt{\pi}$ when $x = 1$.

2. Sketch the graph of the particular solution satisfying the initial condition for the equations in Exercise 1.

3. Verify the indefinite integration formulas

   (a) $\displaystyle \int \frac{du}{u} = \ln |u|,$

   (b) $\displaystyle \int |u|^\alpha \, du = \frac{u|u|^\alpha}{\alpha + 1}$, with $\alpha \neq -1$,

   for the two separate cases $u > 0$ and $u < 0$, and sketch the graphs of the indefinite integral formulas.

4. Newton's law of cooling asserts that the rate of change of the surface temperature $u(t)$ of an object is proportional to the difference between $u(t)$ and the temperature u_0 of the surrounding medium. Thus

$$\frac{du}{dt} = n(u - u_0),$$

   where the constant of proportionality n is negative if $u > u_0$, so that du/dt remains negative. Solve this equation under the assumption that u_0 is constant. Then sketch the graph of $u = u(t)$ under the assumption that $n = -0.5$, $u_0 = 10$, and $u(0)$, the initial temperature of the object, is 20.

5. Each of the following differential equations can be written in one or the other of two forms:

$$\frac{dy}{dx} = f(ax + by + c) \quad \text{or} \quad \frac{dy}{dx} = g\left(\frac{y}{x}\right).$$

   Decide which form applies in each case and use a transformation of the type

$$u = ax + by + c \quad \text{or} \quad u = \frac{y}{x}$$

   to put the equation in separated form. Then solve the equation and eliminate u from the solution.

   (a) $\dfrac{dy}{dx} = 2x + y + 3$.

   (b) $\dfrac{dy}{dt} = (t - y)^2$.

(c) $x^2 \dfrac{dy}{dx} = x^2 + y^2$.

(d) $t\, dx = (t + x)\, dt$.

(e) $\dfrac{dy}{dx} = \dfrac{y - x}{y + x}$.

(f) $\dfrac{dz}{dv} = (2z + v + 1)^2$.

6. Neither of the following equations can be written in the homogeneous form

$$\frac{dy}{dx} = g\!\left(\frac{y}{x}\right).$$

However, a preliminary transformation of the form $x = z + h$, $y = w + k$, where h, k are constant, gives an equation of that form. Solve the equations by first making the transformation with general h and k and then finding h and k.

(a) $\dfrac{dy}{dx} = \dfrac{x + y + 1}{x + 2y + 3}$.

(b) $(x + y)\, dy = (x - y + 1)\, dx$.

7. Show that the differential equation

$$P(x, y)\, dx + Q(x, y)\, dy = 0$$

can be written in the form

$$\frac{dy}{dx} = g\!\left(\frac{y}{x}\right)$$

if both

$$P(tx, ty) = t^n P(x, y) \quad \text{and} \quad Q(tx, ty) = t^n Q(x, y)$$

for some number n. (*Hint:* Let $t = 1/x$.)

8. Can the area between the graph of $y = f(t)$ and the t-axis between $t = a$ and $t = x$ be the same as the arc length of the graph over the same interval, for every a and x? It is easy to verify that this is so for $f(t) = \cosh t$. Find all differentiable functions $f(t)$ for which equality holds over every interval $a \le t \le x$. Recall that $\cosh t = (e^t + e^{-t})/2$.

4 EXACT EQUATIONS

4A Solution

Suppose that we start with an equation

$$F(x, y) = c$$

that implicitly defines $y = y(x)$ as a differentiable function of x for some value of c. Differentiate both sides of the equation with respect to x, using the chain rule.

Denoting partial derivatives with respect to x and y by subscripts, we get

$$F_x(x, y) + F_y(x, y)\,\frac{dy}{dx} = 0.$$

The resulting first-order differential equation is then satisfied by $y = y(x)$. Written in differential form, the equation has the general form

4.1 $$P(x, y)\,dx + Q(x, y)\,dy = 0,$$

where $P = F_x$ and $Q = F_y$. If $F(x, y)$ has continuous mixed partial derivatives, it follows that $F_{xy} = F_{yx}$, in other words that

4.2 $$P_y(x, y) = Q_x(x, y).$$

It is a useful fact that Equation 4.1 always has a solution in any rectangle in the (x, y)-plane throughout which Equation 4.2 holds. Rather than prove this here, we show by example how to find the solution. An equation is called **exact** if it can be written in the form of Equation 4.1 and has an implicit solution $F(x, y) = c$ such that $F_x = P$ and $F_y = Q$. Equation 4.2 is a **consistency condition** that must be satisfied if Equation 4.1 is to be exact.

EXAMPLE 1   The differential equation

$$(e^x \cos y + 2x)\,dx - e^x \sin y\,dy = 0$$

is of the form

$$P(x, y)\,dx + Q(x, y)\,dy = 0.$$

Since $P_y(x, y) = -e^x \sin y = Q_x(x, y)$, we can expect to find a function $F(x, y)$ such that $F_x = P$ and $F_y = Q$. Specifically, we want

$$F_x(x, y) = e^x \cos y + 2x$$

and

$$F_y(x, y) = -e^x \sin y.$$

We can work with whichever equation seems simpler; let's take the second equation and integrate both sides with respect to y. We find

$$F(x, y) = \int^y F_y(x, y)\,dy$$

$$= \int^y -e^x \sin y\,dy$$

$$= e^x \cos y + f(x),$$

where the "constant" of integration $f(x)$ may depend on the other variable x. According to the equation for F_x, we must have

$$F_x(x, y) = \frac{\partial}{\partial x}\left[e^x \cos y + f(x)\right]$$

$$= e^x \cos y + 2x.$$

Thus $e^x \cos y + f'(x) = e^x \cos y + 2x$, so $f'(x) = 2x$. Since $f(x)$ is then of the form $x^2 + c$, we find

$$F(x, y) = e^x \cos y + x^2 + c.$$

Then $F(x, y) = 0$ is the solution in implicit form:

$$e^x \cos y + x^2 + c = 0.$$

4B Justification

If the equation

$$P(x, y)\, dx + Q(x, y)\, dy = 0$$

satisfies the **consistency condition**

$$P_y(x, y) = Q_x(x, y)$$

in some convex region of the (x, y)-plane containing (x_0, y_0), we try to find $F(x, y)$ such that

$$F_x(x, y) = P(x, y), \qquad F_y(x, y) = Q(x, y).$$

We integrate the second equation with respect to the second variable,

$$\int_{y_0}^{y} F_y(x, v)\, dv = \int_{y_0}^{y} Q(x, v)\, dv,$$

getting

$$F(x, y) - F(x, y_0) = \int_{y_0}^{y} Q(x, v)\, dv.$$

If we only knew $F(x, y_0)$, we could solve this equation for $F(x, y)$ and we would have our solution. What we do know is that, in spite of the y's on the right side,

$$F(x, y_0) = F(x, y) - \int_{y_0}^{y} Q(x, v)\, dv,$$

must be a function of x alone, in other words that $\partial F(x, y_0)/\partial y = 0$. But this is true, because if we differentiate the difference first with respect to x and then with respect to y, we get, using $F_x = P$,

$$P(x, y) - \int_{y_0}^{y} Q_x(x, v)\, dv$$

and then

$$P_y(x, y) - Q_x(x, y).$$

The last difference is zero because of the consistency condition; without it we would have $F(x, y_0)$ still depending on the variable y, an impossible situation.

The procedure just described can be carried out in any region in which the integral

$$\int_{y_o}^{y} Q(x, v) \, dv$$

can be effectively computed. Of course, we could interchange the roles of x and y and deal instead with

$$\int_{x_o}^{x} P(u, y) \, du.$$

EXERCISES

1. Most of the following differential equations are exact if written in the form $P \, dx + Q \, dy = 0$. Check the consistency condition $P_y = Q_x$, and solve the equation if possible. Then find the particular solution that satisfies the given initial condition.
   (a) $(y + 2xy^2) \, dx + (x + 2x^2y) \, dy = 0$; $y = 1$ when $x = 1$.
   (b) $\dfrac{dy}{dx} = \dfrac{2x + 1}{2y + 1}$; $y = 0$ when $x = 0$.
   (c) $(y + e^y) \, dx + x(1 + e^y) \, dy = 0$; $y = 0$ when $x = 1$.
   (d) $(\cos y - y \sin x) \, dx + (\cos x - x \sin y) \, dy = 0$; $y = \pi/2$ when $x = \pi/2$.
   (e) $(y + 1) \, dy - xy \, dx = 0$; $y = 1$ when $x = 2$.

2. Determine for which values of the constant k each of the following equations satisfies the consistency condition. Then solve the equation for those values of k.
   (a) $y \, dx + (kx + y^2 + 1) \, dy = 0$.
   (b) $\cos y \, dx + [\cos y + k(x + y)\sin y] \, dy = 0$.
   (c) $\left(1 + \dfrac{1}{x + ky}\right) dx + \left(\dfrac{k}{x + ky}\right) dy = 0$.

3. Show that an equation for which the variables are separable is exact.

4. An **integrating factor** $M(x, y)$ is a function such that

$$M(x, y)P(x, y) \, dx + M(x, y)Q(x, y) \, dy = 0$$

   is exact.
   (a) Show that a **linear** equation of the form

$$\frac{dy}{dx} + f(x)y = g(x)$$

can be written

$$(f(x)y - g(x))\, dx + dy = 0.$$

(b) Show that $M(x) = e^{\int f(x)\, dx}$ is an integrating factor for this last equation.

5 LINEAR EQUATIONS

The first-order differential equations we have looked at so far can nearly all be written in the form

$$\frac{dy}{dx} = F(x,\, y),$$

where F is some fairly simple function of x and y. If we make the requirement that F have the form $F(x,\, y) = -g(x)y + f(x)$, then we get what is called a **linear equation** in its standard form:

5.1 $$\frac{dy}{dx} + g(x)y = f(x).$$

Linear equations are general enough that they have many interesting applications; all the same, they are special enough that we can say significant things about their solutions without knowing much in the way of details about $g(x)$ and $f(x)$. The name "linear" is used because $-gy + f$ is often called a linear function of y. However, there is another reason, which will be explained in Section 5D.

EXAMPLE 1 If f happens to be identically zero, we can find solutions y to equation 5.1 by assuming $y \neq 0$ and writing

$$\frac{y'}{y} = -g(x).$$

Integrating with respect to x, we get

$$\log|y| = -G(x) + c,$$

where G is an indefinite integral of g and c is a constant. Taking the exponential of both sides gives

$$|y| = e^c e^{-G(x)},$$

and removing the absolute value allows us to replace the positive constant e^c by an arbitrary nonzero constant K:

$$y = Ke^{-G(x)}$$
$$= Ke^{-\int g(x)\, dx}.$$

5A Exponential Multipliers

The method of solution used in Example 1 fails if the function f in equation 5.1 is not zero; it also has the technical defect that it forces us to assume $y \neq 0$ (conceivably there are solutions that take on the value zero). Both objections can be avoided at once if we use the following method, suggested by the form of the solution found in Example 1. For the differential equation

$$y' + g(x)y = f(x),$$

written in standard form, we define an **exponential multiplier** to be

$$M(x) = e^{\int g(x)\, dx},$$

where $\int g(x)\, dx$ is an indefinite integral of g. The trick is to multiply the differential equation by M to get

$$e^{\int g(x)\, dx} y' + g(x)e^{\int g(x)\, dx} y = f(x)e^{\int g(x)\, dx}.$$

The whole point is that the left side can now be written as the derivative of $e^{\int g(x)\, dx} y$, because, by the product rule, applied to the factors $e^{\int g(x)\, dx}$ and y,

$$\frac{d}{dx}\left[e^{\int g(x)\, dx} y\right] = e^{\int g(x)\, dx} y' + g(x)e^{\int g(x)\, dx} y.$$

Thus we have rewritten the standard linear differential equation in the form

$$\frac{d}{dx}\left[e^{\int g(x)\, dx} y\right] = e^{\int g(x)\, dx} f(x);$$

it remains only to integrate both sides with respect to x and then to solve for y. The exponential multiplier $M(x)$ is sometimes called an **integrating factor.**

EXAMPLE 2   To find all solutions of the linear differential equation

$$y' = xy + x,$$

we first rewrite the equation in the standard form

$$y' - xy = x.$$

The exponential multiplier is then found by identifying the coefficient function $g(x) = -x$ and computing

$$M(x) = e^{\int g(x)\, dx}$$
$$= e^{-\int x\, dx}$$
$$= e^{-(1/2)x^2}.$$

Multiplying the differential equation by M gives

$$e^{-(1/2)x^2}y' - xe^{-(1/2)x^2}y = xe^{-(1/2)x^2}.$$

But we know from the preceding discussion, or we could verify directly, that this last equation is the same as

$$\frac{d}{dx}\left[e^{-(1/2)x^2}y\right] = xe^{-(1/2)x^2}.$$

Integrating both sides with respect to x gives

$$e^{-(1/2)x^2}y = \int xe^{-(1/2)x^2}\,dx + C$$

$$= -e^{-(1/2)x^2} + C.$$

Then, multiplying by $e^{+(1/2)x^2}$ gives

$$y = -1 + Ce^{(1/2)x^2}$$

for the solution. Figure 8 shows the graph of the particular solution satisfying $y(0) = 0$.

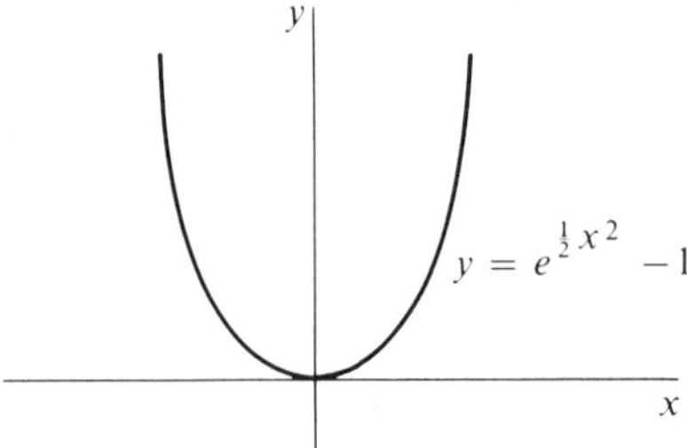

Figure 8

Two points should be emphasized about applying the exponential multiplier method:

1. The linear differential equation should be in the standard form

$$y' + g(x)y = f(x)$$

before identification of the coefficient function g for the purpose of computing $M(x) = e^{\int g(x)\,dx}$.

2. The differential equation

$$y' + g(x)y = f(x)$$

and its multiplied form

$$M(x)y' + M(x)g(x)y = M(x)f(x)$$

are completely equivalent to one another in the sense that any solution of one equation is also a solution of the other. The reason is that the multiplier M, being an exponential function, is never equal to zero, so we can multiply and divide by it as we please.

5B Applications

EXAMPLE 3   Suppose that a 100-gallon vat contains 10 pounds of salt dissolved in water and that a solution of the same salt is being run into the vat at a rate of 3 gallons per minute. The solution being run into the vat has a concentration that increases slowly with time according to the formula

$$C(t) = 1 - e^{-t/100}.$$

The solution is kept thoroughly mixed and the excess is drawn off, also at a rate of 3 gallons per minute. Let $S(t)$ stand for the amount of salt in the tank at time $t \geq 0$. We have

$$\frac{dS}{dt}(t) = 3C(t) - 3\frac{S(t)}{100}$$

$$= 3(1 - e^{-t/100}) - \frac{3}{100}S(t).$$

The resulting first-order equation is linear, and in standard form it is

$$\frac{dS}{dt} + \frac{3}{100}S = 3(1 - e^{-t/100}).$$

An exponential multiplier is given by

$$M(t) = e^{\int (3/100)\,dt} = e^{3t/100}.$$

Multiplying the equation by M puts it in the form

$$\frac{d}{dt}(e^{3t/100}S) = 3(e^{3t/100} - e^{2t/100}),$$

and integration with respect to t gives

$$e^{3t/100}S(t) = \int 3(e^{3t/100} - e^{2t/100})\,dt + K$$

$$= 100e^{3t/100} - 150e^{2t/100} + K.$$

Then multiplication by $e^{-3t/100}$ gives

$$S(t) = 100 - 150e^{-t/100} + Ke^{-3t/100}.$$

To determine the constant K, we recall that the vat initially contains 10 pounds of salt, so $S(0) = 10$. Then setting $t = 0$ in the formula for $S(t)$ gives

$$10 = -50 + K \quad \text{or} \quad K = 60.$$

Thus the desired particular solution is

$$S(t) = 100 - 150e^{-t/100} + 60e^{-3t/100}.$$

Notice that

$$\lim_{t \to \infty} C(t) = 1,$$

so the concentration of the solution being added approaches 1 pound per gallon. From this information we could conclude on physical grounds that the total amount of salt in the 100-gallon tank should approach 100 pounds; indeed, the formula for $S(t)$ shows that

$$\lim_{t \to \infty} S(t) = 100.$$

EXAMPLE 4   Let $f(t)$ be the concentration of a chemical solution on one side of a porous membrane, and let $u(t)$ be the concentration on the other side. Suppose that diffusion takes place through the membrane in such a way that

$$\frac{du}{dt} = 2(f(t) - u),$$

that is, such that the rate of change of u is proportional to the difference in concentrations. If $u(0) = 3$, and f is maintained so that

$$f(t) = \begin{cases} 4, & 0 \le t \le 10 \\ 1, & 10 < t, \end{cases}$$

then we can most easily solve the equation by writing it as

$$\frac{du}{dt} + 2u = 2f(t).$$

An exponential multiplier M is given by

$$M(t) = e^{\int 2\,dt} = e^{2t}.$$

Hence the differential equation can be written

$$\frac{d}{dt}(e^{2t}u) = 2e^{2t}f(t).$$

Integration of both sides from $t = 0$ to $t = s$ gives

$$\int_0^s \frac{d}{dt}[e^{2t}u(t)]\,dt = \int_0^s 2e^{2t}f(t)\,dt$$

or
$$e^{2s}u(s) - u(0) = \int_0^s 2e^{2t}f(t) \, dt.$$

Then

$$u(s) = u(0)e^{-2s} + 2e^{-2s} \int_0^s e^{2t}f(t) \, dt.$$

Using the integral with limits is convenient here because we can write, according to the definition of f,

$$\int_0^s e^{2t}f(t) \, dt = \begin{cases} \int_0^s 4e^{2t} \, dt, & 0 \le s \le 10, \\[2mm] \int_0^{10} 4e^{2t} \, dt + \int_{10}^s e^{2t} \, dt, & 10 \le s, \end{cases}$$

$$= \begin{cases} 2(e^{2s} - 1), & 0 \le s \le 10, \\[2mm] 2(e^{20} - 1) + \frac{1}{2}(e^{2s} - e^{20}), & 10 \le s. \end{cases}$$

$$= \begin{cases} 2(e^{2s} - 1), & 0 \le s \le 10, \\[2mm] \frac{3}{2}e^{20} - 2 + \frac{1}{2}e^{2s}, & 10 \le s. \end{cases}$$

Then

$$u(s) = 3e^{-2s} + \begin{cases} 4 - 4e^{-2s}, & 0 \le s \le 10, \\[2mm] (3e^{20} - 4)e^{-2s} + 1, & 10 \le s. \end{cases}$$

$$= \begin{cases} 4 - e^{-2s}, & 0 \le s \le 10, \\[2mm] (3e^{20} - 1)e^{-2s} + 1, & 10 \le s. \end{cases}$$

We can now return to the original independent variable t. Sketching the graph of the solution $u(t)$ is left as an exercise.

5C Existence and Uniqueness

Quite apart from its utility for writing down solution formulas in particular examples, the exponential multiplier technique answers, for linear equations, the question of when there exists a unique solution with given initial condition of the form

$$y(x_0) = y_0.$$

All we need to assume is that f and g in the equation $y' + gy = f$ are continuous on the relevant intervals. We then know that the factor $M(x)$ exists. We can next integrate both sides of the multiplied equation $(My)' = Mf$ between x_0 and x to get

$$M(x)y(x) - M(x_0)y(x_0) = \int_{x_0}^x M(u)f(u) \, du.$$

Because M is an exponential, it is never zero, so we can divide by it to find $y(x)$. What we have just done shows that $y(x)$, if it exists, must be given (uniquely!) by the resulting equation. And it is an easy matter to check that the resulting equation does in fact provide a solution.

General existence and uniqueness criteria for nonlinear differential equations are more technical and are taken up in Chapter 7.

5D Linearity

The assumption of linearity that we have been making can be looked at in the following way. We can introduce the abbreviation

$$L(y) = y' + g(x)y$$

for the left side of the standard form $y' + g(x)y = f(x)$. We can even write

$$L = \frac{d}{dx} + g(x)$$

with the understanding that L acts on functions $y = y(x)$ as defined previously. L is called an **operator,** acting on functions and producing functions $L(y)$.

EXAMPLE 5 Let $g(x) = x$. Then if $y = \sin x$,

$$L(y) = y' + g(x)y$$

$$= y' + xy$$

$$= \frac{d}{dx}(\sin x) + x \sin x$$

$$= \cos x + x \sin x;$$

in words, the operator $L = (d/dx) + x$ produces $\cos x + x \sin x$ from $\sin x$.

To say that L is a **linear** operator is to say that

$$L(y_1 + y_2) = L(y_1) + L(y_2) \tag{1}$$

and

$$L(cy) = cL(y), \qquad c \text{ constant.} \tag{2}$$

It is easy to check that L as defined satisfies both equations (1) and (2). Equations (1) and (2) can be combined into the single equation

$$L(c_1 y_1 + c_2 y_2) = c_1 L(y_1) + c_2 L(y_2), \qquad c_1, c_2 \text{ constant.} \tag{3}$$

Equation (3) has both equations (1) and (2) as special cases; take $c_1 = c_2 = 1$ to get equation (1), and $c_2 = 0$ to get equation (2). The verification of equation (3) goes like this:

Let y_1, y_2 be differentiable functions on some interval. The definition of L says

$$L(c_1 y_1 + c_2 y_2) = \frac{d}{dx}(c_1 y_1 + c_2 y_2) + g(x)(c_1 y_1 + c_2 y_2)$$

$$= c_1 y_1' + c_2 y_2' + c_1 g(x)y_1 + c_2 g(x)y_2$$

$$= c_1(y_1' + g(x)y_1) + c_2(y_2' + g(x)y_2)$$

$$= c_1 L(y_1) + c_2 L(y_2).$$

EXAMPLE 6   Let $g(x) = x$, $c_1 = 2$, and $c_2 = 5$. The preceding computation simply says that

$$\frac{d}{dx}(2y_1 + 5y_2) + x(2y_1 + 5y_2)$$

is the same as

$$2\left(\frac{d}{dx}y_1 + xy_1\right) + 5\left(\frac{d}{dx}y_2 + xy_2\right).$$

This simple rearrangement of terms does not seem very profound; the point is that operators for which it is possible to make such a rearrangement share many properties, making it possible to deal with large classes of them at once. To fully appreciate this unification, we will study many examples in Chapter 2. For now we will simply make the following observations.

Suppose we have in hand solutions to the two linear differential equations differing only on the right side,

$$y' + g(x)y = f_1(x) \quad \text{and} \quad y' + g(x)y = f_2(x);$$

let's call the respective solutions y_1 and y_2. Suppose then that we want a solution to

$$y' + g(x)y = c_1 f_1(x) + c_2 f_2(x), \tag{4}$$

where c_1, c_2 are constants. It follows from the linearity of the differential equations that

$$c_1 y_1 + c_2 y_2$$

is a solution to the new differential equation. The reason is that, using the abbreviation

$$L(y) = y' + g(x)y,$$

we have

$$L(c_1 y_1 + c_2 y_2) = c_1 L(y_1) + c_2 L(y_2)$$

$$= c_1 f_1 + c_2 f_2.$$

This is the same as equation (4) with $y = c_1 f_1 + c_2 f_2$. Linearity used in this way is sometimes called the **superposition principle.**

EXAMPLE 7   It is easy to check that

$$y_1(x) = \tfrac{1}{3}x^2 \quad \text{is a solution of} \quad y' + \frac{1}{x}y = x, \qquad x > 0,$$

and

$$y_2(x) = \tfrac{1}{4}x^3 \quad \text{is a solution of} \quad y' + \frac{1}{x}y = x^2, \qquad x > 0.$$

It follows from the preceding remarks that to find a solution of

$$y' + \frac{1}{x} = 3x - 5x^2$$

we need only take

$$y(x) = 3(\tfrac{1}{3}x^2) - 5(\tfrac{1}{4}x^3)$$
$$= x^2 - \tfrac{5}{4}x^3.$$

EXERCISES

1. Find an exponential multiplier $M(x)$ for each expression such that $M(x)$ times that expression can be written in the form $(d/dx)(M(x)y)$.

   (a) $y' + 2y.$ $\qquad\qquad\qquad$ (b) $\dfrac{dy}{dx} + y.$

   (c) $\dfrac{dy}{dx} + \dfrac{2}{x}y.$ $\qquad\qquad$ (d) $y' + e^x y.$

   (e) $e^x y' + e^{-x}y.$ $\qquad\qquad$ (f) $y'.$

2. Find the general solution of each of the following first-order equations. Then find the particular solution that satisfies the given initial condition.

   (a) $\dfrac{ds}{dt} + ts = t, \qquad s(0) = 0.$ $\qquad$ (b) $y' = y + 1, \qquad y(0) = 1.$

   (c) $2\dfrac{dy}{dx} = xy, \qquad y(1) = 0.$ $\qquad$ (d) $t\dfrac{dP}{dt} + P = t^3, \qquad P(1) = 0.$

   (e) $\dfrac{dx}{dy} + x = 1, \qquad x(1) = 1.$ $\qquad$ (f) $\dfrac{dx}{dt} = x + t, \qquad x(0) = 0.$

   (g) $y' + y = e^x, \qquad y(0) = 0.$ $\qquad\quad$ (h) $y' + \cos x = 0 \qquad y(0) = 2.$

3. Find the general solution of the differential equation

$$y' + f(x)y = x,$$

where $f(x) = \begin{cases} 1, & 0 \le x < 1, \\ 1/x, & 1 \le x. \end{cases}$ Try $M(x) = \begin{cases} e^x, & 0 \le x < 1, \\ ex, & 1 \le x. \end{cases}$

4. A 100-gallon tank full of salt solution contains initially 10 pounds of salt in solution. At $t = 0$, pure water is added at a rate of 2 gallons per minute, with a resulting overflow of 2 gallons per minute of salt solution. Assuming that the solution is kept thoroughly mixed at all times, find the amount of salt in the tank at time $t > 0$.

5. Salt solution enters a 100-gallon tank of initially pure water from two different sources. One source provides water containing 1 pound of salt per gallon at a rate of 2 gallons per minute. A second source provides 3 gallons of salt solution per minute at a varying concentration $C(t) = 2e^{-2t}$, measured in pounds of salt per gallon. Assume that the contents of the tank are kept thoroughly mixed at all times and that the solution is drawn off at a rate of 5 gallons per minute. Find the amount of salt in the tank at time $t > 0$.

6. The current $I(t)$ in an electric circuit satisfies the differential equation

$$L \frac{dI}{dt} + RI = E(t),$$

where L and R are positive constants representing inductance and resistance, respectively, and $E(t)$ is a variable applied voltage. Show that

$$I(t) = \frac{1}{L} e^{-Rt/L} \int_0^t E(u) e^{Ru/L}\, du + I(0)e^{-Rt/L}.$$

7. A pellet of mass m falling under the influence of gravity through a resisting medium has velocity $v(t)$ at time t, satisfying

$$m \frac{dv}{dt} = mg - kv.$$

Here g is the constant acceleration of gravity, and k is a positive constant that measures the resistance of the medium. Show that

$$v(t) = \left[v(0) - \frac{mg}{k} \right] e^{-kt/m} + \frac{mg}{k}.$$

8. Sketch the graph of the function $u(t)$ found at the conclusion of Example 4 in the text. What is the maximum value of $u(t)$? What is $\lim\limits_{t \to \infty} u(t)$?

9. Verify directly that, if $y_1(x)$ and $y_2(x)$ are solutions of the respective equations

$$y' + gy = f_1 \quad \text{and} \quad y' + gy = f_2,$$

then $c_1 y_1 + c_2 y_2$ is, for every pair of constants c_1, c_2, a solution of

$$y' + gy = c_1 f_1 + c_2 f_2.$$

10. Consider the constant coefficient linear equation $y' + ay = c$, with a, c constant, $a \ne 0$.
   (a) Show that the isoclines of the direction field of this equation are horizontal lines, and that every such line is an isocline.
   (b) Sketch the direction field associated with the differential equation $y' + 2y = 1$.

6 APPLICATIONS

The applications discussed in the earlier sections of this chapter have been kept fairly simple in order not to conflict too much with the discussion of the various types of first-order differential equation. The present section is organized more by area of application than by technique for solution. Also, these examples show that interpreting the solution formula can be more complicated than finding the formula. One reason for this is that the solution often contains parameters that must be adjusted to fit observed data.

6A Population Dynamics

In Example 1 of Section 3, we let $P(t)$ be the size of a population at time t and considered the consequences of the linear growth model

$$\frac{dP}{dt} = kP,$$

where k is a positive constant. The main conclusion is that the population grows exponentially, or that

$$P(t) = P_0 e^{kt},$$

where P_0 is the population size at time $t = 0$. Over short time intervals, these results fit certain populations, mainly bacteria, quite well. However, it is clear that $P(t)$ will tend to infinity as t tends to infinity, so without some restriction on t the model cannot be correct. A more realistic model for some populations is given by the **Verhulst–Pearl differential equation:**

$$\frac{dP}{dt} = kP\left(1 - \frac{P}{P_\infty}\right),$$

where P_∞ is a limiting size for the population, beyond which the growth rate is necessarily zero. As $P(t)$ approaches P_∞, the factor $1 - P/P_\infty$ approaches zero and so has an inhibiting effect on the growth of P. The Verhulst–Pearl equation can be solved by separation of variables:

$$\frac{dP}{P - P^2/P_\infty} = k\,dt, \qquad P \neq P_\infty \neq 0.$$

The partial fraction identity

$$\frac{1}{P} + \frac{1/P_\infty}{1 - P/P_\infty} = \frac{1}{P - P^2/P_\infty}$$

allows us to integrate the left side, and we get

$$\ln|P| - \ln\left|1 - \frac{P}{P_\infty}\right| = kt + c.$$

Then

$$\ln \left| \frac{P}{1 - P/P_\infty} \right| = kt + c$$

or, taking exponentials on both sides,

$$\frac{P}{1 - P/P_\infty} = Ke^{kt}, \qquad K = \pm e^c.$$

Solving for P gives

$$P(t) = \frac{P_\infty K e^{kt}}{P_\infty + K e^{kt}}.$$

Setting $P(0) = P_0$, we can solve the resulting equation for K; we get $K = P_0 P_\infty / (P_\infty - P_0)$. Substituting this value for K into our expression for $P(t)$ and simplifying, we find

$$P(t) = \frac{P_0 P_\infty}{(P_\infty - P_0)e^{-kt} + P_0}.$$

The whole point of the preceding analysis is to arrive at a general formula into which we can put some known information and from which we can then derive something we do not yet know.

EXAMPLE 1   If we know k, P_0, and P_∞ in the formula derived previously, then we can let t tend to infinity, and we see right away that the first term in the denominator tends to 0. Hence

$$\lim_{t \to \infty} P(t) = P_\infty.$$

This result is consistent with our original interpretation of P_∞ as a population size beyond which there is no further increase in size. We now see that, in theory, the value P_∞ is never actually attained in this model, but that the population size $P(t)$ approaches P_∞ asymptotically. Figure 9 shows the graphs of some typical relations for $P = P(t)$.

6B Radioactive Decay

A radioactive substance such as carbon 14 decays in such a way that the amount $y(t)$ present in a sample at time t very nearly satisfies a relation of the form *decay rate is proportional to amount present*, or

$$\frac{dy}{dt} = -ky, \qquad k > 0, \quad t \ge 0,$$

where k is called the **decay constant.** (Note that the *rate* of decay, dy/dt, is not constant in general.) The differential equation is both separable and linear, and we find

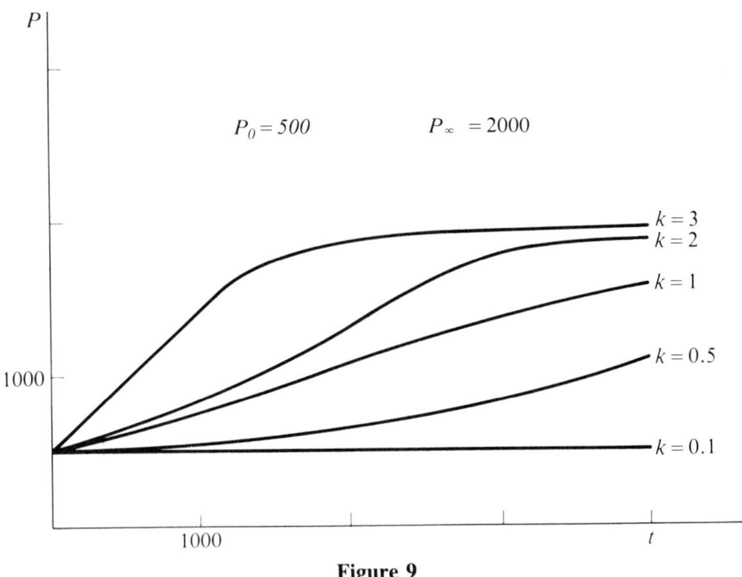

Figure 9

$$y(t) = y_0 e^{-kt}$$

for the general solution where y_0 is the amount present at time $t = 0$. Let $t_1 < t_2$ be two arbitrary times and consider the equation $y(t_2) = y(t_1)/2$, or

$$y_0 e^{-kt_2} = \tfrac{1}{2} y_0 e^{-kt_1};$$

we are asking for the time t_2 at which we have half as much as we did at time t_1. Canceling y_0, we find

$$e^{-k(t_2 - t_1)} = \tfrac{1}{2}.$$

Taking logarithms of both sides gives

$$-k(t_2 - t_1) = -\ln 2,$$

so

$$(t_2 - t_1) = \frac{\ln 2}{k}.$$

Thus the time interval required after t_1 for decay by 50% is the same regardless of what t_1 is. Thus we can speak of this time interval of length $T = t_2 - t_1$ as the **half-life** of the substance. By the previous equation, the half-life is

$$T = \frac{\ln 2}{k},$$

where k is the decay constant. In practice, the half-life is relatively easy to compute, and we can then compute the decay constant from it. For example, carbon 14 has a half-life of about $T \approx 5568$ years, so its decay constant is

$$k = \frac{\ln 2}{T} \approx 1.245 \cdot 10^{-4}.$$

The number of atoms of a substance in a typical sample is very large, on the order of 10^{23} per gram of the substance. However, for some purposes it is more convenient to think of $y(t)$ as a continuous approximation to a number of atoms rather than to a mass; this is so in the next example, in which a decay rate is expressed as a number of atoms per gram per minute.

EXAMPLE 2   In living wood the decay rate of carbon 14 is balanced by absorption of additional carbon 14 from the environment; the resulting decay rate for carbon 14 atoms in living wood is measured empirically to be about 6.68 per minute per gram of wood in recent epochs. When the tree dies, exponential decay takes over because natural absorption stops. If a tree is cut at time $t = 0$, we measure the decay rate at time t later and consider together the two equations

$$\left.\frac{dy}{dt}\right|_{t=0} = -ky_0 = -6.68$$

and

$$\frac{dy}{dt}(t) = -ky(t) = -r(t),$$

where $r(t)$ is the decay rate observed at time t. But $y(t) = y_0 e^{-kt}$, so

$$r(t) = ky_0 e^{-kt}$$
$$= 6.68 e^{-kt}.$$

Solving for t gives

$$t = -\frac{1}{k}\ln\left(\frac{r(t)}{6.68}\right).$$

Thus an observed rate of decay of 6.3 per minute per gram would give an age from cutting time of

$$t = -\frac{1}{1.245 \cdot 10^{-4}}\ln\left(\frac{6.3}{6.68}\right)$$

$$\sim 470 \text{ years.}$$

Note that we have changed the time unit to years in the last equation, since the half-life was measured in years. The decay rates 6.3 and 6.68 can both be measured in any other common time unit since we used only their ratio.

EXAMPLE 3   An atom of a radioactive substance typically decays into an atom of some other radioactive substance with its own decay constant. For example, uranium 238, with a half-life of $4.51 \cdot 10^9$ years and decay constant $m = 1.537 \cdot 10^{-8}$, decays to thorium 234 with a half-life of 0.066 years and decay constant $n = 10 \cdot 498$; after a sequence of 17 decay steps, the result, lead 206, is radioactively stable, and no further decay takes place in this sequence. Some of the longer-lived elements in the uranium series are useful for dating oil paintings that contain lead compounds, and it is important to understand the interplay between two or more elements in the series; we will consider, for example, the first two named here.

Let $U(t)$ be the amount of uranium 238 in a sample at time $t \geq 0$, and let $T(t)$ be the amount of thorium 234 in the same sample. Then

$$\frac{dU}{dt} = -mU$$

and
$$\frac{dT}{dt} = -nT - \frac{dU}{dt};$$

this last term is subtracted, because it is negative but we want to show the positive rate contribution of the decayed radium (i.e., thorium) to the rate of change of thorium. The preceding pair of differential equations is an example of a *system* of first-order equations, to be studied in more detail in Chapter 4. In this example, the system is easy to solve, because the first equation does not contain T; we solve it and get

$$U(t) = U(0)e^{-mt}.$$

The second equation is then

$$\frac{dT}{dt} = -nT + mU(0)e^{-mt}.$$

This is a first-order linear equation, and in standard form it looks like

$$\frac{dT}{dt} + nT = mU(0)e^{-mt}.$$

The integrating factor is e^{nt}, and the solution is

$$T(t) = \frac{mU(0)}{n - m}e^{-mt} + \left(T(0) - \frac{mU(0)}{n - m}\right)e^{-nt}.$$

In the concrete example mentioned previously, $m = 1.537 \cdot 10^{-8}$, and is so small that even over a period of 2000 years the change in $\exp(-mt)$ would be negligible. For this reason, if n is relatively large, the last term in the expression for $T(t)$ tends to zero fairly rapidly as t gets large, and the first term appears to be practically constant. Thus, instead of tending to zero as we would expect, there seems to be a kind of positive *equilibrium* value E for $T(t)$ as t gets moderately large:

$$E \sim \frac{mU(0)}{n - m}.$$

Over a long enough time period, $T(t)$ does tend to zero, of course.

6C Rocket Flight

We begin by analyzing the forces acting in a particular fixed direction, which might, for example, point along a line through the center of mass of the earth. We will let F represent the magnitude of the force acting on some object in that direction, m the mass of the object, and v the magnitude of its velocity in that direction. The **momentum** p of the object is by definition $p = mv$, where both $m = m(t)$ and $v = v(t)$ in general depend on time t. **Newton's second law** then says that

$$F = \frac{dp}{dt}$$

$$= \frac{d(mv)}{dt}.$$

It follows from the definition of momentum that, if the mass m of the object remains constant,

$$F = m\frac{dv}{dt}$$

$$= ma,$$

where $a = dv/dt$ is the acceleration. However, our concern here is with an object, a rocket, that consumes fuel at such a rapid rate that the change in mass due to expulsion of exhaust gas must be taken into account. To that end, we consider the change Δp in momentum over the time interval t to $t + \Delta t$. At time $t + \Delta t$, the rocket's momentum is

$$p + \Delta p = (m + \Delta m)(v + \Delta v) - \Delta m(w + \Delta w),$$

where w represents the velocity of the expelled exhaust. Note that v and w will have opposite signs, and that $\Delta m < 0$, since the rocket's mass is decreasing. If we now subtract $p = mv$, simplify, and divide by Δt, we get

$$\frac{\Delta p}{\Delta t} = m\frac{\Delta v}{\Delta t} + v\frac{\Delta m}{\Delta t} + \Delta m\frac{\Delta v}{\Delta t} - w\frac{\Delta m}{\Delta t} - \Delta m\frac{\Delta w}{\Delta t}.$$

As Δt tends to zero, so does Δm. It follows that

$$F = \frac{dp}{dt} = m\frac{dv}{dt} + (v - w)\frac{dm}{dt}.$$

The difference $v_e = v - w$ represents the relative velocity of separation of the rocket and the exhaust gas.

The principal components that make up the external force F for a rocket leaving the earth's surface are static air pressure, aerodynamic drag or friction, and gravity. With this in mind, we can write the **fundamental rocket equation** as

$$m \frac{dv}{dt} = -v_e \frac{dm}{dt} + F.$$

EXAMPLE 4   Consider a rocket with constant exhaust velocity $v_e > 0$ and constant rate of fuel consumption a. If m_0 is the initial fuel mass and m_1 is the mass of the rocket without fuel, then at time t the total mass of rocket plus fuel is

$$m(t) = m_1 + m_0 - at, \qquad 0 \le t \le \frac{m_0}{a}.$$

When $t > m_0/a$, then $m(t) = m_1$, a constant. Recall that the fundamental equation has the form

$$m \frac{dv}{dt} = -v_e \frac{dm}{dt} + F.$$

Neglecting atmospheric effects, the external force we consider is gravity near the earth's surface:

$$F = -mg.$$

The equation to be studied is thus

$$\frac{dv}{dt} = -\frac{v_e}{m} \frac{dm}{dt} - g.$$

Integration with respect to t gives

$$v(t) = -v_e \ln m(t) - gt + c.$$

The constant c can be determined by supposing the initial velocity is $v(0) = 0$, for example, so $c = v_e \ln m(0)$. Thus

$$v(t) = v_e \ln \left(\frac{m(0)}{m(t)} \right) - gt.$$

So far in the calculation, the rate of fuel consumption has not had any effect, but with our choice for $m(t)$ we get, if $t \le m_0/a$,

$$v(t) = v_e \ln \left(\frac{m_1 + m_0}{m_1 + m_0 - at} \right) - gt.$$

Clearly, the parameters m_0, m_1, and v_e should be chosen so that $v(t) \geq 0$. In particular, when the fuel is all used up at $t = m_0/a$, we want to arrange it so that

$$v_e \geq \frac{g m_0}{a \ln\left(1 + \dfrac{m_0}{m_1}\right)}.$$

EXERCISES

Population Dynamics

1. How should the constants k and P_0 be chosen in the solution

$$P(t) = P_0 e^{kt}$$

to the linear growth equation

$$\frac{dP}{dt} = kP$$

if $P(t)$ is to satisfy $P(10) = 100$ and $P(20) = 150$? With this choice for P_0 and k, what is $P(-10)$?

2. If a solution $P(t)$ for

$$\frac{dP}{dt} = kP, \qquad k > 0,$$

doubles during a time-period of length D, how long does $P(t)$ take to triple?

3. Assume $P(t)$ differentiable for all real t and also that $P'(t) = kP(t)$ for some constant k. Show that the acceleration $P''(t)$ of $P(t)$ is $k^2 P(t)$ and, more generally, that $P^{(n)} = k^n P(t)$.

4. If the population of the United States was 4,000,000 in 1790, 17,000,000 in 1840 and 63,000,000 in 1890, show that the size of the population from 1790 to 1890 cannot be approximated very well by a formula of the form $P(t) = P_0 e^{kt}$.

5. A natural generalization of the Verhulst–Pearl differential equation is

$$\frac{dP}{dt} = kP\left[1 - \left(\frac{P^q}{P_\infty}\right)\right], \qquad q > 0.$$

Note: In doing this problem, the formula

$$\int \frac{dP}{P(1 - P^q)} = \frac{1}{q} \ln\left(\frac{P^q}{1 - P^q}\right) + C$$

is useful; it can be derived using the substitution $P^q = x$.

(a) Show that the solution curves for this equation satisfy

$$\lim_{t \to \infty} P(t) = (P_\infty)^{1/q}.$$

(b) Find $P(t)$ if $P(0) = P_0$, noting the special case $P_0^q = P_\infty$.

6. Show that the Verhulst–Pearl differential equation can be written in the form

$$\frac{dP}{dt} = aP - bP^2,$$

and show how the parameters a and b are related to k and P_∞.

7. Writing the Verhulst–Pearl equation in the form

$$\frac{1}{P}\frac{dP}{dt} = k\left(1 - \frac{P}{P_\infty}\right)$$

shows that for this model the growth rate of the population per individual member decreases with rate $-k/P_\infty$ as P increases. For some insect populations, it turns out that the equation

$$\frac{1}{P}\frac{dP}{dt} = k\left(\frac{P_\infty - P}{P_\infty + cP}\right)$$

is more realistic, where c is a positive constant.

(a) Show that if $c = 0$ we get the Verhulst–Pearl equation.

(b) Find the general solution of this differential equation in implicit form.

(c) What form does the solution take if we determine the constant of integration by $P(0) = P_0$?

Radioactive Decay

8. Thallium 206 has a half-life of 4.19 minutes and decays atom-for-atom to lead 206. If we start with 1 gram of thallium 206, how much is left after 1 minute? How much is left after 1 hour? How much lead 206 will we have after 2 minutes?

9. Suppose that charcoal from an ancient campsite exhibits 0.9 disintegration per minute per gram of carbon 14 as compared with 6.68 disintegrations for live wood. Estimate the age of the charcoal.

10. Suppose that a sample of charcoal is known, for reasons having nothing to do with carbon 14, to be 1200 years old. What rate of carbon 14 decay would the sample be expected to exhibit?

11. If a sample of some radioactive element is reduced from 1.0 gram to 0.91 gram after 37 days, what is the half-life of the element?

12. One of the elements in the uranium series is uranium 234 with a half-life of $h_1 = 2.48 \cdot 10^5$ years. Uranium 234 decays to thorium 230 with a half-life of $h_2 = 80{,}000$ years. Thorium 230 decays to radium 226 with a half-life of $h_3 = 1622$ years.

(a) If a sample consists of pure uranium 234, what fraction will remain after 10,000 years?

(b) If a sample consists of 1 gram of uranium 234, how much thorium 230 will be present after 10,000 years?

(c) If a sample consists of 1 gram of uranium 234 and 1 gram of thorium 230, how much of each will be present after 10,000 years?

(d) If a sample consists of 1 gram of uranium 234, how much radium 225 will be present after 10,000 years?

13. Suppose we have an initial quantity U_0 of a radioactive element, say uranium, with decay constant m. Suppose the uranium decays to another radioactive element, say thorium, with initial quantity T_0 and decay constant n, where $n > m$. Show that if

$$nT_0 < mU_0$$

then, after time $t = 0$, the amount of thorium increases to an absolute maximum and then decays toward zero.

14. Branching sometimes occurs in radioactive decay. For example, bismuth 214 has a half-life of 19.7 minutes, but .9996 of the atoms decay to polonium 214, with a half-life of $1.64 \cdot 10^{-4}$ second, while the remaining .0004 decay to thallium 210, with a half-life 1.32 minutes. The decay of polonium and thallium each leads to the same element, lead 210. Let k_1, k_2, k_3 represent the decay constants for bismuth, polonium, and thallium, and let $0.9996 = p$. Derive a formula for the amount of lead 210 present at time t, in terms of k_1, k_2, k_3 and p, if we were to start at $t = 0$ with 1 gram of pure bismuth 214.

Rocket Flight

15. **(a)** Solve the fundamental rocket equation for constant exhaust velocity $v_e > 0$ under the assumption that external force $F = 0$ and that $v(0) = v_0 \geq 0$. Make no assumption about the form of $m(t)$.

(b) Accept the assumptions in part (a), and let fuel be consumed at a constant rate a so that $m(t) = m_1 + m_0 - at$. Assume that, as a result, exhaust velocity is proportional to $a : v_e = qa$. Show that velocity at burn-out (i.e., when $t = m_0/a$, where m_0 is the total mass of the fuel supply) is also proportional to a.

16. **(a)** Solve the fundamental rocket equation for constant exhaust velocity $v_e \geq 0$ under the assumption that $F = -gm$, where $m(t) = m_1 + m_0 - at$ and that $v(0) = v_0 \geq 0$.

(b) Integrate the result found in part (a) to find a formula for the height of a radially guided rocket starting from the earth's surface.

17. **(a)** Imagine a body of mass m_1 moving through space on a linear path with velocity v_0 and with no forces acting on it. At time $t = 0$, the body begins to change in mass, for example by evaporation, according to a formula $m = m(t)$. Show that after $t = 0$ the velocity satisfies

$$m\frac{dv}{dt} + v\frac{dm}{dt} = 0.$$

(b) Show that after $t = 0$ the velocity of the body in part (a) satisfies

$$v(t) = \frac{m_1 v_0}{m(t)}.$$

(c) Find a formula for the distance traveled after time $t = 0$ by the body described in part (a) if the mass at time t is $m(t) = m_1 + at$, where a is a positive constant.

(d) Find the analogous formula to that in part (c) if $m(t) = m_1 + at^2$, $a > 0$. Then show that the distance traveled after $t = 0$ is bounded above by $v_0\sqrt{m_1/a}$.

18. Aerodynamic friction exerts a force F_1 roughly proportional to velocity: $F = -kv$, $k > 0$. Solve the fundamental rocket equation assuming constant exhaust velocity, initial velocity

$v(0) = 0$, and

$$m(t) = m_1 + m_0 - at, \qquad a > 0.$$

19. A multistage rocket can be thought of as a succession of single-stage rockets in which the initial velocity for each stage is the same as the terminal velocity for the preceding state. Consider a two-stage rocket where the ith stage has rocket mass m_i, total fuel mass f_i, exhaust velocity v_i, and fuel consumption rate a_i, $i = 1, 2$. Assume the first stage ends when its fuel is all gone and that the assembly drops the mass m_1 at that instant. Set up the two differential equations for the flight velocity, assuming no external forces and initial first-stage velocity v_0.

20. Consider the rocket equation

$$m\frac{dv}{dt} = v_e \frac{dm}{dt} + F(t)$$

with constant exhaust velocity $v_e > 0$ and external force $F(t)$ depending only on time.
(a) Assuming $v(t)$ known and $F(t)$ identically zero, derive the formula

$$m(t) = m(0)\, e^{(v(0) - v(t))/v_e}, \; t \geq 0.$$

(b) Derive a formula analogous to the one in part (a) for the case in which $F(t)$ is continuous for $t \geq 0$.

7 NUMERICAL METHODS

A solution

$$y = f(x)$$

for a first-order differential equation

$$\frac{dy}{dx} = F(x, y)$$

can sometimes be expressed by a formula such as the simple example

$$y = ce^{-ax},$$

where a and c are parameters that depend on an initial condition or on some property of $F(x, y)$. The importance of such formulas is that they often show quite easily how the values of the solution depend not only on the independent variable x, but also on the parameters; in our example, it is quite clear that if a is positive, then y tends to 0 as x tends to ∞, regardless of what value c has. However, when solution formulas are not readily available, we rely on numerical approximations. And if our purpose in finding a solution formula is simply to generate numerical values, it may be preferable to bypass the formula and work directly with the differential equation, as we do in this section.

7A Euler Method

Suppose we are given a first-order differential equation together with an initial condition, say

$$y' = \frac{x^2 + y^2}{2}, \qquad y(-2) = -1,$$

and we want to find approximate values for a solution $y = y(x)$. Since we can make only finitely many estimates, we restrict ourselves to some interval of values, say $x_0 \le x \le b$. We then try to approximate $y(x)$ at the points

$$x_1 = x_0 + h, \; x_2 = x_0 + 2h, \; \ldots, \; x_m = x_0 + mh,$$

where m is some chosen number of subdivisions of the interval from x_0 to b, each of the same length

$$h = \frac{b - x_0}{m}.$$

To make our estimates, we approximate the graph of $y = y(x)$ by a sequence of straight-line segments chosen from the direction field of our differential equation

$$y' = F(x, y),$$

starting with the initial condition $y(x_0) = y_0$. The segment of the direction field through (x_0, y_0) has slope $F(x_0, y_0)$, and we extend that segment until it intersects the vertical line $x = x_1$ at a point (x_1, y_1). We then simply repeat the process with (x_1, y_1), replacing (x_0, y_0) to get (x_2, y_2), and so on to (x_m, y_m). Figure 10 shows such a sequence of approximating segments along with the solution curve that they approximate. Each segment is determined by its initial point (x_k, y_k) and the slope $F(x_k, y_k)$; the equation of the line containing the segment is thus

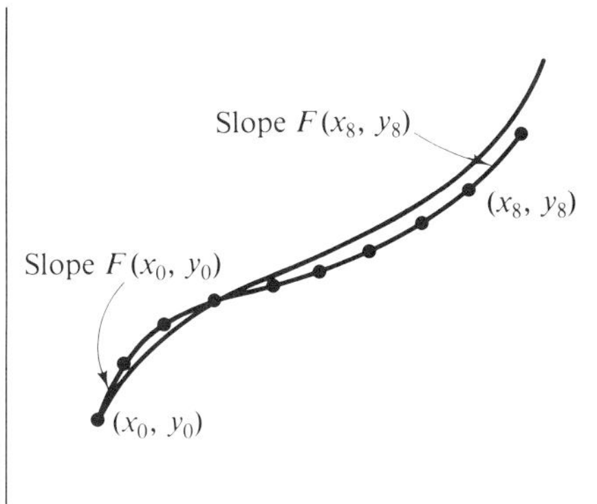

Figure 10

$$y - y_k = F(x_k, y_k)(x - x_k)$$

or
$$y = y_k + F(x_k, y_k)(x - x_k).$$

On this line, the y-value corresponding to $x = x_{k+1}$ is then

$$y_{k+1} = y_k + F(x_k, y_k)(x_{k+1} - x_k),$$

or, since $h = x_{k+1} - x_k$,

$$y_{k+1} = y_k + hF(x_k, y_k).$$

The preceding formulas define the **Euler approximation.** Starting with x_0 and y_0, we get a table of approximations to the solution $y = y(x)$.

k	x_k	$y_{k+1} = y_k + hF(x_k, y_k)$
0	x_0	y_0
1	x_1	$y_1 = y_0 + hF(x_0, y_0)$
2	x_2	$y_2 = y_1 + hF(x_1, y_1)$
3	x_3	$y_3 = y_2 + hF(x_2, y_2)$
.	.	.
.	.	.
.	.	.

A computer program to print 100 approximate values, with step size $h = 0.01$, might look like this:

```
DEF F(X,Y)=X*Y
LET X=0
LET Y=0
LET H=.01
FOR K=1 TO 100
LET Y=Y+H*F(X,Y)
PRINT X,Y
LET X = X+H
NEXT K
```

EXAMPLE 1   The differential equation

$$y' = xy$$

has solutions $y = c\,\exp(x^2/2)$, which we will compare with the results of our numerical approximation method. We have

$$y_{k+1} = y_k + hF(x_k, y_k)$$
$$= y_k + hx_ky_k.$$

With $h = 0.1$, the next table shows the comparison, with the values of the exact solution rounded off. What we get is a fair approximation for some purposes, for example, for drawing a graph. Notice that the accuracy deteriorates as we go down the columns, because the errors compound each other. We have started with initial values $(x_0, y_0) = (0, 1)$.

x_k	y_k	$e^{(1/2)x_k^2}$
0.0	1	1
0.1	1	1.005
0.2	1.01	1.020
0.3	1.03	1.046
0.4	1.06	1.083
0.5	1.10	1.133
0.6	1.16	1.197
0.7	1.23	1.277
0.8	1.31	1.377
0.9	1.42	1.499
1.0	1.55	1.647

To improve the accuracy, we could make h smaller, thus increasing m, the number of subdivisions of the interval from x_o to b. Of course, increasing the number of subdivisions increases the likelihood of significant round-off error in the arithmetic. Rather than recklessly decrease h, we prefer to use a simple modification of the method, described next, that produces a smaller error at each step.

7B Modified Euler Method

To improve the performance of the Euler method we use a process called **prediction–correction**; this assigns a slope to each approximating segment that is an average of the Euler slope and of what the next Euler slope would be if we were to take a single Euler step.

We will now use y_{k+1} to denote our improved approximate value at the $(k + 1)$th step, and we will use p_{k+1} for the corresponding simple Euler approximation, based on a previously computed value y_k. We follow these steps to solve $y' = F(x, y)$, $y(x_0) = y_0$ approximately:

(i) Compute the slope $F(x_k, y_k)$

(ii) Determine p_{k+1} by

$$p_{k+1} = y_k + hF(x_k, y_k).$$

(iii) Compute the average of the two slopes $F(x_k, y_k)$ and $F(x_{k+1}, p_{k+1})$ and use it to determine y_{k+1} by

$$y_{k+1} = y_k + h\left[\frac{F(x_k, y_k) + F(x_{k+1}, p_{k+1})}{2}\right].$$

The formula for y_{k+1} comes from writing the equation for the line through (x_k, y_k) with the average slope and then setting $x = x_{k+1}$ to get the corresponding value $y = y_{k+1}$. The general idea is illustrated in Figure 11. A computer program to implement the method might look like this, with step size $h = 0.01$:

```
DEF F(X,Y) = X*Y
LET X=X0
LET Y=Y0
LET H=.01
FOR K=1 TO 100
LET P=X+H*F(X,Y)
LET Y=Y+.5*H*(F(X,Y)+F(X+H,P))
PRINT X,Y
LET X = X+H
NEXT K
```

At each stage the value p_k is the *prediction* and y_k is the *correction*. If $h < 0$ the approximations move from larger to smaller x values.

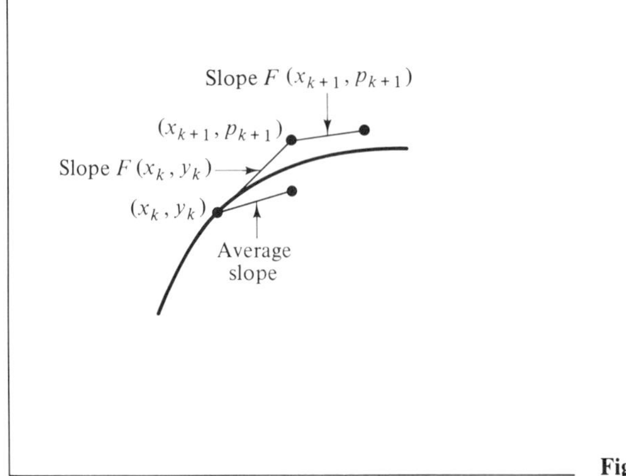

Figure 11

EXAMPLE 2   Here is the modified Euler approximation for the same problem we looked at in Example 1: $y' = xy$ with $(x_0, y_0) = (0, 1)$, and step size

$h = 0.1$. We have $F(x, y) = xy$, so

$$p_{k+1} = y_k + hF(x_k, y_k)$$

$$= y_k + 0.1 x_k y_k$$

and

$$y_{k+1} = y_k + \frac{h}{2}[(F(x_k, y_k) + F(x_{k+1}, p_{k+1})]$$

$$= y_k + 0.05\left[(x_k y_k + x_{k+1}\, p_{k+1})\right].$$

A table of values looks like this:

x_k	p_k	y_k	$e^{(1/2)x_k^2}$
0	1	1	1
0.1	1	1.005	1.005
0.2	1.015	1.020	1.020
0.3	1.041	1.046	1.046
0.4	1.077	1.083	1.083
0.5	1.127	1.133	1.133
0.6	1.190	1.197	1.197
0.7	1.269	1.277	1.277
0.8	1.367	1.377	1.377
0.9	1.487	1.499	1.499
1.0	1.634	1.648	1.649

The values of p_k, y_k and the exact solution are rounded back to three decimal places. Notice that there is a difference between the last two columns only at $x = 1$.

The arithmetic required in the preceding examples can be done quite easily with a hand calculator. However, if you want many thousands of steps, an automatic digital computer is probably essential. It is important to remember that reducing step size h cannot indefinitely improve the accuracy of the approximations, however. The increased amount of arithmetic required will eventually cause enough round-off error such that further reduction of h is self-defeating. It can be shown that the error made at each step by using an approximate formula, called the **local formula error,** is on the order of h^2 for the Euler method and h^3 for the modified Euler method. Thus, taking $h = 0.01$ can be expected to produce about 0.0001 amount of error in one Euler step and about 0.000001 in one modified Euler step. The advantage of the modification is obvious, because to get the latter accuracy from the simple Euler method would require 1000 times as many steps as the modified method over a given interval, thereby increasing by a factor of 1000 the **round-off error** inherent in the arithmetic, in particular in the number of significant digits in the calculations. As a practical matter we often try a reasonable-looking step size and then accept the results if cutting the step size in half produces no significant change in the final outcome.

E X E R C I S E S

1. Each of the following differential equations has a unique solution, expressible in terms of elementary functions, that satisfies the prescribed condition. First find the solution formula; then compare the values given by this formula with approximate values that you compute using the Euler method. Use at least 5 equally spaced steps on the given interval if you compute by hand and at least 20 steps if you use a computer.
   (a) $y' = y$ for $0 \le x \le 0.5$ with $y = 1$ when $x = 0$.
   (b) $y' = y + x$ for $0 \le x \le 0.5$ with $y = 0$ when $x = 0$.

2. Substitute the modified Euler method for the Euler method in each part of Exercise 1.

3. Consider the differential equation

$$y' = 1 + y^2 \qquad \text{for} \qquad -1 \le x \le 1.$$

   (a) Sketch the direction field of the differential equation.
   (b) Sketch a solution curve passing through $(x, y) = (0, 0)$.
   (c) Apply the Euler method to approximate the solution in part (b) at points $x_k = k/5$ for $k = 1$ through $k = 5$.
   (d) Use the modified Euler method to print a table of values for the solution satisfying $y(0) = 0$ for $-1 \le x \le 1$.
   (e) Compare the results of (d) with the solution $y = \tan x$.

4. The initial-value problem

$$y' = f(x), \qquad y(x_0) = y_0$$

   has the solution

$$y = y_0 + \int_{x_0}^{x} f(t)\, dt.$$

   (a) Show that applying the Euler method to the differential equation is equivalent to approximating the solution integral by Riemann sums (i.e., by using the rectangle rule).
   (b) Show that the modified Euler method for the differential equation is equivalent to using the trapezoid rule for approximating the solution integral.

5. Write a computer-plotting program that plots a sketch of the direction field of a first-order differential equation $y' = f(x, y)$ and then plots the solutions of a sequence of initial-value problems determined by conditions of the form $y(x_0) = y_0$. Apply the program to each of the following differential equations, plotting the direction field for points (x, y) in the rectangle $|x| \le 4$, $|y| \le 4$, and plotting the solution satisfying $y(-1) = -1$.
   (a) $y' = -y + x$.   (b) $y' = xy$.
   (c) $y' = \sin y$.   (d) $y' = \sin xy$.
   (e) $y' = \sqrt{1 + y^2}$.   (f) $y' = \sqrt{1 + x^2}$.

6. The fourth-order **Runge–Kutta** approximations to the solution of $y' = F(x, y)$, $y(x_0) = y_0$, are computed as follows:

$$y_{k+1} = y_k + \frac{h}{6}(m_1 + 2m_2 + 2m_3 + m_4),$$

where $m_1 = hF(x_k, y_k)$

$$m_2 = hF\left(x_k + \frac{h}{2}, y_k + \frac{m_1}{2}\right)$$

$$m_3 = hF\left(x_k + \frac{h}{2}, y_k + \frac{m_2}{2}\right)$$

$$m_4 = hF(x_k + h, y_k + m_3).$$

(The local formula error is of order h^5.) Write a computer program to implement the routine. Then compare the results obtained from it for the problem

$$y' = xy, \qquad y(0) = 1,$$

with the results obtained from the modified Euler method, as well as with the solution formula $y = \exp(x^2/2)$. Try step sizes $h = 0.1, 0.01,$ and 0.001.

7. The initial value problem

$$y' = \sqrt{y}, \; y(0) = 0,$$

has two distinct solutions for $x \geq 0$:

$$y_1(x) = 0 \qquad \text{and} \qquad y_2(x) = \frac{x^2}{4}.$$

(a) Show that direct application of the Euler method produces y_1 exactly at the points of approximation.

(b) Show that applying the Euler method to

$$y' = \sqrt{y}, \; y(x_0) = \frac{x_0^2}{4},$$

for small positive x_0 produces an approximation to y_2.

8. Consider the modified Verhulst-Pearl equation

$$\frac{dP}{dt} = kP^r\left(1 - \frac{P^q}{m}\right).$$

(a) Let $k = 1, m = 5$ and $P(0) = 1$. Find numerical approximations to the solution in the range $0 \leq t \leq 10$ for the parameter pairs $(r, q) = (0.5, 1), (0.5, 2), (1.5, 1), (1.5, 2), (2, 2)$.

(b) Estimate a parameter pair (r, q) that yields approximately the values $P(0) = 1$, $P(2) = 2.4$, $P(4) = 2.9$.

9. A 100-gallon mixing tank is full at the time $t = 0$. Salt solution at a concentration of 1 pound per gallon is thereafter added at a decreasing rate of $e\,3^{-0.5t}$ gallons per minute, and solution is drawn from the tank so as to maintain the 100-gallon level.

(a) Find the first order equation satisfied by the amount $S(t)$ of salt in the tank at time t minutes.

(b) Find a numerical approximation to $S(t)$ in the range $0 \leq t \leq 20$, assuming that $S(0) = 0$. Sketch the graph of the solution.

10. The rocket equation

$$m\frac{dv}{dt} = -v_e\frac{dm}{dt} + F$$

becomes a first order equation for $v(t)$ if we specify time dependent mass $m(t) = m_0 + m_1 - at$ for $0 \leq t \leq m_0/a$, and $m(t) = m_1$ for $m_0/a < t$. Let the external force F be given by

$$F(t) = -gm(t) - kv^{\alpha},$$

where g is the acceleration of gravity, and k and α are positive constants.

(a) Find numerical approximations to $v(t)$ in the range $0 \leq t \leq m_0/a$, given that $g = 32.2$, $v(0) = 0$, $m_0 = 25$, $m_1 = 75$, $a = 1$, $v_e = 3500$, and $\alpha = 1.5$. Use successively $k = 2, 1, 0.5, 0.1$, and sketch the corresponding graphs for $v = v(t)$ using an appropriate scale.

(b) Estimate the value of k in part (a) that would produce approximately the values $v(0) = 0$, $v(5) = 17.5$, $v(10) = 40$, $v(15) = 63$, and $v(20) = 87$.

(c) Repeat part (a) for the range $0 \leq t \leq 50$, noting that $m(t)$ cannot be described by a single elementary function.

11. The **logistic equation**

$$\frac{dP}{dt} = kP\left(1 - \frac{P}{L(t)}\right)$$

is a generalization of the Verhulst-Pearl equation discussed in Section 6A. Here $L(t)$ acts to limit population size $P(t)$. Let $k = 1$ and $P(0) = \frac{1}{2}$, and use numerical analysis to sketch the graph of $P(t)$ for $0 \leq t \leq 25$. In each case compare the graph of $P(t)$ with that of $L(t)$.

(a) $L(t) = 3 + \sin t$.   (b) $L(t) = \ln(2 + t)$.   (c) $L(t) = t + 1$.

12. The velocity $v = v(t)$ of a rocket of mass $m(t) = m_1 + m_0 - at$ satisfies

2

Higher-Order Equations

1 INTRODUCTION

Here we move from first-order differential equations, which were treated in Chapter 1, to higher-order equations, such as

$$y'' = f(x, y, y')$$

in which the **order** of the equation is the same as the highest-order derivative in it, but for which the geometric interpretation of the equation is no longer so simple. Fortunately, there is a fairly large class of such equations, with interesting applications, for which there is a widely applicable method for finding solution formulas; these are the linear constant coefficient equations, for example

$$y'' + y = 0,$$
$$y'' - 3y' + 2y = e^x,$$
$$y'' - y = x,$$
$$y''' - y = 1.$$

We will concentrate mainly on the second-order case, partly because those are the ones that most often come up in applied mathematics, and partly because the higher-order extension does not involve any very different ideas.

The standard example of the kind of application considered later in this chapter is

$$m \frac{d^2x}{dt^2} + k \frac{dx}{dt} + hx = F(t),$$

where m is the mass of an oscillating weighted spring, k is a friction coefficient, h is a spring-stiffness coefficient, and F represents externally applied force. The methods of the next few sections are aimed at giving a complete analysis of the initial-value problem for this and many related differential equations, that is, for predicting what will happen to the mechanism in the future if we know the initial position $x(t_0)$ and initial velocity $x'(t_0)$ at some time t_0.

1A Exponential Solutions

The differential equation

$$y' - ry = 0,$$

with r constant, was treated in Sections 3 and 5 of Chapter 1. Using the exponential multiplier method of Section 5, it is easy to show that every solution is of the form

$$y = ce^{rx}, \qquad c \text{ constant.}$$

The reason is that multiplication of the equation by e^{-rx} gives the equivalent equation

$$e^{-rx}y' - re^{-rx}y = 0$$

or
$$(e^{-rx}y)' = 0.$$

It follows that $e^{-rx}y = c$, so $y = ce^{rx}$. It turns out that, with a suitable extension of the definition of the exponential function, every solution of an nth-order constant-coefficient equation of the form

$$a_n y^{(n)} + a_{n-1}y^{(n-1)} + \cdots + a_1 y' + a_0 y = 0, \qquad a_n \neq 0,$$

can be written as a *sum* of simple multiples of exponential functions. Note that the left side is a sum of constant multiples of y and its derivatives. We will often refer to a sum of multiples of functions as a *linear combination* of those functions. All the relevant ideas are explained in the next three sections, and here we will consider only the more accessible examples.

EXAMPLE 1   We look for exponential solutions to

$$y'' - 5y' + 6y = 0,$$

having the form $y = e^{rx}$, r constant. We have $y' = re^{rx}$, $y'' = r^2 e^{rx}$, so to satisfy the equation we need

$$r^2 e^{rx} - 5re^{rx} + 6e^{rx} = 0$$

or
$$(r^2 - 5r + 6)e^{rx} = 0.$$

The exponential factor e^{rx} is never zero, so the last equation is satisfied if and only if

$$r^2 - 5r + 6 = (r - 2)(r - 3) = 0.$$

Letting $r = 2$ and $r = 3$, we get the respective solutions

$$y = e^{2x}, \qquad y = e^{3x}.$$

It is easy to verify directly that these two functions are solutions of

$$y'' - 5y' + 6y = 0.$$

It is also straightforward to verify that, for any constants c_1 and c_2, the function

$$y = c_1 e^{2x} + c_2 e^{3x}$$

is a solution. To find the particular solution satisfying

$$y(0) = 1, \qquad y'(0) = -1,$$

we compute $y' = 2c_1 e^{2x} + 3c_2 e^{3x}$ and substitute $x = 0$ in the expressions for y and y'. We find

$$c_1 + c_2 = 1,$$

$$2c_1 + 3c_2 = -1.$$

Solving this pair of equations for c_1 and c_2 gives $c_1 = 4$, $c_2 = -3$. Hence

$$y = 4e^{2x} - 3e^{3x}$$

is the desired solution.

One thing Example 1 shows is that to find exponential solutions to

$$ay'' + by' + cy = 0$$

the thing to do is solve the associated **characteristic equation**

$$ar^2 + br + c = 0.$$

If the roots $r = r_1$ and $r = r_2$ are distinct real numbers, then the general solution will turn out to be

$$y = c_1 e^{r_1 x} + c_2 e^{r_2 x}.$$

Particular solutions will be determined by imposing two conditions on y in order to fix the values of c_1 and c_2. The roots r_1, r_2 are called **characteristic roots.**

EXAMPLE 2   The differential equation

$$y'' - 2y' - y = 0$$

has the characteristic equation

$$r^2 - 2r - 1 = 0,$$

with roots

$$r = \frac{2 \pm \sqrt{8}}{2} = 1 + \sqrt{2}, \qquad 1 - \sqrt{2}.$$

The corresponding solutions are

$$y_1(x) = e^{(1+\sqrt{2})x}, \qquad y_2(x) = e^{(1-\sqrt{2})x}.$$

The general solution is

$$y = c_1 e^{r_1 x} + c_2 e^{r_2 x}, \qquad r_1 = 1 + \sqrt{2} \qquad r_2 = 1 - \sqrt{2}.$$

Two features of the solution formula in Example 2 are important enough that they have special names. First, the solution is a sum of constant multiples of other solutions, which we express by saying that the solution is a **linear combination** of solutions. Second, neither of the exponentials is a numerical multiple of the other, which we express by saying that the exponentials are **linearly independent;** without linear independence we would have essentially one term in the solution formula, since the constants could be combined into one.

1B Factoring Operators

Differential operator notation will be useful here, as well as later, for solving systems of differential equations, so we will consider some of their basic properties here. We let $D = d/dx$; that is, we let D stand for differentiation with respect to some specified variable, call it x here, and interpret $D - 2$, $D^2 - 1$, and similar expressions, as operators acting on sufficiently often differentiable functions $y = y(x)$. For example,

$$(D + 1)e^{2x} = De^{2x} + e^{2x} = 3e^{2x};$$

$$(D^2 + 2) \sin x = D(D \sin x) + 2 \sin x$$

$$= D \cos x + 2 \sin x = \sin x;$$

$$(D - 2)y = Dy - 2y$$

$$= y' - 2y;$$

$$(D^2 - 1)y = D^2 y - y$$

$$= D(Dy) - y = y'' - y.$$

The most important remark about such operators for our immediate use is that they can be factored if the coefficients are constants. Specifically, if r and s are constants,

$$(D + r)(D + s)y = (D + r)(y' + sy)$$

$$= D(y' + sy) + r(y' + sy)$$

$$= y'' + (r + s)y' + (rs)y$$

$$= [D^2 + (r + s)D + rs]y.$$

Interchanging r and s really changes nothing in this last expression, so we can conclude that

1.1
$$(D + r)(D + s) = (D + s)(D + r)$$

$$= D^2 + (r + s)D + rs,$$

in the sense that all three operators have the same effect on twice-differentiable functions y, assuming r and s are constants. Furthermore, it follows that, if $r = r_1$ and $s = r_2$ are the roots of the characteristic equation of

$$y'' + ay' + by = 0,$$

then

1.2
$$y'' + ay' + by = (D - r_1)(D - r_2)y$$

$$= (D - r_2)(D - r_1)y.$$

EXAMPLE 3   The differential equation

$$y'' + 2y' + y = 0$$

can be written

$$(D^2 + 2D + 1)y = 0$$

or
$$(D + 1)^2 y = 0.$$

Note that the characteristic equation is

$$r^2 + 2r + 1 = 0 \qquad \text{or} \qquad (r + 1)^2 = 0,$$

so there is a double root $r = -1$ with associated solution $y = e^{-x}$. To find another solution, suppose that y is some solution and let $z = (D + 1)y$. Then

$$(D + 1)z = (D + 1)^2 y = 0$$

is a first-order equation satisfied by z:

$$z' + z = 0.$$

The solutions of this equation are all of the form

$$z = c_1 e^{-x}.$$

It follows from $z = (D + 1)y$ that y must satisfy the additional first-order equation

$$y' + y = c_1 e^{-x},$$

for some choice of constant c_1. But this equation can be solved by multiplying by e^x to get $e^x y' + e^x y = c_1$, or

$$(e^x y)' = c_1.$$

Hence $e^x y = c_1 x + c_2$ and

$$y = c_1 x e^{-x} + c_2 e^{-x}.$$

This is the general solution of the differential equation.

The following theorem is proved by applying the method of Example 3.

1.3 Theorem.   Suppose the characteristic equation of

$$y'' + ay' + by = 0$$

has real roots r_1 and r_2. If $r_1 \neq r_2$, the general solution is

$$y = c_1 e^{r_1 x} + c_2 e^{r_2 x}.$$

If $r_1 = r_2$, the general solution is

$$y = c_1 x e^{r_1 x} + c_2 e^{r_1 x}.$$

Proof.   In operator form, the differential equation is

$$(D - r_1)(D - r_2)y = 0.$$

Let $z = (D - r_2)y$ for some solution y. Then solve $(D - r_1)z = 0$ to get

$$z = c_1 e^{r_1 x}.$$

The solution y then satisfies

$$(D - r_2)y = c_1 e^{r_1 x},$$

and multiplication by $e^{-r_2 x}$ gives

$$D(e^{-r_2 x} y) = c_1 e^{(r_1 - r_2)x}.$$

If $r_1 \neq r_2$, we integrate to get

$$e^{-r_2 x} y = \frac{c_1}{r_1 - r_2} e^{(r_1 - r_2)x} + c_2,$$

so

$$y = \frac{c_1}{r_1 - r_2} e^{r_1 x} + c_2 e^{r_2 x}.$$

For neatness, we can rename the constant $c_1/(r_1 - r_2)$ and call it c_1.

If $r_1 = r_2$, we have $D(e^{-r_2 x} y) = c_1$, so integration gives

$$e^{-r_2 x} y = c_1 x + c_2.$$

In that case,

$$y = c_1 x e^{r_2 x} + c_2 e^{r_2 x}.$$

What to do in case the roots r_1, r_2 are complex numbers is taken up in the next section. For now we stay with examples having real roots for the characteristic equation. In any case it will be important for us that the general solution is a **linear combination** of simpler solutions, that is, a *sum of multiples* of simpler solutions.

EXAMPLE 4   Consider the differential equation

$$(D - r)(D - (r + h))y = 0,$$

or

$$y'' - (2r + h)y' + r(r + h)y = 0.$$

If $h = 0$, the characteristic equation has r for a double root. Otherwise, the general solution is

$$y = c_1 e^{(r+h)x} + c_2 e^{rx}.$$

If we let h tend to zero, all we get from the preceding formula is $y = (c_1 + c_2)e^{rx}$, which is only a partial solution. However, the constants c_1, c_2 are at our disposal while h tends to zero, so we can replace them by $c_1 = 1/h$, $c_2 = -1/h$. For each fixed h, the resulting solution is

$$y = \frac{1}{h} e^{(r+h)x} - \frac{1}{h} e^{rx}$$

$$= \frac{1}{h}[e^{(r+h)x} - e^{rx}].$$

The limit of this function as $h \to 0$ is, for each fixed x, the definition of the derivative of e^{rx} *with respect to* r. Thus the limit of the solution is

$$y = \frac{d}{dr} e^{rx} = x e^{rx}.$$

The computation we just did does not by itself prove that $x e^{rx}$ is a solution of

$$y'' - 2ry' + r^2 y = 0$$

(we already know that), but it does show that the apparently abrupt change in the general solution formula as we pass from unequal to equal roots is in fact very natural geometrically. Figure 1 shows the graph of $y = x$ (corresponding to double root $r = 0$), together with the graph of $y = (e^{hx} - 1)/h$ for various values of h getting closer to zero. It is important to understand that, while the

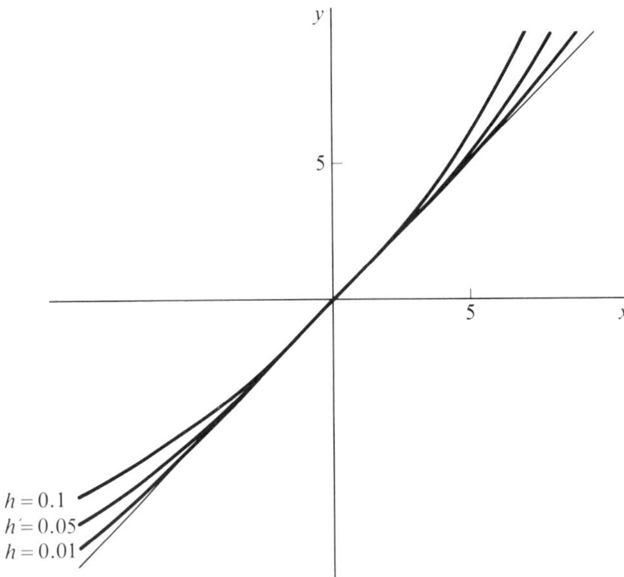

Figure 1

latter graphs get closer to the graph of $y = x$ over any fixed x-interval, the behavior of the exponential is overall very different from that of $f(x) = x$ for $-\infty < x < \infty$.

EXERCISES

1. With $D = \dfrac{d}{dx}$, compute

   (a) $(D + 1)e^{-2x}$. (b) $(D^2 + 1)e^x$. (c) $D^3 e^{3x}$.
   (d) $(D^2 + D - 1) \sin x$. (e) $(D^2 + 1)x \cos x$.

2. Find the characteristic equation of each of the following differential equations. Then solve the characteristic equation and use the roots to write down the general solution of the differential equations.

   (a) $y'' + y' - 6y = 0$. (b) $2y'' - y = 0$.
   (c) $y'' + 2y' + y = 0$. (d) $y'' + 3y' + y = 0$.
   (e) $y'' - y' = 0$. (f) $y'' - 3y' - y = 0$.
   (g) $2y'' - 3y' + y = 0$. (h) $3y'' + 3y' = 0$.

3. For each differential equation in Exercise 2, find the particular solution satisfying the corresponding condition given next. Then sketch the graph of that particular solution.

   (a) $y(0) = 2,\ y'(0) = 2$. (b) $y(0) = 1,\ y'(0) = 0$.
   (c) $y(0) = 1,\ y'(0) = 2$. (d) $y(1) = 1,\ y'(1) = 1$.
   (e) $y(0) = 1,\ y(1) = 0$. (f) $y(0) = 0,\ y(1) = 0$.
   (g) $y(0) = 0,\ y'(0) = 0$. (h) $y(1) = 1,\ y(2) = 2$.

4. Sketch the graph of each function of x given. Then find a differential equation of the form

$$y'' + ay' + by = 0$$

of which each is a solution; write the general solution of the differential equation and verify that the given function is a special case of your general solution.

(a) xe^{-x}.

(b) $e^x + e^{-x}$.

(c) $1 + x$.

(d) $2e^{2x} - 3e^{3x}$.

[*Hint*: What characteristic roots go with each solution?]

5. Write each of the following differential equations in operator form:

$$(aD^2 + bD + c)y = 0.$$

Then factor the operator [e.g., $D^2 - 1 = (D - 1)(D + 1)$].

(a) $y'' + 2y' + y = 0$.

(b) $y'' - 2y = 0$.

(c) $2y'' - y = 0$.

(d) $y'' + 3y' = 0$.

(e) $y'' = 0$.

(f) $y'' - y' = x$.

6. Each of the following equations can be written in factored operator form:

$$(D - r_1)(D - r_2)y = f(x).$$

In each case let $(D - r_2)y = z$ and solve

$$(D - r_1)z = f(x)$$

for the most general possible z. Having found z, solve $(D - r_1)y = z$.

(a) $D(D - 3)y = 0$.

(b) $D^2 y = 1$.

(c) $y'' - y = 1$.

(d) $y'' + 2y' + y = x$.

7. The differential equation

$$y'' + \frac{1}{x}y' - \frac{1}{x^2}y = 0, \qquad x > 0,$$

can be written in operator form as

$$\left(D^2 + \frac{1}{x}D - \frac{1}{x^2}\right)y = 0.$$

(a) Show that the equation can also be written

$$D\left(D + \frac{1}{x}\right)y = 0.$$

(b) Solve the equation in part (a) by letting $z = (D + 1/x)y$ and solving a succession of first-order equations.

(c) Show that

$$D\left(D + \frac{1}{x}\right) \neq \left(D + \frac{1}{x}\right)D.$$

(d) Solve

$$\left(D + \frac{1}{x}\right)Dy = 0.$$

8. The hyperbolic cosine and hyperbolic sine are defined by

$$\cosh x = \tfrac{1}{2}(e^x + e^{-x}), \qquad \sinh x = \tfrac{1}{2}(e^x - e^{-x}).$$

(a) Show that, if constants d_1 and d_2 are suitably chosen in terms of c_1 and c_2, then

$$c_1 e^{rx} + c_2 e^{-rx} = d_1 \cosh rx + d_2 \sinh rx.$$

(b) Express the general solution of

$$y'' - k^2 y = 0$$

in terms of hyperbolic functions.

9. (a) Show that the characteristic equation of

$$Ay'' + By' + Cy = 0,$$

with A, B, C constant, $A \neq 0$, has real roots if and only if $B^2 \geq 4AC$.

(b) Show that when $B^2 > 4AC$, the general solution of the differential equation in part (a) can be written

$$y = e^{\alpha x}(d_1 \cosh \beta x + d_2 \sinh \beta x),$$

where $\alpha = -B/2A$, $\beta = \sqrt{B^2 - 4AC}/2A$.

10. Assume $|A| < \frac{1}{4}$ in the equation

$$Ay'' + y' + y = 0$$

and show that, as A tends to 0, and with proper choice of arbitrary constants, there are solutions of this equation tending, for each fixed x, to solutions of $y' + y = 0$.

2 COMPLEX SOLUTIONS

2A Complex Exponentials

The formal computations carried out in the previous section can all be carried out under the more general assumption that the roots of the characteristic equation

$$r^2 + ar + b = 0$$

associated with

$$y'' + ay' + by = 0$$

are complex numbers. To make sense of these computations, we need to have a **complex exponential** function defined for real numbers x by

$$e^{ix} = \cos x + i \sin x.$$

Plotting the complex number e^{ix} in the complex plane by its real and imaginary parts $\cos x$ and $\sin x$ shows, in Figure 2, that x can be interpreted as an angle measured in radians. Our motivation for this definition of e^{ix} comes from its properties, established next as Equations 2.1, 2.2, and 2.3 (see also Exercise 11).

The numbered formulas listed next are the basic properties of e^{ix} that are routinely used. We begin with the absolute value, recalling that $|\alpha + i\beta| = \sqrt{\alpha^2 + \beta^2}$. We have

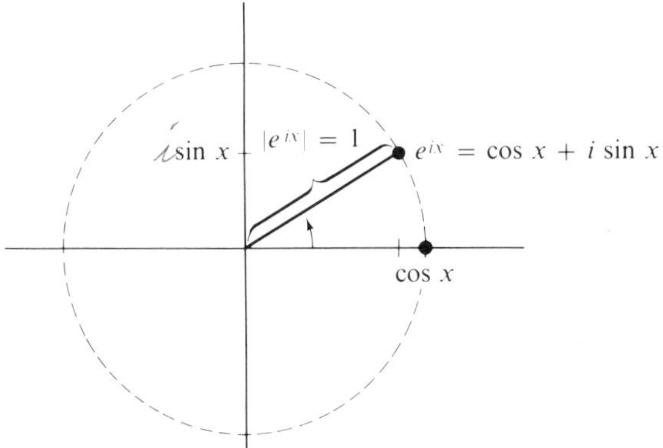

Figure 2

2.1
$$|e^{ix}| = |\cos x + i \sin x|$$
$$= \sqrt{\cos^2 x + \sin^2 x} = 1.$$

Using the addition formulas for sine and cosine shows that

$$(\cos x + i \sin x)(\cos x' + i \sin x')$$
$$= (\cos x \cos x' - \sin x \sin x') + i(\cos x \sin x' + \sin x \cos x')$$
$$= \cos(x + x') + i \sin(x + x').$$

It follows that

2.2
$$e^{ix}e^{ix'} = e^{i(x+x')}.$$

In particular, when $x' = -x$, we get $e^{ix}e^{-ix} = e^0 = 1$, so that

2.3
$$\frac{1}{e^{ix}} = e^{-ix}; \quad \text{that is,} \quad \frac{1}{\cos x + i \sin x} = \cos x - i \sin x.$$

Equations 2.2 and 2.3, aside from their usefulness, serve to justify the use of the exponential notation; the function e^{ix} behaves something like the real-valued exponential e^x, for which $e^x e^{x'} = e^{x+x'}$ and $(e^x)^{-1} = e^{-x}$.

> **EXAMPLE 1**   The function e^{ix} is in general complex valued, and Figure 2 shows that 1 and -1 are the only real values it assumes. In particular, if k is an integer,
>
> $$e^{ik\pi} = \cos k\pi + i \sin k\pi$$
> $$= (-1)^k.$$

However, the real-valued functions $\cos x$ and $\sin x$ can be represented in terms of e^{ix} as follows:

$$\frac{1}{2}(e^{ix} + e^{-ix}) = \frac{1}{2}(\cos x + i \sin x + \cos x - i \sin x) = \cos x,$$

$$\frac{1}{2i}(e^{ix} - e^{-ix}) = \frac{1}{2i}(\cos x + i \sin x - \cos x + i \sin x)$$

$$= \sin x.$$

Using complex exponentials may seem like an awkward way to do trigonometry, but it turns out that Formula 2.2 is often a very natural and convenient substitute for the addition formulas for cosine and sine.

The complex exponential notation is used not only because of the simple formulas described previously, but because it works naturally with differentiation and integration. To differentiate or integrate a complex-valued function $u(x) + iv(x)$, we have only to differentiate or integrate the real and imaginary parts. Here are the precise definitions:

$$\frac{d}{dx}(u(x) + iv(x)) = \frac{du}{dx}(x) + i\frac{dv}{dx}(x),$$

$$\int (u(x) + iv(x))dx = \int u(x)\,dx + i\int v(x)\,dx.$$

Accordingly, we have

$$\frac{d}{dx}e^{ix} = \frac{d}{dx}(\cos x + i \sin x)$$

$$= -\sin x + i \cos x$$

$$= i(\cos x + i \sin x) = ie^{ix}.$$

The end result is

$$\frac{d}{dx}e^{ix} = ie^{ix},$$

just as if i were a real number. Similarly,

$$\int e^{ix}\,dx = \frac{1}{i}e^{ix} + C,$$

where C is an arbitrary real or complex constant. More generally, we can define

$$e^{(\alpha+i\beta)x} = e^{\alpha x}e^{i\beta x}$$

and compute

2.4
$$\frac{d}{dx}e^{(\alpha+i\beta)x} = (\alpha + i\beta)e^{(\alpha+i\beta)x}$$

and

$$\textbf{2.5} \qquad \int e^{(\alpha+i\beta)x}\,dx = \frac{1}{(\alpha+i\beta)}e^{(\alpha+i\beta)x} + C, \qquad C \text{ real or complex,}$$

assuming neither α nor β is zero. These computations are left as exercises, which are simple but nevertheless important to understand.

2B Complex Roots

Given the facts about complex exponentials just described, the derivation of the general solution of

$$y'' + ay' + by = 0,$$

given in Section 1, proceeds word for word the same in case the roots of the characteristic equation are complex. The only thing that remains to be done then is to interpret the solution when a and b are real numbers. Rather than go through the identical general computation again, we will illustrate it with an example and then state the general result, whose proof is formally the same as that of Theorem 1.3.

EXAMPLE 2   The differential equation $y'' + y = 0$ in operator form is $(D^2 + 1)y = 0$, and in factored form this is

$$(D - i)(D + i)y = 0.$$

If we let, for some solution y,

$$(D + i)y = z,$$

we then need to solve

$$(D - i)z = 0.$$

As in the real case previously treated, we multiply by an exponential factor designed to make the left side the derivative of a product. The correct factor is

$$e^{\int -i\,dx} = e^{-ix},$$

and we observe that $e^{-ix}(D - i)z = D(e^{-ix}z)$. To solve

$$D(e^{-ix}z) = 0,$$

all we have to do is integrate with respect to x and then multiply by e^{ix}. Since $e^{ix}e^{-ix} = 1$, we get

$$z = c_1 e^{ix}.$$

Substitution of this result into the equation for y gives

$$(D + i)y = c_1 e^{ix}.$$

This time the multiplier is e^{ix}, and we find

$$D(e^{ix}y) = c_1 e^{2ix}.$$

Since the integral of e^{2ix} is $(\frac{1}{2i})e^{2ix}$, we have

$$e^{ix}y = \frac{1}{2i}c_1 e^{2ix} + c_2.$$

Now multiply by e^{-ix} to get

$$y = c_1 e^{ix} + c_2 e^{-ix},$$

where we have absorbed the factor $1/(2i)$ into the constant c_1. The roots of the characteristic equation $r^2 + 1 = 0$ are $\pm i$, so the same rule for writing down the solution works here as it does in the case of real roots.

It remains only to extract the real-valued solutions from this general complex-valued solution. This is done simply by using the definition of $e^{\pm ix}$ in terms of $\cos x$ and $\sin x$:

$$y = c_1 e^{ix} + c_2 e^{-ix}$$
$$= c_1(\cos x + i \sin x) + c_2(\cos x - i \sin x)$$
$$= (c_1 + c_2) \cos x + i(c_1 - c_2) \sin x.$$

To simplify the formula, we can set $d_1 = c_1 + c_2$ and $d_2 = i(c_1 - c_2)$. This does not involve any change in generality in the constants, because we can always solve for the c's in terms of the d's. In fact,

$$c_1 = \tfrac{1}{2}(d_1 - id_2) \quad \text{and} \quad c_2 = \tfrac{1}{2}(d_1 + id_2).$$

Note that if the d's are to be real numbers, which is the important case in most applications, then the c's will in general be complex. The solution formula

$$y = d_1 \cos x + d_2 \sin x$$

lets us, for example, find a formula for the solution satisfying initial conditions like $y(0) = 1$, $y'(0) = 2$. Since

$$y' = -d_1 \sin x + d_2 \cos x,$$

we see right away that $d_1 = 1$ and $d_2 = 2$. Thus

$$y = \cos x + 2 \sin x$$

is the required solution.

Whenever the equation $y'' + ay' + by = 0$ has coefficients that are real numbers, even though the characteristic equation

$$r^2 + ar + b = 0$$

may have complex roots as in Example 2, these complex roots will be **conjugate** to each other, that is, of the form $r_1 = \alpha + i\beta$ and $r_2 = \alpha - i\beta$. This follows from the quadratic formula

$$r = \frac{-a \pm \sqrt{a^2 - 4b}}{2},$$

and the assumption that $a^2 - 4b < 0$. Thus

$$r_1 = -\frac{a}{2} + \frac{i}{2}\sqrt{4b - a^2}, \qquad r_2 = -\frac{a}{2} - \frac{i}{2}\sqrt{4b - a^2}.$$

It follows that the complex solutions

$$y = c_1 e^{(\alpha + i\beta)x} + c_2 e^{(\alpha - i\beta)x}, \qquad \alpha,\ \beta \text{ real},$$

can always be written

$$
\begin{aligned}
y &= e^{\alpha x}(c_1 e^{i\beta x} + c_2 e^{-i\beta x}) \\
&= e^{\alpha x}[c_1(\cos \beta x + i \sin \beta x) + c_2(\cos \beta x - i \sin \beta x)] \\
&= e^{\alpha x}[(c_1 + c_2)\cos \beta x + i(c_1 - c_2)\sin \beta x] \\
&= d_1 e^{\alpha x} \cos \beta x + d_2 e^{\alpha x} \sin \beta x.
\end{aligned}
$$

This is a form of the solution that is often used in practice, so we include it in the general statement.

2.6 Theorem.   The differential equation

$$y'' + ay' + by = 0, \qquad a,\ b \text{ constant},$$

has for its general solution

$$y = c_1 e^{r_1 x} + c_2 e^{r_2 x}, \qquad r_1 \neq r_2$$

$$y = c_1 x e^{r_1 x} + c_2 e^{r_1 x}, \qquad r_1 = r_2,$$

where r_1, r_2 are the roots of $r^2 + ar + b = 0$. Assuming that a, b are real with $a^2 - 4b < 0$, with $r_1 = \alpha + i\beta$, $r_2 = \alpha - i\beta$, the general solution can be written

$$y = c_1 e^{\alpha x} \cos \beta x + c_2 e^{\alpha x} \sin \beta x.$$

EXAMPLE 3   The solutions of

$$Ay'' + By' + Cy = 0$$

can be classified according to the sign and relative size of the constants A, B, and C. The characteristic equation is

$$Ar^2 + Br + C = 0$$

with roots

$$r_1,\ r_2 = \frac{-B \pm \sqrt{B^2 - 4AC}}{2A}.$$

If $B^2 - 4AC > 0$, the roots are real and unequal, so the solutions are conveniently written

$$y = c_1 e^{r_1 x} + c_2 e^{r_2 x}.$$

If $B^2 - 4AC = 0$, the roots are equal and real, so solutions are all of the form

$$y = c_1 x e^{r_1 x} + c_2 e^{r_1 x}.$$

Finally, with $B^2 - 4AC < 0$, the solutions have the form

$$y = c_1 e^{\alpha x} \cos \beta x + c_2 e^{\alpha x} \sin \beta x$$

where $\alpha = -B/2A$ and $\beta = \sqrt{4AC - B^2}/2A$. Clearly, the case $B^2 - 4AC = 0$ is critical in that it represents the division between oscillatory solutions ($B^2 - 4AC < 0$) and nonoscillatory solutions ($B^2 - 4AC > 0$). The physical significance of this classification is explained in Section 4.

2C Nth-Order Equations

The nth-order equation of the kind we have been looking at can be written

$$L(y) = y^{(n)} + a_{n-1} y^{(n-1)} + \cdots + a_1 y + a_0 = 0,$$

where the a_k are real numbers. The methods used for the second-order case can be successively applied to produce the following theorem.

2.7 Theorem. The nth-order constant-coefficient equation $L(y) = 0$ has for its general solution a linear combination

$$c_1 y_1(x) + \cdots + c_n y_n(x)$$

of solutions $y_k(x)$, where the c_k are arbitrary constants; these constants are uniquely prescribed by initial conditions of the form

$$y(x_0) = y_0, \; y'(x_0) = y_1, \; \ldots, \; y^{(n-1)}(x_0) = y_{n-1}.$$

Let $r_1 \ldots, r_n$ be the roots of the characteristic equation

$$r^n + a_{n-1} r^{n-1} + \cdots + a_1 r + a_0 = 0.$$

The terms $y_k(x)$ in the solutions can all be written in the form

$$x^l e^{r_k x}, \qquad l = 0, 1, \ldots, m - 1,$$

where m is the multiplicity of r_k. If roots $\alpha + i\beta$ and $\alpha - i\beta$ occur in complex pairs, the corresponding pair of solutions can be written

$$x^l e^{\alpha x} \cos \beta x, \qquad x^l e^{\alpha x} \sin \beta x.$$

EXAMPLE 4 The differential equation $y^{(4)} - y = 0$ has the characteristic equation $r^4 - 1 = 0$. To find the roots, we observe that $r^4 = 1$, so $r^2 = 1$ or $r^2 = -1$. Hence the roots are $r_1 = 1$, $r_2 = -1$, $r_3 = i$, and $r_4 = -i$. The first two roots provide solutions e^x and e^{-x}. The second pair provides either e^{ix}, e^{-ix} or the alternative form, $\cos x$, $\sin x$. The general solution is

$$y(x) = c_1 e^x + c_2 e^{-x} + c_3 \cos x + c_4 \sin x.$$

It can be proved that the first sentence in Theorem 2.7 remains true even if the coefficients a_k in the differential equation are continuous functions $a_k(x)$ instead of mere constants. However, we cannot then be specific about the form of the solutions $y_k(x)$.

EXERCISES

1. Show that each of the following complex numbers has absolute value 1. Then find a real number x such that the complex number can be written in the form $e^{ix} = \cos x + i \sin x$; for example,

$$\frac{\sqrt{3} + i}{2} = \cos \frac{\pi}{6} + i \sin \frac{\pi}{6} = e^{i\pi/6}.$$

   (a) i.   (b) $(1 + i)/\sqrt{2}$.   (c) $(1 - i)/\sqrt{2}$.   (d) $(\sqrt{3} - i)/2$.

2. Given a pair d_1, d_2 of real numbers, find complex numbers c_1, c_2 such that

$$c_1 e^{(\alpha + i\beta)x} + c_2 e^{(\alpha - i\beta)x} = e^{\alpha x}(d_1 \cos \beta x + d_2 \sin \beta x).$$

   (a) $d_1 = 1$, $d_2 = 0$.   (b) $d_1 = 4$, $d_2 = -2$.
   (c) $d_1 = 0$, $d_2 = \pi$.   (d) $d_1 = 1$, $d_2 = 1$.

3. Recall that a function f is periodic, with period p, if $f(x + p) = f(x)$ for all x in the domain of f.
   (a) Show that e^{ix} has period 2π.
   (b) Show that $e^{i\beta x}$ is periodic for β real, and find the smallest positive period if $\beta \neq 0$.

4. Show that $e^{-i\beta x} = \cos \beta x - i \sin \beta x$. What properties of $\cos$ and $\sin$ are used here?

5. (a) Verify Equation 2.4.
   (b) Verify Equation 2.5.

6. Solve each of the following differential equations by factoring the differential operator associated with it and then successively solving a pair of first-order linear equations.
   (a) $y'' + y = 1$.   (b) $y'' + 2y' + 2y = 0$.
   (c) $y'' + 2y = 0$.   (d) $y'' + y' = x$.

7. Find all real or complex solutions of $y'' + iy' = 0$.

8. Solve each of the following differential equations by finding the roots of the characteristic equation and then writing down the general solutions as a sum of multiples of functions $e^{\alpha x} \cos \beta x$ and $e^{\alpha x} \sin \beta x$.
   (a) $y'' + 2y = 0$.   (b) $2y'' + 3y' = 0$.
   (c) $y'' - 2y' + 2y = 0$.   (d) $y'' - y' + y = 0$.

(e) $2y'' + y' - y = 0.$ **(f)** $y'' + y' = 2y.$
(g) $2y'' + y' + y = 0.$ **(h)** $3y'' - y' + y = 0.$

9. For each of the differential equations in Exercise 8, find the solution satisfying the following corresponding condition; do this by solving a pair of linear equations for the unknown constants.

(a) $y(0) = 0,\ y'(0) = 1.$ **(b)** $y(0) = 1,\ y'(0) = 0.$
(c) $y(\pi) = 0,\ y'(\pi) = 0.$ **(d)** $y(0) = 2,\ y'(0) = -1.$
(e) $y(0) = 0,\ y'(0) = 2.$ **(f)** $y(0) = 0,\ y(1) = 1.$
(g) $y(0) = 0,\ y'(0) = 0.$ **(h)** $y(0) = 0,\ y(1) = 0.$

10. (a) Show that $(d_1 \cos \beta x + d_2 \sin \beta x)$ can be written in the form $r \cos (\beta x - \theta)$, where $r = \sqrt{c_1^2 + c_2^2}$ and θ satisfies $\cos \theta = c_1/r$ and $\sin \theta = c_2/r.$

(b) The result of part (a) is useful because it shows that

$$c_1 \cos \beta x + c_2 \sin \beta x$$

has a graph that is the same as that of $\cos \beta x$ shifted by a **phase angle** θ and multiplied by an **amplitude** r. Sketch the graph of

$$\cos 2x + \sqrt{3} \sin 2x$$

by first finding θ and r.

11. Separate the real and imaginary terms in the infinite series

$$\sum_{k=1}^{\infty} \frac{(ix)^k}{k!}$$

into two infinite series and use the result to justify the definition

$$e^{ix} = \cos x + i \sin x.$$

12. If f and g are complex-valued differentiable functions and c is a complex number, show that

(a) $(f + g)' = f' + g',$ **(b)** $(cf)' = cf',$
(c) $(fg)' = fg' + f'g,$ **(d)** $(f/g)' = (f'g - fg')/g^2,$

whenever the functions are defined.

13. Find the general solution to each of the following equations.

(a) $y''' - y = 0.$ **(b)** $y^{(4)} - 2y'' + y = 0.$
(c) $y''' - 2y' = 0.$ **(d)** $y^{(4)} + y = 0.$

[*Hint for part* (d): To solve $r^4 + 1 = 0$, note that $r^2 = \pm i = \pm e^{i(\pi/2 + 2k\pi)}$, k an integer. Then $r = \pm e^{i(\pi/4 + k\pi)}$.]

(e) $(D^2 - 4)(D^2 - 1)y = 0.$ **(f)** $D(D - 1)^3 y = 0.$

14. Find constant-coefficient differential equations of minimal order that have the following functions as solutions.

(a) $\sin 2x.$ **(b)** $x \sin 2x.$
(c) $e^x \cos 2x.$ **(d)** $x^2 \sin x.$
(e) $xe^{-x} \cos 3x.$ **(f)** $x^3 \cos 4x.$

[*Hint:* What are the characteristic roots associated with each solution?]

15. If L is a linear operator such that $L(y_1) = w_1$, $L(y_2) = w_2$ and $L(y_3) = w_3$, find a z in terms of the y's such that

(a) $L(z) = 2w_1 - 3w_2.$ **(b)** $L(z) = w_1 + 2w_2 - 4w_3.$
(c) $L(z) = L(2z) + w_1 + w_2.$ **(d)** $L(z) = 0.$

3 NONHOMOGENEOUS EQUATIONS

3A General Solution

A differential operator of the form

$$L = D^2 + aD + b$$

is **linear,** which means that

$$L(c_1 y_1 + c_2 y_2) = c_1 L(y_1) + c_2 L(y_2)$$

for every pair of constants c_1, c_2 and every pair of twice-differentiable functions y_1, y_2. Recall that linearity for first-order operators was discussed in Chapter 1, Section 5, where it was defined in the equivalent form

$$L(cy) = cL(y), \qquad c \text{ constant,}$$

$$L(y_1 + y_2) = L(y_1) + L(y_2).$$

Verification of the linearity of this second-order operator L is straightforward and is left as an exercise. The property of linearity allows us to conclude that $c_1 y_1 + c_2 y_2$ is a solution of the **homogeneous equation** $L(y) = 0$, or

$$y'' + ay' + by = 0,$$

whenever y_1 and y_2 are solutions. (However, more work may be needed to find the most general solution.) The linearity of L plays a part in describing the most general solution of the **nonhomogeneous equation** $L(y) = f$, or

$$y'' + ay' + by = f(x),$$

where $f(x)$ is given on some interval. The function $f(x)$ is often called a **forcing function** because of its interpretation in mechanical and electrical problems.

3.1 Theorem.   Let y_p be some particular solution of the nonhomogeneous equation

$$L(y) = f,$$

where L is linear. Then the most general solution of $L(y) = f$ is of the form

$$y = y_p + y_h,$$

where y_h is the most general solution of the homogeneous equation $L(y) = 0$.

Proof. Assume that $L(y_p) = f$ and that y is some solution of $L(y) = f$. Then

$$L(y - y_p) = L(y) - L(y_p)$$

$$= f - f = 0.$$

Thus $y - y_p$ is equal to some solution y_h of $L(y) = 0$. It follows that $y = y_p + y_h$.

Note that Theorem 3.1 applies to any linear operator at all; it will have several applications later, but for now we will use it to conclude that

$$y'' + ay' + by = f(x)$$

has the general solution

$$y = y_p(x) + c_1 y_1(x) + c_2 y_2(x),$$

where y_p is some particular solution, and y_1 and y_2 are basic exponential solutions of the associated homogeneous equation.

EXAMPLE 1 The equation

$$y'' - y = x$$

has $y'' - y = 0$ for its associated homogeneous equation, with most general solution

$$y_h = c_1 e^x + c_2 e^{-x}.$$

Also it is easy to see by inspection that $y_p = -x$ is a solution of the non-homogeneous equation, since $y_p' = -1$ and $y_p'' = 0$. (We will see shortly how to discover such a solution by integration.) It follows from Theorem 3.1 that every solution has the form

$$y = y_p + y_h$$
$$= -x + c_1 e^x + c_2 e^{-x}.$$

To find the particular solution satisfying conditions of the form

$$y(x_0) = a_0, \qquad y'(x_0) = a_1,$$

just compute $y'(x) = -1 + c_1 e^x - c_2 e^{-x}$ and solve

$$-x_0 + c_1 e^{x_0} + c_2 e^{-x_0} = a_0$$
$$-1 + c_1 e^{x_0} - c_2 e^{-x_0} = a_1$$

for the correct constants c_1 and c_2.

3B Undetermined Coefficient Method

Example 1, although it is simple, is unsatisfactory in that it used a solution $y_p = -x$ that was found by inspection. The remainder of Section 3 is taken up with finding solution formulas for y_p for certain equations. The undetermined coefficient method involves finding trial solutions containing constants to be determined. The correct values of the constants are then found by substitution into the equation

$$y'' + ay' + by = f(x).$$

However, this method works only when the function $f(x)$ is a linear combination, with constant coefficients, of functions that can be written in either of the two forms

$$x^n e^{\alpha x} \cos \beta x \quad \text{or} \quad x^n e^{\alpha x} \sin \beta x,$$

where n is a nonnegative integer. An example is

$$f(x) = 2xe^{2x} + 3 \sin 2x,$$

where for the first term we take $n = 1$, $\alpha = 2$, $\beta = 0$ and for the second we take $n = 0$, $\alpha = 0$, $\beta = 2$. Since the differential operator $L = D^2 + aD + b$ is linear, functions of the form

$$f(x) = a_1 f_1(x) + \cdots + a_n f_n(x),$$

that is, linear combinations, allow us to break the problem down into the solution of several equations

$$L(y_1) = f_1(x), \ \ldots, \ L(y_n) = f_n(x),$$

and then write the particular solution as the linear combination

$$y_p = a_1 y_1 + \cdots + a_n y_n.$$

EXAMPLE 2   To solve $y'' - y = 3e^{2x} + 4e^x$, we note first the homogeneous solution $y_h = c_1 e^x + c_2 e^{-x}$ and then solve

$$y'' - y = e^{2x} \quad \text{and} \quad y'' - y = e^x.$$

Then combine the respective solutions y_1, y_2 of these two equations to get $y_p = 3y_1 + 4y_2$. An obvious trial solution for the first equation is $y_1 = Ae^{2x}$, where A is constant. We compute $y_1' = 2Ae^{2x}$ and $y_1'' = 4Ae^{2x}$ and substitute:

$$y_1'' - y_1 = 4Ae^{2x} - Ae^{2x}$$

$$= 3Ae^{2x} = e^{2x}.$$

We want $3A = 1$, so $A = \frac{1}{3}$ gives the solution $y_1 = \frac{1}{3}e^{2x}$. The trial solution $y_2 = Ae^x$ will not work for the second equation because $y_2'' - y_2 = 0$, regardless of how A is chosen. But $y_2 = Axe^x$ gives $y_2' = Axe^x + Ae^x$ and $y_2'' = Axe^x + 2Ae_x$, and substitution into the differential equation gives

$$y_2'' - y_2 = Axe^x + 2Ae^x - Axe^x$$

$$= 2Ae^x = e^x.$$

Thus $A = \frac{1}{2}$ and $y_2 = (\frac{1}{2})xe^x$. The general solution of the original equation is

$$y = 3(\tfrac{1}{3})e^{2x} + 4(\tfrac{1}{2})xe^x + c_1 e^x + c_2 e^{-x}$$

$$= e^{2x} + 2xe^x + c_1 e^x + c_2 e^{-x}.$$

Here is an outline of the routine for finding the terms in a linear combination for a trial solution y_t to $L(y) = f(x)$, where f has the special form

$$x^n e^{\alpha x} \cos \beta x \quad \text{or} \quad x^n e^{\alpha x} \sin \beta x.$$

(i) Include in y_t the function f itself and all terms in its successive derivatives, discarding any repeated terms.

(ii) If a term included in step (i) happens to be a solution of the homogeneous equation, multiply that term and all terms surviving from (i) by the single lowest power x^k such that the resulting terms are no longer homogeneous solutions.

(iii) Form a linear combination with undetermined constant coefficients of the terms from (ii), and determine the values of the coefficients by substitution into $L(y) = f$.

EXAMPLE 3   Here is a list of functions $f(x)$ and corresponding trial solutions $y_t(x)$, assuming no term in $y_t(x)$ satisfies the homogeneous equation.

$$f(x) = ce^{rx}; \qquad y_t(x) = Ae^{rx}.$$

$$f(x) = cx^2; \qquad y_t(x) = Ax^2 + Bx + C.$$

$$f(x) = cx^2 e^{rx}; \qquad y_t(x) = (Ax^2 + Bx + c)e^{rx}.$$

$$f(x) = c \cos \beta x; \qquad y_t(x) = A \cos \beta x + B \sin \beta x.$$

$$f(x) = cx \sin \beta x; \qquad y_t(x) = (Ax + B) \sin \beta x + (Cx + D) \cos \beta x.$$

$$f(x) = ce^{\alpha x} \cos \beta x; \qquad y_t(x) = (A \cos \beta x + B \sin \beta x)e^{\alpha x}.$$

However, <u>note</u> that in solving

$$y'' + 4y = 3 \cos 2x,$$

which has the associated homogeneous solution $c_1 \cos 2x + c_2 \sin 2x$, the choice $A \cos 2x + B \sin 2x$ for $y_t(x)$ would be inadequate, because substitution into the left side produces a zero. (This is why it is important to be aware of the solutions of the homogeneous equation before finding a trial solution.) In this example, we should take

$$y_t(x) = Ax \cos 2x + Bx \sin 2x.$$

EXAMPLE 4   The equation

$$y'' + 2y' + y = 3e^{-x} + 2x$$

has homogeneous solution

$$y_h(x) = c_1 e^{-x} + c_2 x e^{-x}.$$

Solving first the equation $y'' + 2y' + y = 2x$, we try $y_1(x) = Ax + B$ and find

$y_1'(x) = A$, $y_1''(x) = 0$. Substitution into the equation with $2x$ on the right gives

$$0 + 2A + Ax + B = 2x.$$

Equating coefficients of x gives $A = 2$. The constants satisfy $2A + B = 0$, so $B = -2A = -4$. Thus $y_1(x) = 2x - 4$. To solve $y'' + 2y' + y = 3e^{-x}$, we try

$$y_2(x) = Ax^2 e^{-x}.$$

Then

$$y_2'(x) = A(2x - x^2)e^{-x}$$

$$y_2''(x) = A(2 - 4x + x^2)e^{-x}.$$

Substitution into the equation with $3e^{-x}$ on the right gives

$$A(2 - 4x + x^2)e^{-x} + 2A(2x - x^2)e^{-x} + Ax^2 e^{-x} = 3e^{-x}.$$

The terms with x and x^2 as factors all cancel out and we are left with $2A = 3$. Thus $A = \frac{3}{2}$, so $y_2(x) = (\frac{3}{2})x^2 e^{-x}$. The general solution of the original equation is then

$$y = y_1 + y_2 + y_h$$

$$= 2x - 4 + \frac{3}{2}x^2 e^{-x} + c_1 e^{-x} + c_2 x e^{-x}.$$

We could have found y_p by using a single trial solution

$$y = Ax + B + Cx^2 e^{-x},$$

which would probably be more efficient. However, splitting the problem into two parts does have the effect of reducing the number of undetermined coefficients you need to work with at one time.

The method depends on the observation that if we want to solve

$$L(y) = f(x),$$

where $f(x)$ is itself a solution of a homogeneous equation $My = 0$, then

$$M(L(y)) = M(f(x)) = 0.$$

Thus the desired solution $y(x)$ must be among the solutions of

$$ML(y) = 0.$$

If M and L are linear, constant-coefficient operators, then the solutions of the preceding equation are linear combinations of functions of the form $x^k e^{rx}$, where r may be real or complex. Thus the only problem is to determine the so far "undetermined coefficients" of combination that will actually give a solution of the original equation

$L(y) = f(x)$. For example, the differential equation

$$y'' - y = e^x$$

can be written

$$(D^2 - 1)y = e^x.$$

Since $(D - 1)e^x = 0$, we have for any solution $y(x)$,

$$(D - 1)(D^2 - 1)y = (D - 1)^2(D + 1)y = 0.$$

Hence any particular solution y must have the form

$$y(x) = c_1 e^{-x} + c_2 e^x + c_3 x e^x.$$

Since the first two terms are solutions of the associated homogeneous equation $y'' - y = 0$, we can concentrate on the remaining term $c_3 x e^x$. To find c_3, we substitute into the nonhomogeneous differential equation. The resulting algebraic equation determines c_3.

3C Green's Functions

The undetermined coefficient method is sometimes the most efficient way to solve a specific nonhomogeneous equation whose forcing function $f(x)$ has a special form. However, for $f(x)$ of a more general type and to display the dependence of the solution on $f(x)$ in a general way, other methods are used. The **variation-of-parameters** method and **Laplace transform** method are taken up later. Here we derive a formula, depending on the homogeneous solutions, into which the forcing function can be plugged.

3.2 Theorem.   Let $y_p(x)$ be the solution of

$$y'' + ay' + by = f(x)$$

satisfying $y_p(x_0) = y_p'(x_0) = 0$. If r_1, r_2 are the roots of the characteristic equation $r^2 + ar + b = 0$, then

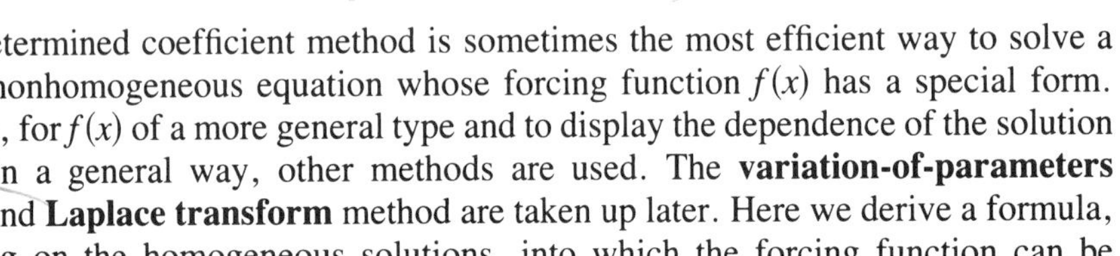

(i) if r_1, r_2 are unequal, $\displaystyle y_p(x) = \frac{1}{r_1 - r_2} \int_{x_0}^{x} (e^{r_1(x-t)} - e^{r_2(x-t)})f(t)\,dt,$

(ii) if $r_1 = r_2$, $\displaystyle y_p(x) = \int_{x_0}^{x} (x - t)e^{r_1(x-t)}f(t)\,dt,$

(iii) if r_1, $r_2 = \alpha \pm i\beta$, $\displaystyle y_p(x) = \frac{1}{\beta} \int_{x_0}^{x} e^{\alpha(x-t)} \sin \beta(x - t)f(t)\,dt.$

In each formula, the factor in the integrand exclusive of $f(t)$ is a function $G(x, t) = g(x - t)$ called the **Green's function** of the operator $D^2 + aD + b$. In general, then, the formula can be written

$$y_p(x) = \int_{x_0}^{x} g(x - t)f(t)\, dt$$

where the function g has one of the three forms

(i) $g(u) = \dfrac{1}{r_1 - r_2}(e^{r_1 u} - e^{r_2 u})$, $r_1 \neq r_2$,

(ii) $g(u) = u e^{r_1 u}$, $r_1 = r_2$,

(iii) $g(u) = \dfrac{1}{\beta} e^{\alpha u} \sin \beta u$, $r_1 = \alpha + i\beta$, $r_2 = \alpha - i\beta$.

Integrals of this form are called **convolution** integrals, and some care is needed not to mix up the variables. Specifically, the variable x that occurs in the upper limit and in the difference $x - t$ must be treated as a constant with respect to integration in the other variable. Before proving Theorem 3.2, we will give some examples.

EXAMPLE 5   Here is an equation that can also be solved by undetermined coefficients:

$$y'' + 2y' + y = e^{-x}.$$

The homogeneous solution is $y_h = c_1 x e^{-x} + c_2 e^{-x}$, corresponding to a double root $r = -1$ of the equation $r^2 + 2r + 1 = 0$. The Green's function is $G(x, t) = (x - t)e^{-(x-t)}$. The particular solution satisfying $y(0) = y'(0) = 0$ is

$$y(x) = \int_{0}^{x} (x - t)e^{-(x-t)}e^{-t}\, dt.$$

Note that integration is with respect to t, so x is temporarily fixed. We find

$$y_p(x) = \int_{0}^{x} x e^{-(x-t)}e^{-t}\, dt - \int_{0}^{x} t e^{-(x-t)}e^{-t}\, dt$$

$$= x e^{-x} \int_{0}^{x} dt - e^{-x} \int_{0}^{x} t\, dt$$

$$= x e^{-x}[t]_0^x - e^{-x}\left[\tfrac{1}{2}t^2\right]_0^x$$

$$= x e^{-x}(x - 0) - e^{-x}(\tfrac{1}{2}x^2 - 0)$$

$$= x^2 e^{-x} - \tfrac{1}{2}x^2 e^{-x} = \tfrac{1}{2}x^2 e^{-x}.$$

The general solution is then

$$y(x) = \tfrac{1}{2}x^2 e^{-x} + c_1 x e^{-x} + c_2 e^{-x}.$$

The next example cannot be done by undetermined coefficients except by breaking the problem into three separate parts and adjusting the initial conditions for each part.

EXAMPLE 6   The equation $y'' - y = f(x)$ has the associated homogeneous solution $y_h = c_1 e^x + c_2 e^{-x}$, so the Green's function is

$$g(x - t) = \tfrac{1}{2}[e^{(x-t)} - e^{-(x-t)}].$$

Suppose $f(x)$ is given by

$$f(x) = \begin{cases} 0, & x < 0, \\ 1, & 0 \le x \le 1, \\ 0, & 1 < x. \end{cases}$$

The solution satisfying $y(0) = y'(0) = 0$ is

$$y(x) = \frac{1}{2} \int_0^x [e^{(x-t)} - e^{-(x-t)}] f(t) \, dt.$$

Since $f(t) = 0$ for $x < 0$, we clearly have $y(0) = 0$ for $x < 0$. When $0 \le x \le 1$, we find

$$y(x) = \frac{1}{2} \int_0^x [e^{(x-t)} - e^{-(x-t)}] \, dt$$

$$= \frac{1}{2} e^x \int_0^x e^{-t} \, dt - \frac{1}{2} e^{-x} \int_0^x e^t \, dt$$

$$= \frac{1}{2} e^x [-e^{-t}]_0^x - \frac{1}{2} e^{-x} [e^t]_0^x$$

$$= \frac{1}{2} e^x - \frac{1}{2} - \frac{1}{2} + \frac{1}{2} e^{-x} = \cosh x - 1.$$

When $1 < x$, we note that $f(t) = 0$ for $1 < t$, so

$$y(x) = \frac{1}{2} e^x \int_0^1 e^{-t} \, dt - \frac{1}{2} e^{-x} \int_0^1 e^t \, dt$$

$$= \frac{1}{2} e^x [-e^{-t}]_0^1 - \frac{1}{2} e^{-x} [e^t]_0^1$$

$$= \frac{1}{2} e^x - \frac{1}{2} e^{x-1} - \frac{1}{2} e^{-(x-1)} + \frac{1}{2} e^{-x} = \cosh x - \cosh(x - 1).$$

Just as $f(x)$ is defined by different formulas on different intervals, so is y:

$$y(x) = \begin{cases} 0, & x < 0, \\ \cosh x - 1, & 0 \le x \le 1, \\ \cosh x - \cosh(x - 1), & 1 < x. \end{cases}$$

The derivation of Theorem 3.2 is an application of the method we used earlier to derive the most general solution to $y'' + ay' + by = 0$. We factor the operator and write

$$(D - r_1)(D - r_2)y = f(x), \qquad (D - r_2)y = z.$$

Then solve $(D - r_1)z = f(x)$ by first multiplying by $e^{-r_1 x}$. The result is

$$e^{-r_1 x}z = \int e^{-r_1 x}f(x)\, dx + c.$$

We choose the constant c so that $z(x_0) = 0$. This can be done by using a definite integral. We get

$$z(x) = e^{r_1 x}\int_{x_0}^{x} e^{-r_1 u}f(u)\, du.$$

Now solve $(D - r_2)y = z$ by the same method:

$$y(x) = e^{r_2 x}\int_{x_0}^{x} e^{-r_2 t}z(t)\, dt$$

$$= e^{r_2 x}\int_{x_0}^{x} e^{(r_1 - r_2)t}\left[\int_{x_0}^{t} e^{-r_1 u}f(u)\, du\right] dt.$$

Here we have replaced x by t in the integral for z. If we perform the t-integration by parts, and assume $r_1 \neq r_2$, we get

$$y(x) = \frac{1}{r_1 - r_2}e^{r_1 x}\int_{x_0}^{x} e^{-r_1 u}f(u)\, du - \frac{1}{r_1 - r_2}e^{r_2 x}\int_{x_0}^{x} e^{-r_2 t}f(t)\, dt$$

(i)

$$= \frac{1}{r_1 - r_2}\int_{x_0}^{x}[e^{r_1(x-t)} - e^{r_2(x-t)}]f(t)\, dt.$$

A similar integration by parts when $r_1 - r_2 = 0$ gives

(ii)
$$y(x) = \int_{x_0}^{x}(x - t)e^{r_1(x-t)}f(t)\, dt.$$

If $r_1 = \alpha + i\beta$ and $r_2 = \alpha - i\beta$, then $r_1 - r_2 = 2i\beta$, and

$$\frac{1}{r_1 - r_2}[e^{r_1(x-t)} - e^{r_2(x-t)}] = \frac{1}{\beta}e^{\alpha(x-t)}\sin\beta(x - t).$$

Substituting this into (i) gives part (iii) of Theorem 3.2. Some of these details are left as exercises.

EXERCISES

General Solution

1. Show that each of the following operators L is linear by verifying that

$$L(c_1 y_1 + c_2 y_2) = c_1 L(y_1) + c_2 L(y_2)$$

for arbitrary constants c_1, c_2 and sufficiently differentiable functions y_1, y_2.

  (a) $L(y) = D^2 y$.   (b) $L(y) = D^2 y + Dy$.
  (c) $L(y) = 2Dy + xy$.   (d) $L(y) = 2D^2 y + xDy$.

2. Which of the following operators L are linear?

  (a) $L(y) = xDy$.   (b) $L(y) = xD^2 y + xy^2$.
  (c) $L(y) = x^2 D^2 y + y^2$.   (d) $L(y) = y^2$.

3. Each of the following linear differential equations has the given function y_p as a solution. Verify this and find the most general solution.

  (a) $y'' + y = x$; $y_p = x$.   (b) $y'' - 2y' + y = 1$; $y_p = 1$.
  (c) $2y'' - y = e^x$; $y_p = e^x$.   (d) $xy' + y = 1$; $y_p = 1$.

4. Find second-order homogeneous differential equations of which the following functions are solutions.

  (a) $e^x + 2e^{2x}$.   (b) $e^x \cos x - e^x \sin x$.
  (c) $x + 1$.   (d) $xe^x - 2e^x$.

Undetermined Coefficient Method

5. Find the general solution of the following equations by first finding the general solution of the associated homogeneous equation and then adding to it a particular solution found by the undetermined coefficient method.

  (a) $y'' - y = e^{2x}$.   (b) $y'' - y = 3e^x$.
  (c) $y'' + 2y' + y = e^x$.   (d) $y'' - 3y = x$.
  (e) $y'' - y = e^x + x$.   (f) $y'' - 2y = \cos 2x$.
  (g) $y'' + y = \cos x$.   (h) $y'' = \cos x + \sin x$.
  (i) $y'' + y = x \cos x$.   (j) $y'' - y = xe^x$.

6. For each of the differential equations in Exercise 5, find the particular solution $y(x)$ satisfying the initial conditions $y(0) = 0$, $y'(0) = 1$. Then sketch the graph of that solution.

Green's Functions

7. For each of the differential equations in Exercise 5, find the Green's function $G(x, t) = g(x - t)$ associated with the equation.

8. For each of the following differential equations, first find the general solution of the associated homogeneous equation; then find the associated Green's function $G(x, t) = g(x - t)$ and use it to find the particular solution satisfying the given initial condition of the special form $y(x_0) = y'(x_0) = 0$.

  (a) $y'' - y = x$; $x_0 = 0$.   (b) $y'' + y = x$; $x_0 = 0$.
  (c) $y'' + 2y' + y = e^{2x}$; $x_0 = 0$.   (d) $y'' + 2y' + y = 0$; $x_0 = 1$.

9. Find the particular solution to

$$y'' - 4y = f(x)$$

satisfying $y(0) = y'(0) = 0$, where

  (a) $f(x) = \begin{cases} 1, & 0 \le x, \\ 0, & x < 0. \end{cases}$ (b) $f(x) = \begin{cases} 0, & 0 \le x, \\ 1, & x < 0. \end{cases}$

(c) $f(x) = \begin{cases} 1, & 0 \le x, \\ -1, & x < 0. \end{cases}$ **(d)** $f(x) = \begin{cases} x, & 0 \le x, \\ 0, & x < 0. \end{cases}$

[*Hint:*   The forcing function f in part (c) is a linear combination of those parts (a) and (b).]

(e) $f(x) = \begin{cases} x - 1, & 1 \le x, \\ 0, & x < 1. \end{cases}$ **(f)** $f(x) = x^2.$

10. Find the general solution of each of the following differential equations by writing the equation in the form

$$(D - r_1)(D - r_2)y = f(x),$$

letting $z = (D - r_2)y$ and then solving in succession the two first-order equations $(D - r_1)z = f(x)$ and $(D - r_2)y = z$.

(a) $y'' - y' - 2y = e^x.$ **(b)** $y'' - 4y = e^{3x}.$

11. Show that if $(D - r)y(x) = f(x)$ then the conditions $y(x_0) = 0, f(x_0) = 0$ hold if, and only if, $y(x_0) = 0$, $y'(x_0) = 0$.

12. **(a)** Show that if $r_1 = \alpha + i\beta$, $r_2 = \alpha - i\beta$ then

$$\frac{1}{r_1 - r_2}(e^{r_1 u} - e^{r_2 u}) = \frac{1}{\beta}e^{\alpha u} \sin \beta u.$$

(b) Show that

$$\lim_{\beta \to 0} \frac{1}{\beta}e^{\alpha u} \sin \beta u = ue^{\alpha u}.$$

(c) Let r_1 and r_2 be real valued. Show that

$$\lim_{r_1 \to r_2} \frac{1}{r_1 - r_2}(e^{r_1 u} - e^{r_2 u}) = ue^{r_2 u}.$$

These limit relations make plausible the Green's function for the case of equal roots.

13. To complete the derivation of the Green's function for the case of equal roots, show that

$$e^{rx} \int_{x_0}^{x} \left[\int_{x_0}^{t} e^{-ru}f(u)\, du \right] dt = \int_{x_0}^{x} (x - t)e^{r(x - t)}f(t)\, dt.$$

[*Hint:*   The left-hand integral is of the form

$$\int_{x_0}^{x} 1[F(t)]\, dt;$$

integrate this by parts.]

14. **(a)** The independent variable in a differential equation is often denoted by t. For example, in

$$\frac{d^2x}{dt^2} + a\frac{dx}{dt} + bx = f(t),$$

t usually stands for time. Show that the Green's function solution can be written

$$x_p(t) = \int_{t_0}^{t} g(t - u)f(u)\, du.$$

(b) Solve

$$\frac{d^2x}{dt^2} + x = \sin t,$$

if $x(0) = 0$ and $dx/dt(0) = 1$.

4 DYNAMICAL APPLICATIONS

Dynamical applications are applications of the second order, linear, constant-coefficient differential equations treated earlier in this chapter, in which the solution $y = y(t)$ that we study will be a function of time t such that $y(t)$ is determined by its relation to its velocity dy/dt and its acceleration d^2y/dt^2. Fundamental to these relations will be Newton's second law of motion, which says that force F is equal to constant mass m times acceleration a. Thus, since $a = d^2y/dt^2$,

$$F = m\frac{d^2y}{dt^2}.$$

This law is a special case of

$$F = \frac{d}{dt}\left(m\,\frac{dy}{dt}\right);$$

that is, force equals the rate of change of momentum, considered in Chapter 1, and follows from it when m is constant. It is customary to write time derivatives using overdots, and we will do so when it is convenient: $\dot{y} = dy/dt$, $\ddot{y}=d^2y/dt^2$.

4A Falling Body

A freely falling body near the surface of the earth, and at distance $y = y(t)$ above it, can be assumed for many purposes to have a constant acceleration g, where the acceleration of gravity g in feet per second is about 32.2. It follows that, with height $y(t)$ measured from some point above the earth's surface, the force F can be written in two equal ways:

$$m\ddot{y} = mg.$$

Both sides have the same sign because gravity acts to increase the velocity $\dot{y}$ and so makes its derivative satisfy $\ddot{y} > 0$. (If we were to measure y up from the earth, we would have $m\ddot{y} = -mg$ instead.) Canceling m and integrating twice, we get

$$\dot{y} = gt + c_1, \qquad y = \frac{1}{2}gt^2 + c_1t + c_2.$$

EXAMPLE 1   If a freely falling body has height $y_0 = 200$ feet below some reference point and initial velocity $\dot{y}_0 = 100$ feet per second at time $t = 0$, the height at time $t \geq 0$ satisfies

$$\dot{y}_0 = g \cdot 0 + c_1 \quad \text{so} \quad c_1 = \dot{y}_0 = 100$$

and

$$y_0 = \frac{1}{2} g \cdot 0 + 100.0 + c_2 \quad \text{so} \quad c_2 = y_0 = 200.$$

Hence

$$y(t) = \frac{1}{2} g t^2 + 100 t + 200.$$

Example 1 could have been treated using the characteristic equation $r^2 = 0$, with double roots $r_1 = r_2 = 0$, but straightforward integration is simpler here.

The assumption that a body is "freely falling" is sometimes admissible, but in practice air resistance often has to be taken into account. One model of somewhat restricted validity (see also Exercise 5) prescribes a **drag force** proportional to the velocity $\dot{y}$:

$$F_d = -k\dot{y}, \qquad k \text{ a positive constant.}$$

The equation of forces is now

$$m\ddot{y} = mg - k\dot{y};$$

the minus sign is needed because the drag decreases the velocity and so tends to push $\ddot{y}$ toward negative values. The solutions of our equation

$$\ddot{y} + \frac{k}{m}\dot{y} = g$$

can be found in two steps. The characteristic equation is $r^2 + (k/m)r = 0$, with roots $r_1 = 0$, $r_2 = -k/m$. The homogeneous solution is

$$y_h = c_1 + c_2 e^{-kt/m}.$$

The trial solution $y_p = At$ shows that $A = mg/k$. The total solution is $y_p + y_h$, or

$$y(t) = \frac{mg}{k} t + c_1 + c_2 e^{-kt/m}.$$

EXAMPLE 2   The **terminal velocity** of a falling body subject to drag force F_d is defined by

$$v_\infty = \lim_{t \to \infty} \dot{y}(t).$$

From the solution $y(t)$ found previously, we compute

$$\dot{y}(t) = \frac{mg}{k} - \frac{k}{m} c_2 e^{-kt/m}.$$

Then $v_\infty = mg/k$. Note that a large drag constant k produces an expectedly small terminal velocity. Note also that the analysis just given does not apply if $k = 0$. But the solution of the freely falling body problem gives

$$v_\infty = \lim_{t \to \infty} y(t)$$
$$= \lim_{t \to \infty} gt + c_1 = \infty.$$

The product mg is the **weight** w of the body, so the formula for terminal velocity can be written $v_\infty = w/k$.

4B Pendulum

A freely swinging or **undamped pendulum** with no outside force but gravity acting on it can be most conveniently described in terms of the angle $\theta = \theta(t)$ that the pendulum makes with a vertical line. Figure 3 shows the setup, together with an

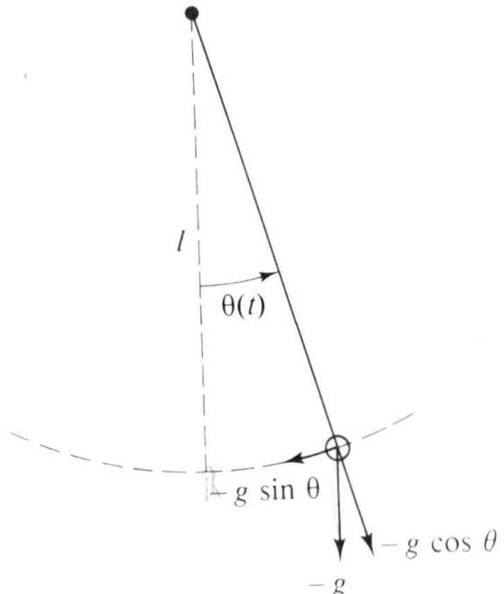

Figure 3

analysis of the gravitational acceleration $-g$ into components parallel and perpendicular to the pendulum. We consider the total mass m of the pendulum as if concentrated at its center of mass, at distance l from the pivot, in the weight at the end. The equation of force is

$$m \frac{d^2}{dt^2}(l\theta) = -mg \sin \theta.$$

The minus sign makes $\dot\theta$ decrease if $0 < \theta < \pi$ and increase if $-\pi < \theta < 0$. Note that $y = l\theta$ is the distance traveled by the weight when the angle changes by θ radians.

Simplification gives

$$\ddot{\theta} = -\frac{g}{l} \sin \theta.$$

The differential equation is nonlinear and we can find numerical approximations to its solutions by the methods of the next section. However, if θ remains small, say $|\theta| < 0.1$, useful results can be obtained by making the approximation $\sin \theta \approx \theta$; this leads to

$$\ddot{\theta} + \frac{g}{l} \theta = 0.$$

EXAMPLE 3   The solutions of the approximate linear equation come from the characteristic equation $r^2 + g/l = 0$ with roots $r_1, r_2 = \pm\sqrt{g/l}$. We get

$$\theta(t) = c_1 \cos \sqrt{\frac{g}{l}}t + c_2 \sin \sqrt{\frac{g}{l}}t.$$

If we arrange it so that $\theta = 0$ when $t = 0$, then $c_1 = 0$, and we can write

$$\theta(t) = c_2 \sin \sqrt{\frac{g}{l}}t.$$

Here c_2 is the maximum absolute angular displacement of the end of the pendulum. The actual displacement of the end of the pendulum is

$$y(t) = l\theta(t) = A \sin \sqrt{\frac{g}{l}}t,$$

where $A = lc_2$ is the amplitude of the pendulum's swing. The velocity of the motion is

$$\dot{y}(t) = A\sqrt{\frac{g}{l}} \cos \sqrt{\frac{g}{l}}t,$$

showing that the maximum velocity, $A\sqrt{g/l}$, occurs when the pendulum is vertical. Also, the velocity is 0 when $\sqrt{g/l}\,t = k\pi/2$, k an integer. This is when the absolute value of the solution reaches its maximum. The period of $y(t)$ is $2\pi\sqrt{l/g}$ (see Exercise 7).

Solution formulas for the nonlinear equation $\ddot{\theta} = -(g/l) \sin \theta$ can be derived using *elliptic functions*, not treated in this book; from these formulas it can be shown that the solutions are periodic as functions of t, just as in the linear model. In practice, we usually have to deal with frictional damping of the motion. This can be allowed for fairly accurately by including a friction term $-k(l\dot{\theta})$, with $k > 0$. On dividing by

lm, we let $K = k/m$ to get

$$\ddot{\theta} = -\frac{g}{l}\sin\theta - K\dot{\theta}.$$

Numerical solution of this equation is taken up in the next section; for now we will look at the **linear damped pendulum equation** that we get when we replace $\sin\theta$ by θ:

$$\ddot{\theta} + K\dot{\theta} + \frac{g}{l}\theta = 0.$$

EXAMPLE 4   The linear pendulum equation has the characteristic equation $r^2 + Kr + g/l = 0$ with roots $r_1, r_2 = \frac{1}{2}(-K \pm \sqrt{K^2 - 4g/l})$. The classification of solutions according to the roots leads to the following list.

(i) $\theta(t) = c_1 e^{\alpha_1 t} + c_2 e^{\alpha_2 t}$ if $K^2 > 4g/l$, where

$$\alpha_1, \alpha_2 = \tfrac{1}{2}(-K \pm \sqrt{K^2 - 4g/l})$$

    are both negative.

(ii) $\theta(t) = c_1 t e^{-Kt/2} + c_2 e^{-Kt/2}$ if $K^2 = 4g/l$.

(iii) $\theta(t) = e^{-Kt/2}(c_1 \cos \beta t + c_2 \sin \beta t)$ if $K^2 < 4g/l$, where

$$\beta = \tfrac{1}{2}\sqrt{4g/l - K^2}.$$

If K should happen to be 0, then only case (iii) can occur, and we get the periodic solution we found in Example 3: $\theta(t) = c_1 \cos \sqrt{g/l}\, t + c_2 \sin \sqrt{g/l}\, t$. For a pendulum, it is usually only the oscillatory case (iii) that is important, so we will illustrate that case. Take $K = 1.0$, $m = 1$, and $l = 1$. Then $\beta \approx 5.65$. The solution

$$y(t) = \theta(t) = e^{-0.5t}(c_1 \cos \beta t + c_2 \sin \beta t)$$

is still subject to initial conditions, for example, $\theta(0) = 0$, $\dot{\theta}(0) = 10$. These imply that $c_1 = 0$ and $c_2\beta = 10$, so we get

$$y(t) = \frac{10}{\beta} e^{-0.5t} \sin \beta t$$

$$\approx 1.8 e^{-0.5t} \sin 5.65t$$

A part of the graph is shown in Figure 4. The damping constant $K = 1.0$ evidently leads to quite significant damping of the oscillations. Of course, the motion is not periodic as in the undamped case, but the points of zero displacement from the vertical are still equally spaced along the time axis, the space between then being about $2\pi/5.65 \approx 1.11$.

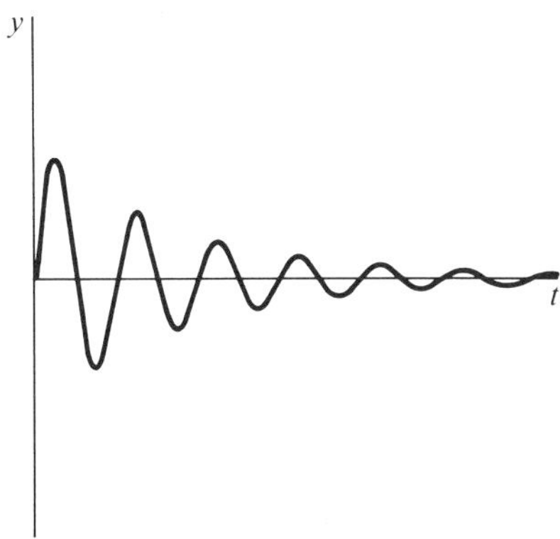

Figure 4

EXERCISES

Falling Body

1. A ball of mass m, and hence weight $w = mg$, is dropped from a point 200 feet above the surface of the earth so that its initial velocity is 0. Assume that the drag force is insignificant.
   (a) When does the ball hit the ground?
   (b) What is the impact velocity at ground level?
   (c) From what height should the ball be dropped to get an impact velocity of 100 feet per second?
   (d) Is the mass of the ball relevant to any of the preceding questions?

2. Suppose in Exercise 1 that the weight of the ball is 32.2 pounds, so we can take $m = 1$, and that the drag force is significant, with drag constant $k = 0.1$.
   (a) Show that the ball hits the ground after time t_1, where t_1 is the positive solution of the equation
   $$t/10 + e^{-t/10} = 171/161, \text{ with } g = 32.2.$$

   (b) By successive approximation, for example, using Newton's method, show that t_1 in part (a) is above 3.74.
   (c) Compare the result obtained in part (b) with the answer obtained by using the free-fall formula $y(t) = gt^2/2$.
   (d) Find the terminal velocity of the ball. [Ans. 322]
   (e) About how long does it take for the ball to reach the velocity of 200 feet per second? [Ans. 9.7 seconds]

3. Suppose a falling body is subject to a drag constant k and has mass m.
   (a) If the initial velocity is 0, show that the distance covered in time t is
   $$y(t) = \frac{mg}{k}t - \frac{m^2 g}{k^2}(1 - e^{-kt/m}).$$

   (b) Show that the formula for $y(t)$ in part (a) satisfies $y(t) \le \frac{1}{2}gt^2$ for $t \ge 0$. [*Hint:* $\ddot{y} = g - (k/m)\dot{y} \le g$. Now integrate from 0 to t.]

(c) Find the analogue of the formula given in part (a) for the case of initial velocity $\dot{y}(0) = v_0$.

4. **(a)** For a body of mass m subject to drag constant k, show that initial velocity v_0 leads to velocity at time t given by

$$\dot{y}(t) = \frac{mg}{k} + \left(v_0 - \frac{mg}{k}\right)e^{-kt/m}.$$

(b) Find the limit as k tends to 0 of the formula for $\dot{y}(t)$ in part (a). Does this agree with the free-fall formula $v_0 + gt$?

5. The drag constant k can be estimated by using an observed value of the terminal velocity $v_\infty = mg/k$.

(a) Find k if a body of weight $w = 100$ pounds achieves terminal velocity $v_\infty = 180$ feet per second.

(b) How far must the body in part (a) fall to obtain the velocity of 150 feet per second?

6. The nonlinear equation

$$m\ddot{y} = mg - k\dot{y}^2$$

is sometimes used to model fall in a resisting medium. Here the retarding action of the drag term is much greater at large velocities than for the linear model. To solve the equation, let $v = \dot{y}$, $\dot{v} = \ddot{y}$ and solve

$$m\dot{v} = mg - kv^2$$

by separation of variables for $v(t)$. Then solve $\dot{y} = v(t)$ by a single integration.

(a) Solve the equation $m\dot{v} = mg - kv^2$ under the assumption that $v(0) = v_0$, and show that the solution can be written

$$v(t) = \sqrt{\frac{mg}{k}}\left(\frac{e^{at} - c_0 e^{-at}}{e^{at} + c_0 e^{-at}}\right),$$

where $c_0 = (\sqrt{mg} - \sqrt{k}\,v_0)/(\sqrt{mg} + \sqrt{k}\,v_0)$, and $a = \sqrt{kg/m}$.

(b) Find the terminal velocity v_∞ in this model and compare it with one obtained from the linear model $m\dot{v} = mg - kv$.

(c) Assuming that $y(0) = 0$, show that

$$y(t) = \frac{v_\infty^2}{g}\ln\left(\cosh\frac{gt}{v_\infty} + \frac{v_0}{v_\infty}\sinh\frac{gt}{v_\infty}\right).$$

Pendulum

7. Determine the amplitude A of the displacement $y(t)$ of an undamped pendulum satisfying the linear equation $\ddot{\theta} + (g/l)\theta = 0$, where $y = l\theta$. What is its period?

8. How long should an undamped pendulum be if it satisfies $\ddot{\theta} + (g/l)\theta = 0$ and it is to have a period of 2 seconds?   [Ans. 3.26 feet]

9. **(a)** Given the relations, $y(t) = l\theta(t)$ and $K = k/lm$, express the differential equation $\ddot{\theta} + K\dot{\theta} + (g/l)\theta = 0$ in terms of y, k, l, and m.

(b) What condition must K satisfy for the pendulum described by the differential equation in part (a) to oscillate?

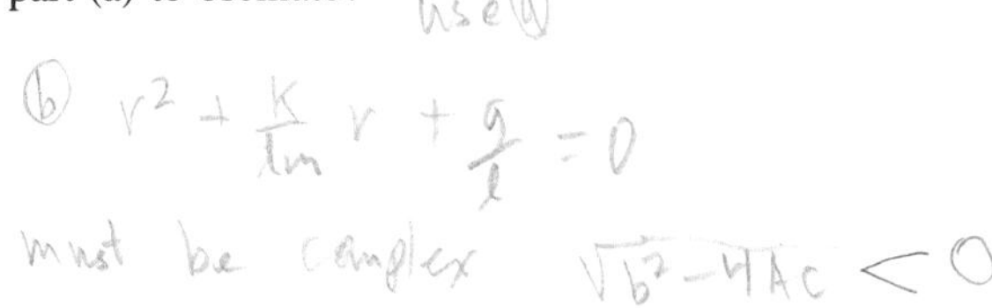

(c) Show that oscillations described by the differential equation in part (a) undergo an increase in the time between successive vertical positions as l is increased. Compare the effect that a corresponding increase in m has.

10. Consider the equation

$$\ddot{\theta} + K\dot{\theta} + \frac{g}{l}\theta = 0,$$

where K, g, and l are positive constants.

(a) Show that if the roots of the characteristic equation are real that they are both negative and that if they are complex then they both have negative real parts.

(b) What can you conclude about the oscillation of a pendulum from part (a)?

(c) If $K^2 \geq 4g/l$, explain why $\theta(t)$ is nonoscillatory.

11. Show that the general solution of

$$\ddot{\theta} + K\dot{\theta} + \frac{g}{l}\theta = 0,$$

in the case $K^2 < 4g/l$, can be written in the form

$$\theta(t) = A e^{-Kt/2} \cos\left(\sqrt{\frac{B}{l}}\, t + \alpha\right).$$

[*Hint:* Use the addition formula for the cosine.]

12. Solve the equation

$$\ddot{y} + \dot{y} + y = a_0 \cos wt$$

for the oscillation of a pendulum with externally applied force $a_0 \cos wt$. First find a particular solution $y_p(t)$ by the method of undetermined coefficients using the trial solution $y_p(t) = A \cos wt + B \sin wt$. This example is treated more generally in Section 4C, Example 10.

13. Let $y = y(t)$ be a periodic solution of

$$y'' = -\frac{g}{l}\sin y,$$

and let y_0 be the maximum value of $y(t)$, such that $0 < y_0 < \pi/2$.

(a) Show that if $y(t)$ satisfies

$$y' = \sqrt{\frac{2g}{l}(\cos y - \cos y_0)}$$

then it satisfies the second-order equation.

(b) Show that the time T required for $y(t)$ to go from $-y_0$ to y_0 is

$$T = \sqrt{\frac{2l}{g}} \int_0^{y_0} \frac{dy}{\sqrt{\cos y - \cos y_0}}.$$

Thus the period of the solution is $p = 2T$.

14. A cylindrical object floating in liquid with its central axis vertical satisfies a differential equation for its vertical displacement $y(t)$ from equilibrium:

$$\ddot{y} + \frac{k}{m}\dot{y} + A\frac{w}{m}y = 0,$$

where m is the mass of the object, A is its circular cross-sectional area, w is the density of the liquid, and k is a positive friction constant.

(a) Show that the differential equation has oscillatory solutions if and only if $k^2 < 4Awm$.

(b) Discuss the effect on $y(t)$ of increasing m and of increasing w, assuming that

$$k^2 < 4Awm.$$

(c) Discuss the analogy with solutions of the (linearized) damped pendulum equation

$$\ddot{y} + \frac{k}{lm}\dot{y} + \frac{g}{l}y = 0.$$

4C Oscillations

A second-order, constant-coefficient linear differential equation

$$a\frac{d^2x}{dt^2} + b\frac{dx}{dt} + cx = f(t)$$

often has a physical interpretation that allows a neat classification of the solutions and equations into distinct types, depending on the relations between the constants a, b, c and the function f. Equations of this kind are important not only because of their direct physical applicability, but also because of the insight they yield about related nonlinear phenomena. A typical mechanism that can be analyzed using a constant-coefficient equation is shown in Figure 5, in cross section. The working parts consist

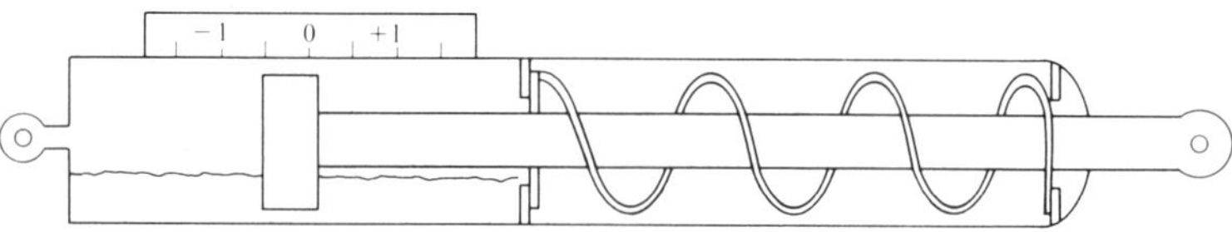

Figure 5

of a piston that travels in a cylinder containing fluid, and a spring that can expand and compress. A spring usually exerts a force roughly proportional to its extension or compression from its equilibrium position, denoted by 0 on the fixed scale. Thus, if x is the amount of displacement from 0, then the force f_s exerted by the spring is representable by

$$f_s = -hx, \qquad h > 0.$$

For small displacements, it is assumed that h is constant. Similarly, we assume that the frictional force f_F in the mechanism due to the viscosity of the fluid is proportional to the velocity:

$$f_F = -k\frac{dx}{dt}, \qquad k > 0.$$

The time-dependent external force $f_E = f(t)$ acts independently of f_s and f_F, which are, in turn, assumed to act independently of one another, so the total force acting parallel to the scale is

$$f_s + f_F + f_E = -hx - k\frac{dx}{dt} + f(t).$$

On the other hand, general physical principles assert that this force must also be equal to the mass m of the moving parts times the acceleration d^2x/dt^2. Thus

4.1
$$m\frac{d^2x}{dt^2} + k\frac{dx}{dt} + hx = f(t).$$

Next, we make various assumptions about k, h, and f. The mass m will be a fixed positive constant. We consider first **free oscillations,** with external force f identically zero.

Harmonic Oscillation.   We assume that $k = 0$ and $f = 0$. These assumptions represent an ideal situation that can only be approximated by the mechanism shown in Figure 5. The differential equation becomes

$$\frac{d^2x}{dt^2} + \frac{h}{m}x = 0.$$

The solutions all have the form

$$x(t) = c_1 \cos \sqrt{\frac{h}{m}}\,t + c_2 \sin \sqrt{\frac{h}{m}}\,t,$$

but to interpret a given solution easily it is good to rewrite it as follows. We choose an α such that

$$\cos \alpha = \frac{c_1}{\sqrt{c_1^2 + c_2^2}}, \qquad \sin \alpha = \frac{c_2}{\sqrt{c_1^2 + c_2^2}};$$

such an α can always be found if c_1 and c_2 are not both zero. Then, letting $A = \sqrt{c_1^2 + c_2^2}$ gives

$$x(t) = A\left(\cos \alpha \cos \sqrt{\frac{h}{m}}\,t + \sin \alpha \sin \sqrt{\frac{h}{m}}\,t\right)$$

$$= A \cos \left(\sqrt{\frac{h}{m}}\,t - \alpha\right).$$

This last formula shows that A is the amplitude of an oscillation about the equilibrium position. The graph of the displacement as a function of time is shown in Figure 6 for the choice $\alpha = \pi/2$, $A = 2$, and $h = m = 1$. Changing the number α shifts the graph to the right or left. The frequency of the oscillation depends only on the ratio h/m; increasing h or decreasing m increases the frequency of the oscillation. The fact that the oscillation is periodic is a direct consequence of the assumption that there is no friction.

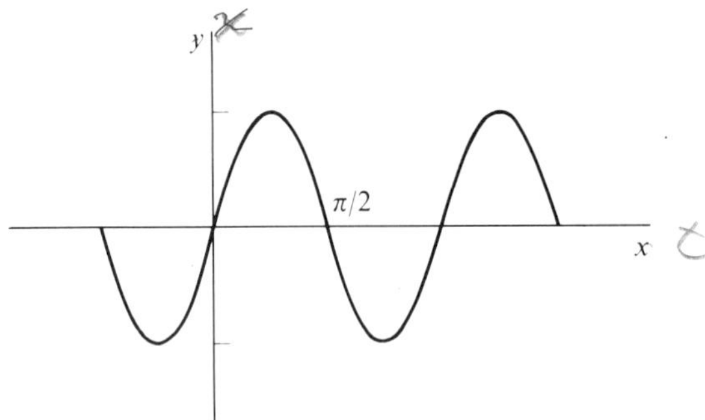

Figure 6

EXAMPLE 5   If in Equation 4.1 we take $m = h = 1$, $k = 0$, and $f = 0$, then the resulting differential equation

$$\frac{d^2x}{dt^2} + x = 0$$

has the solution

$$x(t) = A \cos(t - \alpha).$$

The initial conditions

$$x(0) = x_0, \qquad \frac{dx}{dt}(0) = v_0,$$

require that

$$A \cos(-\alpha) = x_0, \qquad -A \sin(-\alpha) = v_0.$$

We can solve these equations for A and α in terms of x_0 and v_0 to get

$$A = \sqrt{x_0^2 + v_0^2}, \qquad \alpha = \arctan \frac{v_0}{x_0}.$$

In the special case when $x_0 = 0$, the initial displacement from equilibrium is zero, and we find

$$A = |v_o|, \qquad \alpha = \pm \frac{\pi}{2}.$$

Thus the solution is

$$x(t) = |v_0| \cos\left(t \pm \frac{\pi}{2}\right)$$

$$= \pm |v_0| \sin t.$$

The sign has to be chosen so that $\pm |v_0| = v_0$. Other special cases are treated in the exercises.

Damped Oscillation.   The piston in the view of the mechanism shown in Figure 5 exerts a damping force that can be varied by changing the viscosity of the medium in which the piston moves. If we continue to assume that $f = 0$ in Equation 4.1, then we have to deal with the differential equation

$$\frac{d^2x}{dt^2} + \frac{k}{m}\frac{dx}{dt} + \frac{h}{m}x = 0.$$

The characteristic equation is

$$r^2 + \frac{k}{m}r + \frac{h}{m} = 0,$$

which has the roots

$$r_1 = \frac{1}{2m}(-k + \sqrt{k^2 - 4mh}), \qquad r_2 = \frac{1}{2m}(-k - \sqrt{k^2 - 4mh}).$$

We can distinguish three distinct cases depending on the discriminant $k^2 - 4mh$.

Overdamping, $k^2 - 4mh > 0$. Because we have assumed that k and h are both positive, this case occurs when $k > 2\sqrt{mh}$. Physically, this inequality means that the friction constant k exceeds the constant $\sqrt{mh}$, depending on the spring stiffness h and the mass m, by a factor more than 2. The effect of the assumption $k > 2\sqrt{mh}$ is to make the roots of the characteristic equation satisfy

$$r_2 < r_1 < 0.$$

As a result, the general solution has the form

$$x(t) = c_1 e^{r_1 t} + c_2 e^{r_2 t},$$

where the exponentials decrease as t increases.

EXAMPLE 6   We assume $m = 2$, $h = 1$, and $k = 3$, so that $k > 2\sqrt{mh}$. Then

$$r_1 = -\frac{1}{2}, \qquad r_2 = -1,$$

so that the displacement from equilibrium at time t is

$$x(t) = c_1 e^{-t/2} + c_2 e^{-t}.$$

A typical graph is shown in Figure 7. The maximum displacement occurs at just one point, after which the displacement tends steadily to 0 as t increases.

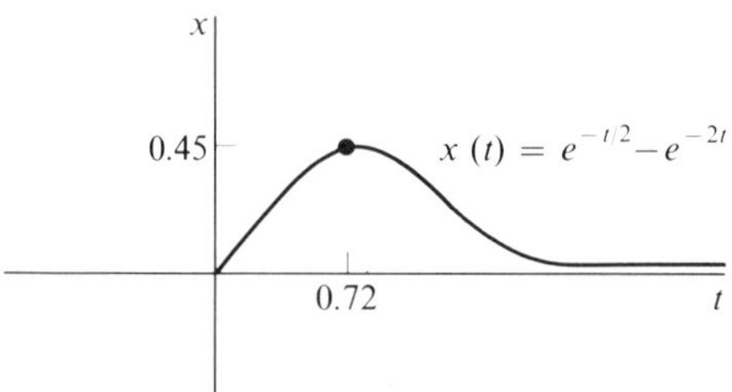

Figure 7

Underdamping, $k^2 - 4mh < 0$. This case occurs when $k < 2\sqrt{mh}$, so that, relative to $\sqrt{mh}$, the friction constant k is small. The characteristic roots are now complex conjugates of one another:

$$r_1 = \frac{1}{2m}(-k + i\sqrt{4mh - k^2}), \qquad r_2 = \frac{1}{2m}(-k - i\sqrt{4mh - k^2}).$$

The general form of the displacement function is then

$$x(t) = e^{-kt/2m}\left(c_1 \cos \frac{\sqrt{4mh - k^2}}{2m} t + c_2 \sin \frac{\sqrt{4mh - k^2}}{2m} t \right)$$

$$= Ae^{-kt/2m} \cos \left(\frac{\sqrt{4mh - k^2}}{2m} t - \alpha \right),$$

where $A = \sqrt{c_1^2 + c_2^2}$ and $\alpha = \arctan (c_2/c_1)$.

EXAMPLE 7   Take $h = 2$, $m = 1$, and $k = 2$. Then $k < 2\sqrt{mh}$ and the displacement at time t is

$$x(t) = Ae^{-t} \cos (t - \alpha).$$

Figure 8 shows the graph of such a function with $A = 1$ and $\alpha = 0$. It is easy

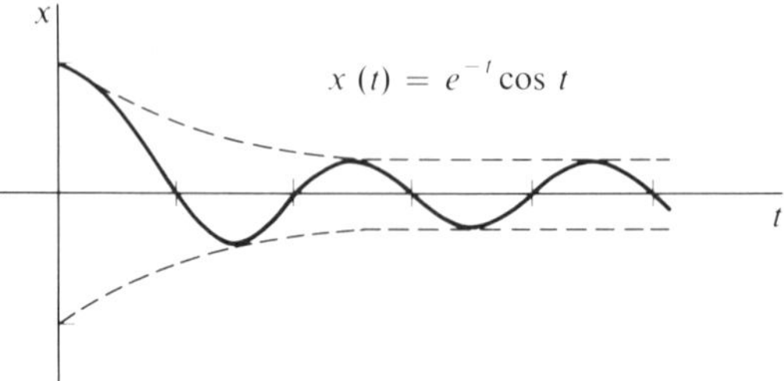

Figure 8

to see that this choice for the constants A and α gives a solution satisfying the initial conditions

$$x(0) = 1, \qquad \frac{dx}{dt}(0) = -1.$$

Critical Damping, $k = 2\sqrt{mh}$. This case lies between overdamping and underdamping, and it is unstable in the sense that an arbitrarily small change in one of the parameters k, m or h will disturb the equality $k = 2\sqrt{mh}$ and produce one of the other two cases. Numerically, the case of critical damping is distinguished by the equality of the characteristic roots: $r_1 = r_2 = -k/2m$. It follows that the displacement function is given by

$$x(t) = c_1 t e^{-kt/2m} + c_2 e^{-kt/2m}.$$

EXAMPLE 8   Take $m = h = 1$ and $k = 2$. Then

$$x(t) = (c_1 t + c_2)e^{-t}.$$

If $x(0) = x_0$ and $dx/dt\,(0) = v_0$, then

$$c_1 = (v_0 + x_0), \qquad c_2 = x_0,$$

so that

$$x(t) = [(v_0 + x_0)t + x_0]e^{-t}.$$

Figure 9 shows four possibilities, depending on the size of v_0, the initial velocity.

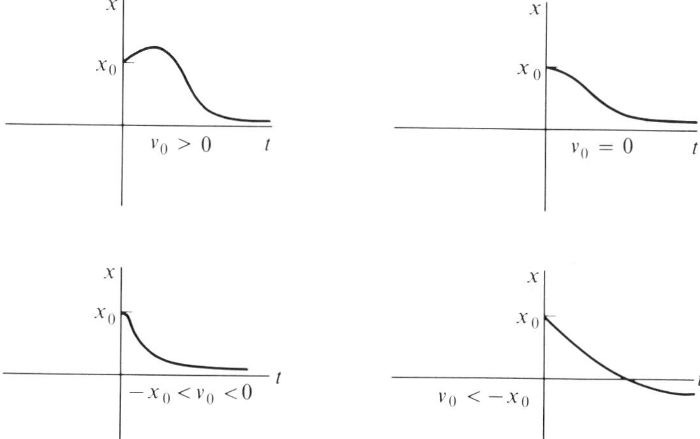

Figure 9

A critically damped displacement is like an overdamped one in that there is no oscillation from one side to the other of the equilibrium position. But with fixed mass m and spring constant h, the critical viscosity value $k = 2\sqrt{mh}$ produces the most rapid return toward equilibrium from an initial position in which $v_0 = 0$. The physical reason is that a higher viscosity produces a more sluggish return, and lower viscosity allows oscillation.

Forced Oscillation.   In the specific instances considered so far, the differential equation

$$m \frac{d^2x}{dt^2} + k \frac{dx}{dt} + hx = f(t)$$

has been subject to initial conditions $x(0) = x_0$, $dx/dt\,(0) = v_0$, but the external force function f has been assumed to be identically zero. The resulting free oscillation is described by a solution of a homogeneous differential equation. (Note that "free" does not mean "undamped.") When f is not identically zero, we speak of **forced oscillation.** From a purely mathematical point of view, there is no reason why the function f on the right side of the preceding differential equation cannot be chosen to be, say, an arbitrary continuous function for $t \geq 0$. However, a force function f that assumes large values could easily drive the oscillations quite outside the range in which we can maintain the original assumptions used to derive the differential equation. (For example, stretching a spring too far might change its characteristics to the point of destroying its elasticity altogether.) For this reason, the function f is chosen in the examples and exercises to have a rather restricted range of values. In every example, of course, we can use the decomposition of a solution $x(t)$ into homogeneous and particular parts,

$$x(t) = x_h(t) + x_p(t).$$

Fortunately, $x_h(t)$ has already been discussed earlier in this section for various choices of m, k and h in the homogeneous differential equation. What remains to be done, then, is to discuss the effect of adding a solution of the nonhomogeneous equation. If $k > 0$, the analysis given in the earlier examples shows that every homogeneous solution tends to zero like an exponential of the form $e^{-kt/2m}$. Thus, for values of t that make $(kt/2m)$ moderately large, the addition of the homogeneous solution has a negligible effect. Such an effect is called **transient,** and the particular solution is therefore called the **steady-state** solution.

EXAMPLE 9   If $f(t) = a_0 \cos \omega t$, then the differential equation

$$m \frac{d^2x}{dt^2} + k \frac{dx}{dt} + hx = a_0 \cos \omega t$$

has a particular solution of the form

$$x_p(t) = A \cos \omega t + B \sin \omega t$$

that can be found by the method of undetermined coefficients. Substitution of x_p into the equation yields

$$(h - \omega^2 m)A + \omega k B = a_0,$$

$$\omega k A - (h - \omega^2 m)B = 0.$$

It follows that

$$A = \frac{(h - \omega^2 m)a_0}{(h - \omega^2 m)^2 + \omega^2 k^2}, \qquad B = \frac{\omega k a_0}{(h - \omega^2 m)^2 + \omega^2 k^2},$$

so the particular solution we get is

$$x_p(t) = \frac{a_0}{(h - \omega^2 m)^2 + \omega^2 k^2}[(h - \omega^2 m) \cos \omega t + \omega k \sin \omega t]$$

$$= \frac{a_0}{\sqrt{(h - \omega^2 m)^2 + \omega^2 k^2}} \cos(\omega t - \alpha),$$

where $\alpha = \arctan[\omega k / (h - \omega^2 m)]$. What we have found is just one of many solutions, each satisfying a different set of initial conditions. However, our particular solution is the steady-state solution. Notice that $x_p(t)$ is very much like the external force $f(t)$; in fact, since $f(t) = a_0 \cos \omega t$,

$$x_p(t) = \frac{1}{\sqrt{(h - \omega^2 m)^2 + \omega^2 k^2}} f\left(t - \frac{\alpha}{\omega}\right).$$

The choice $\omega^2 = h/m$ makes the amplitude of x_p equal to $a_0\sqrt{m}/k\sqrt{h}$. This choice of the frequency ω is called **resonant** because it produces a response of large amplitude for values of the system parameters k and h that are small relative to m. Notice that in this example of resonance we have $\alpha = \arctan(\infty) = \pi/2$, so that

$$x_p(t) = \frac{a_0}{\omega k} \cos\left(\omega t - \frac{\pi}{2}\right)$$

$$= \sqrt{\frac{m}{h}} \cdot \frac{a_0}{k} \cos\left(\sqrt{\frac{h}{m}} t - \frac{\pi}{2}\right).$$

Thus, in a system with $\sqrt{m}$ large relative to $k\sqrt{h}$, a small external force may produce virbrations of large amplitude if the external force oscillates at the resonant frequency. For this reason, resonance can completely upset an operating system, even though the external force remains small in magnitude.

EXERCISES

1. Each of the following differential equations represents a free oscillation. Classify them according to type: harmonic, overdamped, underdamped, or critically damped. (*Note:* $\dot{x} = dx/dt$, $\ddot{x} = d^2x/dt^2$.)

(a) $\dfrac{d^2x}{dt^2} + 2\dfrac{dx}{dt} + x = 0.$ (b) $\dfrac{d^2x}{dt^2} + 2\dfrac{dx}{dt} + 2x = 0.$

(c) $\ddot{x} + p^2x = 0.$ (d) $\ddot{x} + 3\dot{x} + x = 0.$

(e) $\ddot{x} + a\dot{x} + a^2x = 0,\ a > 0.$ (f) $\dfrac{d^2x}{dt^2} + \dfrac{1}{4}\dfrac{dx}{dt} + \dfrac{1}{8}x = 0.$

2. Find the general solution for each differential equation in Exercise 1.

3. For each of the following general solutions of a second-order, constant-coefficient equation, find the choice of the arbitrary constants that satisfies the corresponding initial conditions. Sketch the solution that you find. Also find a differential equation that the solution satisfies.

(a) $x(t) = c_1 \cos 2t + c_2 \sin 2t;\ x(0) = 0,\ dx/dt(0) = 1.$

(b) $x(t) = A\cos(3t - \alpha);\ x(0) = 1,\ dx/dt(0) = -1.$

(c) $x(t) = c_1 t e^{-2t} + c_2 e^{-2t};\ x(0) = 0,\ dx/dt(0) = -1.$

(d) $x(t) = Ae^{-t}\cos(2t - \alpha);\ x(0) = 2,\ dx/dt(0) = 2.$

(e) $x(t) = c_1 e^{-2t} + c_2 e^{-4t};\ x(0) = -1,\ dx/dt(0) = -1.$

4. Find the steady-state solution to each of the following differential equations. Also estimate the earliest time beyond which the transient solution remains less than 0.01, if we are given the initial conditions $x(0) = 1,\ \dot{x}(0) = 0.$

(a) $\dfrac{d^2x}{dt^2} + 2\dfrac{dx}{dt} + 2x = 2\cos 3t.$ (b) $\dfrac{d^2x}{dt^2} + \dfrac{dx}{dt} + x = e^{-t}.$

5. There is a well-known analogy between the behavior of a damped mass–spring system and of the simple *RLC* circuit illustrated in Figure 10. Here L is the inductance (analogue of mass) of a coil, R is the resistance (analogue of friction constant) in the circuit, and C is the capacitance (analogue of reciprocal of spring stiffness) or ability of a capacitor to store a charge. The differential equation satisfied by the charge $Q(t)$ on the capacitor at time t is

$$L\frac{d^2Q}{dt^2} + R\frac{dQ}{dt} + \frac{1}{C}Q = E(t),$$

where $E(t)$ is the voltage impressed on the circuit from an external source. [The charge $Q(t)$ is related to the current flow $I(t)$ by $I = dQ/dt.$

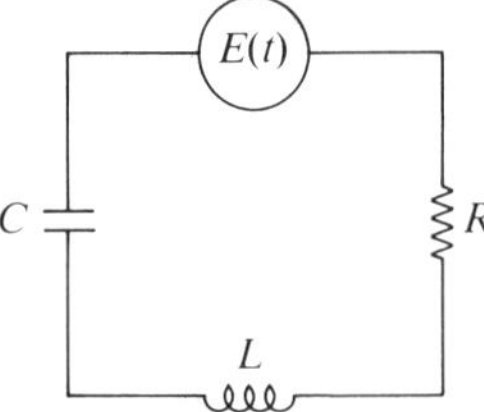

Figure 10

(a) Derive the relations that must hold between R, L, and C in order that the response of $Q(t)$ should be respectively underdamped, critically damped, and overdamped.

(b) Show that if $C = \infty$ (capacitor is absent) the equation for $I(t)$ is

$$L\frac{dI}{dt} + RI = E(t).$$

(c) Solve the equation in part (b) when $E(t) = E\sin \omega t$ and $I(0) = 0$, and show that, if t is large enough, the current response differs negligibly from

$$\frac{E}{Z}\sin(\omega t - \theta),$$

where $Z = \sqrt{R^2 + \omega^2 L^2}$ and $\cos\theta = R/Z$. The function $Z(\omega)$ is called the **impedance** of the circuit in response to the sinusoidal input of frequency ω.

(d) Show that, if $0 < C < \infty$, the long-term response to input voltage $E(t) = E\sin\omega t$ is $E(t)/Z(\omega)$, where $Z(\omega) = \sqrt{R^2 + [\omega L - 1/\omega C]^2}$.

5 NUMERICAL METHODS

Second-order differential equations are either linear, of the form

$$\frac{d^2 y}{dt^2} + a(t)\frac{dy}{dt} + b(t)y = f(t),$$

or nonlinear, for example, the damped pendulum equation

$$\frac{d^2 y}{dt^2} + k\frac{dy}{dt} + \frac{g}{l}\sin y = 0.$$

Equations of both types can be written in the very general form

$$\frac{d^2 y}{dt^2} = F\!\left(t,\, y,\, \frac{dy}{dt}\right).$$

It is this form that we will treat here along with **initial conditions** of the form

$$y(t_0) = y_0, \qquad \frac{dy}{dt}(t_0) = z_0,$$

where t_0, y_0, and z_0 are given constants.

It is important to realize that the only type of problem for which we so far have any general method of solution is the linear equation with $a(t)$ and $b(t)$ both constant. Even then, if $f(t)$ is not a function that can be integrated formally, we may have trouble finding a formula for a particular solution. What we may then settle for is a numerical approximation y_k to the value $y(t_k)$ of the true solution at a discrete set of points t_k. Such approximations were treated in Chapter 1 for first-order equations, and the methods used here for second-order equations are simple modifications of the first-order methods.

If the purpose in solving a differential equation is just to obtain numerical values for some particular solution, then a purely numerical approach may be more efficient than first finding a solution formula and then digging the desired numbers, or graph, out of the formula. On the other hand, if what you want is to display the nature of a solution's dependence on certain parameters in the differential equation, or on initial conditions, then solution by formula is preferable if at all possible. Beyond that, detailed properties of a solution, such as whether it is periodic or only approximately so, can be hard to get from a numerical approximation and easy to get from a formula like $y(t) = 2\cos 3t$.

5A Euler Methods

Numerical methods for first-order equations can be motivated by using the interpretation of the first derivative as a slope. Rather than trying to make something of the interpretation of the second derivative in a second-order equation, what we will do is to find a pair of first-order equations equivalent to a given second-order equation and then apply first-order methods to the simultaneous solution of the pair of equations. The principle is easiest to understand in the general case

$$y'' = F(t, y, y'), \qquad y(t_0) = y_0, \qquad y'(t_0) = z_0.$$

There are many ways to find an equivalent pair of first-order equations, but the most natural is usually to introduce the first derivative y' as a new unknown function z. We write $z = y'$, $z' = y''$, so we can replace $y'' = F(t, y, y')$ by $z' = F(t, y, z)$. The pair to be solved numerically is then

$$y' = z, \qquad y(t_0) = y_0,$$
$$z' = F(t, y, z), \qquad z(t_0) = z_0.$$

Since $y'(t_0) = z(t_0) = z_0$, there is an initial condition that goes naturally with each equation. To find an approximate solution, we can do what we would do with a single first-order equation, except that at each step we find new approximate values for both unknown functions y and z, and then use these values to compute new approximations in the next step. The iterative formulas are as follows for the simple **Euler method,** with step size h:

$$y_{k+1} = y_k + hz_k,$$
$$z_{k+1} = z_k + hF(t_k, y_k, z_k),$$

where $t_k = t_0 + kh$. The starting values y_0, z_0 come from the initial conditions. We will often use overdots to indicate derivatives with respect to t.

EXAMPLE 1   The initial-value problem

$$\ddot{y} = -\sin y, \qquad y(0) = 2, \qquad \dot{y}(0) = 1,$$

describes the motion of a pendulum with fairly large amplitude, so large that the linear approximation $\ddot{y} = -y$ would be inadequate. We go ahead to solve the system

$$\dot{y} = z, \qquad y(0) = 2,$$
$$\dot{z} = -\sin y, \qquad z(0) = 1.$$

The essence of the computation can be expressed in the following basic loop.

```
DEF F(T,Y,Z)=-SIN(Y)
LET T=0
LET Y=2
```

```
LET Z=1
LET H=.01
FOR K=1 TO 2000
LET W=Y
LET Y=Y+H*Z
LET Z=Z+H*F(T,W,Z)
LET T=T+H
PRINT T,Y
NEXT K
```

Note the line LET W=Y that saves the current value of Y for use two lines later; without it the advanced value Y+H*Z would be used, which is not correct. The printout results in 2000 values of t from 0.01 to 20 by steps of size 0.01 along with the corresponding Y-values. We could also get the Z-values, which are approximations to the values of the derivative $\dot{y}$ at the same t points. Rather than show a table with 2000 entries, Figure 11 shows the corresponding graph for the initial condition $y(0) = 2$, $y'(0) = 1$. The approximation $\sin y \approx y$ is inappropriate for this problem, because the values of y that occur are too large to make the approximation a good one. The resulting equation $\ddot{y} = -y$ has solutions of the form $A \cos(t - b)$, whose graphs have the same general shape as the graph in Figure 11. But closer examination shows these graphs really vary significantly from the shape and period shown in Figure 11 (see Exercise 7).

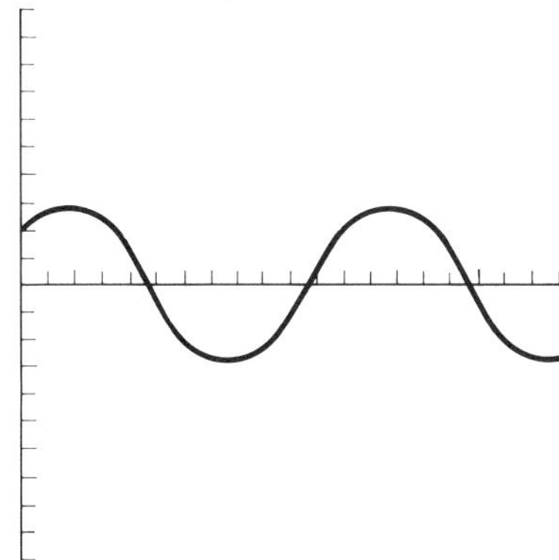

Figure 11

The **modified Euler method** follows this routine:

$$p_{k+1} = y_k + hz_k$$

$$q_{k+1} = z_k + hF(t_k, y_k, z_k)$$

$$y_{k+1} = y_k + \frac{h}{2}(z_k + q_{k+1})$$

$$z_{k+1} = z_k + \frac{h}{2}[F(t_k, y_k, z_k) + F(t_{k+1}, p_{k+1}, q_{k+1})].$$

Here p_k and q_k provide the simple Euler estimates that are then used to compute the final estimates for y_k and $z_k = \dot{y}_k$; $t_k = t_0 + kh$ as before. (As with the simple Euler method, the value y_k has to be kept for use in computing z_{k+1} and cannot be replaced by y_{k+1} without significant error.) The advantage of the modification is that the error in the final estimates is substantially reduced without adding much complexity to the computation on a digital computer.

EXAMPLE 2   The initial-value problem

$$y'' + y = 0, \qquad y(0) = 0, \qquad y'(0) = 1$$

has solution $y = \sin x$. Here is a table of values that compares the Euler method, the modified Euler method, and the correct value to as many decimal places as are given. The value of h used is 0.001, but only every hundredth value is given. The only discrepancies between the last two columns occur in the sixth and the last entries.

x	Euler y	Mod-Euler y	$y = \sin x$
0.1	0.099838	0.099834	0.099834
0.2	0.198689	0.198669	0.198669
0.3	0.295564	0.29552	0.29552
0.4	0.389496	0.389418	0.389418
0.5	0.479545	0.479426	0.479426
0.6	0.564812	0.564643	0.564642
0.7	0.644443	0.644218	0.644218
0.8	0.717643	0.717356	0.717356
0.9	0.783679	0.783327	0.783327
1.0	0.841892	0.841471	0.841471
1.1	0.891697	0.891208	0.891207
1.2	0.932598	0.932039	0.932039
1.3	0.964185	0.963558	0.963558
1.4	0.98614	0.98545	0.98545
1.5	0.998243	0.997495	0.997495
1.6	1.00037	0.999574	0.999574
1.7	0.992508	0.991665	0.991665
1.8	0.974725	0.973848	0.973848
1.9	0.9472	0.9463	0.9463
2.0	0.910207	0.909297	0.909297
2.1	0.864117	0.863209	0.863209
2.2	0.809387	0.808496	0.808496
2.3	0.746564	0.745705	0.745705
2.4	0.676275	0.675463	0.675463
2.5	0.599221	0.598472	0.598472
2.6	0.516173	0.515501	0.515501
2.7	0.427958	0.427379	0.42738
2.8	0.335458	0.334988	0.334988
2.9	0.239597	0.239249	0.239249
3.0	0.141333	0.141119	0.14112

A computer program to produce the first, third, and fourth columns in the previous table might look like this.

```
DEF F(T,Y,Z)=−Y
LET T=0
LET Y=0
LET Z=1
LET H=.001
FOR J=1 TO 30
  FOR K=1 TO 100
  LET P=Y+H*Z
  LET Q=Z+H*F(T,Y,Z)
  LET W=Y
  LET Y=Y+.5*H*(Z+Q)
  LET Z=Z+.5*H*(F(T,W,Z)+F(T+H,P,Q))
  LET T=T+H
  NEXT K
  PRINT T,Y,COS(T)
NEXT J
```

Recall that in this program we are dealing with a pair of equations of the form

$$\dot{y} = z$$

$$\dot{z} = F(t, y, z)$$

The line LET W=Y saves the current value of Y for use two lines later.

5B Stiffness

We remarked in Section 7 of Chapter 1 that the local formula errors due to using the Euler and modified Euler methods are respectively of order h^2 and h^3. In addition to formula error and the round-off error that we have to deal with, there occurs for some second-order equations an additional difficulty due to what is called **stiffness** of a differential equation. The problem is that in a two-parameter family of solutions a small error at some point may lead to approximation of a solution that is very different from the one we started on. Here is a simple example.

EXAMPLE 3   The differential equation $y'' - 100y = 0$ has the general solution

$$y(t) = c_1 e^{10t} + c_2 e^{-10t}.$$

The initial conditions $y(0) = 1$, $y'(0) = -10$ single out the solution $y_1(t) = e^{-10t}$. But the "nearby" initial conditions $y(0) = 1$, $y'(0) = -10 + 2\epsilon$, for small $\epsilon > 0$, single out the solution

$$y_2(t) = 0.1\epsilon e^{10t} + (1 - 0.1\epsilon)e^{-10t}.$$

The behavior of these two solutions is radically different for only moderately large values of t; $y_1(t)$ tends to zero, but $y_2(t)$ gets very large unless the "error" ϵ is required to be small to an impractical degree. Figure 12 shows comparative behaviors, with $\epsilon = 0.00001$. Solutions of nonlinear equations may exhibit the same kind of behavior, and one way to deal with the problem is suggested in the exercises.

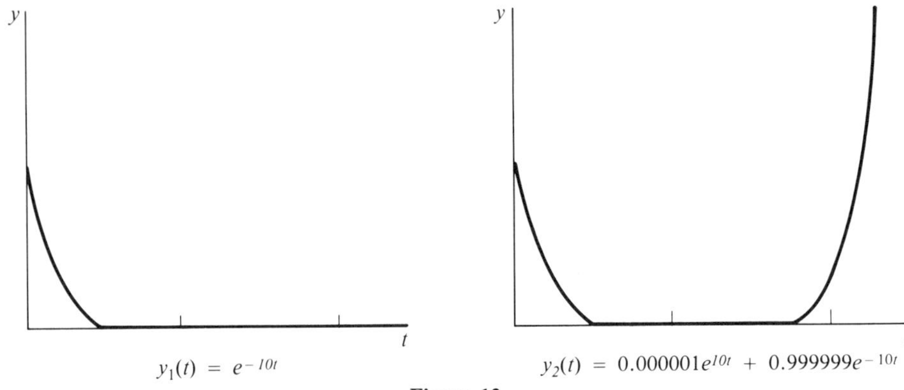

$$y_1(t) = e^{-10t} \qquad\qquad y_2(t) = 0.000001e^{10t} + 0.999999e^{-10t}$$

Figure 12

EXERCISES

1. **(a)** Devise a program to implement the Euler method for

$$y'' = F(t, y, y'), \qquad y(t_0) = y_0, \ y'(t_0) = z_0,$$

and apply it to the pendulum equation with $F(t, y, y') = -16 \sin y$, $y(0) = 0$, $y'(0) = 0.5$.
   (b) Do part (a) using the modified Euler method.

2. For small oscillations of y, the approximation $\sin y \approx y$ is fairly good, leading to the replacement of the pendulum equation $y'' + (g/L)\sin y = 0$ by the linearized equation $y'' + (g/L)y = 0$, with solutions of the form

$$y_l(t) = c_1 \cos\sqrt{\frac{g}{L}}t + c_2 \sin\sqrt{\frac{g}{L}}t.$$

Assuming $g/L = 16$, compare $y_l(t)$ with the modified Euler approximation to the solution of the nonlinear equation under initial conditions.
   (a) $y(0) = 0$, $y'(0) = 0.1$.
   (b) $y(0) = 0$, $y'(0) = 4$.

3. Improved accuracy in the Euler methods can, within reasonable limits, be obtained by

decreasing the step size h. This requires more steps to reach a given value of t and may produce more approximate solution values than is convenient. Show how to modify a program to print results only after M steps of calculation. For example, $h = 0.001$ and $M = 10$ would produce approximate values with argument differences of 0.01.

4. Consider the **Airy equation** $\ddot{y} + ty = 0$ for $-20 \leq r \leq 20$.
   (a) Estimate the location of the zero values of the solution satisfying $y(0) = 0$ and $\dot{y}(0) = 2$. [*Hint*: Look for successive approximations to y with opposite sign.]
   (b) Estimate the location of the maxima and minima of the solution satisfying $y(0) = 0$ and $\dot{y}(0) = 2$. [*Hint*: Look for successive approximations to $z = \dot{y}$ with opposite sign.]

5. The **Bessel equation** of order zero is

$$t^2 y'' + ty' + t^2 y = 0.$$

For $1 \leq t \leq 40$, estimate the location of the zero values of a solution satisfying $y(1) = 1$, $y'(1) = 0$.

6. Consider the damped pendulum equation

$$\ddot{\theta} = -\frac{g}{l} \sin \theta - \frac{k}{m} \dot{\theta}$$

with $g = 32.2$, $l = 20$, $k = 0.03$, and $m = 5$. For the solution with $y(0) = 0$, $y(0) = 0.2$:
   (a) Estimate the maximum angles θ for $0 \leq t \leq 30$.
   (b) Estimate the successive times between occurrences of the value $\theta = 0$ for $0 \leq t \leq 30$.

7. Make a numerical comparison of the solution $\ddot{y} = -\sin y$, $y(0) = 0$, $y(0) = 1$, with the solution of $y = -y$ using the same initial conditions.

Stiff Equations

8. The initial-value problem

$$y'' = 25y, \qquad y(0) = 1, \qquad y'(0) = -5$$

has solution $y(t) = e^{-5t}$. Try to approximate this solution using the modified Euler method with step size $h = 0.001$ over the interval $0 \leq t \leq 10$. Interpret your results.

Note: The following exercises deal with second-order equations of somewhat different forms from those that occur in the previous exercises. Computer programs may have to be altered accordingly.

9. The second-order initial-value problem of Exercise 8 can be converted to a pair of first-order equations by means of substitutions other than $y' = z$.
   (a) Let $y' = -5y + u$ [i.e., $(D + 5)y = u$]. Then $u' = 5u$ [i.e., $(D - 5)u = 0$] gives $y'' = 25y$. Show that the initial conditions $y(0) = 1$, $y'(0) = -5$ are converted to $y(0) = 1$, $u(0) = 0$. Then approximate y and u by the modified Euler method over the interval $0 \leq t \leq 10$ and compare the results with those of Exercise 8.
   (b) Let $y' = wy$, which is called a **Riccati substitution,** and show that $y'' = 25y$ becomes $w' = 25 - w^2$, provided that $y \neq 0$. Then show that $y(0) = 1$, $y'(0) = -5$ is equivalent to $y(0) = 1$, $w(0) = -5$, and approximate the solution to the system

$$y' = wy, \qquad y(0) = 1,$$

$$w' = 25 - w^2, \quad w(0) = -5.$$

10. Show that the **Riccati substitution** $y' = wy$ reduces the general linear equation

$$\frac{d^2y}{dt^2} + a(t)\frac{dy}{dt} + b(t)y = 0$$

to the system of equations

$$y' = wy,$$

$$w' = -a(t)w - b(t) - w^2$$

provided that $y \neq 0$. Show also that the initial conditions $y(t_0) = y_0$, $y'(t_0) = z_0$ get converted by the same substitution into $y(t_0) = y_0$, $w'(t_0) = z_0/y_0$.

Applications

11. Suppose that the displacement $y(t)$ of a falling body is subject to a nonlinear friction force

$$m\ddot{y} = mg - k(\dot{y})^\alpha.$$

(a) Find numerical approximations to $y(t)$ in the range $0 \leq t \leq 20$, with $g = 32.2$, $y(0) = 0$, $\dot{y}(0) = 0$, $M = 1$ and $\alpha = 1.5$. Use the values $k = 0, 0.1, 0.5, 1$, and sketch the graphs of $y = y(t)$ using an appropriate scale.

(b) Estimate the value of k that produces the values $y(0) = 0$, $y(5) = 66$, $y(10) = 137$ and $y(15) = 208$, approximately.

12. An unforced oscillator displacement $x = x(t)$ satisfies

$$m\ddot{x} + k\dot{x} + hx = 0,$$

where k and h may depend on time t.

(a) Suppose that $k(t) = 0.2(1 - e^{-0.1t})$, $h = 5$, $m = 1$ and that $x(0) = 0$, $\dot{x}(0) = 5$. Compute a numerical approximation to $x(t)$ on the range of $0 \leq t \leq 20$. Then sketch the graph of $x = x(t)$.

(b) In part (a) make $k = 0$ and $h(t) = 5(1 - e^{-0.2t})$.

13. Consider the nonlinear oscillator equation

$$m\ddot{x} + k\dot{x}|\dot{x}|^\beta + hx = 0, \qquad 0 \leq \beta = \text{const}.$$

Let $m = 1$, $k = 0.2$, $h = 5$ and suppose that $x(0) = 0$, $\dot{x}(0) = 5$. Compute numerical approximations to $x(t)$ on the range $0 \leq t \leq 20$ for $\beta = 0, 0.5, 1$. Then sketch the resulting graphs.

14. The nonlinear equation

$$\frac{d^2y}{dx^2} - m\frac{dy}{dx} + ky^2 = 0, \qquad m, k \text{ constant},$$

arises in the study of certain chemical reactions. Compare the behavior of numerical solutions of the initial value problem $y(0) = 0, y'(0) = 1$ with the corresponding behavior when the nonlinear term ky^2 is replaced by a linear term ky.

3

Laplace Transforms

1 BASIC PROPERTIES

The techniques described in Chapter 2 can be used to find formulas for the solution of differential equations such as

$$y^{(n)} + a_{n-1}y^{(n-1)} + \cdots + a_0 y = f(t),$$

with initial conditions at $t = 0$ of the form

$$y(0) = c_0, \; y'(0) = c_1, \; \ldots, \; y^{(n-1)}(0) = c_{n-1}.$$

If we are in a position to assume that $f(t)$ is defined for all t satisfying $t \geq 0$, and that $f(t)$ does not grow too rapidly as $t \to \infty$, we can use an alternative method that has achieved widespread popularity in electrical engineering and control theory. Experience with exponential multipliers shows that it is natural to multiply a solution $y(t)$ by a factor e^{-st}, where s is some real or complex number, to get a product $e^{-st}y(t)$. What seems less natural, but nevertheless turns out to be effective, is to integrate with respect to t between 0 and ∞. In particular, the equation

1.1
$$\int_0^\infty e^{-st}e^{-at}\, dt = \frac{1}{s - a}$$

holds when $(s - a) > 0$, and is used repeatedly. To prove Equation 1.1, we compute the partial integral

$$\int_0^T e^{-(s-a)t}\, dt = \left[-\frac{1}{s-a} e^{-(s-a)t} \right]_0^T$$

$$= -\frac{1}{s-a} e^{-(s-a)T} + \frac{1}{s-a}.$$

When $T \to \infty$, the exponential factor tends to zero, so letting $T \to \infty$ on both sides gives Equation 1.1. A convergent integral of the form

$$\mathcal{L}[f](s) = \int_0^\infty e^{-st} f(t)\, dt$$

produces a function $\mathcal{L}[f]$, defined for certain real or complex numbers s, called the **Laplace transform** of the function f. The key to the Laplace transform method is the following formula:

1.2
$$\int_0^\infty e^{-st} y'(t)\, dt = -y(0) + s \int_0^\infty e^{-st} y(t)\, dt,$$

which we prove under the assumptions (i) the improper integral on the left is convergent, and (ii) $\lim_{t \to \infty} e^{-st} y(t) = 0$. Indeed, integration by parts of the partial integral on the left gives

$$\int_0^T e^{-st} y'(t)\, dt = \left[e^{-st} y(t) \right]_0^T + s \int_0^T e^{-st} y(t)\, dt$$

$$= e^{-sT} y(T) - y(0) + s \int_0^T e^{-st} y(t)\, dt.$$

Because of assumption (ii), the first term on the right tends to 0 as $T \to \infty$. Equation 1.2 follows by letting $T \to \infty$ in the preceding equation.

EXAMPLE 1   The example

$$y' + 2y = 0, \qquad y(0) = 3,$$

is too simple to show the real advantages of using Laplace transforms, but it does illustrate the general principles involved. We form the Laplace transform of both sides of the differential equation by multiplying both sides by e^{-st} and then integrating from 0 to ∞ with respect to t. The result can be written

$$\int_0^\infty e^{-st} y'\, dt + 2 \int_0^\infty e^{-st} y\, dt = 0.$$

Equation 1.2 allows us to rewrite the first integral, obtaining

$$-3 + s \int_0^\infty e^{-st} y\, dt + 2 \int_0^\infty e^{-st} y\, dt = 0.$$

Here we have used the assumption that $y(0) = 3$. We also rely on our knowledge of the exponential nature of the solution to justify the assumptions (i) and (ii) needed for the application of Equation 1.2. The previous equation can now be solved for the Laplace transform of the solution $y(t)$ in the form

$$\int_0^\infty e^{-st} y \, dt = \frac{3}{s + 2}. \tag{1}$$

Thus we have found not the solution $y(t)$, but its Laplace transform. However, if we set $a = -2$ in Equation 1.1 and then multiply by 3, we get

$$\int_0^\infty e^{-st} 3e^{-2t} \, dt = \frac{3}{s + 2}. \tag{2}$$

Since we already know from the general theory of Chapter 2 that $y(t)$ must be an exponential solution, there remains only the question of the constants involved, and we see by comparing equations (1) and (2) that the solution

$$y(t) = 3e^{-2t}$$

satisfies our requirements. Even though we really do not need to use the fact here, it is worth commenting that a function $f(t)$ defined for $t > 0$ is determined at its points of continuity by its Laplace transform $\mathcal{L}[f]$.

To apply the Laplace transform to differential equations with order higher than one, we need a simple extension of Equation 1.2. To simplify the notation, we write 1.2 in the form

$$\mathcal{L}[y'](s) = -y(0) + s\mathcal{L}[y](s).$$

Applying this equation to $\mathcal{L}[y''](s)$, the Laplace transform of y'', gives

$$\mathcal{L}[y''](s) = -y'(0) + s\mathcal{L}[y'].$$

Applying the equation gives

$$\mathcal{L}[y''](s) = -y'(0) + s\{-y(0) + s\mathcal{L}[y](s)\}$$

$$= -y'(0) - sy(0) + s^2\mathcal{L}[y](s).$$

The same formal routine leads after n steps to the formula

1.3 $\mathcal{L}[y^{(n)}](s) = -y^{(n-1)}(0) - sy^{(n-2)}(0) - \ldots - s^{n-1}y(0) + s^n\mathcal{L}[y](s).$

Of course, the assumptions (i) and (ii) needed for Equation 1.2 have to be increased to

(a). The integrals $\int_0^\infty e^{-st} y^{(k)}(t) \, dt$ are convergent for $k = 1, 2, \ldots, n$.

(b). $\lim\limits_{t \to \infty} e^{-st} y^{(k)}(t) = 0$ for $k = 0, 1, \ldots, n - 1$.

EXAMPLE 2   The differential equation

$$y'' - y' - 2y = 3e^t,$$

with initial conditions

$$y(0) = 1, \qquad y'(0) = 0,$$

can be solved by applying the Laplace transform to both sides. For simplicity, we denote the Laplace transform of y by Y; that is, $Y(s) = \mathscr{L}[y](s)$. Using Equation 1.3 for $n = 1$ and 2, together with the initial conditions, we find

$$\mathscr{L}[y] = Y(s),$$

$$\mathscr{L}[y'] = -y(0) + sY(s)$$

$$= -1 + sY(s),$$

$$\mathscr{L}[y''] = -y'(0) - sy(0) + s^2Y(s)$$

$$= -s + s^2Y(s).$$

Because integration from 0 to ∞ and multiplication by e^{-st} are both linear operations, the equation

$$\mathscr{L}[y'' - y' - 2y] = \mathscr{L}[3e^t]$$

simplifies to

$$\mathscr{L}[y''] - \mathscr{L}[y'] - 2\mathscr{L}[y] = 3\mathscr{L}[e^t].$$

The expressions found for $\mathscr{L}[y]$, $\mathscr{L}[y']$, and $\mathscr{L}[y'']$, together with Equation 1.1, allow us to write the equation as

$$[-s + s^2Y(s)] - [-1 + sY(s)] - 2[Y(s)] = 3\,\frac{1}{s-1}.$$

Rearrangement gives

$$(s^2 - s - 2)Y(s) = \frac{3}{s-1} + s - 1 = \frac{s^2 - 2s + 4}{s-1}$$

or

$$Y(s) = \frac{s^2 - 2s + 4}{(s-1)(s^2 - s - 2)}.$$

Having found an expression for $Y(s)$, our problem is now to identify precisely the solution $y(t)$ that satisfies $\mathscr{L}[y](s) = Y(s)$. Because $Y(s)$ is a rational function, it can theoretically always be broken down according to the partial fraction decomposition usually associated with the computation of indefinite integrals. In our example, the decomposition can, in fact, be carried out because the denominator of Y can be factored. We need to determine the coefficients A, B, and C in

$$\frac{s^2 - 2s + 4}{(s - 1)(s + 1)(s - 2)} = \frac{A}{s - 1} + \frac{B}{s + 1} + \frac{C}{s - 2}.$$

Multiplying through by $(s - 1)$, and then setting $s = 1$, gives $A = -\frac{3}{2}$. Similarly, we multiply by $(s + 1)$ and then set $s = -1$ to get $B = \frac{7}{6}$. Finally, we multiply by $(s - 2)$ and then set $s = 2$ to get $C = \frac{4}{3}$. As a result, we have

$$Y(s) = -\frac{\frac{3}{2}}{s - 1} + \frac{\frac{7}{6}}{s + 1} + \frac{\frac{4}{3}}{s - 2}.$$

Equation 1.1 now allows us to identify $y(t)$ as

$$y(t) = -\tfrac{3}{2}e^t + \tfrac{7}{6}e^{-t} + \tfrac{4}{3}e^{2t}.$$

As a check on the computation, we can verify that $y(0) = 1$ and $y'(0) = 0$.

In the examples given previously, we used the linearity of $\mathscr{L}$, the Laplace transform operator. The property is formally expressed by the two equations

1.4

$$\mathscr{L}[y_1 + y_2] = \mathscr{L}[y_1] + \mathscr{L}[y_2],$$

$$\mathscr{L}[cy] = c\mathscr{L}[y], \qquad c = \text{constant}.$$

These equations, together with Equation 1.3 for the Laplace transform of a derivative, need to be supplemented in practice by the calculation of Laplace transforms of specific functions, as is done in Equation 1.1, which asserts that $\mathscr{L}[e^{at}](s) = 1/(s - a)$. Table I contains more than enough entries to do all the problems in this section, although more elaborate tables may contain several hundred entries. Such tables are meant to be used in both directions, so that while the entry

$$\mathscr{L}[t^n] = \frac{n!}{s^{n+1}}, \qquad n = 0, 1, 2, \ldots$$

provides the transform of $f(t) = t^n$, it also provides, after division by $n!$, the inverse transform

$$\mathscr{L}^{-1}\left[\frac{1}{s^{n+1}}\right] = \frac{t^n}{n!}, \qquad n = 0, 1, 2 \ldots$$

For the proof that for every Laplace transform Y there is a *unique* function y such that $\mathscr{L}[y] = Y$, we can refer to more theoretical accounts of the subject. All the entries in Table I can be computed using elementary integration techniques.

In finding inverse transforms, it is sometimes essential to decompose a rational function $P(s)/Q(s)$, with the degree of P less than the degree of Q, according to the following rules:

TABLE I. TABLE OF LAPLACE TRANSFORMS

$f(t)$	$\mathcal{L}[f](s) = \int_0^\infty e^{-st} f(t)\, dt$
1. 1	$\dfrac{1}{s}$
2. t	$\dfrac{1}{s^2}$
3. t^n	$\dfrac{n!}{s^{n+1}}, \qquad n = 0, 1, 2, \ldots$
4. e^{at}	$\dfrac{1}{s-a}$
5. te^{at}	$\dfrac{1}{(s-a)^2}$
6. $t^n e^{at}$	$\dfrac{n!}{(s-a)^{n+1}}, \qquad n = 0, 1, 2, \ldots$
7. $\sin bt$	$\dfrac{b}{s^2 + b^2}$
8. $\cos bt$	$\dfrac{s}{s^2 + b^2}$
9. $t \sin bt$	$\dfrac{2bs}{(s^2 + b^2)^2}$
10. $t \cos bt$	$\dfrac{s^2 - b^2}{(s^2 + b^2)^2}$
11. $e^{at} \sin bt$	$\dfrac{b}{(s-a)^2 + b^2}$
12. $e^{at} \cos bt$	$\dfrac{s-a}{(s-a)^2 + b^2}$
13. $e^{at} - e^{bt}$	$\dfrac{a-b}{(s-a)(s-b)}$
14. $ae^{at} - be^{bt}$	$\dfrac{(a-b)s}{(s-a)(s-b)}$
15. $(b-c)e^{at} + (c-a)e^{bt} + (a-b)e^{ct}$	$\dfrac{(a-b)(b-c)(a-c)}{(s-a)(s-b)(s-c)}$
16. $\sin bt - bt \cos bt$	$\dfrac{2b^3}{(s^2 + b^2)^2}$

1. If the denominator $Q(s)$ has the factor $(s - a)^m$ as the highest power of $s - a$ that divides $Q(s)$, then include in the decomposition of $P(s)/Q(s)$ the fractions of the form

$$\frac{A_j}{(s-a)^j}, \qquad j = 1, 2, \ldots, m.$$

2. If the denominator $Q(s)$ has the factor $(s^2 + ps + q)^n$ as the highest power of $s^2 + ps + q$ that divides $Q(s)$, then include in the decomposition of $P(s)/Q(s)$ the fractions of the form

$$\frac{B_k s + C_k}{(s^2 + ps + q)^k}, \qquad k = 1, 2, \ldots, n.$$

EXAMPLE 3   To find the function $f(t)$ having Laplace transform

$$F(s) = \frac{s + 1}{(s - 1)^2(s^2 + 1)},$$

we decompose the function into a sum of fractions as follows:

$$\frac{s + 1}{(s - 1)^2(s^2 + 1)} = \frac{A}{s - 1} + \frac{B}{(s - 1)^2} + \frac{Cs + D}{s^2 + 1}.$$

To compute B, we can multiply through by $(s - 1)^2$ and then set $s = 1$. We get $B = 1$. The same kind of trick does not apply directly to the other coefficients, but if we subtract $1/(s - 1)^2$ from both sides, we find we can cancel $(s - 1)$ on the left to get

$$\frac{-s^2 + s}{(s - 1)^2(s^2 + 1)} = \frac{-s}{(s - 1)(s^2 + 1)} = \frac{A}{s - 1} + \frac{Cs + D}{s^2 + 1}.$$

Now multiply by $(s - 1)$ and then set $s = 1$ to get $A = -\frac{1}{2}$. As a result,

$$\frac{-s}{(s - 1)(s^2 + 1)} = \frac{-\frac{1}{2}}{s - 1} + \frac{Cs + D}{s^2 + 1}.$$

To find C and D, we can multiply through by $(s - 1)(s^2 + 1)$ to get

$$-s = -\tfrac{1}{2}(s^2 + 1) + (s - 1)(Cs + D).$$

Rearranging the powers of s gives

$$\tfrac{1}{2}s^2 - s + \tfrac{1}{2} = Cs^2 + (D - C)s - D.$$

We equate coefficients of like powers on both sides and find that $C = \frac{1}{2}$, whereas $D = -\frac{1}{2}$. The result is that

$$\frac{s + 1}{(s - 1)^2(s^2 + 1)} = \frac{-\frac{1}{2}}{s - 1} + \frac{1}{(s - 1)^2} + \frac{\frac{1}{2}s}{s^2 + 1} - \frac{\frac{1}{2}}{s^2 + 1}.$$

From the table of transforms, we conclude that

$$\mathcal{L}^{-1}[F(s)] = -\tfrac{1}{2}e^t + te^t + \tfrac{1}{2}\cos t - \tfrac{1}{2}\sin t.$$

EXERCISES

1. (a) Verify that

$$\mathcal{L}[e^{at}](s) = \int_0^\infty e^{-st} e^{at} \, dt = \frac{1}{s-a}, \qquad \text{for } s > a.$$

   (b) Use integration by parts to verify that

$$\mathcal{L}[t](s) = \int_0^\infty t e^{-st} \, dt = \frac{1}{s^2}, \qquad \text{for } s > 0.$$

2. Use Equation 1.3 to show that, if $y(0) = 2$ and $y'(0) = 3$, then

$$\int_0^\infty e^{-st} y''(t) \, dt = s^2 Y(s) - 2s - 3,$$

   where $Y(s)$ is the Laplace transform of $y(t)$.

3. By computing the appropriate integral, or by using the table of Laplace transforms, compute $\mathcal{L}[f](s)$, where $f(t)$ is as follows:

   (a) $t \sin 2t$.   (b) $\cos t + 2 \sin t$.   (c) $t^2 + 2t - 1$.

   (d) $\cos(t + a)$.   (e) $(2t + 1) e^{3t}$.   (f)) $e^t + e^{-t}$.

4. Use Table I to find the *inverse* Laplace transforms of the following functions $F(s)$; that is, find $y(t)$ such that $\mathcal{L}[y](s) = F(s)$.

   (a) $\dfrac{1}{s^2 - 1}$.   (b) $\dfrac{s}{s^2 - 4}$.

   (c) $\dfrac{2}{s^2 + 4}$.   (d) $\dfrac{4s}{(s^2 + 4)^2}$.

   (e) $\dfrac{1}{(s - 2)^2 + 9}$.   (f) $\dfrac{1}{s^2} - \dfrac{1}{(s - 1)^2}$.

5. Use the Laplace transform to solve the initial-value problems. Check by substitution.

   (a) $y' - y = t$, $y(0) = 2$.
   (b) $y' + 2y = 1$, $y(0) = 1$.
   (c) $y' + 3y = \cos 2t$, $y(0) = 0$.
   (d) $y'' + y = e^{-t} + 1$, $y(0) = -1$, $y'(0) = 1$.
   (e) $2y'' - y' = 2 \cos 3t$, $y(0) = 0$, $y'(0) = 2$.

6. (a) Define

$$H_a(t) = \begin{cases} 0, & \text{if } 0 \le t \le a \\ 1, & \text{if } a < t. \end{cases}$$

   Show that $\mathcal{L}[H_a](s) = (1/s)e^{-as}$.

   (b) Show that if $g(t) = H_a(t)f(t)$ for $0 \le t$, then

$$\mathcal{L}[g](s) = e^{-as} \mathcal{L}[f](s).$$

   (c) Sketch the graph of $H_a(t)$ for $a = -1$.

   (d) Solve the differential equation

$$y'' = H_a(t), \qquad 0 < a$$

   with initial condition $y(0) = 1$, $y'(0) = 0$.

7. Let $P(D)$ be an nth-order, linear, constant-coefficient, differential operator. Show that

$$\mathscr{L}[P(D)y](s) = P(s)\mathscr{L}[y](s) + Q(s),$$

for some polynomial Q of degree $n - 1$. Use induction.

8. (a) Show that if $f(t)$ is defined and differentiable only for $t > 0$ (instead of $t \geq 0$), then

$$\mathscr{L}[f'](s) = -f(0+) + s\mathscr{L}[f](s),$$

where $f(0+) = \lim_{t \to 0+} f(t)$.

   (b) Show that, if the limits $f^{(k)}(0+)$ all exist, then Formula 1.3 can be generalized to

$$\mathscr{L}[f^{(n)}](s) = -f^{(n-1)}(0+) - \cdots - s^{(n-1)}f(0+) + s^n\mathscr{L}[f](s).$$

9. Show that if $p > 0$ and $f(t + p) = f(t)$ for $t \geq 0$, then

$$\mathscr{L}[f](s) = (1 - e^{-ps})^{-1} \int_0^p e^{-st}f(t)\, dt.$$

2 CONVOLUTION

Let us review the solution of the second-order differential equation

$$y'' + py' + qy = f(t), \qquad y(0) = y_0, \quad y'(0) = y_1.$$

Taking the Laplace transform of both sides gives

$$(s^2 + ps + q)\, Y(s) = F(s) + y'(0) + (s + p)y(0).$$

The polynomial factor $P(s) = s^2 + ps + q$ on the left is the **characteristic polynomial** of the operator $D^2 + pD + q$. The reciprocal $Q(s) = 1/P(s)$ is called the **transfer function** of the operator, and if we multiply by $Q(s)$, or divide by $P(s)$, we get the formula

$$Y(s) = \frac{F(s) + y'(0) + (s + p)y(0)}{P(s)}$$

for the Laplace transform $Y = \mathscr{L}[y]$. The remaining step is to find the inverse transform $y(t) = \mathscr{L}^{-1}[y](t)$. The essence of the method is to use the Laplace transform to reduce the solution of the problem to some routine algebraic manipulations.

In addition to the table of specific Laplace transforms in the previous section (Table I), a number of general formulas, such as Formula 1.3, are useful in solving problems. The most important of these answers the following question: If $F(s)$ and $G(s)$ are the Laplace transforms of $f(t)$ and $g(t)$, respectively, what function has Laplace transform equal to the product $F(s)G(s)$? It turns out that under rather general hypotheses there is an answer, given by the **convolution integral**

$$f * g(t) = \int_0^t f(u)g(t - u)\, du.$$

The function $f * g(t)$ is called the **convolution** of the functions $f(t)$ and $g(t)$ and is defined for $t \geq 0$, provided that f and g are integrable on every finite interval. The convolution $f * g$ is to be thought of as a kind of product of f and g and, in fact, it turns out that $f * g = g * f$, although this is not obvious from the definition. The basic information about convolutions is summarized as follows.

2.1 Theorem   Let $f(u)$ and $g(u)$ be integrable on $0 \leq u \leq t$ for every positive t; then $f * g$ and $g * f$ both exist and are equal; that is, convolution is commutative:

$$f * g = g * f.$$

If $\mathcal{L}[|f|](s)$ and $\mathcal{L}[|g|](s)$ are both finite, then

$$\mathcal{L}[f * g](s) = (\mathcal{L}[f](s))(\mathcal{L}[g](s)).$$

Proof. The first statement is easily proved by changing the variable in the definition of $f * g$. We have, on replacing u by $t - v$,

$$f * g(t) = \int_0^t f(u)g(t - u)\, du = -\int_t^0 f(t - v)g(v)\, dv$$

$$= \int_0^t g(v)f(t - v)\, dv = g * f(t).$$

Proving the second statement involves an important technical point for which we will not give a proof, but the rest of the argument is complete. To simplify the writing of limits of integration, we can extend both $f(t)$ and $g(t)$ to have the value 0 for $t < 0$. Then we can write

$$\mathcal{L}[f](s)\mathcal{L}[g](s) = \int_{-\infty}^{\infty} e^{-su}f(u)\, du \int_{-\infty}^{\infty} e^{-sv}g(v)\, dv$$

$$= \int_{-\infty}^{\infty} f(u)\left[\int_{-\infty}^{\infty} e^{-s(u + v)}g(v)\, dv\right] dv.$$

We can make the change of variable $v = t - u$ in the inner integral to get

$$\mathcal{L}[f](s)\mathcal{L}[g](s) = \int_{-\infty}^{\infty} f(u)\left[\int_{-\infty}^{\infty} e^{-st}g(t - u)\, dt\right] du.$$

Under the hypotheses of the theorem, we can interchange the order of integration. (We will assume the required theorem, called Fubini's theorem.) Then we have

$$\mathcal{L}[f](s)\mathcal{L}[g](s) = \int_{-\infty}^{\infty} e^{-st}\left[\int_{-\infty}^{\infty} f(u)g(t - u)\, du\right] dt.$$

Because we have assumed $f(t)$ and $g(t)$ are zero for $t < 0$, the inner integral is zero for $t < 0$. It follows that the t integration is needed only for $0 \leq t < \infty$. Similarly, the u integration is needed only for $0 \leq u \leq t$. Hence

$$\mathcal{L}[f](s)[g](s) = \int_0^\infty e^{-st} \left[\int_0^t f(u)g(t - u)\, du \right] dt$$

$$= \mathcal{L}[f * g](s),$$

which is what we wanted to prove.

EXAMPLE 1.   From the table in Section 1, we see that $\mathcal{L}[t](s) = 1/s^2$ and $\mathcal{L}[\sin t](s) = 1/(s^2 + 1)$. It follows from Theorem 2.1 that

$$\frac{1}{s^2} \cdot \frac{1}{s^2 + 1} = \mathcal{L}\left[\int_0^t (t - u)\sin u\, du \right](s).$$

But holding t fixed, we can use integration by parts to show that

$$\int_0^t (t - u)\sin u\, du = \left[-(t - u)\cos u \right]_0^t - \int_0^t \cos u\, du$$

$$= t - \sin t.$$

We could have obtained the same result by computing a partial fraction decomposition of the form

$$\frac{1}{s^2} \cdot \frac{1}{s^2 + 1} = \frac{A}{s} + \frac{B}{s^2} + \frac{Cs + D}{s^2 + 1},$$

and then finding the inverse transform of each term.

Table II lists the most frequently used general properties of the Laplace transform. The entries that have not already been discussed can easily be proved using elementary calculus techniques. The table omits the precise conditions under which each formula holds. The distinction between Formulas 2 and 3 and Equations 1.2 and

TABLE II.   GENERAL PROPERTIES OF THE LAPLACE TRANSFORM

1. $\mathcal{L}[af + bg] = a\mathcal{L}[f] + b\mathcal{L}[g], \quad a, b \text{ constant}$

2. $\mathcal{L}[f'](s) = s\mathcal{L}[f](s) - f(0+), \quad f(0+) = \lim_{t \to 0+} f(t)$

3. $\mathcal{L}[f^{(n)}](s) = s^n\mathcal{L}[f](s) - s^{n-1}f(0+) - s^{n-2}f'(0+) - \cdots - f^{(n-1)}(0+)$

4. $\mathcal{L}\left[\int_0^t f(u)\, du \right](s) = \frac{1}{s}\mathcal{L}[f](s)$

5. $\mathcal{L}[e^{at}f(t)](s) = \mathcal{L}[f](s - a)$

6. $\mathcal{L}[H_a(t)f(t - a)](s) = e^{-as}\mathcal{L}[f](s), \quad H_a(t) = \begin{cases} 0, & t < a, \\ 1, & a \leq t \end{cases}.$

7. $\mathcal{L}[tf(t)](s) = -\dfrac{d}{ds}\mathcal{L}[f](s)$

8. $\mathcal{L}[f * g](s) = \mathcal{L}[f](s)\mathcal{L}[g](s)$

1.3 of the previous section occurs because in the latter we assumed that we were dealing with solutions of differential equations and that these solutions had continuous derivatives at $t = 0$. The corresponding formulas in Table II are valid under the weaker assumption that $\lim_{t \to 0+} f^{(k)}(t)$ exists, but is not necessarily equal to $f^{(k)}(0)$. (See Exercise 8 of Section 1.)

EXAMPLE 2.   From Table I, we find that

$$\mathcal{L}[\sin t](s) = \frac{1}{s^2 + 1}.$$

Taking $f(t) = \sin t$ in Formula 4 of Table II gives

$$\frac{1}{s(s^2 + 1)} = \frac{1}{s} \mathcal{L}[\sin t](s)$$

$$= \mathcal{L}\left[\int_0^t \sin u \, du\right](s) = \mathcal{L}[-\cos t + 1].$$

Hence

$$\mathcal{L}^{-1}\left[\frac{1}{s(s^2 + 1)}\right] = -\cos t + 1.$$

Repeating the application of Formula 4 gives

$$\frac{1}{s^2(s^2 + 1)} = \mathcal{L}\left[\int_0^t (-\cos u + 1) \, du\right]$$

$$= \mathcal{L}[-\sin t + t].$$

This establishes the formula

$$\mathcal{L}^{-1}\left[\frac{1}{s^2(s^2 + 1)}\right] = -\sin t + t,$$

which was derived in Example 1 using convolution.

EXAMPLE 3   Starting with the formula

$$\mathcal{L}[\cos t](s) = \frac{s}{s^2 + 1},$$

we can apply Formula 7 in Table II to get

$$\mathcal{L}[t \cos t](s) = -\frac{d}{ds} \frac{s}{s^2 + 1}$$

$$= \frac{s^2 - 1}{(s^2 + 1)^2}.$$

Another application of the same formula gives

$$\mathcal{L}[t^2 \cos t](s) = -\frac{d}{ds}\frac{s^2 - 1}{(s^2 + 1)^2}$$

$$= \frac{2s^2 - 6s}{(s^2 + 1)^3}.$$

EXAMPLE 4   To apply Formula 6 of Table II to the function $f(t) = t$, we define $f(t) = 0$ for $t < 0$. The graphs of $f(t)$ and $f(t - a)$ are shown in Figure 1 for $a = 1$ and $a = 2$. Each function is zero where it is not positive. From Formula 6 we find

$$\mathcal{L}[f(t - 1)](s) = e^{-s}\mathcal{L}[f(t)](s)$$

$$= e^{-s}\mathcal{L}[t](s) = e^{-s}\frac{1}{s^2}.$$

Similarly

$$\mathcal{L}[f(t - 2)](s) = e^{-2s}\mathcal{L}[f(t)](s)$$

$$= e^{-2s}\mathcal{L}[t](s) = e^{-2s}\frac{1}{s^2}.$$

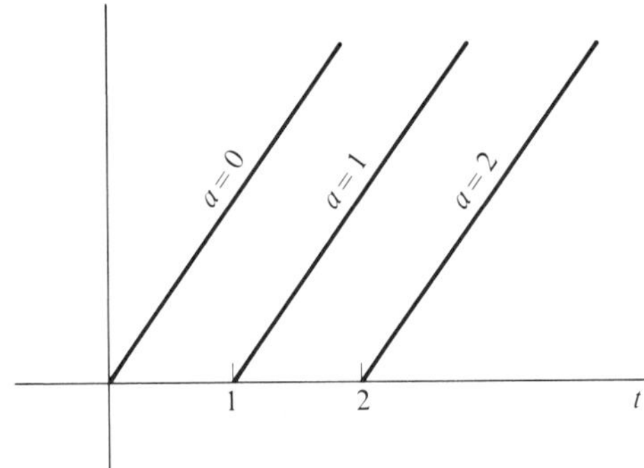

Figure 1

EXERCISES

1. Find the convolution $f * g$ of the following pairs of functions:
   (a) $f(t) = t$, $g(t) = e^{-t}$, $t \geq 0$.
   (b) $f(t) = t^2$, $g(t) = (t^2 + 1)$, $t \geq 0$.
   (c) $f(t) = 1$, $g(t) = t$, $t \geq 0$.

2. Use the convolution of two functions to find the inverse Laplace transform of each of the following products of Laplace transforms:
   (a) $\dfrac{1}{s^2(s + 1)}.$
   (b) $\dfrac{1}{(s - 1)(s - 2)}.$
   (c) $\dfrac{1}{s^3}e^{-2s}.$

3. By using the formulas in Tables I and II, find the inverse Laplace transform of each of the following functions.

(a) $\dfrac{1}{s(s+3)^2}$.

(b) $\dfrac{e^{-2s}}{s(s^2+4)}$.

(c) $\dfrac{1}{s^2+2s+2}$.

(d) $\dfrac{1}{s^2+1}$.

(e) $\dfrac{(e^{-s}+1)}{s}$.

(f) $\dfrac{s}{s^2+10}$.

4. Solve the following initial-value problems. Check by substitution.

   (a) $y'' - y = \sin 2t + 1$, $y(0) = 1$, $y'(0) = -1$.

   (b) $y'' + 2y = t$, $y(0) = 0$, $y'(0) = 1$.

   (c) $y'' + y' = t + e^{-t}$, $y(0) = 2$, $y'(0) = 1$.

5. Solve the equation

$$y' + y = \int_0^t y(u)\,du + t,$$

given that $y(0) = 1$.

6. Use integration by parts to show that

$$\int_0^t (t-u)^n f(u)\,du = n! \int_0^t\left(\int_0^{t_1}\left(\cdots\int_0^{t_n} f(t_{n+1})\,dt_{n+1}\right)\ldots dt_2\right)dt_1.$$

7. The gamma function, denoted by Γ, can be defined by

$$\Gamma(z) = \int_0^\infty t^{z-1}e^{-t}\,dt, \qquad z > 0.$$

   (a) Use integration by parts to show that

$$\Gamma(z+1) = z\Gamma(z).$$

   (b) Deduce from part (a) that $\Gamma(n+1) = n!$, for $n = 0, 1, 2, \ldots$.

   (c) Show that if $a > -1$ then

$$\mathcal{L}[t^a](s) = \frac{\Gamma(a+1)}{s^{a+1}}.$$

3 GENERALIZED FUNCTIONS

3A The δ Function

The solution of the initial-value problem

$$y'' + ay' + by = f(t), \qquad y(0) = y_0, \quad y'(0) = z_0,$$

is often wanted when $f(t)$ is an **impulse function,** that is, a function that is identically zero outside some small interval and that assumes relatively large values on the

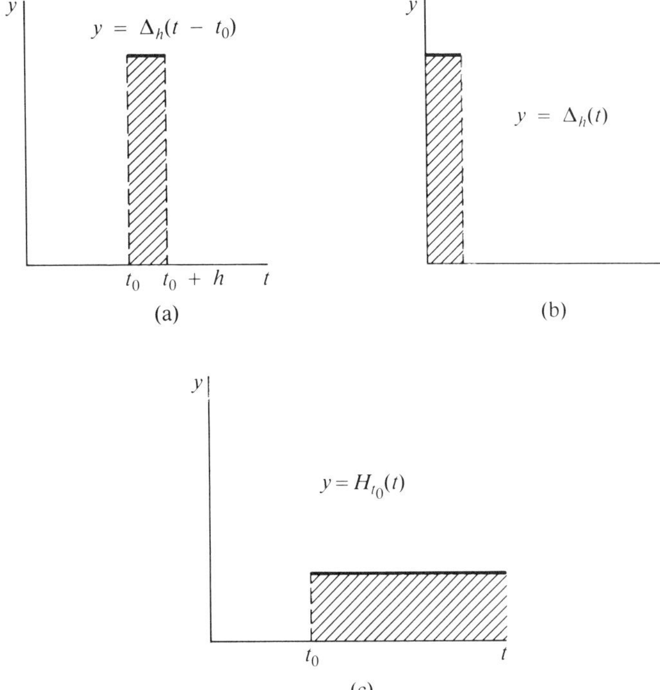

Figure 2

interval itself. Figure 2(a) shows the graph of just such a function, one that assumes the value $1/h$ on the interval $t_0 < t < t_0 + h$. The integral of this function is

$$\int_{t_0}^{t_0 + h} \frac{1}{h}\,dt = 1 \,.$$

As h tends to zero, the value of the function on the decreasing interval $t_0 \le t \le t_0 + h$ tends to infinity while the value of the integral remains 1. When $t_0 = 0$, we denote the function by $\Delta_h(t)$ so that the more general function is $\Delta_h(t - t_0)$.

We now compute

$$\lim_{h \to 0} \int_0^\infty \Delta_h(t - t_0)g(t)dt$$

under the assumption that g is continuous at t_0. If we define

$$G(t) = \int_0^t g(u)\,du,$$

then

$$\int_0^\infty \Delta_h(t - t_0)g(t)dt = \int_{t_0}^{t_0+h} \frac{1}{h} g(t)dt$$

$$= \frac{1}{h}\left[\int_0^{t_0+h} g(t)dt - \int_0^{t_0} g(t)dt\right]$$

$$= \frac{1}{h}(G(t_0 + h) - G(t_0)).$$

As $h \to 0$, this last expression tends to $G'(t_0)$. By one of the two fundamental theorems of calculus, $G'(t_0) = g(t_0)$. Hence

$$\lim_{h \to 0} \int_0^\infty \Delta_h(t - t_0)g(t)dt = g(t_0).$$

If we were to take the limit under the integral sign, we would get

$$\lim_{h \to 0} \Delta_h(t - t_0) = \begin{cases} 0, & t \neq t_0, \\ +\infty, & t = t_0. \end{cases}$$

The resulting function is not integrable in an ordinary sense that is helpful here, but the way we arrived at it does suggest that we *define* the **(Dirac) $\delta-$ function δ** to be such that

$$\int_0^\infty \delta(t - t_0)g(t)\, dt = g(t_0)$$

whenever $g(t)$ is continuous at $t = t_0$. It is important to realize that not only the definition of *function*, but also the definition of *integral* has been extended beyond its usual meaning here. We will sometimes use the notation $\delta_a(t) = \delta(t - a)$.

EXAMPLE 1 $\displaystyle\int_0^\infty \delta(t - 2)e^{-t}\, dt = e^{-2}.$ More generally, we can compute the Laplace transform

$$\mathcal{L}[\delta_{t_0}](s) = \int_0^\infty \delta(t - t_0)e^{-st}\, dt = e^{-t_0 s}.$$

In particular, with $t_0 = 0$, we get $\mathcal{L}[\delta](s) = 1$. The convolution of δ with f turns out to be f:

$$\int_0^t f(t - u)\delta(u)\, du = f(t), \qquad t > 0.$$

Thus the rule $\mathcal{L}[f * \delta](s) = \mathcal{L}[f](s)\mathcal{L}[\delta](s)$, shown already when δ is an ordinary function, still holds.

The formal definition of $\delta(t - t_0)$, given previously, can be interpreted as providing a large instantaneous impulse at $t = t_0$. To give some meaning to the word "large" in this context, consider the following. We define the **Heaviside function** H_{t_0} by

$$H_{t_0}(t) = \begin{cases} 0, & t < t_0, \\ 1, & t \geq t_0. \end{cases}$$

Its graph for a particular value of t_0 is shown in Figure 2(c). Note that H_{t_0} is a perfectly well defined function in the ordinary sense of the word function. It does have the property, however, that its derivative H'_{t_0} fails to exist at t_0, even though $H'_{t_0}(t)$ exists for all values of t except t_0. Given the abruptness of the jump in H_{t_0} at t_0, we might guess that $H'_{t_0}(t) = \delta(t - t_0)$. To justify this equation, we need to show that it is consistent with the usual computations in calculus. So let $g(t)$ be a differentiable function tending rapidly enough to zero as $t \to \infty$ that the improper integral

$$\int_0^\infty g(t)\, dt$$

exists. For example, $g(t) = e^{-t}$ is such a function. We now do the following formal integration by parts, assuming $t_0 > 0$:

$$\int_0^\infty H'_{t_0}(t)g(t)\, dt = H_{t_0}(t)g(t)\Big|_0^\infty - \int_0^\infty H_{t_0}(t)g'(t)\, dt$$

$$= H_{t_0}(\infty)g(\infty) - H_{t_0}(0)g(0) - \int_{t_0}^\infty g'(t)\, dt.$$

Here we have used the property that $H_{t_0}(0) = 0$ for $t < t_0$. But $H_{t_0}(\infty) = \lim_{t \to \infty} H_{t_0}(t) = 1$, and $g(\infty) = \lim_{t \to \infty} g(t) = 0$. Also $H_{t_0}(0) = 0$ since $t_0 > 0$. Hence

$$\int_0^\infty H'_{t_0}(t)g(t)\, dt = -\int_{t_0}^\infty g'(t)\, dt = -g(\infty) + g(t_0) = g(t_0).$$

Comparing this result with our previous definition,

$$\int_0^\infty \delta(t - t_0)g(t)\, dt = g(t_0),$$

we see that $H'_{t_0}(t)$ behaves just like $\delta(t - t_0)$. So we define the **derivative of the Heaviside function** by $H'_{t_0} = \delta_{t_0}$, so that

$$H'_{t_0}(t) = \delta_{t_0}(t) = \delta(t - t_0).$$

EXAMPLE 2

$$\int_0^\infty H'_{t_0}(t)e^{-st}\, dt = \int_0^\infty \delta(t - t_0)e^{-st}\, dt = e^{-t_0 s}.$$

Thus

$$\mathcal{L}[H'_{t_0}](s) = e^{t_0 s}.$$

On the other hand,

$$\mathcal{L}[H_{t_0}](s) = \int_0^\infty H_{t_0}(t) e^{-st}\, dt = \int_{t_0}^\infty e^{-st}\, dt = -\frac{1}{s} e^{-st} \Big|_{t_0}^\infty$$

$$= \frac{1}{s} e^{-t_0 s}.$$

It follows that

$$\mathcal{L}[H'_{t_0}](s) = s\mathcal{L}[H_{t_0}](s) - H_{t_0}(0+) = s\mathcal{L}[H_{t_0}](s).$$

This rule has already been shown to hold with H_{t_0} replaced by an honestly differentiable function f for which $f(0+) = 0$. It is this kind of consistency with the rules of ordinary calculus that facilitates operating with $\delta(t - t_0)$ or its equivalent, $H'_{t_0}(t)$.

3B Solution of Equations

Here is a direct solution of $y'' = \delta(t - t_0)$. We recall that $H'_{t_0}(t) = \delta(t - t_0)$, so

$$y' = H_{t_0}(t) + c_1,$$

$$y = \int_0^t H_{t_0}(u)\, du + c_1 t + c_2$$

$$y = \begin{cases} 0 + c_1 t + c_2, & t < t_0 \\ (t - t_0) + c_1 t + c_2, & t \ge t_0. \end{cases}$$

Since $H_{t_0}(t) = 0$ for $t < t_0$, $y(t)$ can be written

$$y = (t - t_0)H_{t_0}(t) + c_1 t + c_2.$$

If we also require $y(0) = y'(0) = 0$, we get the solution

$$y = (t - t_0)H_{t_0}(t).$$

As a check, we calculate, using the product rule,

$$y' = H_{t_0}(t) + (t - t_0)H'_{t_0}(t)$$

$$= H_{t_0}(t) + (t - t_0)\delta(t - t_0).$$

But the second term simply acts like 0, because

$$\int_0^\infty (t - t_0)\delta(t - t_0) g(t)\, dt = (t - t_0)g(t)\big|_{t=t_0} = 0,$$

for every g. It follows that $y'' = H'_{t_0}(t) = \delta(t - t_0)$, as we wanted to show.

EXAMPLE 3 We can use the Laplace transform to solve the problem $y'' = \delta(t - t_0)$, $y(0) = 0$, $y'(0) = 0$, more quickly. We find

$$\mathcal{L}[y''](s) = s^2\mathcal{L}[y](s) - sy(0+) - y'(0+) = s^2\mathcal{L}[y](s).$$

Since $\mathcal{L}[\delta(t - t_0)](s) = e^{-t_0 s}$, we have

$$\mathcal{L}[y](s) = \frac{1}{s^2}e^{-t_0 s}.$$

Since $\mathcal{L}[t](s) = 1/s^2$, reference to Table II (6) shows that

$$y(t) = (t - t_0)H_{t_0}(t).$$

The interpretation of $\delta(t - t_0)$ in the equation $y'' = \delta(t - t_0)$ follows from the formula for y' we derived previously:

$$y' = H_{t_0}(t) + c_1.$$

Thus the presence of $\delta(t - t_0)$ produces a jump of height $+1$ in y' at the point t_0.

EXAMPLE 4   To solve the initial-value problem

$$y'' + 2y' + y = 2\delta(t - 1), \qquad y(0) = 1, \qquad y'(0) = 1,$$

apply $\mathcal{L}$ to both sides:

$$s^2\mathcal{L}[y](s) + 2s\mathcal{L}[y] + \mathcal{L}[y] - 3 - s = 2e^{-s}.$$

Then

$$\mathcal{L}[y](s) = \frac{2}{(s + 1)^2} + \frac{1}{s + 1} + 2\frac{e^{-s}}{(s + 1)^2}.$$

From Tables I and II, we find

$$y(t) = 2te^{-t} + e^{-t} + 2(t - 1)e^{-(t-1)}H_1(t).$$

Notice that $y(t)$ is continuous at $t = 1$. However, its derivative $y'(t)$ will jump up by 2 at $t = 1$.

EXAMPLE 5   It may seem that the advantage of using the Laplace transform in Example 4 lies partly in the ease with which it handles the δ-function. The real advantage, however, still lies in the way the initial conditions are handled. To see this, note that we can easily find a particular solution of the differential equation using a Green's function. The roots of the characteristic equation $r^2 + 2r + 1 = 0$ are $r_1 = r_2 = -1$.   Hence   the   Green's   function   is $g(t - u) = (t - u)e^{-(t-u)}$. The solution satisfying $y(0) = y'(0) = 0$ is

$$y_p(t) = \int_0^t (t - u)e^{-(t-u)}2\delta(u - 1)\, du.$$

The integral can be written

$$y_p(t) = \int_0^\infty (t - u)e^{-(t-u)}H_u(t) \cdot 2\delta(u - 1)\, du,$$

since $H_u(t) = 1$ when $u < t$ and 0 otherwise. It follows right away from the definition of δ that

$$y_p(t) = 2(t - 1)e^{-(t-1)}H_1(t).$$

To finish the problem, we would still have to find the additional homogeneous solution, te^{-t}, that satisfies the initial conditions.

EXERCISES

1. Sketch the graphs of $\Delta_1(t)$, $\Delta_1(t - 2)$, and $H_2(t - 4)$.
2. Show that, if $\Delta_h(t) = 1/h$ for $0 \le t < h$ and $\Delta_h(t) = 0$ otherwise, then

$$\Delta_h(t) = [H_0(t) - H_h(t)]/h.$$

   Conclude then that

$$\Delta_h(t - t_0) = H_0(t - t_0) + H_h(t - t_0).$$

3. Consider the convolution integral with $g(u) = 0$ if $u < 0$:

$$\Delta_h * g(t) = \int_0^t \Delta_h(u)g(t - u)\, du.$$

   (a) By making simple changes of variable in the integral, show that this convolution can be written

$$\Delta_h * g(t) = \frac{1}{h}\int_{t-h}^h g(u)\, du, \qquad \text{if } t \ge h \ge 0.$$

   (b) Show that the convolution can also be written

$$\Delta_h * g(t) = \frac{1}{h}\int_0^t g(u)\, du, \qquad \text{if } h \ge t \ge 0.$$

4. (a) Solve $y' = \Delta_h(t - t_0)$ under the initial condition $y(0) = 0$.
   (b) Sketch the graph of the solution $y_p(t)$ found in part (a) when $t_0 = 1$.
   (c) Find $\lim_{h \to 0+} y_p(t)$ and compare the result with $H_{t_0}(t)$.
5. Express the solution of the initial-value problem

$$y'' = \delta_{t_0}(t), \qquad y(0) = y'(0) = 0,$$

   in terms of $H_{t_0}(t)$ if $t_0 > 0$.
6. (a) Solve the initial-value problems

$$y'' + y = \delta_\pi(t), \qquad y(0) = y'(0) = 0,$$

and

$$y'' + y = \delta_{2\pi}(t), \qquad y(0) = y'(0) = 0.$$

(b) Use the results of part (a) to solve

$$y'' + y = c\delta_{\pi}(t) + d\delta_{2\pi}(t), \qquad y(0) = y'(0) = 0,$$

where c and d are constants.

(c) Show that if $c = d$ in part (b) then the solution is not only 0 for $0 < t < \pi$, but also for $2\pi < t$.

7. Solve the following initial-value problems using transforms or Green's functions.
   (a) $y'' + y' = \delta(t - 1)$, $y(0) = 0$, $y'(0) = 1$.
   (b) $y'' + y' = \delta(t - 1)$, $y(0) = 1$, $y'(0) = 1$.
   (c) $y'' - y = 2\delta(t - 2)$, $y(0) = 0$, $y'(0) = 0$.
   (d) $y'' + 4y = \delta(t - \pi)$, $y(0) = 0$, $y'(0) = 0$.

8. Verify that

$$\mathcal{L}[\delta(t - t_0)f(t)] = e^{-t_0 s}f(t_0).$$

Then use this result to solve

$$y'' + 2y' + y = \delta(t - 1)t, \qquad y(0) = 0, \qquad y'(0) = 0.$$

9. The Heaviside function is useful for representing functions that are defined piecewise on intervals. Sketch the graphs of the following functions $f(t)$.
   (a) $f(t) = H_1(t) - H_2(t)$.
   (b) $f(t) = H_0(t) + 2H_1(t)$.
   (c) $f(t) = H_0(t) + H_1(t) - 2H_2(t)$.
   (d) $f(t) = t[H_0(t) - H_1(t)] + H_1(t)$.
   (e) $f(t) = t[H_0(t) - H_1(t)] - t[H_1(t) - H_2(t)]$.

10. Derive the following formulas:
    (a) $\mathcal{L}[H_a(t)](x) = \dfrac{1}{s}e^{-as}$, $a \geq 0$.

    (b) $\mathcal{L}[tH_a(t)](s) = \dfrac{1 + as}{s^2}e^{-as}$, $a \geq 0$.

11. Use the formulas in Exercise 10 to compute the Laplace transforms of the functions in Exercise 9.

4

First-Order Systems

The earlier chapters are about differential equations whose solutions are real-valued functions of a real variable. A useful generalization is to consider vector differential equations (or, equivalently, systems of real-valued differential equations) whose solutions are vector-valued functions of a real variable. There are two main reasons for making this generalization. One is that many phenomena in applied mathematics are most naturally presented in vector form. The other reason is that, even if the original problem is not in vector form, there may be a valuable technical trade-off in replacing a model of higher derivative order by a vector model with first-order derivatives. Both of these statements will be illustrated in the present chapter. We will follow the convention of writing tuples of coordinates for vectors. Thus (x, y) means the same thing as $x\mathbf{i} + y\mathbf{j}$ and (x, y, z) means the same thing as $x\mathbf{i} + y\mathbf{j} + z\mathbf{k}$.

1 VECTOR EQUATIONS

1A Geometric Setting

A solution of a single first-order differential equation of the form

$$\frac{dx}{dt} = F(t, x)$$

is a real-valued function $x = x(t)$ defined on some interval $a < t < b$ and such that

$$\frac{dx}{dt}(t) = F(t, x(t)), \qquad a < t < b.$$

For example, $dx/dt = x + t$ has the general solution $x = ce^t - t - 1$, defined for all real t. The pair of equations

$$\frac{dx}{dt} = F_1(t, x, y)$$

$$\frac{dy}{dt} = F_2(t, x, y)$$

is called a **system** of **dimension 2**, and a solution will have the form

$$x = x(t)$$

$$y = y(t), \qquad a < t < b.$$

As t increases from a to b, the point $(x(t), y(t))$ in the x, y-plane will trace out some path, perhaps like the one in Figure 1(b). Such a path is called a **trajectory** of the system. It is important to realize that the trajectory of a solution fails by itself to give a complete geometric description of the solution; the correspondence between t values and points on the trajectory path is not made explicit by the picture. The variable t often represents time in applied problems, and we may use dots instead of primes to indicate time derivatives: $dx/dt = \dot{x}$, $dy/dt = \dot{y}$.

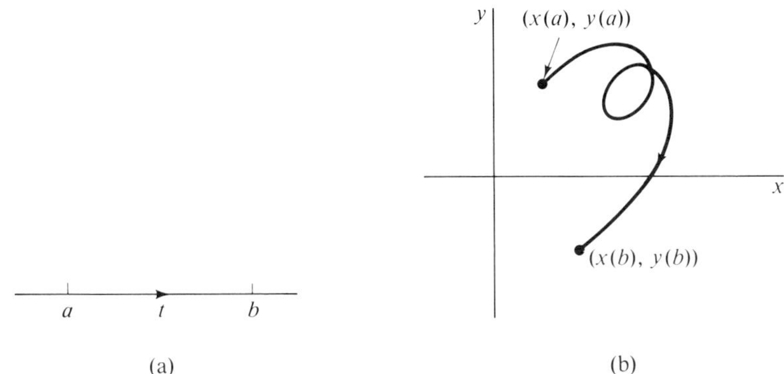

Figure 1

EXAMPLE 1   The system

$$\dot{x} = x,$$

$$\dot{y} = 2y$$

is particularly simple because each unknown function occurs in just one equation. This means that we can solve each equation by itself to get the general solution

$$x = c_1 e^t,$$

$$y = c_2 e^{2t}.$$

At $t = 0$, we have $x(0) = c_1$, $y(0) = c_2$, so the initial conditions $x(0) = 3$, $y(0) = 1$ leave us with

$$x = 3e^t,$$

$$y = e^{2t}.$$

The trajectory then satisfies $y = e^{2t} = (x/3)^2$, with the restriction that $x > 0$ and $y > 0$. Figure 2(a) shows the trajectory for $-\infty < t < \ln 2$. The labels show the points corresponding to some t-values. Including a t-axis perpendicular to the x, y-axes allows us to draw a graph of the solution for $-10 < t < \ln 2$, shown in Figure 2(b). This picture shows the complete correspondence between t and $(x(t), y(t))$ for the interval in question. However, in practice it may be that the trajectory is easier to interpret if it represents, for example, the path taken by some physical object.

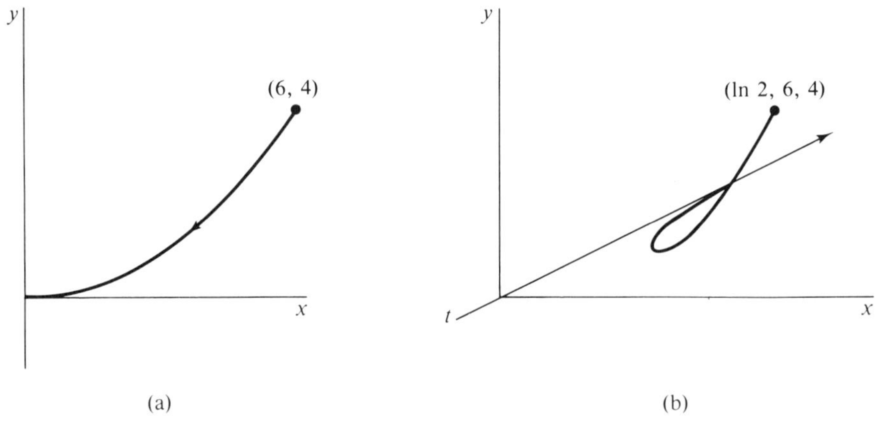

Figure 2

The system of differential equations in Example 1 can be written using vector notation by letting $\mathbf{x} = (x, y)$, $d\mathbf{x}/dt = (\dot{x}, \dot{y})$, and $F(x, y) = (x, 2y)$. Then equating corresponding coordinates gives

$$(\dot{x}, \dot{y}) = (x, 2y) \quad \text{or} \quad \frac{d\mathbf{x}}{dt} = F(\mathbf{x}).$$

EXAMPLE 2   The system

$$\frac{dx}{dt} = x + y + t,$$

$$\frac{dy}{dt} = x - y - t$$

could be written $d\mathbf{x}/dt = F(t, \mathbf{x})$, where

$$F(t, x, y) = (x + y + t, x - y - t).$$

While the systems encountered in practice are always of some specific dimension, it is important to realize that systems of very high dimension are important. For example, the motion of four bodies subject only to their own mutual gravitational attraction is described by the solution of a 24-dimensional system. And much larger systems come up often. When we speak of a **first-order system of dimension** n, we mean a system of the form

$$\frac{dx_1}{dt} = F_1(t, x_1, x_2, \ldots, x_n),$$

$$\frac{dx_2}{dt} = F_2(t, x_1, x_2, \ldots, x_n),$$

$$\vdots$$

$$\frac{dx_n}{dt} = F_n(t, x_1, x_2, \ldots, x_n).$$

Using the vector notations

$$\mathbf{x} = (x_1, x_2, \ldots, x_n), \qquad \frac{d\mathbf{x}}{dt} = \left(\frac{dx_1}{dt}, \frac{dx_2}{dt}, \ldots, \frac{dx_n}{dt}\right),$$

we can write the system more concisely for some purposes as

$$\frac{d\mathbf{x}}{dt} = F(t, \mathbf{x}), \qquad \text{where} \quad F(t, \mathbf{x}) = (F_1(t, \mathbf{x}), \ldots, (F_n(t, \mathbf{x})).$$

A solution is then a vector-valued function $\mathbf{x} = \mathbf{x}(t)$ defined on the interval $a < t < b$. Verifying that $\mathbf{x}(t)$ really is a solution amounts to checking that

$$\frac{d\mathbf{x}(t)}{dt} = F(t, \mathbf{x}(t)), \qquad a < t < b.$$

Trajectory curves for n-dimensional systems can be pictured only for $n = 2$ and $n = 3$, but the terminology is still very useful in higher dimensions. A three-dimensional trajectory is shown in Figure 3(a).

One advantage of the vector interpretation of a system is that a solution $\mathbf{x}(t)$ can be interpreted as a position, at time t, of a point in a space of some dimension, and we can even draw a picture of this space in the case of dimension 2 or 3. Another advantage is that the derivative $d\mathbf{x}/dt$ has a geometric meaning that the derivatives dx_k/dt do not have when considered separately: $d\mathbf{x}/dt$ is a tangent vector to the trajectory $\mathbf{x} = \mathbf{x}(t)$. Furthermore, if t is interpreted as a time variable, then $d\mathbf{x}/dt$ is a velocity vector in the sense that its length is the speed of traversal of the trajectory. The following discussion shows how the formal definitions of tangent and velocity are motivated by the intuitive ideas behind them.

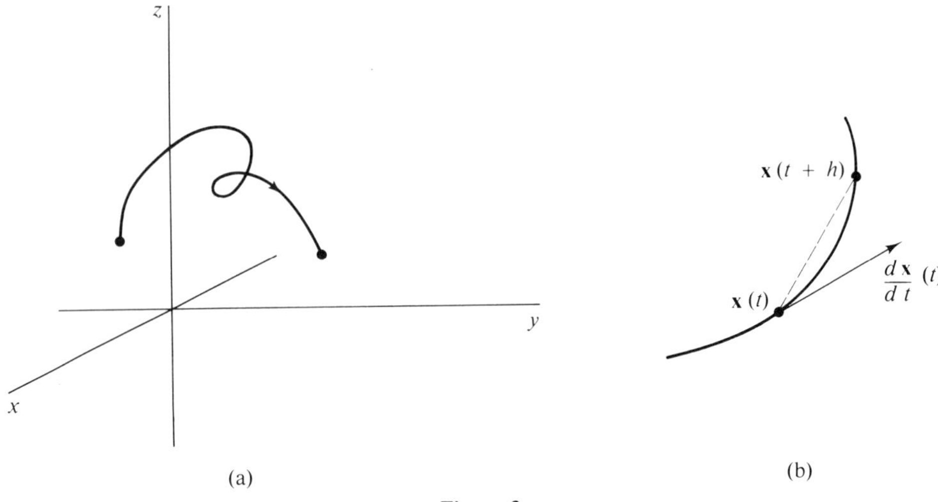

(a)

(b)

Figure 3

Figure 3(b) shows the points $\mathbf{x}(t)$ and $\mathbf{x}(t + h)$. The parallelogram law for addition of vectors shows that, if $\mathbf{z}$ is the vector (arrow) joining $\mathbf{x}(t)$ to $\mathbf{x}(t + h)$, then $\mathbf{z} + \mathbf{x}(t) = \mathbf{x}(t + h)$. It follows that $\mathbf{z} = \mathbf{x}(t + h) - \mathbf{x}(t)$. Now multiply this vector by the number $1/h$ to get a parallel vector that approaches the tangent direction to the trajectory at $\mathbf{x}(t)$ as $h \to 0$. The result is easier to read if we write the entries in $\mathbf{x}(t)$ in a column:

$$\lim_{h \to 0} \frac{\mathbf{x}(t + h) - \mathbf{x}(t)}{h} = \begin{pmatrix} \lim\limits_{h \to 0} \dfrac{x_1(t + h) - x_1(t)}{h} \\ \cdot \\ \cdot \\ \cdot \\ \lim\limits_{h \to 0} \dfrac{x_n(t + h) - x_n(t)}{h} \end{pmatrix}$$

$$= \begin{pmatrix} \dfrac{dx_1(t)}{dt} \\ \cdot \\ \cdot \\ \cdot \\ \dfrac{dx_n(t)}{dt} \end{pmatrix} = \frac{d\mathbf{x}(t)}{dt}.$$

We define $d\mathbf{x}/dt\,(t)$ to be the **tangent vector** or **velocity vector** to the trajectory at $\mathbf{x}(t)$.

The **speed** of $\mathbf{x}(t)$ at time t is the length of $d\mathbf{x}/dt$ at t, defined by

$$\left|\frac{d\mathbf{x}}{dt}(t)\right| = \sqrt{\left(\frac{dx_1}{dt}(t)\right)^2 + \cdots + \left(\frac{dx_n}{dt}(t)\right)^2}.$$

EXAMPLE 3   The pair of equations

$$\dot{x} = x - y$$
$$\dot{y} = x + y$$

has for one of its solutions $x(t) = e^t \cos t$, $y(t) = e^t \sin t$. This is so because

$$\dot{x}(t) = e^t \cos t - e^t \sin t = x(t) - y(t),$$
$$\dot{y}(t) = e^t \cos t + e^t \sin t = x(t) + y(t).$$

The trajectory of this solution is a spiral curve satisfying the initial condition $(x(0), y(0)) = (1, 0)$. See Figure 4(a). Notice that the distance of the point $(x(t), y(t))$ from the origin increases exponentially, since

$$|(x(t), y(t))| = \sqrt{(e^t \cos t)^2 + (e^t \sin t)^2} = e^t.$$

The speed at time t can be found by using the differential equations:

$$|(\dot{x}, \dot{y})| = \sqrt{(x - y)^2 + (x + y)^2}$$
$$= \sqrt{2x^2 + 2y^2} = \sqrt{2}e^t.$$

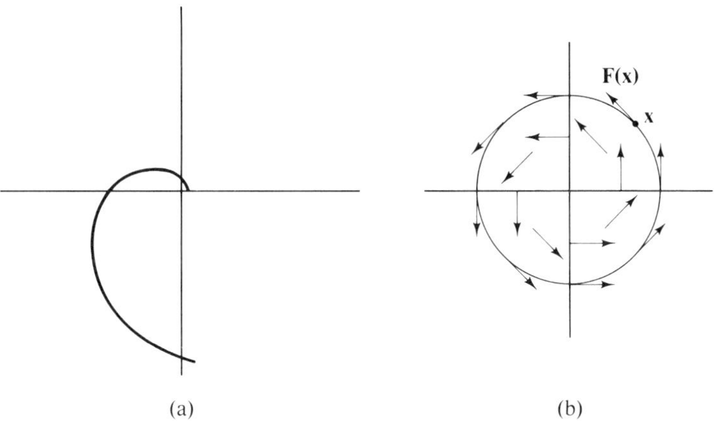

(a)   (b)

Figure 4

1B Vector Fields

In the general first-order system

$$\frac{d\mathbf{x}}{dt} = F(t, \mathbf{x}),$$

the vector-valued function F is understood to depend explicitly on t, so different tangent vectors may be assigned to a trajectory if it happens to pass through the same point $\mathbf{x}$ at two different times, as in Figure 1(b). If $F(t, \mathbf{x})$ is independent of t, the system is called **autonomous** and can be written

$$\frac{d\mathbf{x}}{dt} = F(\mathbf{x}).$$

Examples 1 and 3 are about autonomous systems, whereas Example 2 is time dependent. For an autonomous system, the tangent vector $F(\mathbf{x})$, which is the tangent vector at $\mathbf{x}$, is always the same regardless of what time it is when the trajectory passes through $\mathbf{x}$. Such an assignment of vectors (i.e., arrows) $F(\mathbf{x})$ to points $\mathbf{x}$ is called a **vector field.** Figure 4(b) shows a sketch of one. Note that each vector $F(\mathbf{x})$ is translated parallel to itself so that its tail is located at $\mathbf{x}$. In other words, *the arrow is drawn from the point* $\mathbf{x}$ *to the point* $\mathbf{x} + F(\mathbf{x})$.

EXAMPLE 4   The system $dx/dt = -y$, $dy/dt = x$ can be written $d\mathbf{x}/dt = F(\mathbf{x})$, where $\mathbf{x} = (x, y)$ and $F(\mathbf{x}) = (-y, x)$. The system can be looked at as saying that the tangent vector to a solution trajectory through the point (x, y) has the direction of the vector $(-y, x)$. The tangent then has slope $-x/y$, while the line joining the origin to the point (x, y) has slope y/x. It follows that these two directions are perpendicular. Figure 4(b) shows a sketch of a few tangent vectors. The picture suggests that the trajectories themselves are circular in shape, which turns out to be true. [Using the method in the next section, we can show that the trajectories all come from solutions of the form $x(t) = r \cos(t - t_0)$, $y(t) = r \sin(t - t_0)$.]

A sketch of a vector field plays somewhat the same role for a system that a direction field does for a single equation. In a direction field the segments, all of the same length, are tangent to the *graph* of a solution, and speed is depicted by the slopes of the segments. In a vector field, the arrows are tangent to the *trajectory* (i.e., image curve) of a solution, and speed is depicted by the lengths of the arrows.

For the general time-dependent system $d\mathbf{x}/dt = F(t, \mathbf{x})$, the function $F(t, \mathbf{x})$ specifies a tangent vector (to a trajectory through $\mathbf{x}$) that may be different for different t. The stationary vector field of the kind pictured in Figure 4(b) is no longer appropriate, but it can be replaced by a sequence of "snapshots" taken at different times. Each snapshot will be a sketch of a single vector field, but will show changes in the arrows as time t varies.

EXAMPLE 5   The system

$$\frac{dx}{dt} = (1 - t)x - ty,$$

$$\frac{dy}{dt} = tx + (1 - t)y$$

is determined by the time-dependent vector field

$$F(t, x, y) = ((1 - t)x - ty, \ tx + (1 - t)y).$$

See Figure 5 for sketches of the vector field for three values of t.

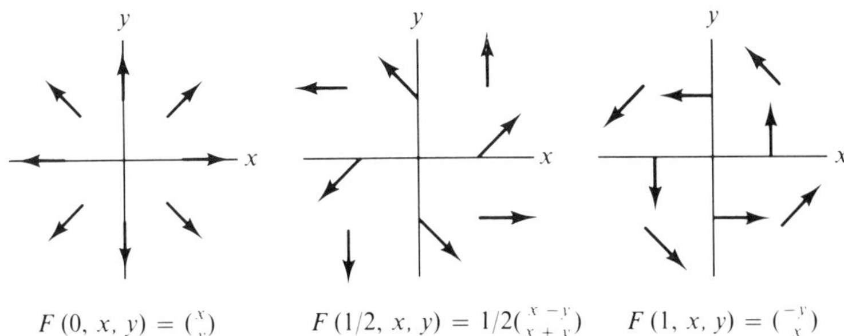

$$F(0, x, y) = \begin{pmatrix} x \\ y \end{pmatrix} \qquad F(1/2, x, y) = 1/2\begin{pmatrix} x - y \\ x + y \end{pmatrix} \qquad F(1, x, y) = \begin{pmatrix} -y \\ x \end{pmatrix}$$

Figure 5

1C Order Reduction

It is often a useful trade-off to replace a differential equation of order higher than one by a system with more unknown functions. This is done, for example, in Chapter 2, Section 5, and in Section 3D of this chapter.

EXAMPLE 6   In the second-order equation

$$\frac{d^2 y}{dt^2} = f\left(t, \ y, \ \frac{dy}{dt}\right)$$

we can let $dy/dt = x$. Then $d^2y/dt^2 = dx/dt$, so the two equations

$$\frac{dx}{dt} = f(t, \ y, \ x), \qquad \frac{dy}{dt} = x$$

form a system equivalent to the original second-order equation. Notice that a solution of the system, being a pair of functions, produces not only a solution $y(t)$ of the given equation but also its derivative $dy/dt = x(t)$. This idea will be pursued further in Section 3; for now we simply use it as a source of examples of systems. Here are some equations and their companion systems.

$$\text{(a)} \ \ y'' + y = 0; \qquad \begin{cases} y' = x, \\ x' = -y. \end{cases}$$

$$\text{(b)} \ \ y'' = -\sin y; \qquad \begin{cases} y' = x \\ x' = -\sin y. \end{cases}$$

$$\text{(c)} \ \ y'' + ty' + t^2 y = 0; \qquad \begin{cases} y' = x, \\ x' = -tx - t^2 y. \end{cases}$$

EXERCISES

1. The following systems can be solved by treating each equation separately. Find the general solution, and then find the particular solution that satisfies the given initial conditions.

(a) $\dfrac{dx}{dt} = x + 1,\; x(0) = 1,$

   $\dfrac{dy}{dt} = y,\; y(0) = 2.$

(b) $\dfrac{dx}{dt} = t,\; x(1) = 0,$

   $\dfrac{dy}{dt} = y,\; y(1) = 0.$

(c) $\dfrac{dx}{dt} = x,\; x(0) = 0,$

   $\dfrac{dy}{dt} = \dfrac{1}{2}y,\; y(0) = 1,$

   $\dfrac{dz}{dt} = \dfrac{1}{3}z,\; z(0) = -1.$

(d) $\dfrac{dx}{dt} = x + t,\; x(0) = 0,$

   $\dfrac{dy}{dt} = y - t,\; y(0) = 0,$

   $\dfrac{dz}{dt} = z,\; z(0) = 1.$

2. For each of the systems in Exercise 1, there is a vector-valued function $F(t, \mathbf{x})$, with $\mathbf{x} = (x, y)$ or (x, y, z) such that the system can be written in the form $d\mathbf{x}/dt = F(t, \mathbf{x})$.
   (i)   Find F in each case.
   (ii) Find the speed of a trajectory though $\mathbf{x}$ at time t.

3. Sketch the following vector fields, associated with the systems of Exercise 1, by drawing a few arrows for $F(\mathbf{x})$ or $F(t, \mathbf{x})$ with their tails at selected points $\mathbf{x}$ of the form (x, y) or (x, y, z). In parts (b) and (d), make separate sketches for $t = -1, t = 0$, and $t = 1$.
   (a) $F(x, y) = (x + 1, y).$
   (b) $F(t, x, y) = (t, y).$
   (c) $F(x, y, z) = (x, \tfrac{1}{2}y, \tfrac{1}{3}z).$
   (d) $F(t, x, y, z) = (x + t, y - t, z).$

4. (a) Write a computer graphics program to sketch a vector field of the form $F(x, y) = (F_1(x, y), F_2(x, y))$.
   (b) Do the same for the fields of the form

$$F(x, y, z) = (F_1(x, y, z), F_2(x, y, z), F_3(x, y, z)).$$

   In this case, the program is essentially no harder to write, but care is needed to keep the sketch from getting too crowded.

5. Sketch the vector field $F(x, y) = (-y, x)$. Then sketch the trajectory curve tangent to arrows in the field sketch, starting at $(x, y) = (1, 0)$.

6. By letting $\dot{y} = x$, express each of the following second-order differential equations as a first-order system of dimension 2. Also find the corresponding initial conditions for $x(0)$ and $y(0)$.
   (a) $\ddot{y} + \dot{y} + y = 0,\; y(0) = 1,\; \dot{y}(0) = 1.$
   (b) $\ddot{y} + t\dot{y} = t,\; y(0) = 0,\; \dot{y}(0) = 1.$

7. For each part of Exercise 6, solve either the given differential equation or the associated system, whichever seems easier.

8. Find a first-order system equivalent to the following differential equations:
   (a) $\dfrac{d^2x}{dt^2} + \left(\dfrac{dx}{dt}\right)^2 + x^2 = e^t.$ $\quad\left(\text{Let } \dfrac{dx}{dt} = y.\right)$
   (b) $\dfrac{d^2x}{dt^2} = x\left(\dfrac{dx}{dt}\right).$

(c) $\dfrac{d^3x}{dt^3} = \left(\dfrac{d^2x}{dt^2}\right)^2 - x\dfrac{dx}{dt} - t.$ $\left(\text{Let } \dfrac{dx}{dt} = y, \dfrac{dy}{dt} = z\right).$

(d) $\dfrac{d^3x}{dt^3} = 12x\dfrac{dx}{dt}.$

9. Suppose the two-dimensional system $dx/dt = F(t, x, y)$, $dy/dt = G(t, x, y)$ is such that the ratio $G(t, x, y)/F(t, x, y) = P(x, y)$ happens to be independent of t. This would occur in particular if neither F nor G depended explicitly on t. Since the chain rule allows us to write

$$\frac{dy}{dx} = \frac{\dfrac{dy}{dt}}{\dfrac{dx}{dt}}$$

under fairly general conditions, we can sometimes conclude that there are trajectory curves of the system satisfying the differential equation

$$\frac{dy}{dx} = P(x, y).$$

If this equation can be solved, we have a way to plot trajectories without finding solutions $x(t)$, $y(t)$. For example, $dx/dt = ty$, $dy/dt = -tx$ leads us to consider

$$\frac{dy}{dx} = -\frac{x}{y},$$

which has solutions $x^2 + y^2 = c$, representing circular trajectories. Using the method described, sketch some trajectories for the following systems. It may help to first sketch a *direction* field.

(a) $dx/dt = x - y$, $dy/dt = x^2 - y^2$.
(b) $dx/dt = e^{2y}$, $dy/dt = e^{x+y}$.
(c) $dx/dt = e^t y$, $dy/dt = e^t x$.
(d) $dx/dt = xy + y^2$, $dy/dt = x + y$.

2 LINEAR SYSTEMS

2A Definition

A differential equation for a single unknown function $y(t)$ is called a **linear equation** if it can be written in the form

$$a_n\frac{d^n y}{dt^n} + \cdots + a_1\frac{dy}{dt} + a_0 y = f$$

where $a_0, \ldots, a_n, f$ are in general functions of t. Defining the **linear operator** $L = a_n D^n + \cdots + a_1 D + a_0$, such an equation can be written $L(y) = f$. A **linear system** for unknown functions $x(t)$, $y(t)$, $\ldots$, is one in which each equation can be written so that its left side is a sum of terms $L_1(x)$, $L_2(y)$, . . . , where the L_k are linear operators.

EXAMPLE 1   Here are some linear systems.

(a) $\dfrac{dx}{dt} + \dfrac{dy}{dt} = e^t,$

$\qquad \dfrac{dx}{dt} - \dfrac{dy}{dt} = e^{-t}.$

(b) $\dfrac{dx}{dt} - ty = t,$

$\qquad \dfrac{dy}{dt} + tx = t^2.$

(c) $\dfrac{d^2x}{dt^2} + \dfrac{dx}{dt} + \dfrac{dy}{dt} = 0,$

$\qquad x + \dfrac{dy}{dt} + y = 0.$

(d) $\dfrac{dx}{dt} = x + y + z,$

$\qquad \dfrac{dy}{dt} = x + y - y,$

$\qquad \dfrac{dz}{dt} = x - y + z.$

2B Elimination Method

By adding nonzero multiples of one equation to another, we can sometimes rewrite a system so that it becomes easier to interpret and solve. For example, the sum and difference respectively of the two equations in Example 1(a) are, respectively,

$$\frac{dx}{dt} = \cosh t, \qquad \frac{dy}{dt} = \sinh t.$$

The solutions are then $x(t) = \sinh t + c_1$, $y(t) = \cosh t + c_2$. Differential operators are helpful in making such modifications if the coefficients in the system are constant.

EXAMPLE 2   The vector equation

$$\frac{d}{dt}(x, y) = (2x + 4y + 2, x - y + 4)$$

represents the system

$$\frac{dx}{dt} = 2x + 4y + 2,$$

$$\frac{dy}{dt} = x - y + 4.$$

Using $D = d/dt$, we can write the system as

$$(D - 2)x - 4y = 2,$$

$$-x + (D + 1)y = 4.$$

Since the desired $y(t)$ must of course be at least once differentiable, the first equation shows that $x(t)$ must be twice differentiable. A similar remark applies to $y(t)$. If D were a number, we could add multiples of one equation to another so as to eliminate either x or y and then substitute back to get the other variable. The algebra of differential operators allows us to do something similar here. Operate on the second equation with $(D - 2)$:

$$(D - 2)x - 4y = 2,$$

$$-(D - 2)x + (D - 2)(D + 1)y = (D - 2)4.$$

Noting that $(D - 2)4 = -8$, we add the first equation to the second to get

$$(D - 2)(D + 1)y - 4y = -6.$$

But this equation is the same as

$$(D^2 - D - 6)y = -6.$$

The characteristic equation is $r^2 - r - 6 = (r + 2)(r - 3) = 0$. The roots are $r_1 = -2$, $r_2 = 3$; so the solution of the homogeneous equation for y is

$$y_h = c_1 e^{-2t} + c_2 e^{3t}.$$

By inspection, we find a particular solution $y_p = 1$, so $y(t)$ must be of the form

$$y(t) = c_1 e^{-2t} + c_2 e^{3t} + 1.$$

Now solve the second of the two equations for x:

$$x = (D + 1)y - 4.$$

Substitution of the formula for $y(t)$ gives

$$x(t) = -c_1 e^{-2t} + 4c_2 e^{3t} - 3.$$

The vector solution is then

$$(x, y) = (-c_1 e^{-2t} + 4c_2 e^{3t} - 3, \qquad c_1 e^{-2t} + c_2 e^{3t} + 1).$$

Initial conditions $(x(0), y(0)) = (2, 1)$ require that

$$(-c_1 + 4c_2 - 3, c_1 + c_2 + 1) = (2, 1)$$

or $-c_1 + 4c_2 = 5$, $c_1 + c_2 = 0$. Thus $c_2 = 1$ and $c_1 = -1$ for the particular solution

$$(x_p, y_p) = (e^{-2t} + 4e^{3t} - 3, -e^{-2t} + e^{3t} + 1).$$

A careful analysis of Example 2 would show that we have indeed found the most general solution and that it contains exactly the right number of arbitrary constants. However, it can happen that extraneous constants turn up in a solution formula. This may happen, for example, because applying an operator increases the order of the system and so increases the number of constants. To find relations among the extra constants, substitute the solution back into the given system. For another remark on this point, see the last paragraph in Section 2C.

EXAMPLE 3   The second-order system

$$\frac{d^2x}{dt^2} = x + 2y + t,$$

$$\frac{d^2y}{dt^2} = 3x + 2y$$

can be written in operator form as

$$(D^2 - 1)x - 2y = t,$$

$$-3x + (D^2 - 2)y = 0.$$

If we multiply the second equation by 2 and operate on the first with $(D^2 - 2)$, then adding the resulting equations eliminates y:

$$(D^2 - 2)(D^2 - 1)x - 6x = -2t \quad \text{or} \quad (D^4 - 3D^2 - 4)x = -2t.$$

The characteristic equation is

$$r^4 - 3r^2 - 4 = (r^2 + 1)(r^2 - 4) = 0,$$

and has roots $r_1 = i$, $r_2 = -i$, $r_3 = 2$, $r_4 = -2$. The homogeneous solution is then

$$x_h(t) = c_1 \cos t + c_2 \sin t + c_3 e^{2t} + c_4 e^{-2t}.$$

By inspection, we find a particular solution $x_p = \frac{1}{2}t$. Solving the first of the given equations for y, a straightforward computation gives

$$y(t) = \frac{1}{2}(D^2 - 1)x - \frac{1}{2}t$$

$$= -c_1 \cos t - c_2 \sin t + \frac{3}{2}c_3 e^{2t} + \frac{3}{2}c_4 e^{-2t} - \frac{3}{4}t.$$

The complete solution is

$$x(t) = c_1 \cos t + c_2 \sin t + c_3 e^{2t} + c_4 e^{-2t} + \frac{1}{2}t,$$

$$y(t) = -c_1 \cos t - c_2 \sin t + \frac{3}{2}c_3 e^{2t} + \frac{3}{2}c_4 e^{-2t} - \frac{3}{4}t.$$

The four constants can be determined by imposing four initial conditions, for example, $x(0) = x_0$, $\dot{x}(0) = x_1$, $y(0) = y_0$, and $\dot{y}(0) = y_1$.

2C Standard Form

For doing computations, it is important to be able to reduce a system to the **standard form**

$$\frac{d\mathbf{x}}{dt} = F(t, \mathbf{x})$$

containing only first-order derivatrives, even if the original system contains higher-order derivatives.

EXAMPLE 4   The systems

$$\begin{cases} \dfrac{dx}{dt} = y + t, \\[2mm] \dfrac{dy}{dt} = x + t, \end{cases} \qquad \begin{cases} \dfrac{dx}{dt} = xy, \\[2mm] \dfrac{dy}{dt} = x^2 + t \end{cases}$$

are both in standard form, although only the first is linear.

EXAMPLE 5   To convert the second-order system

$$\frac{d^2x}{dt^2} + 2\frac{dy}{dt} = t,$$

$$\frac{d^2y}{dt^2} - \frac{dx}{dt} + y = 0,$$

let $dx/dt = u$, $dy/dt = v$. Then $d^2x/dt^2 = du/dt$ and $d^2y/dt^2 = dv/dt$. Substitution gives us altogether

$$\frac{du}{dt} = -2v + t,$$

$$\frac{dv}{dt} = u - y,$$

$$\frac{dx}{dt} = u, \qquad \frac{dy}{dt} = v.$$

This reduction to standard form depended only on our ability to solve for the second derivatives.

EXAMPLE 6   The first-order system

$$\frac{dx}{dt} + \frac{dy}{dt} = 2x + 4y,$$

$$2\frac{dx}{dt} + 3\frac{dy}{dt} = 2x + 6y$$

can be put in standard form by applying simple elimination to the derivatives. Multiply the first equation by 2 and subtract from the second to get $dy/dt = -2x - 2y$. Now subtract this equation from the first to get $dx/dt = 4x + 6y$. The result is a system in standard form:

$$\frac{dx}{dt} = 4x + 6y,$$

$$\frac{dy}{dt} = -2x - 2y.$$

The advantage to having the system in this form before proceeding with elimination using operators is that, having found x or y from a second-order equation, we can then get the other one easily. For example, if we know y, then $x = y - \frac{1}{2}(dy/dt)$, from the second equation.

The theory of Chapters 6 and 7 shows that the general solution of a first-order linear system in standard form contains a number of arbitrary constants equal to the number of equations in the system. Thus, to determine the appropriate number of constants for the solution of a linear system, try to put the system in standard form and count the equations. Exercise 8 shows, however, that not every linear system is equivalent to one in standard form. In particular, a nonstandard system may be **inconsistent**, having no solutions at all.

EXERCISES

1. Classify each of these first- or second-order systems as linear or nonlinear.

(a) $\dfrac{dx}{dt} = t + x^2 + y,$

$\dfrac{dy}{dt} = t^2 + x + y.$

(b) $\dfrac{dy}{dt} = t^2 + z,$

$\dfrac{dz}{dt} = t^3 + y.$

(c) $\dfrac{dx}{dt} = t^2x - y + e^t,$

$\dfrac{dy}{dt} = 1.$

(d) $\dfrac{d^2x}{dt^2} = tx + y,$

$\dfrac{dy}{dt} = x + ty.$

(e) $\dfrac{dx}{dt} + \dfrac{dz}{dt} = 1,$

$\ \dfrac{dx}{dt} - t\dfrac{dz}{dt} = x.$

(f) $\dfrac{d^2x}{dt^2} + \dfrac{dy}{dt} = 0,$

$\ \ y + t\,\dfrac{d^2y}{dt^2} = t.$

2. Use elimination by operator multiplication to get rid of one of the dependent variables. Solve the resulting equation for the remaining variable, and then determine the general solution $(x(t),\ y(t))$.

(a) $\dfrac{dx}{dt} = 6x + 8y,$

$\ \dfrac{dy}{dt} = -4x - 6y.$

(b) $\dfrac{dx}{dt} = x + 2y,$

$\ \dfrac{dy}{dt} = -2x + y.$

(c) $\dfrac{dx}{dt} = x + 2y,$

$\ \dfrac{dy}{dt} = x + y + t.$

(d) $\dfrac{dx}{dt} = -y - t,$

$\ \dfrac{dy}{dt} = x + t.$

3. It is in general not possible to find closed-form solutions for linear systems with nonconstant-coefficient functions. Here is one that can nevertheless be solved fairly easily by solving a second-order equation for y.

$$x' = (t^{-1} - t)x - t^2 y, \qquad t > 0,$$

$$y' = x + ty.$$

$$[\text{Ans. } x = -c_1 t^3 + (2c_1 - c_2)t,\ y = c_1 t^2 + c_2.]$$

4. Reduce each system to the standard form with first derivatives on the left side: $d\mathbf{x}/dt = F(t, \mathbf{x})$.

(a) $\dfrac{dx}{dt} + \dfrac{dy}{dt} = t,$

$\ \dfrac{dx}{dt} - \dfrac{dy}{dt} = x.$

(b) $\dfrac{dx}{dt} + \dfrac{dy}{dt} = y,$

$\ \dfrac{dx}{dt} + 2\dfrac{dy}{dt} = x.$

(c) $2\dfrac{dx}{dt} + \dfrac{dy}{dt} + x + 5y = t,$

$\ \dfrac{dx}{dt} + \dfrac{dy}{dt} + 2x + 2y = 0.$

(d) $\dfrac{dx}{dt} + \dfrac{dy}{dt} = \sin t,$

$\ \dfrac{dx}{dt} - \dfrac{dy}{dt} = \cos t.$

5. After putting it in standard form, solve each system in Exercise 4. Then determine the constants in your solution so as to satisfy the corresponding initial conditions given here.

(a) $x(0) = 1,\ y(0) = -1.$

(b) $x(0) = 0,\ y(0) = 5.$

(c) $x(0) = -1,\ y(0) = 0.$

(d) $x(\pi) = 1,\ y(\pi) = 2.$

6. Use elimination by operator multiplication to get rid of one of the dependent variables. Solve the resulting equation for the remaining variable and then determine the general solution of the system. Substitution may be necessary to find relations among constants. Then determine the constants so that the initial conditions are satisfied.

(a) $\dfrac{d^2x}{dt^2} - x + \dfrac{dy}{dt} + y = 0,$

$\dfrac{dx}{dt} - x + \dfrac{d^2y}{dt^2} + y = 0,$

$x(0) = y(0) = 0,\ \dot{x}(0) = 0,\ \dot{y}(0) = 1.$

(b) $\dfrac{d^2x}{dt^2} - \dfrac{dy}{dt} = 0,$

$\dfrac{dx}{dt} + \dfrac{d^2y}{dt^2} = 0,$

$x(0) = 1,\ y(0) = 0,\ \dot{x}(0) = \dot{y}(0) = 0.$

(c) $\dfrac{d^2x}{dt^2} - \dfrac{dy}{dt} = t,$

$\dfrac{dx}{dt} + \dfrac{dy}{dt} = x + y,$

$x(0) = y(0) = 0,\ \dot{x}(0) = 1.$

(d) $\dfrac{d^2x}{dt^2} - y = e^t,$

$\dfrac{d^2y}{dt^2} + x = 0,\ x(0) = y(0) = \dot{x}(0) = \dot{y}(0) = 0.$

7. By introducing new independent variables, $u = \dot{x}$, $v = \dot{y}$, attempt to reduce each system in Exercise 6 to first-order standard form.

8. None of these linear systems is equivalent to a first-order system in standard form. Discuss their solutions, or lack thereof.

(a) $\dfrac{dx}{dt} + \dfrac{dy}{dt} = 0,$

$\dfrac{dx}{dt} + \dfrac{dy}{dt} = 1.$

(b) $\dfrac{dx}{dt} + \dfrac{dy}{dt} = 0,$

$\dfrac{dx}{dt} + \dfrac{dy}{dt} = x.$

(c) $\dfrac{dx}{dt} + \dfrac{dy}{dt} = t,$

$\dfrac{dx}{dt} + \dfrac{dy}{dt} = x.$

(d) $\dfrac{dx}{dt} + \dfrac{dy}{dt} = y,$

$\dfrac{dx}{dt} + \dfrac{dy}{dt} = x.$

9. Suppose L_1, L_2, L_3, and L_4 are constant-coefficient linear differential operators, and consider the system

$$L_1 x + L_2 y = 0,$$

$$L_3 x + L_4 y = 0.$$

Suppose that the operator $L_1 L_4 - L_2 L_3$ is not identically zero.

(a) Show that coordinates $z = x(t)$ and $z = y(t)$ of a solution must both satisfy the differential equation

$$(L_4 L_1 - L_2 L_3)z = 0.$$

(b) Show that applying the Laplace transform to both sides of the differential equations we get equations of the form

$$P_1(s)\mathscr{L}[x](s) + P_2(s)\mathscr{L}[y](s) = Q_1(s),$$

$$P_3(s)\mathscr{L}[x](s) + P_4(s)\mathscr{L}[y](s) = Q_2(s),$$

where the P_k are polynomials that represent the L_k (that is, $P_k(D) = L_k$) and the Q_k are polynomials that contain initial conditions.

(c) Show that the degree d of $P_1P_4 - P_2P_3$, which is the same as the order of $L_1L_4 - L_2L_3$, is equal to the number of arbitrary constants in the solutions $x(t)$, $y(t)$.

10. (a) Find a first-order system of dimension 4 equivalent to the second-order system

$$\ddot{y} - 3x - 2y = 0,$$

$$\ddot{x} - y + 2x = 0.$$

(b) Write the system found in part (a) in standard form.

(c) Solve the system in part (a).

11. Find a two-dimensional system of order 2 equivalent to the

$$\dot{x} = u + v,$$

$$\dot{y} = u - v,$$

$$\dot{u} = x - y,$$

$$\dot{v} = x + y.$$

Then solve the second-order system and use its solution to solve the given system.

3 APPLICATIONS

The examples in this section are all of a type that arise frequently in applied mathematics. For some, we will be able to give complete solutions, whereas the others are examples for which we need the numerical methods described in Section 4.

3A Mixing

EXAMPLE 1   Figure 6 shows two 50-gallon tanks connected by flow pipes and with inlets and outlets all having the rates of flow as marked in gallons per minute (g/m). The flow rates are arranged so that each tank is maintained at its capacity at all times. We suppose that each tank initially contains salt solution at a concentration in pounds per gallon that we leave unspecified for the moment, that the left-hand tank is receiving salt solution at a concentration of 1 pound per gallon, and that the right-hand tank is receiving pure water. The problem is to find out what happens to the amount of salt, in pounds, as time goes on. We assume that each tank is kept thoroughly mixed at all times so that the concentration of salt is the same throughout the whole tank at any time. In the left-hand tank, with salt content $x(t)$, the rate of change of the amount of salt is, of course,

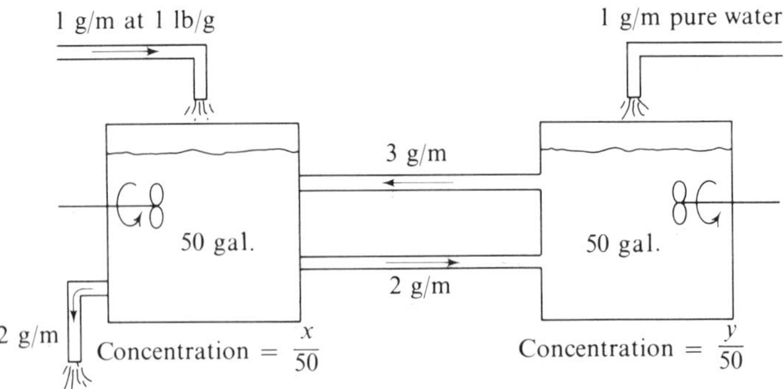

Figure 6

dx/dt. On the other hand, because of the various flow rates, we can break this rate of change into three parts

$$\frac{dx}{dt} = -4\left(\frac{x}{50}\right) + 3\left(\frac{y}{50}\right) + 1,$$

where $x/50$ is the concentration of salt in the left tank and $y/50$ the concentration in the right tank, both in pounds per gallon. The term $-4(x/50)$ is the rate of outflow of salt, and the other two terms represent the rate of inflow. Similarly,

$$\frac{dy}{dt} = 2\left(\frac{x}{50}\right) - 3\left(\frac{y}{50}\right).$$

Thus we have a system of differential equations that we can write as

$$\frac{dx}{dt} = -\frac{4}{50}x + \frac{3}{50}y + 1,$$

$$\frac{dy}{dt} = \frac{2}{50}x - \frac{3}{50}y.$$

To solve it, we can use the elimination method, first writing the system in the form

$$\left(D + \frac{4}{50}\right)x - \frac{3}{50}y = 1,$$

$$-\frac{2}{50}x + \left(D + \frac{3}{50}\right)y = 0.$$

We multiply the first equation by $\frac{2}{50}$ and operate on the second by $(D + \frac{4}{50})$. Addition of the two equations then gives

$$\left(D + \frac{4}{50}\right)\left(D + \frac{3}{50}\right)y - \frac{6}{(50)^2}y = \frac{2}{50},$$

or
$$\left[D^2 + \frac{7}{50}D + \frac{6}{(50)^2}\right] y = \frac{2}{50}.$$

The roots of the characteristic equation can be found in this case by the factorization

$$r^2 + \frac{7}{50}r + \frac{6}{(50)^2} = \left(r + \frac{1}{50}\right)\left(r + \frac{6}{50}\right)$$

to be $r_1 = -\frac{1}{50}$ and $r_2 = -\frac{6}{50}$. A particular solution is clearly $y_p(t) = \frac{50}{3}$, a constant. Thus, in general,

$$y(t) = c_1 e^{-(1/50)t} + c_2 e^{-(6/50)t} + \frac{50}{3}.$$

Using the second equation of the system to write $x(t)$ in terms of $y(t)$, we find

$$x(t) = \tfrac{50}{2}(D + \tfrac{3}{50})y(t)$$

$$= c_1 e^{-(1/50)t} - \tfrac{3}{2}e^{-(6/50)t} + \tfrac{50}{2}.$$

Thus the general solution is

$$x(t) = c_1 e^{-(1/50)t} - \tfrac{3}{2}c_2 e^{-(6/50)t} + \tfrac{50}{2},$$

$$y(t) = c_1 e^{-(1/50)t} + c_2 e^{-(6/50)t} + \tfrac{50}{3}.$$

From these equations, we see immediately that

$$\lim_{t\to\infty} x(t) = \tfrac{50}{2},$$

$$\lim_{t\to\infty} y(t) = \tfrac{50}{3}.$$

In other words, the concentration, in pounds per gallon, in the left tank approaches $\frac{1}{2}$, and in the right tank approaches $\frac{1}{3}$.

The constants c_1 and c_2 depend on the initial values $x(0)$ and $y(0)$. Thus the equations

$$x(0) = c_1 - \tfrac{3}{2}c_2 + \tfrac{50}{2},$$

$$y(0) = c_1 + c_2 + \tfrac{50}{3}$$

determine c_1 and c_2 when $x(0)$ and $y(0)$ are known. In fact, the constants are determined if we know $x(t_1)$ and $y(t_1)$ at any time t_1. We leave these details as an exercise.

3B Oscillations

EXAMPLE 2   Consider two weights of mass m_1 and m_2 separated by springs from each other and from fixed walls. Suppose the springs have stiffness constants k_1, k_2, and k_3 as shown in Figure 7; thus the restoring force to rest position for the ith spring is proportional to k_i. Let x and y be the displacements from

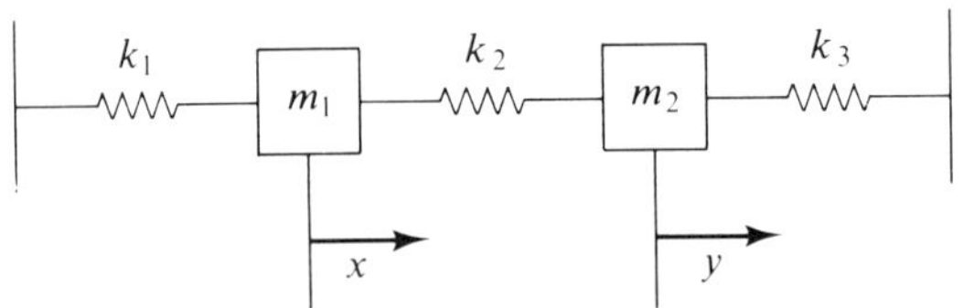

Figure 7

equilibrium of the first and second weights. The force acting on the first weight is, of course, equal to $m_1(d^2x/dt^2)$, but we also have

$$m_1 \frac{d^2x}{dt^2} = -k_1 x + k_2(y - x).$$

The choice of signs is dictated by whether a positive displacement causes an increase or decrease in velocity. Similarly,

$$m_2 \frac{d^2y}{dt^2} = -k_2(y - x) - k_3 y.$$

In deriving both equations, we have neglected frictional forces. We can rewrite the system in the form

$$\frac{d^2x}{dt^2} = -\frac{k_1 + k_2}{m_1} x + \frac{k_2}{m_1} y,$$

$$\frac{d^2y}{dt^2} = \frac{k_2}{m_2} x - \frac{k_2 + k_3}{m_2} y.$$

For example, if the weights are equal, say $m_1 = m_2 = 1$, and $k_1 = k_2 = k_3 = 1$, then the system can be written

$$(D^2 + 2)x - y = 0,$$

$$-x + (D^2 + 2)y = 0.$$

Operating on the second equation with $(D^2 + 2)$ and adding gives

$$(D^4 + 4D^2 + 3)y = (D^2 + 1)(D^2 + 3)y = 0.$$

The general solution of this equation is

$$y(t) = c_1 \cos t + c_2 \sin t + c_3 \cos \sqrt{3}\, t + c_4 \sin \sqrt{3}\, t.$$

Using the second of the pair of equations to find x gives

$$x(t) = (D^2 + 2)y(t)$$

$$= c_1 \cos t + c_2 \sin t - c_3 \cos \sqrt{3}\, t - c_4 \sin \sqrt{3}\, t.$$

The constants c_1, c_2, c_3 and c_4 would be determined by initial displacements and velocities: $x(0)$, $y(0)$, $\dot{x}(0)$, $\dot{y}(0)$.

EXERCISES

1. Suppose that two 100-gallon tanks of salt solution contain amounts of salt $y(t)$ and $z(t)$ at time t. Suppose that the solution in the y tank is flowing to the z tank at a rate of 1 gallon per minute, and that the solution in the z tank is flowing to the y tank at the rate of 4 gallons per minute. Suppose also that the overflow from the y tank goes down the drain, whereas the z tank is kept full by the addition of fresh water. Assume that each tank is kept thoroughly mixed at all times.
   (a) Find a linear system satisfied by y and z.
   (b) Find the general solution of the system in part (a) and then determine the constants in it so that the initial values will be $y(0) = 10$ and $z(0) = 20$.
   (c) Draw the graphs of the particular solutions found in part (b) and interpret the results.

2. In Example 1 of the text, the general solution to a system of differential equations is found to be

$$x(t) = c_1 e^{-(1/50)t} - \tfrac{3}{2} c_2 e^{-(6/50)t} + \tfrac{50}{2},$$

$$y(t) = c_1 e^{-(1/50)t} + c_2 e^{-(6/50)t} + \tfrac{50}{3}.$$

   (a) Find values for the constants c_1 and c_2 so that the initial conditions $x(0) = 25$, $y(0) = \tfrac{2}{3}$ are satisfied.
   (b) Show that it is possible to choose c_1 and c_2 so that an arbitrary initial condition $(x(0), y(0)) = (x_0, y_0)$ is satisfied. Is this a reasonable state of affairs from a physical standpoint?

3. In Example 2 of the text, the system

$$(D^2 + 2)x - y = 0$$

$$-x + (D^2 + 2)y = 0$$

   is shown to have the general solution

$$x(t) = c_1 \cos t + c_2 \sin t - c_3 \cos \sqrt{3}\, t - c_4 \sin \sqrt{3}\, t,$$

$$y(t) = c_1 \cos t + c_2 \sin t + c_3 \cos \sqrt{3}\, t + c_4 \sin \sqrt{3}\, t,$$

   where $x(t)$ and $y(t)$ are interpreted as the displacements at time t of two masses in a mass–spring physical system.
   (a) Show that the initial conditions

$$x(0) = 0, \qquad \dot{x}(0) = 1$$

$$y(0) = 1, \qquad \dot{y}(0) = 0$$

   can be satisfied by choosing the constants properly in the general solution.
   (b) Show that general initial conditions of the form $x(0) = x_0$, $y(0) = y_0$, $\dot{x}(0) = u_0$, $\dot{y}(0) = v_0$ can always be satisfied.

4. Two points start from $x_1 = 0$ and $x_2 = 1$ on a line and move with positions $x_1(t)$ and $x_2(t)$ at time $t \geq 0$. Suppose that the x_1-point always maintains its velocity at exactly 10 units per second greater than that of the x_2-point. Suppose also that the sum of the two velocities is e^{-t} for $t \geq 0$.

(a) Express the relation between the velocities as a first-order system.

(b) Describe the motion of the two points. Are they ever at the same position at the same time?

3C Inverse Square Law

Let $\mathbf{x}_1 = \mathbf{x}_1(t)$ and $\mathbf{x}_2 = \mathbf{x}_2(t)$ represent the positions at time t of two bodies in space such that each acts on the other by the **inverse square law** of gravitational attraction, with no other forces considered. If m_1 and m_2 are the respective masses of the two bodies, the magnitude of the mutually attractive force is then

$$F = \frac{Gm_1 m_2}{r^2},$$

where $r = |\mathbf{x}_1 - \mathbf{x}_2|$ is the distance between $\mathbf{x}_1$ and $\mathbf{x}_2$ (i.e., the length of the vector between them). The **gravitational constant** G is about $6.673 \cdot 10^{-11}$ if the relevant units are meters, kilograms, and seconds. The normalized vectors

$$\mathbf{u}_2 = \frac{\mathbf{x}_1 - \mathbf{x}_2}{|\mathbf{x}_1 - \mathbf{x}_2|}, \qquad \mathbf{u}_1 = -\frac{\mathbf{x}_1 - \mathbf{x}_2}{|\mathbf{x}_1 - \mathbf{x}_2|}$$

have length 1 and point respectively from the second body to the first, and vice versa. Thus the vectors that describe the force acting on each body are the product of magnitude times normalized direction vector; $F\mathbf{u}_1$ acts on the first and $F\mathbf{u}_2$ acts on the second. Since these forces can also be described by Newton's second law as mass times acceleration, we have

$$m_1\ddot{\mathbf{x}}_1 = F\mathbf{u}_1, \qquad m_2\ddot{\mathbf{x}}_2 = F\mathbf{u}_2.$$

Figure 8 shows the positions and normalized vectors. Written out in more detail these **Newton equations** are

3.1 $$\ddot{\mathbf{x}}_1 = -\frac{Gm_2}{r^3}(\mathbf{x}_1 - \mathbf{x}_2), \qquad \ddot{\mathbf{x}}_2 = \frac{Gm_1}{r^3}(\mathbf{x}_1 - \mathbf{x}_2),$$

where m_1 has been canceled from the first equation and m_2 from the second. Subtracting the second equation from the first gives

$$\ddot{\mathbf{x}}_1 - \ddot{\mathbf{x}}_2 = -\frac{G(m_1 + m_2)(\mathbf{x}_1 - \mathbf{x}_2)}{r^3}.$$

Equations 3.1 form a system of vector equations for the motions of the two bodies relative to some coordinate system. If a moving coordinate system has its origin maintained at the center of mass of one of the bodies, say the second, we can let $\mathbf{x} = \mathbf{x}_1 - \mathbf{x}_2$ and consider only the equation of relative motion for the first one:

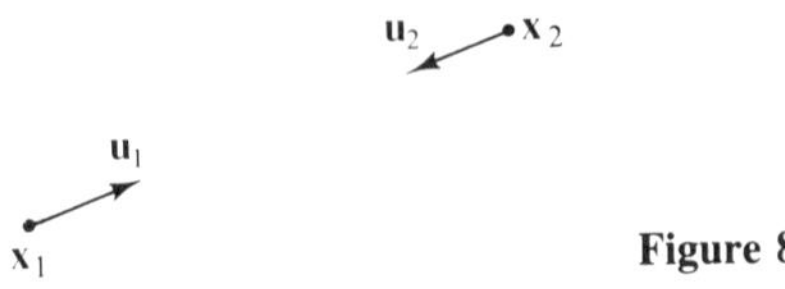

Figure 8

3.2
$$\ddot{\mathbf{x}} = -\frac{G(m_1 + m_2)}{|\mathbf{x}|^3}\,\mathbf{x}.$$

Writing $\mathbf{x} = (x, y, z)$ and $k = G(m_1 + m_2)$, we get three numerical equations:

3.3
$$\ddot{x} = \frac{-kx}{(x^2 + y^2 + z^2)^{3/2}}, \quad \ddot{y} = \frac{-ky}{(x^2 + y^2 + z^2)^{3/2}}, \quad \ddot{z} = \frac{-kz}{(x^2 + y^2 + z^2)^{3/2}}.$$

A solution of this nonlinear system will describe a trajectory of the first body relative to the second, or vice versa. We can eliminate the third equation from consideration by choosing (x, y, z) coordinates so that initial conditions on z are $z(0) = \dot{z}(0) = 0$. The solutions $z = z(t)$ of that equation are then identically zero, so the solution is viewed in an x, y-plane.

There are no simple formulas for the solution of the remaining two equations in 3.3. The classical approach to the problem is to derive characteristic properties of the solutions. These properties are usually stated as **Kepler's laws** of planetary motion, laws that were discovered empirically for closed trajectories before Newton's work.

1. The path described by a solution $(x(t), y(t))$ is an ellipse with the sun (fixed body) at one focus.
2. The radius from the sun to the planet sweeps out equal areas in equal time periods.
3. If T is the time required to complete one orbit and a is half the major axis of the orbit, then

$$T^2 = \frac{4\pi^2}{G(m_1 + m_2)}\,a^3.$$

These beautiful laws are derived from Newton's equations in many physics texts and in some calculus texts. (Among the latter are R. Courant, *Differential and Integral Calculus*, Vol. 2, Interscience Publishers, New York, 1936, and C. H. Edwards and D. E. Penny, *Calculus and Analytic Geometry*, Prentice-Hall, Inc., Englewood Cliffs, New Jersey, 1982.) In these derivations, it is usually assumed that one of the bodies has negligible mass relative to the other, so only one mass appears in the equations. For example, the sun is about 333,434 times as massive as the earth, so for some purposes the mass of the earth can be neglected.

EXAMPLE 3   The closed orbits predicted by the first Kepler law occur only if the speed of separation of the two bodies is not too large. To keep the analysis simple, suppose the relative motion of two bodies is restricted to a fixed line, which we take to be the x-axis. Thus y and z are always zero, so that all that survives of Equations 3.3 is

$$\frac{d^2x}{dt^2} = -\frac{k}{x^2}, \qquad k = G(m_1 + m_2).$$

Let $dx/dt = v$, the velocity of separation. By the chain rule,

$$\frac{d^2x}{dt^2} = \frac{dv}{dt} = \frac{dv}{dx}\frac{dx}{dt} = v\frac{dv}{dx}.$$

The second-order equation then becomes

$$v\frac{dv}{dx} = -\frac{k}{x^2}.$$

Integration of both sides with respect to x gives

$$\int_{v_0}^{v} v\,dv = -k\int_{x_0}^{x}\frac{dx}{x^2},$$

or

$$\frac{v^2}{2} - \frac{v_0^2}{2} = \frac{k}{x} - \frac{k}{x_0},$$

where v_0 is the speed at distance $x = x_0$. This relation between speed and distance allows us to find the **escape speed** of the two bodies relative to each other, that is, the speed v_0 that must be attained at distance x_0 so that speed v always remains strictly positive thereafter. To achieve this effect, we must have

$$0 < v^2 = v_0^2 + \frac{2k}{x} - \frac{2k}{x_0}.$$

Since $k/x \to 0$ as $x \to \infty$, the only way the inequality can hold forever is to have

$$v_0^2 - \frac{2k}{x_0} > 0.$$

The critical escape speed that must be exceeded at distance x_0 is thus

$$v_0 = \sqrt{\frac{2G(m_1 + m_2)}{x_0}},$$

since $k = G(m_1 + m_2)$. Considerations involving energy can be used to show that the formula for escape speed is correct even if the relative motion is not restricted to a linear path.

The same sort of argument that leads to Equations 3.1 for two bodies can be applied to the case of three or more bodies of mass m_k located at $\mathbf{x}_k(t)$, $k = 1, 2, 3, \ldots$. Let $r_{ij} = r_{ji} = |\,\mathbf{x}_i - \mathbf{x}_j\,|$. The analogous vector equations for three bodies are

$$\ddot{\mathbf{x}}_1 = \frac{Gm_2}{r_{21}^3}(\mathbf{x}_2 - \mathbf{x}_1) + \frac{Gm_3}{r_{31}^3}(\mathbf{x}_3 - \mathbf{x}_1)$$

3.4
$$\ddot{\mathbf{x}}_2 = \frac{Gm_1}{r_{12}^3}(\mathbf{x}_1 - \mathbf{x}_2) + \frac{Gm_3}{r_{32}^3}(\mathbf{x}_3 - \mathbf{x}_2)$$

$$\ddot{\mathbf{x}}_3 = \frac{Gm_1}{r_{13}^3}(\mathbf{x}_1 - \mathbf{x}_3) + \frac{Gm_2}{r_{23}^3}(\mathbf{x}_2 - \mathbf{x}_3).$$

There is no known set of characteristic laws describing the solutions of these equations except for very special cases. However, the numerical methods described in the next section can generally be used to give good graphical and numerical descriptions for an initial-value problem associated with initial conditions on the three-dimensional vectors $\mathbf{x}_k(0)$ and $\dot{\mathbf{x}}_k(0)$, $k = 1, 2, 3, \ldots$. Note that there are 18 such conditions for three bodies in three-dimensional space. The main problem in numerical solution of the Newton equations occurs when trajectories of separate bodies come close to each other at the same time. The resulting small denominators can be hard to deal with in the equations.

EXERCISES

Inverse Square Law

1. If position as a function of time is given by the vector $\mathbf{x}(t) = (x(t), y(t), z(t))$, then the magnitude of the acceleration vector is $a = \sqrt{(\ddot{x})^2 + (\ddot{y})^2 + (\ddot{z})^2}$. Use Equations 3.3 to show that $a = k/r^2$, where $r = \sqrt{x^2 + y^2 + z^2}$.

2. The radius of the earth's atmospheric shell is about $6500 \cdot 10^3$ meters, and the earth's mass is about $5976 \cdot 10^{21}$ kilograms. With $G = 6.673 \cdot 10^{-11}$, estimate the escape speed required near the surface of the shell for a projectile of mass 100 kilograms. How is your answer affected if the projectile mass becomes 1000 kilograms? How about 10^{22} kilograms?

3. Letting $z = 0$ in Equations 3.3, we get a system of two second-order equations for x and y. Find the equivalent first-order system obtained by letting $\dot{x} = \dot{u}$, $y = v$.

4. Equations 3.4 for three bodies are equivalent to nine second-order equations for the coordinates (x_1, y_1, z_1), (x_2, y_2, z_2), (x_3, y_3, z_3) of the three bodies. Write out these equations in terms of x_k, y_k, z_k and $r_{ij} = \sqrt{(x_i - x_j)^2 + (y_i - y_j)^2 + (z_i - z_j)^2}$.

5. Write the four second-order vector equations for four bodies of mass m_k at $\mathbf{x}_k(t)$, $k = 1, 2, 3, 4$.

6. The **conservation of momentum** law states that for the total of all forces acting on a system of bodies the sum of the momenta is a constant, where the **momentum** vector is defined by $m\dot{\mathbf{x}}(t)$, with $m = $ mass, $\mathbf{x}(t) = $ position. Verify that conservation of momentum holds for (a) Equations 3.1 and (b) Equations 3.4.

7. Derive the analogue of Equation 3.2 assuming an inverse pth power gravitation law.

Miscellaneous Applications

8. Two points start from $x_1 = 0$ and $x_2 = 1$ on a line and move with positions $x_1(t)$ and $x_2(t)$ at time $t \geq 0$. Suppose that the first point always has a velocity of e^{-t} units per second greater than the second point and that the sum of the two velocities is $f(t)$, arbitrary.
   (a) Express the relation between the velocities as a first-order system.
   (b) Describe the motion of the two points. Are they ever at the same point at the same time?

9. The **Lotka–Volterra equations**

$$\frac{dH}{dt} = (a - bP)H$$

$$\frac{dP}{dt} = (cH - d)P, \qquad a, b, c, d > 0,$$

are sometimes used to describe the size relationship of parasite $P(t)$ and host $H(t)$ populations at time t.

(a) Show that if $P(t) > a/b$, then $H(t)$ decreases, and that if $H(t) < d/c$, then $P(t)$ decreases.

(b) Show that the parameterized solution curves $(H, P) = (H(t), P(t))$ satisfy

$$\frac{dH}{dP} = \frac{(a - bP)H}{(cH - d)P}.$$

Then solve this equation by separation of variables to get

$$H^d P^a = k e^{cH + bP}, \qquad k \text{ constant.}$$

(c) Show that for $k > 0$ the curves found in part (b) are closed circuits in the first quadrant of the P, H-plane. [*Hint:* Analyze the graph of $f(x) = x^a/e^{bx}$.] Thus the Lotka–Volterra model can be used to account for the cyclic variation in the sizes of certain populations.

(d) Show that the closed circuits identified in part (c) are convex by showing that

$$\frac{d^2H}{dP^2} = \frac{1}{d - cH}\left[\frac{aH}{P^2} + \frac{d}{H}\left(\frac{dH}{dP}\right)^2\right].$$

Thus the sign of the second derivative is determined by the first factor: negative when $H > d/c$, positive when $H < d/c$.

10. A point in the (x, y)-plane starts at the origin and moves out along the positive x-axis dragging behind it (as if attached by a string of length a) another point that starts at the point $(0, a)$; thus the distance between the two points remains constantly equal to a, and the velocity vector of the second point always points directly at the first point.

(a) Show that the trajectory of the second point, called a **tractrix**, satisfies the differential equation

$$\frac{dy}{dx} = \frac{-y}{\sqrt{1 - y^2}}.$$

(b) Express the velocity vector $(\dot{x}, \dot{y})$ of the second point in terms of $\dot{x}$, the velocity of the first point along the x-axis.

(c) Let $a = 1$ and sketch the diection field of the differential equation in part (a); then sketch the trajectory of the second point.

(d) Show that the trajectory can be parametrized by

$$x = a \ln(\sec u + \tan u) - a \sin u$$

$$y = a \cos u$$

with $0 \leq u < \pi/2$.

3D Phase Space

The differential equation for the displacement $y = y(t)$ of an oscillating mechanism is

$$m\ddot{y} + k\dot{y} + hy = f(t).$$

Our original treatment in Chapter 2 assumed that k, h, and m were constants. More generally, it may be that k, h, and even m vary with time as the characteristics of a spring change with time $[h = h(t)]$ or the frictional forces change $[k = k(t)]$. Whether these coefficients are variable or not, it is often useful to make a graphic display of $y(t)$ and its derivative $\dot{y}(t) = z(t)$ in the y, z-plane. For one thing, it may be that the velocity $\dot{y}$ is just as interesting as the displacement and so deserves to be displayed along with y. For another, we can very often get a good picture in the y, z,-plane without actually solving the differential equation. The technique is simply to replace a second-order equation by an equivalent first-order system, and then either plot a solution of the system or else its associated vector field. A general second-order equation has the form

$$\ddot{y} = f(t, y, \dot{y}).$$

Letting $z = \dot{y}$, the associated first-order system is

$$\dot{y} = z,$$

$$\dot{z} = f(t, y, z).$$

EXAMPLE 4   The pendulum equation

$$\ddot{y} = -2 \sin y$$

is equivalent to

$$\dot{y} = z,$$

$$\dot{z} = -2 \sin y.$$

The associated vector field is autonomous, so we can easily make a sketch of it. Since the length of the vector (y, z) is not particularly significant, we normalize the arrows that represent it so that they all have the same length. This produces a less cluttered looking picture. See Figure 9. The normalized vector field has y and z coordinates $z/\sqrt{z^2 + 4 \sin^2 y}$ and $-2 \sin y/\sqrt{z^2 + 4 \sin^2 y}$, respectively, if the segments are to have length 1. The continuous curves shown there are plots of $(y(t), z(t))$ for two different sets of initial conditions. One of them gives a closed loop, indicating that the solution $y(t)$ is periodic and so is its derivative $z(t) = \dot{y}(t)$. In the other curve, only $z(t) = \dot{y}(t)$ is periodic and represents a motion in which the pendulum swings over the top, with displacement angle $y(t)$ increasing indefinitely.

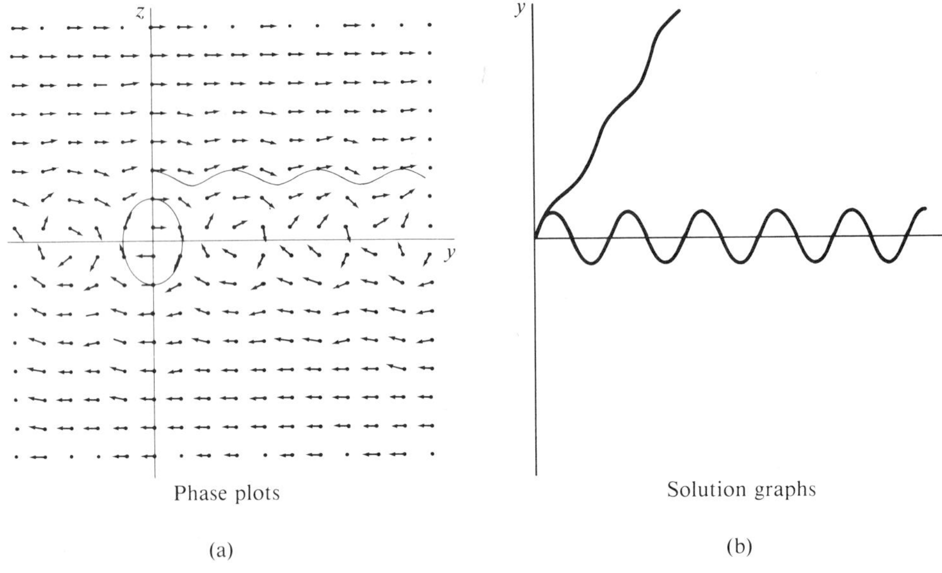

Phase plots

Solution graphs

(a)

(b)

Figure 9

The y, z-plane to which a second-order equation is referred is called the **phase space** or **state space** of the differential equation. The reason is that the pair (y, my), consisting of position and momentum, constitute the **state** of a particle of mass in the position y.

EXERCISES

1. The following differential equations are equivalent to first-order systems in y and z. Find the systems.
   (a) $\ddot{y} + y = 0.$ (b) $\ddot{y} - \dot{y} + 2y = 0.$
   (c) $\ddot{y} - ty = t.$ (d) $\ddot{y} = ty.$

2. Make a sketch in the (y, z)-phase space of the solution of each of the following equations that satisfies the given initial conditions.
   (a) $\ddot{y} + y = 1,$
       $y(0) = 2, \dot{y}(0) = 0.$
   (b) $\ddot{y} - y = 0,$
       $y(0) = 1, \dot{y}(0) = 0.$
   (c) $\ddot{y} = 1,$
       $y(0) = \dot{y}(0) = 0.$
   (d) $\ddot{y} + y = 0,$
       $y(0) = 2, \dot{y}(0) = -1.$

3. For each equation in Exercise 2, find the vector field $(F(y, z), G(y, z))$ associated with its corresponding first-order system. Then normalize the field by dividing by a suitable constant multiple of $\sqrt{F(y, z)^2 + G(y, z)^2}$ and sketch the resulting normalized field.

4. Write a computer graphics program to do the work of Exercise 3.

5. A differential equation of the form $\ddot{x} = f(x)$ expresses the acceleration of a point on a line as a function of position x on the line.

(a) Show that the equation is equivalent to the system

$$\dot{x} = z, \ \dot{z} = f(x)$$

(b) Show that the trajectories of the system in part (a) satisfy

$$\frac{dz}{dx} = \frac{f(x)}{z},$$

having solutions $\frac{1}{2}z^2 - F(x) = \text{const.}$, where $F'(x) = f(x)$.

(c) Assume $f(x) = x^{-2}$ and sketch the trajectory through $(x, z) = (2, 1)$. [The function $-F(x)$ is called the potential energy of the point, which, when added to the kinetic energy $\frac{1}{2}\dot{x}^2$ yields a constant total energy.]

6. The analogue of phase space variables can be introduced for equations of order higher than two. For example, in $\dddot{y} = f(t, y, \dot{y}, \ddot{y})$ let $\dot{y} = z$ and $\dot{z} = w$ so that $\ddot{y} = w$ and $\dddot{y} = \dot{w}$. Find the particular system corresponding to each of the following equations.

(a) $\dddot{y} = y + \ddot{y}.$ **(b)** $\dddot{y} = y\ddot{y} + \dot{y}.$ **(c)** $y^{(4)} = y + t.$

3E Electric Networks

We can analyze an electric network by using a combination of differential and algebraic equations. To understand the analysis it will be enough for our purposes to become familiar with the following ideas. An RLC-network contains resistors (with resistance R_j), inductors (with inductance L_j), and capacitors (with capacitance C_j) joined by conductors in such a way that any two points of the network can be included in a closed loop, or circuit, contained in the network. A value of R_j, L_j, or C_j describes the essential character of the corresponding element of a network. In addition a network typically contains voltage sources characterized by voltages E_j together with the direction in which each source E_j would by itself cause current to flow. A point of the network at which entering current can flow out on more than one conductor is called a **junction.** Two junctions are shown in Figures 10(a) and (b), and four in Figure 10(c). The segment of a network joining two successive junctions is called a **branch**. Six branches are shown in Figure 10(c), while there are only three each in (a) and (b). Each branch will contain at most one voltage source and at most one of each type of network element. In practice there is some resistance, however small, present in every branch.

The problem usually posed is to find the current flowing in the jth branch at time t; we'll denote this current by $I_j(t)$, measured in amperes. We follow the convention that if current is flowing in an arbitrary fixed direction in the jth branch then $I_j > 0$, while flow in the opposite direction corresponds to $I_j < 0$. In analyzing a network we assign an arbitrary positive direction to each branch and indicate our choices by arrows in the network diagram.

Relations between the currents $I_j(t)$ sufficient to determine their values can be derived from initial conditions together with the two Kirchhoff laws:

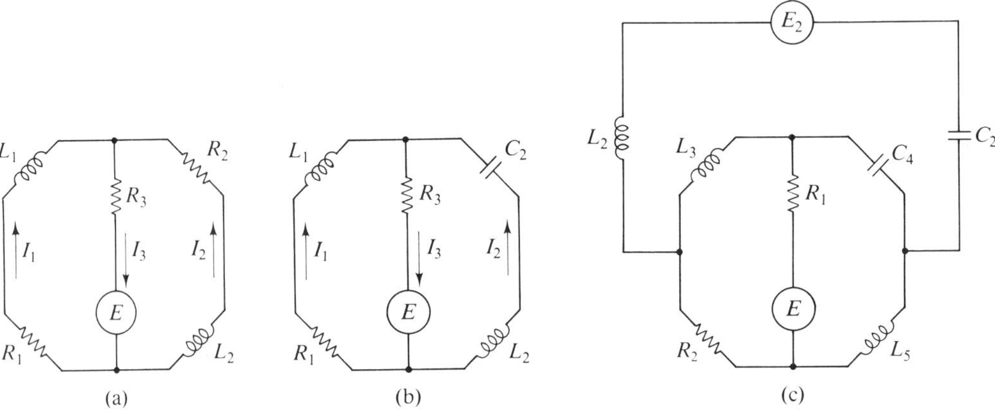

Figure 10

Junction law. The sum of currents directed toward a junction equals the sum of currents directed away from it.

Loop law. The sum of voltage differences across the elements in a closed loop equals the sum of the source voltages in the loop.

The voltage differences in a loop, referred to in the loop law, are caused by the presence of one or more of resistance, inductance, or capacitance in the loop, and are computed for the current I_j in the branch containing such a network element as follows.

Voltage difference $R_j I_j$ is due to resistance R_j.

Voltage difference $L_j \, dI_j/dt$ is due to inductance L_j.

Voltage difference Q_j/C_j is due to capacitance C_j, where Q_j is the charge on the capacitor and charge is related to current by $I_j = dQ_j/dt$.

In computing the sum of voltage differences in a loop, we assign a fixed direction of traversal to the loop, either clockwise or counterclockwise, and attach a minus sign to a difference for which this direction of traversal is counter to the preassigned direction of the branch that contains the element causing the difference. A voltage source is designated **positive** if by itself it would cause current to flow in the direction of traversal of a loop and **negative** otherwise.

A network that doesn't contain any capacitors gives rise immediately to a system of equations to be satisfied by the currents $I_j(t)$. (If the network doesn't contain any inductors either, the system will consist entirely of purely algebraic relations between the $I_j(t)$.) If one or more inductors are also present the system will consist of a combination of algebraic equations and of differential equations of order one. The following example is of this type.

EXAMPLE 1   Figure 10(a) shows a network that contains just resistors and inductors in addition to a single voltage source. Applying the junction law gives the same equation regardless of which of the two junctions it is applied to. Either way we get

$$I_1 + I_2 = I_3.$$

Suppose that the voltage source causes current to flow in the direction shown for I_3 when $E > 0$. The loop-law applied to the left-hand loop then gives

$$L_1 \dot{I}_1 + R_1 I_1 + R_3 I_3 = E.$$

From the right-hand loop we get an equation of the same form relating I_2 and I_3, with L_2 replacing L_1 and R_2 replacing R_1:

$$L_2 \dot{I}_2 + R_2 I_2 + R_3 I_3 = E.$$

Replacing I_3 by $I_1 + I_2$ gives two differential equations for I_1 and I_2:

$$L_1 \dot{I}_1 + (R_1 + R_3)I_1 + R_3 I_2 = E$$
$$L_2 \dot{I}_2 + (R_2 + R_3)I_2 + R_3 I_1 = E.$$

The unit of inductance L is the **henry,** and that of resistance R is the **ohm.** Suppose that $L_1 = L_2 = 0.1$ and that $R_1 = 10$, $R_2 = 20$ and $R_3 = 30$. Suppose also that at the time we're looking at the network the voltage source has been switched off and replaced by a conductor. Then $E = 0$, and our differential equations can be written

$$\dot{I}_1 + 400 I_1 + 300 I_2 = 0$$

$$300 I_1 + \dot{I}_2 + 500 I_2 = 0.$$

In operator form these equations are

$$(D + 400)I_1 + 300 I_2 = 0$$

$$300 I_1 + (D + 500)I_2 = 0.$$

The elimination method leads to the characteristic equation

$$r^2 + 900r + 110000 = 0,$$

with roots (to the nearest integer) $r_1 = -754$, $r_2 = -146$. The general solution then has the form

$$I_1 = 150 c_1 e^{-754t} + 150 c_2 e^{-146t}$$

$$I_2 = 177 c_1 e^{-754t} - 127 c_2 e^{-146t}.$$

Initial values $I_1(0)$ and $I_2(0)$ can be used to determine c_1 and c_2.

We've already seen the relation $dQ/dt = I$ that relates the current in a branch to the charge on a capacitor in the branch. Equations containing a charge Q_j are usually differentiated once with respect to t in order to eliminate Q_j and get equations entirely in terms of currents I_j. If this is done, the equation for a loop containing both an inductor and a capacitor will become a second-order differential equation for I_j, because it already contains a term of the form $L_j dI_j/dt$, whose derivative is $L_j d^2 I_j/dt^2$.

EXAMPLE 2   The network shown in Figure 10(b) contains circuit elements of all three types, R, L and C. Applying the junction law to either junction gives the same equation:

$$I_1 + I_2 = I_3.$$

Suppose that the voltage source causes current to flow in the direction shown for I_3 when $E > 0$. Applying the loop law to the left-hand loop then gives

$$L_1 \dot{I}_1 + R_1 I_1 + R_3 I_3 = E.$$

If $Q_2(t)$ represents the charge on the capacitor, so that $\dot{Q}_2 = I_2$, the right hand loop yields

$$L_2 \dot{I}_2 + Q_2/C_2 + R_3 I_3 = E.$$

If we differentiate this last equation with respect to t and replace $\dot{Q}_2$ by I_2, we get

$$L_2 \ddot{I}_2 + (1/C_2)I_2 + R_3 \dot{I}_3 = \dot{E}.$$

Finally we can replace I_3 by $I_1 + I_2$ to get a pair of differential equations for I_1 and I_2:

$$L_1 \dot{I}_1 + (R_1 + R_3)I_1 + R_3 I_2 = E$$

$$L_2 \ddot{I}_2 + R_3 \dot{I}_2 + (1/C_2)I_2 + R_3 \dot{I}_1 = \dot{E}.$$

The form of these two differential equations suggests that specifying initial values for I_1, I_2 and $\dot{I}_2$ will be enough to determine $I_1(t)$ and $I_2(t)$, and hence also their sum $I_3(t)$. (Note that $\dot{I}_2(0)$ is determined by the equation for the right-hand loop if we know $Q_2(0)$, the initial charge on the capacitor, as well as $E(0)$ and $I_3(0)$.) If E represented a constant voltage source of size V_0 that caused current to flow upon the central branch, then E should be replaced by $-V_0$, because the upward direction is counter to the way the two loops were traversed. For a downward directed source, we would set $E = V_0$. In either case $\dot{E} = 0$, because E is constant. For a variable voltage source, similar remarks apply. For example, if $E(t) = \sin t$, then $\dot{E}(t) = \cos t$, and in the time interval between 0 and π the source would be causing current to flow down in the central branch, followed by an upward flow in the next time interval of length π, *etc.*

E X E R C I S E S

The three networks shown below differ only in the elements they contain; each one results from the one to its left by including a new element. In these diagrams let the symbol E stand for a constant voltage source of E volts that would cause current to flow down relative to the diagram. (Such a source could be provided by a battery with its $(+)$ terminal attached to the junction below it and its $(-)$ terminal attached to the resistor above it.)

1. In the network shown in Fig. 11(a), let $R_1 = 5$, $R_2 = 10$, $R_3 = 15$ and $E = 110$.
   (a) Find a system of equations that determines the currents in the three branches and solve the system.
   (b) What happens to the solution in part (a) if the voltage source is applied upward instead of down?
   (c) What happens to the solution if $R_3 = 0$ instead of 10?
   (d) What happens to the solution if $R_1 = R_3 = 0$?

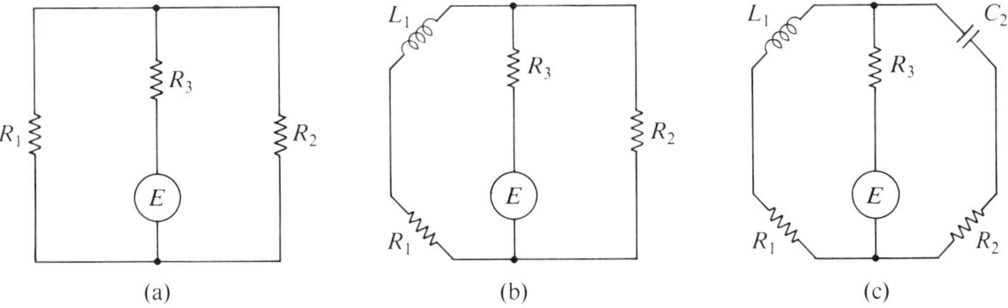

Figure 11

2. In the network shown in Figure 11(b) let $R_1 = 5$, $R_2 = 10$, $R_3 = 15$, $L_1 = 0.1$, and $E = 110$.
   (a) Find a system of equations that determines the currents in the three branches.
   (b) Find the general solution of the system of equations found in part (a).
   (c) Determine the constant in the general solution by imposing the condition that at time $t = 0$ a current of 3 amperes is flowing upward in the left branch.
   (d) Find the limit as t tends to infinity of the solutions found in part (c).

3. In the network shown in Figure 11(c) let $R_1 = 5$, $R_2 = 10$, $R_3 = 15$, $L_1 = 0.1$, $C_2 = 1/250 = 0.004$, $E = $ const.
   (a) Find a system of equations that determines the currents in the three branches.
   (b) Find the general solution of the system of equations found in part (a).
   (c) Determine the constants in the general solution by imposing the condition that at time $t = 0$ currents of 2 and 3 amperes respectively are flowing upward in the right and left branches and that the initial charge on the capacitor is $Q_2(0) = 5$.

4. In the network in Figure 11(c) let $R_1 = 5$, $R_2 = 10$, $R_3 = 15$, $L_1 = 0.1$, $C_2 = 1/1250 = 8 \cdot 10^{-4}$, $E = $ const.
   (a) Find a system of equations that determines the currents in the three branches.

(b) Find the general solution of the system of equations found in part (a), and show that they exhibit damped oscillation.

(c) Determine the constants in the general solution by imposing the condition that at time $t = 0$ currents of 2 and 3 amperes respectively are flowing upward in the right and left branches and that the initial charge on the capacitor is $Q_2(0) = 5$.

5. Find a system of equations for the currents in the six branches of the network shown in Figure 10(c), assuming general constant values for the voltages and the characteristics of the circuit elements.

4 NUMERICAL METHODS

Our numerical methods apply to a first-order vector equation

$$\frac{d\mathbf{x}}{dt} = f(t, \mathbf{x})$$

with initial condition

$$\mathbf{x}(t_0) = \mathbf{x}_0.$$

4A Euler's Method

We choose a step of size h and find successive approximations $\mathbf{x}_k$ to the true values $\mathbf{x}(t_0 + kh)$ of the solution $\mathbf{x}(t)$. The idea is to use the derivative approximation

$$\frac{\mathbf{x}(t + h) - \mathbf{x}(t)}{h} \approx f(t, \mathbf{x}),$$

in the form

$$\mathbf{x}(t + h) \approx \mathbf{x}(t) + hf(t, \mathbf{x}).$$

Thus having found $\mathbf{x}_k$ corresponding to $t_k = t_0 + kh$, we define the approximation $\mathbf{x}_{k+1}$ at t_{k+1} by

$$\mathbf{x}_{k+1} = \mathbf{x}_k + hf(t_k, \mathbf{x}_k).$$

For a two-dimensional system,

$$\dot{x} = F(t, x, y), \qquad x(t_0) = x_0,$$

$$\dot{y} = G(t, x, y) \qquad y(t_0) = y_0,$$

the 0th step starts with x_0 and y_0. Then

$$x_1 = x_0 + hF(t_0, x_0, y_0),$$

$$y_1 = y_0 + hG(t_0, x_0, y_0).$$

Next, with $t_1 = t_0 + h$,

$$x_2 = x_1 + hF(t_1, x_1, y_1),$$

$$y_2 = y_1 + hG(t_1, x_1 \, y_1).$$

In general, with $t_k = t_{k-1} + h = t_0 + kh$, we get

$$x_{k+1} = x_k + hF(t_k, x_k, y_k),$$

$$y_{k+1} = y_k + hG(t_k, x_k, y_k).$$

The basic loop for computer implementation is then:

$$X = X + H * F(T, X, Y),$$

$$Y = Y + H * G(T, X, Y),$$

$$T = T + H,$$

where the letters on the left represent the new values and the letters on the right represent the values computed in the previous step of the loop.

EXAMPLE 1   The system

$$\dot{x} = ty + 1, \qquad x(0) = 1,$$

$$\dot{y} = x, \qquad\qquad y(0) = -1$$

is equivalent to the equation $\ddot{y} - ty = 1$. With step size $h = 0.1$, the loop

$$X = X + H * (T*Y + 1),$$

$$Y = Y + H * X,$$

$$T = T + H$$

produces the following table.

t	x	y	$\dot{x} = ty + 1$	$\dot{y} = x$
0	1	-1	1	1
0.1	1.1	-0.89	1	1.1
0.2	1.19	-0.77	0.92	1.19
0.3	1.28	-0.64	0.87	1.28
0.4	1.35	-0.51	0.85	1.35
0.5	1.44	-0.36	0.85	1.44
0.6	1.52	-0.21	0.89	1.52
0.7	1.61	-0.05	0.97	1.61
0.8	1.70	0.12	1.08	1.70
0.9	1.81	0.30	1.24	1.81
1	1.94	0.49	1.44	1.94
1.1	2.09	0.70	1.70	2.09
1.2	2.26	0.93	2.02	2.26

(continued)

t	x	y	$\dot{x} = ty + 1$	$\dot{y} = x$
1.3	2.48	1.18	2.41	2.48
1.4	2.73	1.45	2.88	2.73
1.5	3.03	1.75	3.45	3.03
1.6	3.39	2.09	4.14	3.39
1.7	3.83	2.47	4.96	3.83
1.8	4.35	2.91	5.95	4.35
1.9	4.97	3.41	7.13	4.97
2	5.72	3.98	8.56	5.72
2.1	6.62	4.64	10.28	6.62
2.2	7.69	5.41	12.36	7.69
2.3	8.98	6.31	14.88	8.98
2.4	10.53	7.36	17.93	10.53
2.5	12.40	8.60	21.64	12.40

4B Modified Euler Method

The next simplest numerical method for a vector equation is a modification of the Euler method in which, instead of using the tangent vector $f(t_k, \mathbf{x}_k)$ to find the next value, we use the vector average

$$\tfrac{1}{2}[f(t_k, \mathbf{x}_k) + f(t_k + h, \mathbf{p}_{k+1})],$$

where $\mathbf{p}_{k+1}$ is the value that the Euler method would have predicted, that is,

$$\mathbf{p}_{k+1} = \mathbf{x}_k + hf(t_k, \mathbf{x}_k).$$

Using the predicted value $\mathbf{p}_{k+1}$ a **modified Euler approximation to** $\mathbf{x}(t_k, \mathbf{x}_k)$ is given by

$$\mathbf{x}_{k+1} = \mathbf{x}_k + \frac{h}{2}[f(t_k, \mathbf{x}_k) + f(t_k + h, \mathbf{p}_{k+1})].$$

For a two-dimensional system,

$$\dot{x} = F(t, x, y), \qquad x(t_0) = x_0,$$

$$\dot{y} = G(t, x, y), \qquad y(t_0) = y_0,$$

we start with (x_0, y_0). Then letting $\mathbf{p} = (p, q)$, we compute the Euler approximation

$$p_1 = x_0 + hF(t_0, x_0, y_0),$$

$$q_1 = y_0 + hG(t_0, x_0, y_0),$$

followed by the modified approximation

$$x_1 = x_0 + \frac{h}{2}[F(t_0, x_0, y_0) + F(t_0 + h, p_1, q_1)],$$

$$y_1 = y_0 + \frac{h}{2}[G(t_0, x_0, y_0) + G(t_0 + h, p_1, q_1)].$$

At the $(k + 1)$th step, we compute $t_k = t_{k-1} + h = t_0 + kh$, and

$$p_{k+1} = x_k + hF(t_k, x_k, y_k),$$

$$q_{k+1} = y_k + hG(t_k, x_k, y_k),$$

$$x_{k+1} = x_k + \frac{h}{2}[F(t_k, x_k, y_k) + F(t_{k+1}, p_{k+1}, q_{k+1})],$$

$$y_{k+1} = y_k + \frac{h}{2}[G(t_k, x_k, y_k) + G(t_{k+1}, p_{k+1}, q_{k+1})].$$

Note that the value x_{k+1} assigned to x in the third of these equations is *not* to be used in the fourth equation; the previous value $x_k = w_k$ should be kept and used instead.

EXAMPLE 2 Consider the second-order equation

$$\ddot{y} + y = 0; \qquad y(0) = 0, \quad \dot{y}(0) = 1.$$

Reducing the equation to a first-order system by letting $\dot{y} = x$ gives

$$\dot{x} = -y, \qquad x(0) = 1,$$

$$\dot{y} = x, \qquad y(0) = 0.$$

The recursive formulas for the basic loop have the form

$$P = X - H * Y,$$

$$Q = Y + H * X,$$

$$W = X$$

$$X = X + H/2 * (-Y - Q),$$

$$Y = Y + H/2 * (W + P).$$

The following table is made with step size $h = 0.01$, but records only every tenth step. It includes also the values of $\cos t$ and $\sin t$ for comparison, because clearly $(x(t), y(t)) = (\cos t, \sin t)$ is the correct elementary formula for the solution.

T	X	Y	$\cos T$	$\sin T$
0	1	0	0.1	0.0
0.1	0.99505	0.09983	0.99500	0.09983
0.2	0.98011	0.19868	0.98006	0.19867
0.3	0.95538	0.29553	0.95533	0.29552
0.4	0.92110	0.38944	0.92105	0.38942
0.5	0.87762	0.47945	0.87757	0.47943
0.6	0.82537	0.56467	0.82533	0.56465
0.7	0.76487	0.64425	0.76483	0.64422
0.8	0.69673	0.71740	0.69669	0.71736
0.9	0.62163	0.78337	0.62159	0.78333
1	0.54031	0.84152	0.54028	0.84148
1.1	0.45360	0.89126	0.45358	0.89121
1.2	0.36235	0.93209	0.36233	0.93204
1.3	0.26749	0.96361	0.26747	0.96356
1.4	0.16995	0.98550	0.16994	0.98545
1.5	0.07071	0.99754	0.07071	0.99749
1.6	0.02922	0.99962	0.02922	0.99957

The distinction between error at a single step and accumulated error is such that the two can be treated separately. If $F(t, \mathbf{x})$ is continuously differentiable in $\mathbf{x}$, then the one-step error using the Euler method is proportional to h^2, where h is the step size. Using the modified Euler method, the one-step error is proportional to h^3, provided $F(t, \mathbf{x})$ is twice continuously differentiable in $\mathbf{x}$. For either method, the size of the accumulated error is controlled by the size of the constant L if F satisfies a **Lipschitz inequality**

$$|F(t, \mathbf{x}) - F(t, \mathbf{y})| \le L|\mathbf{x} - \mathbf{y}|.$$

Thus, if there is a Lipschitz constant L, then the accumulated error over a range of length T is at most

$$\frac{E}{L}(e^{LT} - 1),$$

where E is the sum of the one-step errors.*

EXAMPLE 3   Recall that Newton's equations of planetary motion for a star with one planet are

$$\ddot{x} = -\frac{GMx}{(x^2 + y^2)^{3/2}}, \qquad x(0) = x_0, \quad \dot{x}(0) = u_0,$$

*These ideas are introduced here only as a rough guide (for a more detailed discussion see, for example, G. Birkhoff and G.-C. Rota, *Ordinary Differential Equations,* Ginn & Company, Boston, 1962, Chapter VII).

$$\ddot{y} = -\frac{GMy}{(x^2 + y^2)^{3/2}}, \qquad y(0) = y_0, \quad \dot{y}(0) = v_0.$$

These equations are derived in the Section 3C, where we remarked that the second-order system is equivalent to the first-order system

$$\dot{x} = u, \qquad\qquad x(0) = x_0,$$

$$\dot{y} = v, \qquad\qquad y(0) = y_0,$$

$$\dot{u} = -\frac{GMx}{(x^2 + y^2)^{3/2}}, \qquad u(0) = u_0,$$

$$\dot{v} = -\frac{GMy}{(x^2 + y^2)^{3/2}}, \qquad v(0) = v_0.$$

It is customary to choose units of measurement so that $GM = 4\pi^2$, although for some purposes we could just as well choose them so that $GM = 1$. In choosing initial values for position and velocity, recall that we get an elliptic orbit only if the orbital speed is less than the escape speed, $v_e = \sqrt{2GM/r}$. In other words, we want

$$u_0^2 + v_0^2 < \frac{2GM}{\sqrt{x_0^2 + y_0^2}}.$$

Thus, if $GM = 1$ and $(x_0, y_0) = (1, 0)$, we should choose (u_0, v_0) so that

$$u_0^2 + v_0^2 < 2$$

to get a closed trajectory. The following table was made using the modified Euler method, having chosen $GM = 1$, $(x_0, y_0) = (1, 0)$, and $(u_0, v_0) = (0, 1)$. The step size was $h = 0.01$, but the result is printed only for every 50 steps.

t	x	y
0	1	0
0.5	0.877588	0.479435
1.0	0.540327	0.841496
1.5	0.070792	0.997561
2.0	-0.416073	0.909448
2.5	-0.801107	0.598749
3.0	-0.990088	0.141517
3.5	-0.936777	-0.350347
4.0	-0.654219	-0.756473
4.5	-0.211552	-0.977463
5.0	0.282893	-0.959211
5.5	0.708087	-0.706161
6.0	0.959937	-0.280239

The initial conditions we have chosen in these examples are satisfied by the solutions $(x(t), y(t)) = (\cos t, \sin t)$, which has a circular orbit for its trajectory. Hence we can use this solution as a check on the accuracy of our method of numerical approximation.

The remarks about error accumulation at the end of Chapter 1, Section 7, apply to systems also. However, one outstanding source of error in any numerical method for the approximate solution of differential equations is the evaluation of functions. If, for a single first-order equation $y' = F(x, y)$, the Euler method requires a single evaluation of $F(x, y)$ at each step, the modified Euler method applied to the same equation requires two evaluations, $F(x_k, y_k)$ and $F(x_{k+1}, p_{k+1})$, at the kth step. For an n-dimensional system, the modified Euler method typically requires $2n$ evaluations at the kth step:

$$F_j(t_k, \mathbf{x}_k), \ F_j(t_{k+1}, \mathbf{p}_{k+1}), \qquad j = 1, 2, \ldots, n,$$

where $F_j(t, \mathbf{x})$ is the jth coordinate function of the system. Since errors made at each step are then fed back into the functions F_j at the next step, it is clear that the possibility for error accumulation will be significantly enhanced for a large system. Nevertheless, it can be shown that the local formula errors for the Euler and modified Euler methods remain proportional to h^2 and h^3, respectively; it is only the proportionality factors that may increase as the dimension increases.

EXERCISES

1. The first-order autonomous system

$$\dot{x} = y$$

$$\dot{y} = x$$

is equivalent to the single equation $\ddot{y} = y$, via the relation $\dot{y} = x$.

(a) Show that the system has solutions

$$x(t) = c_1 e^t + c_2 e^{-t},$$

$$y(t) = c_1 e^t - c_2 e^{-t}.$$

(b) Find the particular solution satisfying $x(0) = 1$, $y(0) = 2$.

(c) Compute a table of numerical approximations to the particular solution found in part (b). Do this computation on the interval $0 \le t \le \frac{1}{2}$ in steps of size $h = 0.1$, both by computing from the explicit exponential solution formula and by a direct numerical solution of the system using either the Euler method or its corrected modification.

2. Find a table of approximations to the solution $x(t), y(t)$ of the system

$$\dot{x} = x + y^2,$$

$$\dot{y} = x^2 + y + t,$$

with initial condition $x(0) = 1$, $y(0) = 2$. Use a step of size $h = 0.1$ on the interval $0 \le t \le \frac{1}{2}$, and make the approximation with (a) the Euler method, and (b) the modified Euler method.

3. The Lotka–Volterra equations discussed on page 156 (Exercise 9) contain the special case

$$\frac{dH}{dt} = (3 - 2P)H,$$

$$\frac{dP}{dt} = (\tfrac{1}{2}H - 1)P.$$

   (a) Using the initial conditions $H(0) = 3$, $P(0) = \frac{3}{2}$, compute sufficiently many approximations to the solutions $(H(t), P(t))$ so that the values very nearly return to the initial values.

   (b) Using the approximate data obtained in part (a), sketch the graphs of $H(t)$ and $P(t)$ using the same vertical axis and the same horizontal t-axis.

   (c) Sketch an approximate trajectory in the H, P-plane for the solution found in part (a).

4. The system of Newton equations

$$\ddot{x} = \frac{-x}{(x^2 + y^2)^{3/2}},$$

$$\ddot{y} = \frac{-y}{(x^2 + y^2)^{3/2}},$$

with initial conditions $x(0) = 1$, $y(0) = 0$, $\dot{x}(0) = 0$, $\dot{y}(0) = v_0$, has a solution with a closed trajectory if $v_0 < \sqrt{2}$. Use the modified Euler method to make an approximate computation of the trajectory of a single orbit if
   (a) $v_0 = 0.35$.   **(b)** $v_0 = 0.7$.   **(c)** $v_0 = 1.4$.

5. (a) Show that the second-order equation

$$\ddot{y} - 2\dot{y} + y = t,$$

   with initial conditions $y(0) = 1$, $\dot{y}(0) = 2$, is equivalent to the first-order system

$$\dot{x} = 2x - y + t, \qquad x(0) = 2,$$

$$\dot{y} = x \qquad y(0) = 1.$$

   (b) Find a numerical approximation to the solution of the system in part (a) for the interval $0 \le t \le 1$.

   (c) Solve the given second-order equation by using its characteristic equation, and compare the solution with the numerical results of part (b).

6. (a) Find a first-order system equivalent to

$$y'' - ty' + y = t \qquad y(0) = 1, \quad y'(0) = 2.$$

   (b) Find a numerical approximation to the solution of the system in part (a) for the interval $0 \le t \le 1$.

7. Apply the modified Euler method to the single equation

$$\frac{dy}{dt} = t^2 + y^2, \qquad y(0) = 1.$$

8. Write a program that applies the modified Euler method to the three body problem with all trajectories confined to a single fixed plane. It is convenient here to choose the origin of the coordinate system at the center of mass of the physical system:

$$m_1\mathbf{x}_1 + m_2\mathbf{x}_2 + m_3\mathbf{x}_3 = 0.$$

Differentiating with respect to t in this equation gives zero momentum:

$$m_1\dot{\mathbf{x}}_1 + m_2\dot{\mathbf{x}}_2 + m_3\dot{\mathbf{x}}_3 = 0.$$

Thus if m_1 is the largest mass it is appropriate to let the initial values for $\mathbf{x}_1(0)$ and $\dot{\mathbf{x}}_1(0)$ be determined by the corresponding initial values for the other two bodies. Because of the law of conservation of momentum the previous two equations will in theory be maintained forever.

9. The **van der Pol equation** is

$$\ddot{x} - \alpha(1 - x^2)\dot{x} + x = 0,$$

where α is a positive constant.

(a) Let $y = \dot{x}$ and write down the first-order system in x and y that is equivalent to the van der Pol equation.

(b) Use the modified Euler method to plot numerical solutions to the system found in part (a), using initial values $x(0) = 2$, $y(0) = 0$, while successively letting $\alpha = 0.1$, 1.0, and 2.0. Plot for $0 < t < t_1$, where t_1 is in each plot large enough that the trajectory of $(x(t), y(t))$ appears to be a closed loop.

(c) Repeat the three experiments in part (b) with initial values $x(0) = 1$, $y(0) = 0$, and with $x(0) = 3$, $y(0) = 0$. [The closed loops being approximated in part (b) are each an example of a **limit cycle.**]

5 STABILITY FOR AUTONOMOUS SYSTEMS

When an explicit solution formula $\mathbf{x}(t)$ is available for a vector differential equation

$$\frac{d\mathbf{x}}{dt} = f(t, \mathbf{x}), \qquad t_0 \le t < \infty,$$

it may be possible to determine whether $\lim_{t \to \infty} \mathbf{x}(t)$ exists or, alternatively, exhibits some other kind of behavior. Sometimes, even when solution formulas are not available, we can still get some information about the long-term behavior of solutions if we make the simplifying assumption that $F(t, \mathbf{x}) = F(\mathbf{x})$ does not depend explicitly on t; such a system is called **autonomous.** We will only deal with the case in which $\mathbf{x} = (x, y)$ is two-dimensional, because a detailed analysis of higher-dimensional systems is fairly complicated, and the two-dimensional case is interesting enough by itself.

5A Linear Systems

Consider the autonomous system

5.1
$$\frac{dx}{dt} = ax + by,$$

$$\frac{dy}{dt} = cx + dy, \qquad a,\ b,\ c,\ d \text{ real numbers.}$$

Such a system can be reduced by the elimination method to the solution of a single second-order equation. (In fact, the system might have arisen initially from a phase-space analysis of a second-order equation.) We have seen that the solutions are all derived from characteristic roots λ_1, λ_2 and have the form

5.2
$$x(t) = c_1 e^{\lambda_1 t} + c_2 e^{\lambda_2 t},$$

$$y(t) = c_3 e^{\lambda_1 t} + c_4 e^{\lambda_2 t},$$

unless $\lambda_1 = \lambda_2$, in which case $e^{\lambda_2 t}$ may be replaced by $te^{\lambda_1 t}$. (There will in general be some relation between the constants c_k.) The long-term behavior of a constant solution $x(t) = x_0$, $y(t) = y_0$ is obvious: it simply remains constant; and such a solution is called an **equilibrium solution.** Viewed as a single point, (x_0, y_0) is called an **equilibrium point.** The linear system we are looking at always has an equilibrium solution: the zero equilibrium solution $x(t) = 0$, $y(t) = 0$. The following theorem describes some conditions under which a solution trajectory may or may not tend to the zero equilibrium point as $t \to \infty$.

5.3 Theorem.   Let λ_1 and λ_2 be the characteristic roots that occur in Solution 5.2 to System 5.1. Depending on how the c_k's are chosen, we have the following possibilities.

(i) If the real parts of both λ_1 and λ_2 are negative, then $\lim_{t \to \infty} (x(t), y(t)) = (0, 0)$, for all possible choices of the constants c_k.

(ii) If λ_1 and λ_2 are conjugate and purely imaginary ($\lambda_1 = iq$, $\lambda_2 = -iq$, $q \neq 0$), then all Solutions 5.2 are periodic of period $2\pi/q$.

(iii) If either λ_1 or λ_2 has positive real part, then there are choices of the c_k for which $x(t)$ and $y(t)$ are unbounded.

(iv) If either (or both) of λ_1 and λ_2 is zero, the solution trajectory lies on a straight line or else reduces to a single point.

Proof. The proof consists in examining the form that the solutions take in the four cases:

(i) With $\lambda_1 = -p + iq$, $\lambda_2 = -p - iq$, $p > 0$, the solutions are linear combinations of $e^{-pt} \cos qt$ and $e^{-pt} \sin qt$, or else of e^{-pt} and te^{-pt}. Since these functions tend to zero as $t \to \infty$, the solutions do also.

(ii) With $\lambda_1 = iq$, $\lambda_2 = -iq$, $q \neq 0$, the solutions are linear combinations of cos qt and sin qt and so are either periodic or zero.

(iii) With $\lambda_1 = p + iq$, $\lambda_2 = p - iq$, $p < 0$, the solutions are linear combinations of $e^{pt} \cos qt$ and $e^{pt} \sin qt$, or else of e^{pt} and te^{pt}. Thus nonzero linear combinations can assume arbitrarily large values as $t \to \infty$; if the coefficients are zero, the solutions are identically zero.

(iv) With $\lambda_1 = 0$ and $\lambda_2 = r \neq 0$, the solutions are of the form $x = c_1 + c_2 e^{rt}$, $y = c_3 + c_4 e^{rt}$. If $r = 0$ or if both c_2 and c_3 are zero, the trajectory is a single point. Otherwise, if for instance $c_2 \neq 0$, we can eliminate e^{rt} and find that there is a straight-line trajectory.

The description that Theorem 5.3 gives for the behavior of solutions of System 5.1 fails to capture the interesting variety of behaviors that solutions can exhibit; the best way to show this is to look at some pictures of what can happen. Since everything depends on the roots λ_1, λ_2 of the characteristic equation, it is convenient to derive the equation once and for all. Writing the system in operator form, we have

$$(D - a)x - by = 0,$$

$$-cx + (D - d)y = 0.$$

Eliminating y gives

$$(D - d)(D - a)x - bcx = 0,$$

or $\qquad [D^2 - (a + d)D + (ad - bc)]x = 0.$

Thus the **characteristic roots** λ_1, λ_2 are the roots of

$$\lambda^2 - (a + d)\lambda + (ad - bc) = 0;$$

this is easy to recall from an equation containing a 2-by-2 determinant:

$$\begin{vmatrix} a - \lambda & b \\ c & d - \lambda \end{vmatrix} = (a - \lambda)(d - \lambda) - bc$$

$$= \lambda^2 - (a + d)\lambda + (ad - bc) = 0.$$

EXAMPLE 1   The system

$$\frac{dx}{dt} = -3x + 2y,$$

$$\frac{dy}{dt} = -4x + 3y$$

has the characteristic equation

$$\lambda^2 - 1 = 0.$$

The roots are $\lambda_1 = -1$, $\lambda_2 = 1$. The general solution is

$$x(t) = c_1 e^{-t} + c_2 e^t,$$

$$y(t) = c_1 e^{-t} + 2c_2 e^t.$$

The choice $c_1 = 0$, $c_2 = 1$ gives us

$$x(t) = e^t, \qquad y(t) = 2e^t.$$

The trajectory satisfies $2x = y$, which is the equation of a line. Since $x(t) \to \infty$ and $y(t) \to \infty$ as $t \to \infty$, we are on the part that is in the first quadrant, shown in Figure 12(a). On the other hand, choosing $c_1 = c_2 = 1$ gives us

$$x(t) = e^{-t} + e^t, \qquad y(t) = e^{-t} + 2e^t.$$

Since $y(t) - 2x(t) = -e^{-t}$ and $y(t) - x(t) = e^t$, the trajectory satisfies $(y - x)(y - 2x) = -1$, which is a hyperbola. Again $x(t) \to \infty$ and $y(t) \to \infty$ as $t \to \infty$. Introducing new coordinates $(\bar{x}, \bar{y})$ by $\bar{x} = y - x$, $\bar{y} = y - 2x$, the hyperbola becomes $\overline{xy} = -1$, shown in Figure 12(b). The directed trajectory is shown for $t > 0$. If we choose $c_1 = 1$ and $c_2 = 0$, the trajectory satisfies $x = y$ and is directed toward the origin, as shown in Figure 12(a).

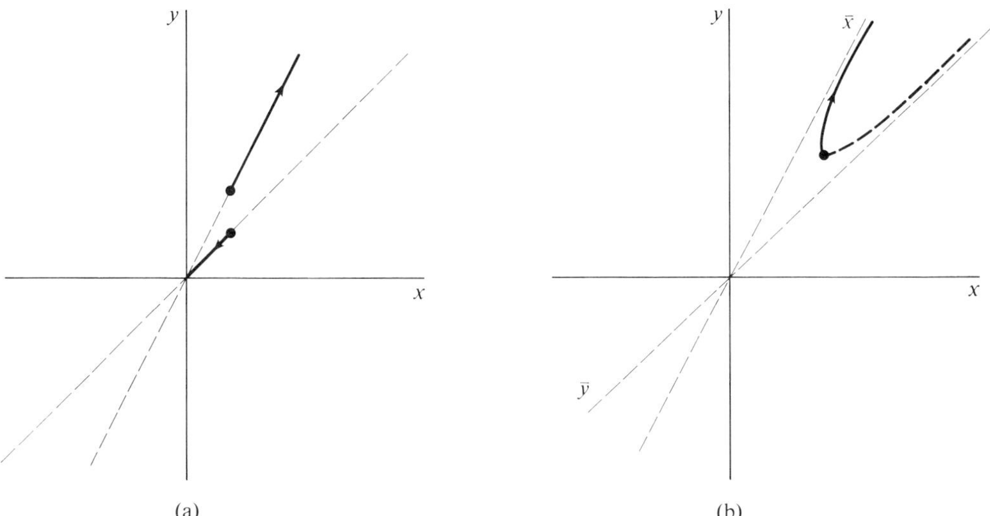

(a)   (b)

Figure 12

What follows is a file of trajectory sketches for some typical two-dimensional systems. The first sketch in each case shows the simplest possible behavior relative to (x, y)-axes. Also included are some trajectory sketches relative to tilted $(\bar{x}, \bar{y})$-axes, where $\bar{x} = a_1 x + a_2 y$, $\bar{y} = a_3 x + a_4 y$, for appropriate choices of the constants a_k.

I. Unstable Node: $0 < \lambda_1 < \lambda_2$

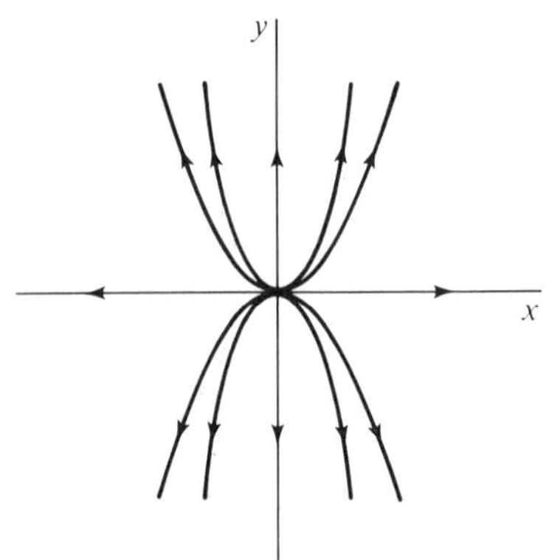 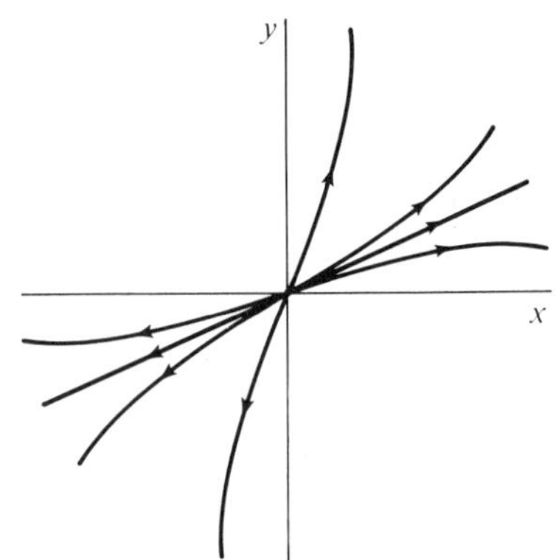

Example: $x = 2e^t$,

$$y = e^{3t};\ 8y = x^3.$$

II. Saddle (Unstable): $\lambda_1 < 0 < \lambda_2$

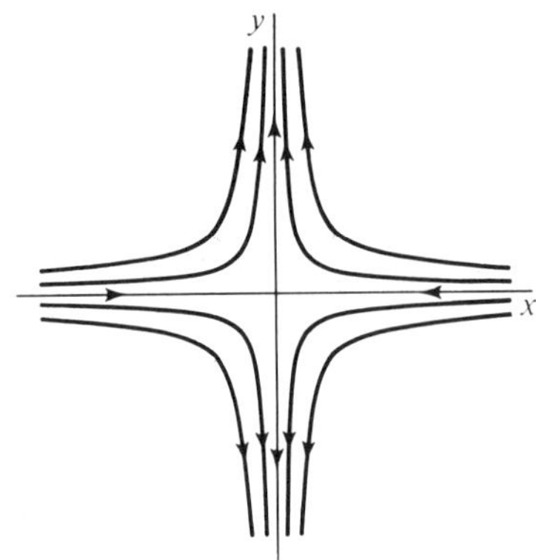 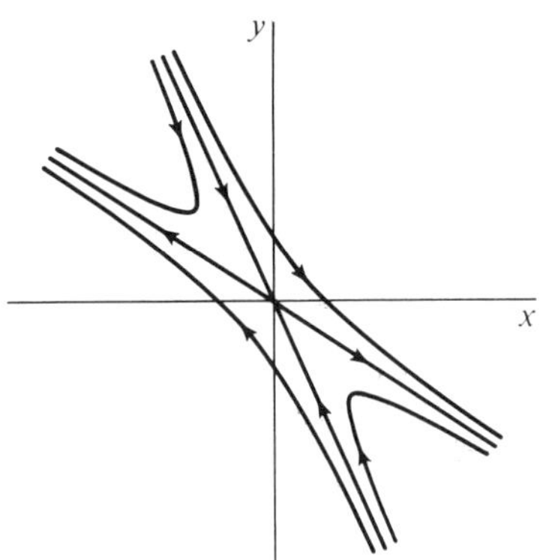

Example: $x = e^{-t}$,

$$y = 3e^t;\ xy = 3.$$

III. Asymptotically Stable Node: $\lambda_1 < \lambda_2 < 0$

Example: $x = 2e^{-3t}$,

$$y = e^{-t};\ x = 2y^3.$$

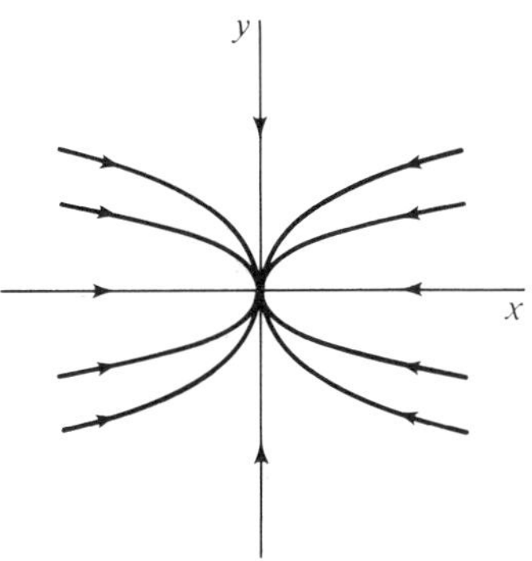 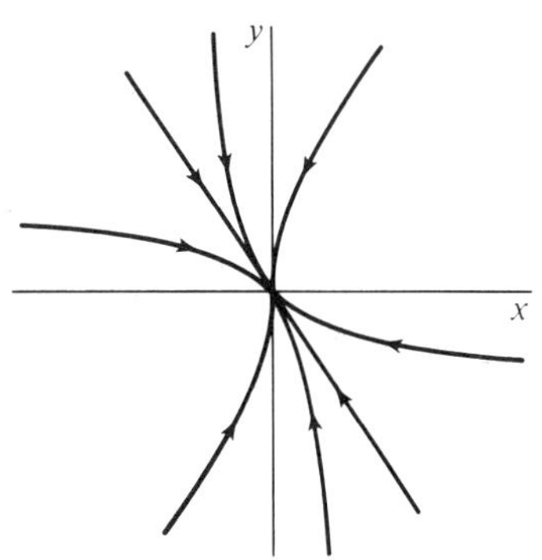

IV. Unstable Spiral: $\lambda_1 = p + iq, \lambda_2 = p - iq, p > 0, q \neq 0$

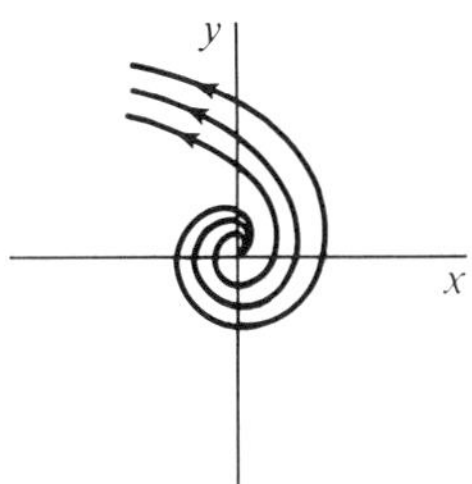

Example: $x = e^t \cos 2t,$

$$y = e^t \sin 2t; x^2 + y^2 = e^{\arctan y/x}.$$

V. Stable Center: $\lambda_1 = iq, \lambda_2 = -iq, q \neq 0.$

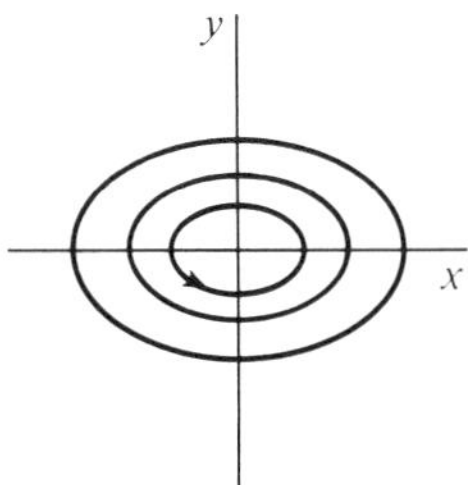

Example: $x = 2 \cos t,$

$$y = \sin t; x^2 + 4y^2 = 4.$$

VI. Asymptotically Stable Spiral: $\lambda_1 = -p + iq,\ \lambda_2 = -p - iq,\ p > 0,\ q \neq 0$

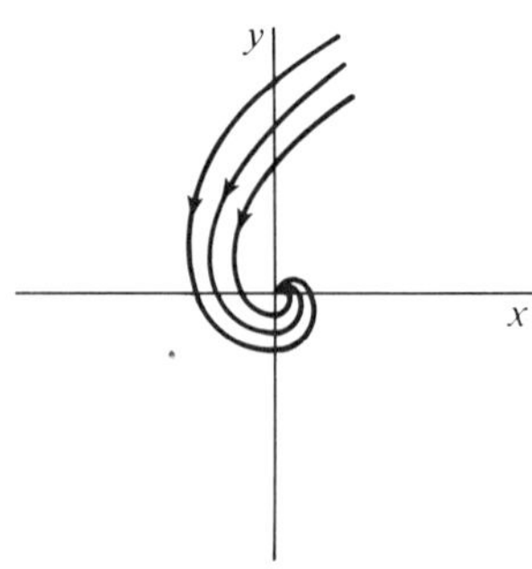

Example: $x = e^{-t} \cos t$,

$$y = e^{-t} \sin t;\ x^2 + y^2 = e^{-2\,\arctan y/x}.$$

VII. Unstable Star: $\lambda_1 = \lambda_2 > 0$

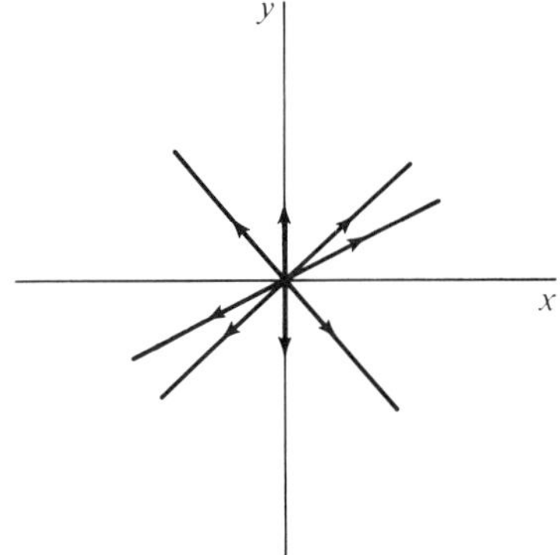 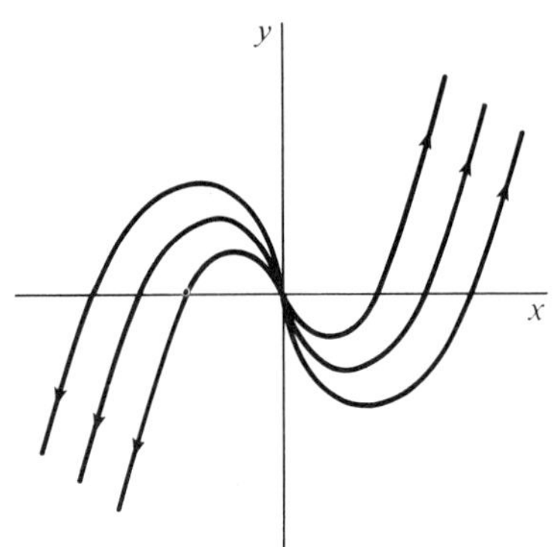

Example: $x = 2e^t$,

$$y = 3e^t,\ 3x = 2y.$$

Example: $x = e^t$,

$$y = te^t,\ y = x \ln x.$$

VIII. Asymptotically Stable Star: $\lambda_1 = \lambda_2 < 0$

Example: $x = e^{-t}$,

$$y = 3e^{-t},\ 3x = y.$$

Example: $x = e^{-t}$,

$$y = te^{-t},\ y = -x \ln x.$$

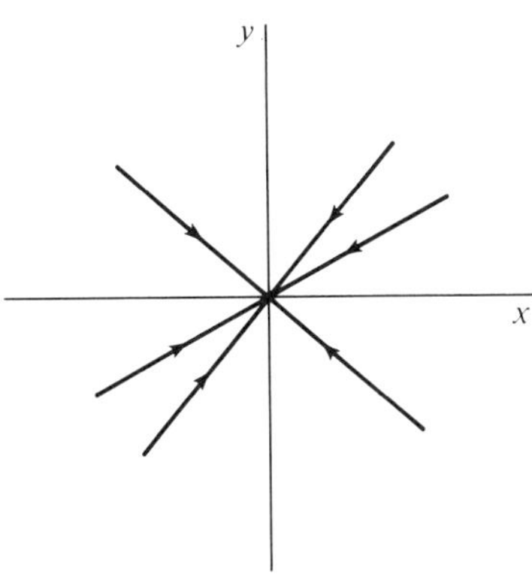 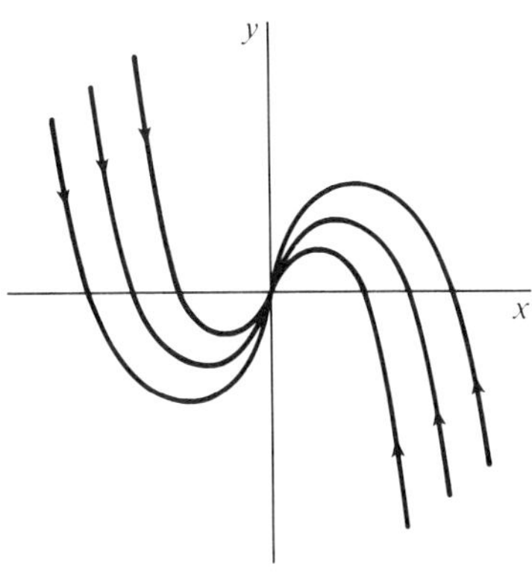

IX. Degenerate: $\lambda_1 = 0$ or $\lambda_2 = 0$

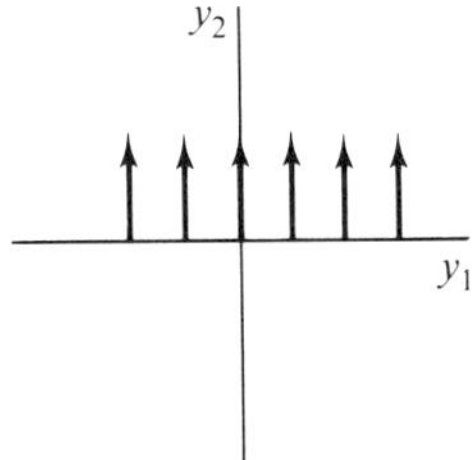

Example: $x = 1$,

$$y = e^t; x = 1.$$

The characteristic equation

$$\lambda^2 - (a + d)\lambda + (ad - bc) = 0$$

has $\lambda_1 = 0$ for a root precisely when $ad - bc = 0$. This case is described as **degenerate** in the above classification for two reasons. One is that the linear combination $c_1 + c_2 e^{\lambda_2 t}$ contains at most one nonconstant function (or none if $a + d = 0$, so that $\lambda_2 = 0$ also). The other reason is that $ad - bc = 0$ precisely when there are infinitely many equilibrium points for the system. The points (x, y) occur when

$$ax + by = 0$$

$$cx + dy = 0$$

both hold. It is a straightforward application of the elimination routine to this algebraic system to see that the equilibrium points constitute a straight line when, and only when, $ad - bc = 0$.

EXAMPLE 2   The system

$$\frac{dx}{dt} = x + y,$$

$$\frac{dy}{dt} = 2x + 2y$$

has equilibrium points (x, y) satisfying

$$x + y = 0,$$

$$2x + 2y = 0.$$

The solutions (x, y) lie on the line $x + y = 0$. The characteristic roots satisfy $\lambda^2 - 3\lambda = 0$, and are $\lambda_1 = 0$, $\lambda_2 = 3$. The solutions of the system are

$$x(t) = c_1 + c_2 e^{3t},$$

$$y(t) = -c_1 + 2c_2 e^{3t}.$$

Elimination of e^t between these two equations shows that the trajectories of the solutions are always straight lines.

EXERCISES

1. Find the characteristic roots λ_1, λ_2 associated with each of the following systems. Then classify each system according to the preceding list.

(a) $dx/dt = -3x + 2y,$
   $dy/dt = -4x + 3y.$

(b) $dx/dt = x + 4y,$
   $dy/dt = 5y.$

(c) $dx/dt = 2x - y,$
   $dy/dt = x + 2y.$

(d) $dx/dt = x,$
   $dy/dt = 3x + y.$

(e) $dx/dt = 2x - 3y,$
   $dy/dt = 2x - 2y.$

(f) $dx/dt = -x + y,$
   $dy/dt = -4x - y.$

(g) $dx/dt = x,$
   $dy/dt = 2x.$

(h) $dx/dt = x + y,$
   $dy/dt = x + y.$

2. Each of the following second-order equations is equivalent to a first-order system obtained by letting $dx/dt = y$. Classify each equation (or equivalently each system) by finding the characteristic roots of the given second-order equation directly.

(a) $d^2x/dt^2 - x = 0.$

(b) $d^2x/dt^2 + dx/dt - x = 0.$

(c) $d^2x/dt^2 + dx/dt + x = 0.$

(d) $d^2x/dt^2 + x = 0.$

3. The following second-order equations can be thought of as determining the development $x(t)$ of a spring system subject to a frictional force kdx/dt, where $k \geq 0$ is constant. Find conditions on k under which the equation belongs to any or all of the classes in the standard list.

(a) $d^2x/dt^2 + kdx/dt + x = 0.$

(b) $d^2x/dt^2 + kdx/dt + 3x = 0.$

4. Sketch typical phase plots in the (x, y)-plane for each of the systems in Exercise 1.

5. Sketch typical phase plots in the (x, y)-plane, where $y = dx/dt$, for each of the second-order equations in Exercise 2.

6. Show that the system

$$\frac{dx}{dt} = ax + by, \qquad \frac{dy}{dt} = cx + dy$$

has infinitely many equilibrium solutions if and only if $ad - bc = 0$.

7. Consider the general solution 5.2 of a linear system in which $\lambda_1 < \lambda_2 < 0$, for which $(0, 0)$ is an asymptotically stable node.

   (a) Show that, if c_2 and c_4 are not both zero in Equation 5.2, then as $t \to \infty$ the slope $dy/dx = (dy/dt)/(dx/dt)$ of the associated trajectory approaches c_4/c_2, interpreted as a vertical slope if $c_2 = 0$.

   (b) Show that if $c_2 = c_4 = 0$, but c_1 and c_3 are not both zero, then the slope of the associated trajectory approaches a limit independent of c_1 and c_3 as $t \to \infty$.

8. **(a)** Show that the nonlinear system

$$\frac{dx}{dt} = A - Bx - x + x^2 y, \, A \neq 0,$$

$$\frac{dy}{dt} = Bx - x^2 y$$

has a single constant solution $x(t) = A$, $y(t) = B/A$. The trajectory of a constant solution is called an **equilibrium point.**

   (b) Assume that $A > 0$ and $B > A^2 + 1$. Trajectories starting sufficiently near, but not at, the equilibrium point exhibit **limit cycle** behavior in that they approach a closed trajectory. Investigate this claim by using a graphical–numerical method.

5B Nonlinear Systems

Methods for determining the long term $(t \to \infty)$ stability or instability of solutions of the nonlinear system

$$\textbf{5.4} \qquad\qquad \frac{dx}{dt} = f(x, y), \qquad \frac{dy}{dt} = g(x, y)$$

are particularly important, because explicit solution formulas are usually impossible to find, and approximate numerical solutions are necessarily always limited to some finite time interval $t_0 \leq t \leq t_1$. In what follows we will consider the behavior of solution trajectories near an **equilibrium point** of the system, that is, a point (x_0, y_0) having the property that $f(x_0, y_0) = g(x_0, y_0) = 0$. Associated with each such point is a constant **equilibrium solution** $x(t) = x_0$, $y(t) = y_0$. We will be concerned with finding out whether or not a solution that passes near such a point tends asymptotically to the point or exhibits some other kind of behavior. Given the nature of the criteria we will develop, it will be necessary to restrict ourselves to considering an **isolated equilibrium point,** that is, an equilibrium point that does not have other equilibrium points arbitrarily close to it. Thus the kind of solution behavior described as "degenerate" in Section 5A will not be included.

Let (x_0, y_0) be an isolated equilibrium point of System 5.4. Since we are interested in solution behavior near (x_0, y_0), we can make a change of variable by replacing x by $x + x_0$ and y by $y + y_0$. The system then looks like

5.5
$$\frac{dx}{dt} = f(x + x_0, y + y_0), \qquad \frac{dy}{dt} = g(x + x_0, y + y_0),$$

and the equilibrium point of interest is now at $x = 0$, $y = 0$. Assuming that f and g are continuously differentiable, we then replace System 5.5 by the **linearized system,** in terms of partial derivatives f_x, f_y, g_x, g_y of f and g:

5.6
$$\frac{dx}{dt} = f_x(x_0, y_0)x + f_y(x_0, y_0)y,$$

$$\frac{dy}{dt} = g_x(x_0, y_0)x + g_y(x_0, y_0)y.$$

It can be shown that System 5.6 is in a precise sense the best of all linear systems to use near (x_0, y_0) instead of the nonlinear System 5.5. (Of course, if 5.5 were itself a linear system, then 5.6 would be exactly the same system as 5.5. for instance, if $f(x, y) = ax + by$, then $f_x(x_0, y_0) = a$ and $f_y(x_0, y_0) = b$, for all x_0, y_0.)

EXAMPLE 3   The equilibrium points of the nonlinear system

$$\frac{dx}{dt} = 2x - y^2,$$

$$\frac{dy}{dt} = x + y$$

are found by solving the algebraic system

$$2x - y^2 = 0,$$

$$x + y = 0.$$

The only solutions are $(x_0, y_0) = (0, 0)$ and $(x_1, y_1) = (2, -2)$, so we have two isolated equilibrium points. Near (x_0, y_0) we consider the linearized system

$$\frac{dx}{dt} = 2x - (2y_0)y$$

$$\frac{dy}{dt} = x + y.$$

When $y_0 = 0$, the system has solutions

$$x(t) = c_1 e^{2t}$$

$$y(t) = c_1 e^{2t} + c_2 e^t.$$

The two positive characteristic roots show that the linear system has an unstable node at $(0, 0)$. Near $(2, -2)$ the linear system has characteristic roots $\lambda_1 = (3 + \sqrt{17})/2$, $\lambda_2 = (3 - \sqrt{17})/2$. Thus the solutions exhibit saddle-type behavior as $t \to \infty$, since $\lambda_1 > 0$ and $\lambda_2 < 0$.

The next important step in Example 3 would be to draw some conclusion about the behavior of solutions of the given nonlinear system from what we know about the solutions of the two linearized systems. To state the connection clearly and succinctly, we will use some standard terminology. Let $x_0(t) = x_0$, $y_0(t) = y_0$ be an equilibrium solution of an autonomous system of differential equations, linear or nonlinear. Then (x_0, y_0) is **asymptotically stable** if every solution $(x(t), y(t))$ of the system that has a trajectory point close enough to (x_0, y_0) satisfies

$$\lim_{t \to \infty} (x(t), y(t)) = (x_0, y_0).$$

For the linear systems in Section 5A, the examples numbered III, VI, and VIII show asymptotically stable behavior at $(0,0)$. Special types of asymptotic stability are the **spiral,** in which the trajectory winds around the equilibrium point as it approaches it, and the **node,** in which the tangent to each trajectory approaches a limiting position. A less strict condition on the system at (x_0, y_0) is that (x_0, y_0) be simply **stable,** which means that every trajectory that has a point close enough to (x_0, y_0) will remain within a preassigned distance of (x_0, y_0). Clearly, if (x_0, y_0) is asymptotically stable for a system, then it is also stable. Number V in Section 5A is stable without being asymptotically stable. All other equilibrium points are called **unstable;** these include types I, II, IV, and VII of Section 5A. In particular, type II, the **saddle point,** can be characterized for the nonlinear system as having two lines through it, one of which is tangent to a trajectory approaching the point and the other of which is tangent to a trajectory that moves away from the point.

5.7 Theorem.   Let $f(x, y)$ and $g(x, y)$ be continuously differentiable in a region R of the (x, y)-plane, and suppose the point (x_0, y_0) in R is an isolated equilibrium point of the system

$$\frac{dx}{dt} = f(x, y), \qquad \frac{dy}{dt} = g(x, y).$$

Suppose that the associated linearized system at (x_0, y_0),

$$\frac{dx}{dt} = f_x(x_0, y_0)x + f_y(x_0, y_0)y, \qquad \frac{dy}{dt} = g_x(x_0, y_0)x + g_y(x_0, y_0)y,$$

is **nondegenerate:**

$$f_x(x_0, y_0)g_y(x_0, y_0) - f_y(x_0, y_0)g_x(x_0, y_0) \neq 0.$$

Then, at (x_0, y_0):

(i) Asymptotic stability for the linearized system implies asymptotic stability for the given system, with nodal and spiral behavior corresponding similarly.

(ii) Instability for the linearized system implies instability for the given system. In particular, if (x_0, y_0) is a saddle point for the linearized system, it is a saddle point for the given system, while nodal and spiral behavior correspond similarly.

(iii) If (x_0, y_0) is a stable center for the linearized system, then (x_0, y_0) may be an unstable spiral, an asymptotically stable spiral, or may be merely stable for the nonlinear system.

The proof of parts (i) and (ii) involves some technical analysis that we will not undertake. Proving part (iii) just amounts to looking at some appropriate examples, and this is done in Exercise 4 at the end of the section.

EXAMPLE 4   Continuing with Example 1, we conclude from Theorem 5.7 (ii), stated just before this, that the nonlinear system has an unstable equilibrium at $(x_0, y_0) = (0, 0)$, since $(0, 0)$ is an unstable node for the linearized system. The same conclusion holds for the point $(2, -2)$, at which the linearized system has a saddle point. More specifically, $(2, -2)$ is a saddle point for the nonlinear system since $\lambda_1 = (3 - \sqrt{17})/2 < 0 < \lambda_2 = (3 + \sqrt{17})/2$.

EXAMPLE 5   The undamped nonlinear pendulum equation $\ddot{\theta} = -(g/l) \sin \theta$ is derived in Section 4B of Chapter 2. Here $\theta = \theta(t)$ is the displacement angle from the downward-oriented vertical. Letting $x = \theta$ and $y = \dot{\theta}$, the equivalent system is

$$\frac{dx}{dt} = y, \qquad \frac{dy}{dt} = -\left(\frac{g}{l}\right) \sin x.$$

Mechanical intuition suggests that the equilibrium solutions are those for which $y = \dot{\theta} = 0$ and $x = \theta = n\pi$, where n is an integer. When n is even, the pendulum hangs straight down, and when n is odd, it is precisely balanced in a straight-up position, which is obviously mechanically unstable. Solving the equations $y = 0$, $-(g/l) \sin x = 0$ for $(x, y) = (0, n\pi)$ confirms at least a part of the intuition. As to stability *versus* instability, consider the linearized system at $(x_n, y_n) = (nl, 0)$:

$$\frac{dx}{dt} = y,$$

$$\frac{dy}{dt} = -(g/l)(\cos n\pi)x.$$

The characteristic roots satisfy

$$\lambda^2 + (-1)^n(g/l) = 0,$$

since $\cos n\pi = (-1)^n$. If n is even (pendulum down), the roots are $\pm\, i\sqrt{g/l}$. If n is odd, the roots are $\pm\, \sqrt{g/l}$. In the first case, we have a center for the linear system, so Theorem 5.7 fails to confirm or deny our intuition. It takes a more detailed analysis (see Exercise 3) to show that the expected stability does occur near $(x_0, y_0) = (2m\pi, 0)$. When n is odd, the linear system has a saddle point at $((2m + 1)\pi, 0)$. This tells us that, while the vertical positions with zero velocity are unstable, there is in principle a path in the (x, y) phase space leading to a straight-up balanced position. Figure 13 shows the result of some numerical computations of phase-space trajectories. The effect of frictional damping on trajectories is taken up in the exercises.

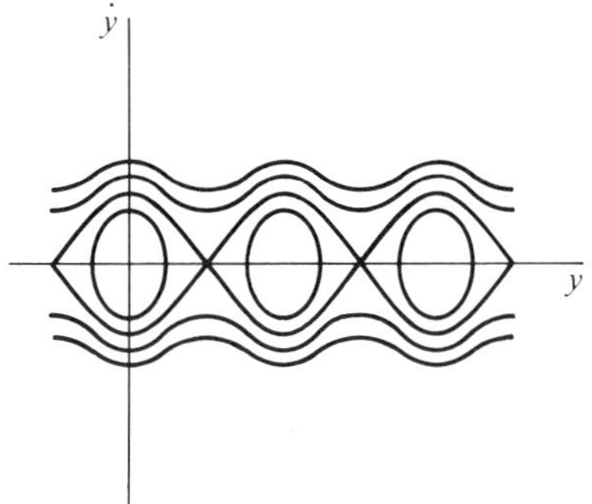

Figure 13

EXERCISES

1. Find all the equilibrium points (x_0, y_0) for each of the following systems. Then find the associated linearized system for each such point. Use Theorem 5.7 to draw what conclusions you can about the behavior of solution trajectories near the equilibrium point. *Warning:* If the linearized system is degenerate $(ad = bc)$, Theorem 5.7 does not apply.

    (a) $dx/dt = y,$ (b) $dx/dt = y,$
       $dy/dt = x + y^2.$ $dy/dt = y + x^2.$
    (c) $dx/dt = -y,$ (d) $dx/dt = x + y^2,$
       $dy/dt = x + x^2.$ $dy/dt = x^2 - y.$
    (e) $dx/dt = x - xy^2,$ (f) $dx/dt = e^x - 1,$
       $dy/dt = x - y^3.$ $dy/dt = x + y.$

2. Each of the following systems has infinitely many equilibrium points. In each case, single out all isolated equilibrium points for which the linearized system in nondegenerate, and draw what conclusions you can about solution trajectories near the equilibrium points.

(a) $dx/dt = 1 - y,$ (b) $dx/dt = x(1 - x^2 - y^2),$
$\quad dy/dt = \sin x.$ $\quad dy/dt = y(1 - x^2 - y^2).$

***3.** In Example 5 of the text we looked at the system

$$\frac{dx}{dt} = y, \qquad \frac{dy}{dt} = -\left(\frac{g}{l}\right)\sin x,$$

derived from the nonlinear pendulum equation. Since $dy/dx = (dy/dt)/(dx/dt)$, the solutions are related by

$$\frac{dy}{dx} = \frac{-g \sin x}{ly}.$$

(a) Use separation of variables to show that solution trajectories satisfy

$$y^2 = (2g/l)\cos x + c, \text{ where } c \text{ is constant.}$$

(b) Show that, if the constant c in part (a) satisfies $|c| < 2g/l$, then the trajectories are closed circuits about the equilibrium points $(x, y) = (2m\pi, 0)$, corresponding to stable oscillations.

(c) Show that if $c > 2g/l$ the trajectories are periodic graphs, unbounded in the x-direction. These trajectories correspond to a perpetual winding motion of the pendulum.

(d) Show that if $c = 2g/l$, the trajectories represent critical paths of motion that approach the unstable equilibrium points.

4. Consider the family of nonlinear systems

$$\frac{dx}{dt} = y + \alpha x(x^2 + y^2),$$

$$\frac{dy}{dt} = -x + \alpha y(x^2 + y^2),$$

where α is constant. Linearized analysis is inadequate for this system.

(a) Show that the only equilibrium point is $(x_0, y_0) = (0, 0)$, regardless of the value of α.

(b) Show that the linearized system associated with $(0, 0)$ is

$$\frac{dx}{dt} = y, \qquad \frac{dy}{dt} = -x,$$

and that $(0, 0)$ is a stable center as $t \to \infty$. Note that $(0, 0)$ is also a stable center for the given system when $\alpha = 0$.

(c) Show that in polar coordinates the given system takes the form

$$\frac{dr}{dt} = \alpha r^3, \qquad \dot{\theta} = -1.$$

[*Hint*: Apply d/dt to the equations $x = r\cos\theta$, $y = r\sin\theta$.]

(d) Solve the polar-form system in part (c), and show that if $\alpha > 0$ the equilibrium point is unstable, and that if $\alpha < 0$ it is stable.

5. The nonlinear pendulum with frictional damping has a displacement angle $\theta = \theta(t)$ that satisfies

$$\ddot{\theta} + \frac{k}{lm}\dot{\theta} + \frac{g}{l}\sin\theta = 0,$$

where $k < 0$ is constant.

(a) Show that the equation for θ is equivalent to the first-order system

$$\dot{x} = y$$

$$\dot{y} = -\frac{g}{l}\sin x - \frac{k}{lm}y.$$

(b) Show that the equilibrium points of the system in part (a) are independent of k.

(c) Show that the unstable equilibrium points are all saddles.

(d) Show that the stable equilibrium points are nodes if $k^2 > 4glm^2$ and asymptotic spirals if $k^2 < 4glm^2$. Why does this distinction make sense physically?

6. The **Lotka–Volterra equations** are discussed in Exercise 9 of the Miscellaneous Applications following Section 3C:

$$\frac{dH}{dt} = (a - bP)H,$$

$$\frac{dP}{dt} = (cH - d)P, \qquad a, b, c, d > 0.$$

(a) Show that for $H > 0$ and $P > 0$, the only equilibrium point is $(H_0, P_0) = (d/c, a/b)$, and find the associated linearized system.

(b) Show that the equilibrium solution of the linearized system is a stable center.

(c) Explain why the result of part (b) is consistent with the results of Exercise 9 referred to above.

(d) Discuss the equilibrium solution of the nonlinear system at $(H, P) = (0, 0)$. Do your conclusions make sense, given the interpretation of P and H as sizes of parasite and host populations, respectively?

7. The **van der Pol equation**

$$\ddot{x} + a(1 - x^2)\dot{x} + x = 0$$

is equivalent to the system

$$\dot{x} = y,$$

$$\dot{y} = -x - a(1 - x^2)y.$$

Discuss the behavior of solutions near $(x_0, y_0) = (0, 0)$ and its dependence on the constant $a > 0$. What happens if $a = 0$ or if $a < 0$?

8. Discuss the nature of the equilibrium point of the system in Exercise 8 of Section 5A, under the assumptions $A > 0$, $B > A^2 + 1$.

5

Linear Systems

1 MATRIX ALGEBRA

Matrices are important in many areas of mathematics, but their use in this book is for solving systems of linear differential equations and systems of linear algebraic equations. The systems

$$\begin{cases} \dfrac{dx}{dt} = 2x + 5y \\[2mm] \dfrac{dy}{dt} = 2x - y \end{cases} \quad \text{and} \quad \begin{cases} 5x + y - z = 0 \\[2mm] x + 2y + 3z = 0 \end{cases}$$

are examples of the two kinds of system we are concerned with. Once we know the general form of these systems, it is clear that they are completely determined by the respective rectangular arrays of numbers

$$\begin{pmatrix} 2 & 5 \\ 2 & -1 \end{pmatrix} \quad \text{and} \quad \begin{pmatrix} 5 & 1 & -1 \\ 1 & 2 & 3 \end{pmatrix}.$$

Any such rectangular array is called a **matrix**. Thus

$$\begin{pmatrix} 1 & 0 \\ 2 & 1 \\ 1 & 1 \end{pmatrix}, \quad \begin{pmatrix} 1 & 2 & 1 & 7 \\ 0 & 0 & 0 & 1 \\ 0 & 1 & 2 & 5 \end{pmatrix}, \quad \begin{pmatrix} 0 \\ 4 \\ 8 \end{pmatrix}, \quad \begin{pmatrix} 0.111 & 0.325 & 5.002 \\ 0.200 & 0.007 & 3.125 \\ 3.001 & 0.555 & 1.007 \end{pmatrix}$$

are examples of matrices. We will sometimes want the entries in a matrix to be functions rather than constants; but since the underlying algebra is the same in both instances, we will restrict ourselves at first to the simpler looking case of constant entries. The horizontal lines of numbers in a matrix are called **rows** and the vertical lines are called **columns**. The numbers of rows and columns determine the **dimensions** of a matrix. Thus the four previous examples have dimensions 3-by-2, 3-by-4, 3-by-1, and 3-by-3. The number of rows is always listed before the number of columns. The 1-by-n matrices are called **n-dimensional row vectors**, and n-by-1 matrices are called **n-dimensional column vectors**. A matrix is **square** if it has the same number of rows as columns.

Two matrices are **equal** if they have the same dimensions and corresponding entries are equal. We will use capital letters to denote matrices and corresponding small letters to denote a typical entry. Thus we may write

$$A = \begin{pmatrix} a_{11} & a_{12} & a_{13} \\ a_{21} & a_{22} & a_{23} \end{pmatrix} \quad \text{or} \quad A = (a_{ij}),$$

where a_{ij}, the ijth entry, is the entry in the ith row and jth column. If there is only one row or one column, we usually omit the unnecessary subscript and use a boldface letter to denote such a row or column:

$$\mathbf{x} = \begin{pmatrix} x_1 \\ x_2 \\ x_3 \end{pmatrix}, \qquad \mathbf{a} = (a_1 \quad a_2 \quad a_3).$$

The set of all n-tuples, whether written vertically or horizontally, is usually denoted by $\mathscr{R}^n$.

1A Sum and Numerical Multiple

Addition and multiplication by a number can be defined for vectors and matrices. If A and B have the same dimensions, then the **sum** $A + B$ is defined to be the matrix having the same dimensions as A and B and with ijth entry equal to $a_{ij} + b_{ij}$ where $A = (a_{ij})$ and $B = (b_{ij})$.

EXAMPLE 1

$$\begin{pmatrix} 1 & 1 \\ 0 & 2 \end{pmatrix} + \begin{pmatrix} -1 & 1 \\ 1 & 2 \end{pmatrix} = \begin{pmatrix} 0 & 2 \\ 1 & 4 \end{pmatrix}.$$

$$\begin{pmatrix} 1 & 2 & 1 \\ -1 & 1 & 0 \end{pmatrix} + \begin{pmatrix} -1 & -2 & -1 \\ 1 & -1 & 0 \end{pmatrix} = \begin{pmatrix} 0 & 0 & 0 \\ 0 & 0 & 0 \end{pmatrix}.$$

Addition is not defined for matrices with different dimensions so we can't add

$$\begin{pmatrix} 1 & 1 \\ 0 & 2 \end{pmatrix} \quad \text{and} \quad \begin{pmatrix} 1 & 2 & 1 \\ -1 & 1 & 0 \end{pmatrix}.$$

For any matrix A and any number r, the **numerical multiple** or **scalar multiple** rA is defined to be the matrix with the same dimensions as A and with ijth the entry ra_{ij}.

EXAMPLE 2

$$-2\begin{pmatrix} 1 & 1 \\ 0 & 2 \end{pmatrix} = \begin{pmatrix} -2 & -2 \\ 0 & -4 \end{pmatrix}.$$

$$3\begin{pmatrix} 1 & 2 & 1 \\ -1 & 1 & 0 \end{pmatrix} = \begin{pmatrix} 3 & 6 & 3 \\ -3 & 3 & 0 \end{pmatrix}.$$

$$\begin{pmatrix} 1 \\ 2 \\ 1 \end{pmatrix} + \begin{pmatrix} 2 \\ 0 \\ -1 \end{pmatrix} - \begin{pmatrix} 3 \\ 3 \\ 3 \end{pmatrix} = \begin{pmatrix} 0 \\ -1 \\ -3 \end{pmatrix}.$$

$$2\begin{pmatrix} 3 \\ -5 \end{pmatrix} - 3\begin{pmatrix} 2 \\ 1 \end{pmatrix} = \begin{pmatrix} 6 \\ -10 \end{pmatrix} + \begin{pmatrix} -6 \\ -3 \end{pmatrix} = \begin{pmatrix} 0 \\ -13 \end{pmatrix}.$$

A sum of numerical multiples of matrices is called a **linear combination** of matrices. Because of the way the matrix operations of addition and numerical multiplication are defined, a linear combination can of course always be written as a single matrix. However, it often happens that we want to expand a matrix into a linear combination. For example, if

$$\mathbf{e}_1 = \begin{pmatrix} 1 \\ 0 \\ 0 \end{pmatrix}, \qquad \mathbf{e}_2 = \begin{pmatrix} 0 \\ 1 \\ 0 \end{pmatrix}, \qquad \mathbf{e}_3 = \begin{pmatrix} 0 \\ 0 \\ 1 \end{pmatrix},$$

then any three-dimensional column can be written as a linear combination of $\mathbf{e}_1$, $\mathbf{e}_2$, $\mathbf{e}_3$:

$$\begin{pmatrix} c_1 \\ c_2 \\ c_3 \end{pmatrix} = c_1\begin{pmatrix} 1 \\ 0 \\ 0 \end{pmatrix} + c_2\begin{pmatrix} 0 \\ 1 \\ 0 \end{pmatrix} + c_3\begin{pmatrix} 0 \\ 0 \\ 1 \end{pmatrix}$$

$$= c_1\mathbf{e}_1 + c_2\mathbf{e}_2 + c_3\mathbf{e}_3.$$

Once it is clearly understood which dimension you are working in, you can use the $\mathbf{e}$-notation without confusion. Let $\mathbf{e}_k$ be the n-dimensional column with 1 in the kth entry and 0 elsewhere. Then any n-dimensional column can be represented uniquely as a linear combination of $\mathbf{e}_1, \ldots, \mathbf{e}_n$. For this reason, $\mathbf{e}_1, \ldots, \mathbf{e}_n$ are called a **basis** for the n-dimensional vectors.

EXAMPLE 3   With $\mathbf{e}_1$, $\mathbf{e}_2$ of dimension 2, we have

$$\begin{pmatrix} 3 \\ 7 \end{pmatrix} = 3\begin{pmatrix} 1 \\ 0 \end{pmatrix} + 7\begin{pmatrix} 0 \\ 1 \end{pmatrix}$$

$$= 3\mathbf{e}_1 + 7\mathbf{e}_2.$$

EXAMPLE 4   A vector function can have a basis representation. For example,

$$\begin{pmatrix} t \\ e^t \\ e^{-t} \end{pmatrix} = t\begin{pmatrix} 1 \\ 0 \\ 0 \end{pmatrix} + e^t\begin{pmatrix} 0 \\ 1 \\ 0 \end{pmatrix} + e^{-t}\begin{pmatrix} 0 \\ 0 \\ 1 \end{pmatrix}$$

$$= t\mathbf{e}_1 + e^t\mathbf{e}_2 + e^{-t}\mathbf{e}_3.$$

Using both addition and numerical multiplication, we can write a linear combination of matrices with the same dimensions. For example, using 2-by-2 matrices,

$$2\begin{pmatrix} 1 & -1 \\ 2 & 3 \end{pmatrix} - 3\begin{pmatrix} 0 & 1 \\ 2 & 1 \end{pmatrix} = \begin{pmatrix} 2 & -2 \\ 4 & 6 \end{pmatrix} + \begin{pmatrix} 0 & -3 \\ -6 & -3 \end{pmatrix}$$

$$= \begin{pmatrix} 2 & -5 \\ -2 & 3 \end{pmatrix}.$$

We write $-A$ for $(-1)A$ and $A - B$ for $A + (-1)B$. Also, for any shape n-by-m, there is a **zero matrix**, denoted by 0, with all entries equal to zero, such that $A + 0 = A$.

Notational warning. When just a single 0 is used to denote the zero matrix, it must be made clear from the context what dimension is intended. For example, if $A = \begin{pmatrix} 1 & 2 \\ 4 & 3 \end{pmatrix}$, then the 0 in $A + 0$ must mean $\begin{pmatrix} 0 & 0 \\ 0 & 0 \end{pmatrix}$, because no other dimensions would allow addition to a 2-by-2 matrix.

As a practical matter, it's important that if vectors $\mathbf{v}_1, \ldots, \mathbf{v}_n$ are used to express a vector $\mathbf{x}$ as a linear combination

$$\mathbf{x} = c_1\mathbf{v}_1 + \ldots + c_n\mathbf{v}_n$$

then the coefficients $c_1, \ldots, c_n$ should be uniquely determined by the vector $\mathbf{x}$. In particular, the zero vector $\mathbf{x} = 0$ should be uniquely expressible, in which case $c_1 = c_2 = \ldots = c_n = 0$. It's easy to see that uniqueness in general follows from this special case because

$$c_1\mathbf{v}_1 + \ldots + c_n\mathbf{v}_n = d_1\mathbf{v}_1 + \ldots + d_n\mathbf{v}_n$$

is equivalent to $(c_1 - d_1)\mathbf{v}_1 + \ldots + (c_n - d_n)\mathbf{v}_n = 0$; thus $c_k - d_k = 0$ for $k = 1, 2, \ldots, n$. Another way of expressing the uniqueness condition on $\mathbf{v}_1, \ldots, \mathbf{v}_n$ is to say that no one of them should be expressible as a linear combination of the remaining ones. If the vectors $\mathbf{v}_1, \ldots, \mathbf{v}_n$ satisfy any one of these equivalent conditions, then the vectors are said to be **linearly independent**; otherwise they are called **linearly dependent**.

EXAMPLE 5   The vectors $\mathbf{e}_1$, $\mathbf{e}_2$, $\mathbf{e}_3$, in $\mathbf{R}^3$ are linearly independent, because

$$c_1\begin{pmatrix} 1 \\ 0 \\ 0 \end{pmatrix} + c_2\begin{pmatrix} 0 \\ 1 \\ 0 \end{pmatrix} + c_3\begin{pmatrix} 0 \\ 0 \\ 1 \end{pmatrix} = \begin{pmatrix} 0 \\ 0 \\ 0 \end{pmatrix}$$

implies $c_1 = c_2 = c_3 = 0$. On the other hand

$$2 \begin{pmatrix} 1 \\ 1 \\ 1 \end{pmatrix} + 3 \begin{pmatrix} 0 \\ 1 \\ 2 \end{pmatrix} - \begin{pmatrix} 2 \\ 5 \\ 8 \end{pmatrix} = \begin{pmatrix} 0 \\ 0 \\ 0 \end{pmatrix}$$

so the three vectors on the left are linearly dependent; the third one is in fact 2 times the first plus 3 times the second.

Additional computations like the one in Example 5 are deferred until Section 6. The underlying idea is referred to occasionally in intervening sections.

1B Matrix Products

First consider an m-by-n matrix A and n-dimensional column $\mathbf{x}$. The **product** $A\mathbf{x}$ is defined to be the m-dimensional column vector equal to the linear combination of the columns of A with coefficients from $\mathbf{x}$, taken in the same order. For example,

$$\begin{pmatrix} a_{11} & a_{12} \\ a_{21} & a_{22} \end{pmatrix} \begin{pmatrix} x_1 \\ x_2 \end{pmatrix} = x_1 \begin{pmatrix} a_{11} \\ a_{21} \end{pmatrix} + x_2 \begin{pmatrix} a_{12} \\ a_{22} \end{pmatrix}$$

$$= \begin{pmatrix} a_{11}x_1 + a_{12}x_2 \\ a_{21}x_1 + a_{22}x_2 \end{pmatrix}.$$

Thus the product $A\mathbf{x} = \mathbf{b}$ represents the system of linear equations

$$a_{11}x_1 + a_{12}x_2 = b_1,$$

$$a_{21}x_1 + a_{22}x_2 = b_2$$

where b_1, b_2 are the entries in the column vector $\mathbf{b}$. Note also that if $\mathbf{x} = \mathbf{e}_j$, with 1 in the jth entry and 0 elsewhere, then $A\mathbf{e}_j$ is just the jth column of A:

$$A\mathbf{e}_j = \begin{pmatrix} a_{1j} \\ \cdot \\ \cdot \\ \cdot \\ a_{mj} \end{pmatrix}$$

where $A = (a_{ij})$.

If B is an n-by-p matrix, the **matrix product** AB is the m-by-p matrix whose columns are the respective products of A with the columns of B. For example,

$$\begin{pmatrix} a_{11} & a_{12} \\ a_{21} & a_{22} \end{pmatrix} \begin{pmatrix} b_{11} & b_{12} \\ b_{21} & b_{22} \end{pmatrix} = \begin{pmatrix} a_{11}b_{11} + a_{12}b_{21} & a_{11}b_{12} + a_{12}b_{22} \\ a_{21}b_{11} + a_{22}b_{21} & a_{21}b_{12} + a_{22}b_{22} \end{pmatrix}.$$

EXAMPLE 6   If

$$A = \begin{pmatrix} 4 & 3 \\ -1 & 2 \\ 1 & -1 \end{pmatrix} \quad \text{and} \quad B = \begin{pmatrix} b_{11} & b_{12} \\ b_{21} & b_{22} \end{pmatrix},$$

then

$$AB = \begin{pmatrix} 4 & 3 \\ -1 & 2 \\ 1 & -1 \end{pmatrix}\begin{pmatrix} b_{11} & b_{12} \\ b_{21} & b_{22} \end{pmatrix} = \begin{pmatrix} 4b_{11} + 3b_{21} & 4b_{12} + 3b_{22} \\ -b_{11} + 2b_{21} & -b_{12} + 2b_{22} \\ b_{11} - b_{21} & b_{12} - b_{22} \end{pmatrix}.$$

If

$$C = \begin{pmatrix} c_{11} & c_{12} \\ c_{21} & c_{22} \end{pmatrix} \quad \text{and} \quad D = \begin{pmatrix} 1 & 2 \\ 4 & 5 \end{pmatrix},$$

then

$$CD = \begin{pmatrix} c_{11} & c_{12} \\ c_{21} & c_{22} \end{pmatrix}\begin{pmatrix} 1 & 2 \\ 4 & 5 \end{pmatrix} = \begin{pmatrix} c_{11} + 4c_{12} & 2c_{11} + 5c_{12} \\ c_{21} + 4c_{22} & 2c_{21} + 5c_{22} \end{pmatrix}.$$

Note that the only requirement on the dimensions of matrices in a product is that the number of columns in the first matrix be the same as the number of rows in the second. Hence we can form the product

$$AD = \begin{pmatrix} 4 & 3 \\ -1 & 2 \\ 1 & -1 \end{pmatrix}\begin{pmatrix} 1 & 2 \\ 4 & 5 \end{pmatrix} = \begin{pmatrix} 16 & 23 \\ 7 & 8 \\ -3 & -3 \end{pmatrix}.$$

The entry in the first row and second column of AD is the ordinary dot product $(4)(2) + (3)(5) = 23$; you should check that the other entries are correct as shown. Matrix multiplication is sometimes called **row-by-column multiplication**, and schematically the process can be illustrated as follows:

$$\begin{pmatrix} * & * & * & * \\ \boxed{* \;\; * \;\; * \;\; *} \\ * & * & * & * \end{pmatrix}\begin{pmatrix} * & * & * & \boxed{*} & * \\ * & * & * & * & * \\ * & * & * & * & * \\ * & * & * & * & * \end{pmatrix} = \begin{pmatrix} * & * & * & * & * & * \\ * & * & * & \boxed{*} & * \\ * & * & * & * & * \end{pmatrix}.$$

EXAMPLE 7   The system of linear algebraic equations

$$\begin{aligned} 3x + 7y &= 1, \\ 2x + 5y &= 2, \end{aligned} \quad \text{or} \quad \begin{pmatrix} 3 & 7 \\ 2 & 5 \end{pmatrix}\begin{pmatrix} x \\ y \end{pmatrix} = \begin{pmatrix} 1 \\ 2 \end{pmatrix}$$

can be written $A\mathbf{x} = \mathbf{c}$, where

$$A = \begin{pmatrix} 3 & 7 \\ 2 & 5 \end{pmatrix}, \quad \mathbf{x} = \begin{pmatrix} x \\ y \end{pmatrix}, \quad \mathbf{c} = \begin{pmatrix} 1 \\ 2 \end{pmatrix}.$$

Similarly, the system of differential equations

$$\frac{dx}{dt} = 3x + 7y - 1, \qquad \text{or} \qquad \begin{pmatrix} \dfrac{dx}{dt} \\ \dfrac{dy}{dt} \end{pmatrix} = \begin{pmatrix} 3 & 7 \\ 2 & 5 \end{pmatrix}\begin{pmatrix} x \\ y \end{pmatrix} - \begin{pmatrix} 1 \\ 2 \end{pmatrix},$$
$$\frac{dy}{dt} = 2x + 5y - 2,$$

can be written $\dfrac{d\mathbf{x}}{dt} = A\mathbf{x} - \mathbf{c}$, since the vector derivative is computed in terms of coordinates by

$$\frac{d\mathbf{x}}{dt} = \frac{d}{dt}\begin{pmatrix} x \\ y \end{pmatrix} = \begin{pmatrix} \dfrac{dx}{dt} \\ \dfrac{dy}{dt} \end{pmatrix}.$$

It is worth remarking that the algebraic system becomes a special case of the differential system if we make the restriction that the solutions be constant, in which case $d\mathbf{x}/dt = 0$. It is easy to see that the algebraic system has solution $x = -9$, $y = 4$. It follows that the differential system has those constant solutions also. Of course, it has many nonconstant solutions in addition. (The constant solutions are called **equilibrium solutions**, and they are important in the discussion of stability in Section 5 of Chapter 4.)

For general questions about matrix products, it is sometimes useful to write the entries in the product of $A = (a_{ij})$ and $B = (b_{ij})$ by using the summation notation. The ijth entry in AB is

$$\sum_{k=1}^{n} a_{ik}b_{kj}.$$

Note that the summation index k runs over the column index of A (that is, along a row) and over the row index of B (that is, along the column).

The next theorem lists the basic properties of matrix products. Of course, we assume that the various matrices have the proper shapes so that the combinations are well defined. The **derivative of a matrix**, and in particular of a vector, is computed entry by entry.

1.1 Theorem

1. $(A + B)C = AC + BC$.
2. $C(A + B) = CA + CB$.
3. $(rA)B = r(AB) = A(rB)$.
4. $A(BC) = (AB)C$.
5. $\dfrac{dAB}{dt} = A\dfrac{dB}{dt} + \dfrac{dA}{dt}B$.

Proof. We prove only property 4 since it is the most complicated; the others are left as exercises. We let A be m-by-n, let B be n-by-p, and let C be p-by-q. Then the ijth entry of BC is $\sum_{k=1}^{p} b_{ik}c_{kj}$, so the ijth entry of $A(BC)$ is

$$\sum_{l=1}^{n} a_{il}\left(\sum_{k=1}^{p} b_{lk}c_{kj}\right) = \sum_{l=1}^{n}\sum_{k=1}^{p} a_{il}b_{lk}c_{kj}.$$

Similarly, the ijth entry of $(AB)C$ is

$$\sum_{k=1}^{p}\left(\sum_{l=1}^{n} a_{il}b_{lk}\right)c_{kj} = \sum_{k=1}^{p}\sum_{l=1}^{n} a_{il}b_{lk}c_{kj}.$$

The two double sums on the right consist of the same terms added in different orders, so the ijth entries in $A(BC)$ and $(AB)C$ are the same.

Formulas 1 and 2 in Theorem 1.1 are called the **distributive laws** for matrix multiplication. Two of them are needed because *matrix multiplication is not in general commutative*; that is, there are square matrices A and B, for which AB and BA both make sense but such that $AB \neq BA$. Exercise 7 gives an example. Formula 4 is the **associative law** for matrix multiplication, and it shows that it makes sense to write the product ABC since the result is independent of the order in which the products are formed. Properties 1, 2 and 3 of Theorem 1.1 express the **linearity** of matrix multiplication; in particular, if $\mathbf{x}$ and $\mathbf{y}$ are n-dimensional column vectors and A is an m-by-n matrix, the two equations

$$A(\mathbf{x} + \mathbf{y}) = A\mathbf{x} + A\mathbf{y}, \qquad A(r\mathbf{x}) = rA\mathbf{x}$$

express the fact that A acts as a **linear function** from $\mathcal{R}^n$ to $\mathcal{R}^m$.

EXAMPLE 8   We verify here that $(A + B)C = AC + BC$ for

$$A = \begin{pmatrix} 1 & 2 \\ -1 & 2 \end{pmatrix}, \qquad B = \begin{pmatrix} 0 & 2 \\ 2 & 1 \end{pmatrix}, \qquad C = \begin{pmatrix} -1 & 1 & 2 \\ 1 & 2 & 1 \end{pmatrix}.$$

We have

$$\left(\begin{pmatrix} 1 & 2 \\ -1 & 2 \end{pmatrix} + \begin{pmatrix} 0 & 2 \\ 2 & 1 \end{pmatrix}\right)\begin{pmatrix} -1 & 1 & 2 \\ 1 & 2 & 1 \end{pmatrix} = \begin{pmatrix} 1 & 4 \\ 1 & 3 \end{pmatrix}\begin{pmatrix} -1 & 1 & 2 \\ 1 & 2 & 1 \end{pmatrix}$$

$$= \begin{pmatrix} 3 & 9 & 6 \\ 2 & 7 & 5 \end{pmatrix}.$$

Also,

$$\begin{pmatrix} 1 & 2 \\ -1 & 2 \end{pmatrix}\begin{pmatrix} -1 & 1 & 2 \\ 1 & 2 & 1 \end{pmatrix} + \begin{pmatrix} 0 & 2 \\ 2 & 1 \end{pmatrix}\begin{pmatrix} -1 & 1 & 2 \\ 1 & 2 & 1 \end{pmatrix} = \begin{pmatrix} 1 & 5 & 4 \\ 3 & 3 & 0 \end{pmatrix} + \begin{pmatrix} 2 & 4 & 2 \\ -1 & 4 & 5 \end{pmatrix}$$

$$= \begin{pmatrix} 3 & 9 & 6 \\ 2 & 7 & 5 \end{pmatrix}.$$

1C Identity Matrices

A square matrix of the form

$$I = \begin{pmatrix} 1 & 0 \\ 0 & 1 \end{pmatrix} \quad \text{or} \quad I = \begin{pmatrix} 1 & 0 & 0 \\ 0 & 1 & 0 \\ 0 & 0 & 1 \end{pmatrix} \quad \text{or} \quad I = \begin{pmatrix} 1 & 0 & \cdots & 0 & 0 \\ 0 & 1 & \cdots & 0 & 0 \\ \cdot & \cdot & & \cdot & \cdot \\ \cdot & \cdot & & \cdot & \cdot \\ \cdot & \cdot & & \cdot & \cdot \\ 0 & 0 & \cdots & 0 & 1 \end{pmatrix}$$

that has 1's on its **main diagonal** and zeros elsewhere is called an **identity matrix**. It has the property that

$$IA = A, \qquad BI = B$$

for any matrices A, B such that the products are defined. Thus it is an identity element for matrix multiplication somewhat as the number 1 is an identity for multiplication of numbers. There is an $n \times n$ identity matrix for every value of n, but as with the zero matrices, it must be made clear from the context what the dimension of an identity matrix is.

EXAMPLE 9   You can check the following.

$$\begin{pmatrix} 1 & 0 \\ 0 & 1 \end{pmatrix}\begin{pmatrix} 1 & 2 & 3 \\ 4 & 5 & 6 \end{pmatrix} = \begin{pmatrix} 1 & 2 & 3 \\ 4 & 5 & 6 \end{pmatrix}\begin{pmatrix} 1 & 0 & 0 \\ 0 & 1 & 0 \\ 0 & 0 & 1 \end{pmatrix} = \begin{pmatrix} 1 & 2 & 3 \\ 4 & 5 & 6 \end{pmatrix}.$$

$$\begin{pmatrix} 1 & 0 & 0 \\ 0 & 1 & 0 \\ 0 & 0 & 1 \end{pmatrix}\begin{pmatrix} 1 & 2 \\ 3 & 4 \\ 5 & 6 \end{pmatrix} = \begin{pmatrix} 1 & 2 \\ 3 & 4 \\ 5 & 6 \end{pmatrix}\begin{pmatrix} 1 & 0 \\ 0 & 1 \end{pmatrix} = \begin{pmatrix} 1 & 2 \\ 3 & 4 \\ 5 & 6 \end{pmatrix}.$$

Notice that, while we use the letter I to denote the identity matrix when it occurs on either side of a given matrix A, these identity matrices will be of different sizes unless A itself is a square matrix. However, we do have, for example,

$$\begin{pmatrix} 1 & 0 \\ 0 & 1 \end{pmatrix}\begin{pmatrix} a & b \\ c & d \end{pmatrix} = \begin{pmatrix} a & b \\ c & d \end{pmatrix}\begin{pmatrix} 1 & 0 \\ 0 & 1 \end{pmatrix} = \begin{pmatrix} a & b \\ c & d \end{pmatrix}.$$

If $\mathbf{x}$ is an n-dimensional column vector, then $I\mathbf{x} = \mathbf{x}$ looks like this when $n = 3$:

$$\begin{pmatrix} 1 & 0 & 0 \\ 0 & 1 & 0 \\ 0 & 0 & 1 \end{pmatrix}\begin{pmatrix} x_1 \\ x_2 \\ x_3 \end{pmatrix} = \begin{pmatrix} x_1 \\ x_2 \\ x_3 \end{pmatrix}.$$

1D Invertibility

If A is a given square matrix and B is a square matrix of the same size such that

$$AB = BA = I,$$

then A is said to be an **invertible** matrix and B is an **inverse** of A. In fact, Exercise 18(c) shows that a matrix A can have at most one inverse, so we can speak of *the* inverse of A and denote it by A^{-1}. Thus, if A is invertible,

$$AA^{-1} = A^{-1}A = I.$$

EXAMPLE 10   If $A = \begin{pmatrix} 1 & 2 \\ 3 & 7 \end{pmatrix}$, then $A^{-1} = \begin{pmatrix} 7 & -2 \\ -3 & 1 \end{pmatrix}$, because

$$AA^{-1} = \begin{pmatrix} 1 & 2 \\ 3 & 7 \end{pmatrix}\begin{pmatrix} 7 & -2 \\ -3 & 1 \end{pmatrix} = \begin{pmatrix} 1 & 0 \\ 0 & 1 \end{pmatrix}$$

and

$$A^{-1}A = \begin{pmatrix} 7 & -2 \\ -3 & 1 \end{pmatrix}\begin{pmatrix} 1 & 2 \\ 3 & 7 \end{pmatrix} = \begin{pmatrix} 1 & 0 \\ 0 & 1 \end{pmatrix}.$$

In general, any 2-by-2 matrix $\begin{pmatrix} a & b \\ c & d \end{pmatrix}$ is invertible if $ad - bc$, called the **determinant** of the matrix, is not zero. In that case

1.2
$$\begin{pmatrix} a & b \\ c & d \end{pmatrix}^{-1} = \frac{1}{ad - bc}\begin{pmatrix} d & -b \\ -c & a \end{pmatrix}.$$

In particular,

$$\begin{pmatrix} 1 & 3 \\ -1 & 2 \end{pmatrix}^{-1} = \tfrac{1}{5}\begin{pmatrix} 2 & -3 \\ 1 & 1 \end{pmatrix} = \begin{pmatrix} \tfrac{2}{5} & -\tfrac{3}{5} \\ \tfrac{1}{5} & \tfrac{1}{5} \end{pmatrix}.$$

Formula 1.2 is worth remembering, and we leave it as an exercise to show that the formula is correct.

If A is an n-by-n matrix, then the matrix equation $A\mathbf{x} = \mathbf{b}$ is equivalent to a system of n linear equations in n unknowns. If A happens to be an invertible matrix with inverse A^{-1}, then we can solve the system in matrix form by multiplying both sides on the left by A^{-1} to get

$$A^{-1}A\mathbf{x} = A^{-1}\mathbf{b}.$$

Since $A^{-1}A = I$, we have $A^{-1}A\mathbf{x} = I\mathbf{x} = \mathbf{x}$, so the equation becomes

$$\mathbf{x} = A^{-1}\mathbf{b}.$$

In other words, $\mathbf{x} = A^{-1}\mathbf{b}$ is a solution to the given equation, and in fact is the only solution, because $\mathbf{x}$ could have been an arbitrary vector satisfying $A\mathbf{x} = \mathbf{b}$.

EXAMPLE 11   The system

$$\begin{array}{rcr} x + 2y &=& 3 \\ 3x + 7y &=& -4 \end{array} \quad \text{is equivalent to} \quad \begin{pmatrix} 1 & 2 \\ 3 & 7 \end{pmatrix}\begin{pmatrix} x \\ y \end{pmatrix} = \begin{pmatrix} 3 \\ -4 \end{pmatrix}.$$

By Formula 1.2,

$$\begin{pmatrix} 1 & 2 \\ 3 & 7 \end{pmatrix}^{-1} = \begin{pmatrix} 7 & -2 \\ -3 & 1 \end{pmatrix},$$

so we multiply the left side of the equation by the inverse to get

$$\begin{pmatrix} 7 & -2 \\ -3 & 1 \end{pmatrix}\begin{pmatrix} 1 & 2 \\ 3 & 7 \end{pmatrix}\begin{pmatrix} x \\ y \end{pmatrix} = \begin{pmatrix} 1 & 0 \\ 0 & 1 \end{pmatrix}\begin{pmatrix} x \\ y \end{pmatrix} = \begin{pmatrix} x \\ y \end{pmatrix}.$$

Hence multiplying also on the right gives

$$\begin{pmatrix} x \\ y \end{pmatrix} = \begin{pmatrix} 7 & -2 \\ -3 & 1 \end{pmatrix}\begin{pmatrix} 3 \\ -4 \end{pmatrix} = \begin{pmatrix} 29 \\ -13 \end{pmatrix}.$$

Thus $(x, y) = (29, -13)$ is the unique solution.

1E Powers of a Square Matrix

If A is a square matrix, it can be multiplied by itself repeatedly, and we define

$$A^2 = AA, \quad A^3 = AAA = AA^2, \ldots, A^n = AA^{n-1}.$$

These powers of A all have the same dimension as A, but note that if B is a nonsquare matrix then B^2 never makes sense.

EXAMPLE 12   If $A = \begin{pmatrix} 2 & 1 \\ 0 & 3 \end{pmatrix}$, then

$$A^2 = \begin{pmatrix} 2 & 1 \\ 0 & 3 \end{pmatrix}\begin{pmatrix} 2 & 1 \\ 0 & 3 \end{pmatrix} = \begin{pmatrix} 4 & 5 \\ 0 & 9 \end{pmatrix},$$

$$A^3 = \begin{pmatrix} 2 & 1 \\ 0 & 3 \end{pmatrix}\begin{pmatrix} 4 & 5 \\ 0 & 9 \end{pmatrix} = \begin{pmatrix} 8 & 19 \\ 0 & 27 \end{pmatrix}, \ etc.$$

EXERCISES

1. Given that

$$A = \begin{pmatrix} -1 & 2 \\ 0 & 1 \end{pmatrix}, \quad B = \begin{pmatrix} 1 & 4 \\ 1 & 1 \end{pmatrix}, \quad \mathbf{x} = \begin{pmatrix} 1 \\ 2 \end{pmatrix}, \quad \mathbf{b} = \begin{pmatrix} 1 \\ 2 \end{pmatrix},$$

$$C = \begin{pmatrix} 1 & 1 & 1 \\ 0 & 1 & 1 \\ 0 & 0 & 1 \end{pmatrix}, \quad D = \begin{pmatrix} 1 & 0 & 1 \\ 1 & 0 & 1 \\ 2 & 2 & 2 \end{pmatrix}, \quad \mathbf{y} = \begin{pmatrix} 1 \\ 2 \\ 3 \end{pmatrix}, \quad \mathbf{d} = \begin{pmatrix} 1 \\ 1 \\ 1 \end{pmatrix},$$

compute

(a) AB.

(b) $A\mathbf{x} + \mathbf{b}$.

(c) $BA + B^2$.

(d) CD.

(e) $C\mathbf{y} + \mathbf{d}$.

(f) $DC + C^2$.

2. For each of the following systems find a matrix A, a column vector $\mathbf{x}$, and a column vector $\mathbf{b}$ such that the system can be written in the form $A\mathbf{x} = \mathbf{b}$.

(a) $x - y = 1,$
$\quad x + y = 2.$

(b) $2x + 3y = 0,$
$\quad x + 3y = 0.$

(c) $x + y = 1,$
$\quad\quad\quad y = 1.$

(d) $x + y + z = 0,$
$\quad x + y - z = 1,$
$\quad x - y - z = 0.$

(e) $x + y = 0,$
$\quad y + z = 0,$
$\quad x - z = 1.$

(f) $u + v - w = 1,$
$\quad u - v + w = 2.$

3. For each of the following systems of differential equations, find a matrix A, a column vector $\mathbf{x}$, and a column vector $\mathbf{b}$ such that the system can be written in the form $d\mathbf{x}/dt = A\mathbf{x} + \mathbf{b}$. Note that in general the entries in A, $\mathbf{b}$, and $\mathbf{x}$, will depend on t.

(a) $dx/dt = 2x + 3y,$
$\quad dy/dt = 2x - 4y.$

(b) $dx/dt = x + y + t,$
$\quad dy/dt = x - y - t.$

(c) $dx/dt = x + y - z,$
$\quad dy/dt = x - y + z,$
$\quad dz/dt = x + ty.$

(d) $du/dt = tu - tv + w,$
$\quad dv/dt = u + v,$
$\quad dw/dt = u + v + t.$

4. Find matrices A and vectors $\mathbf{b}$ such that each of the following systems of equations can be written in the form $\dot{\mathbf{x}} = A\mathbf{x} + \mathbf{b}$, where $\mathbf{x}$ is an n-by-1 column vector of the appropriate dimension.

(a) $\quad \dot{x} - 2y = 1,$
$\quad x + 3\dot{y} = 2.$

(b) $\quad\quad \dot{x} = 1,$
$\quad\quad\quad \dot{y} = 2.$

(c) $\dot{x} - \dot{y} = 0,$
$\quad \dot{x} + \dot{y} = 0.$

(d) $\dot{x} - y + z = 1,$
$\quad x - \dot{y} - z = 0,$
$\quad x + y + \dot{z} = 0.$

(e) $\dot{x} - 2y = 0,$
$\quad \dot{y} + z = 0,$
$\quad x + \dot{z} = 1.$

5. Express each of the following vector functions as a linear combination of $\mathbf{e}_1$ and $\mathbf{e}_2$, in the case of dimension 2, or $\mathbf{e}_1$, $\mathbf{e}_2$, and $\mathbf{e}_3$ in the case of dimension 3.

(a)
$$f(t) = \begin{pmatrix} e^t \\ e^t \\ t^2 \end{pmatrix}.$$

(b)
$$f(t) = \begin{pmatrix} t^2 \\ t^3 \\ t^4 \end{pmatrix}.$$

(c)
$$f(t) = \begin{pmatrix} t - 1 \\ t - 1 \\ t + 2 \end{pmatrix}.$$

(d) $f(t) = \begin{pmatrix} 1 \\ t \end{pmatrix}.$

(e) $f(t) = \begin{pmatrix} \cos t \\ \sin t \end{pmatrix}.$

(f) $f(t) = \begin{pmatrix} 1 \\ 1 \end{pmatrix}.$

6. Recall that the derivative of a vector-valued function is computed by elementwise differentiation of the entries. The same is true for matrices, for example,

$$\frac{d}{dt} \begin{pmatrix} 1 & t \\ t & 2 \end{pmatrix} = \begin{pmatrix} 0 & 1 \\ 1 & 0 \end{pmatrix}.$$

Compute the derivative $dA(t)/dt$ for each of the following matrices $A(t)$.

(a) $A(t) = \begin{pmatrix} e^t & te^t \\ 0 & e^t \end{pmatrix}.$

(b) $A(t) = \begin{pmatrix} t^2 & 2t \\ 3 & 4 \end{pmatrix}\begin{pmatrix} t & 2t \\ t & 3t \end{pmatrix}.$

(c) $A(t) = \begin{pmatrix} e^t & 0 & 0 \\ 0 & e^{2t} & 0 \\ 0 & 0 & e^{3t} \end{pmatrix}.$

(d) $A(t) = \begin{pmatrix} t & t^2 & t^3 \\ 1 & t^3 & t^4 \\ 1 & 1 & t^5 \end{pmatrix}.$

7. Let $U = \begin{pmatrix} -1 & 2 \\ 2 & -4 \end{pmatrix}$, $V = \begin{pmatrix} 2 & 6 \\ 1 & 3 \end{pmatrix}$. Compute UV and VU. Are they the same? Is it possible for the product of two matrices to be zero without either factor being zero?

8. (a) Using the matrix C of Exercise 1, compute the vectors $C\mathbf{e}_1$, $C\mathbf{e}_2$, $C\mathbf{e}_3$, where the $\mathbf{e}_k$ are the natural basis vectors in $\mathcal{R}^3$.

(b) Show that if A is an m-by-n matrix, then $A\mathbf{e}_j$ is the jth column of A, where the $\mathbf{e}_k$ are the natural basis vectors in $\mathcal{R}^n$.

9. Compute the products.

(a) $\begin{pmatrix} 1 & 2 & 3 \\ 4 & 5 & 6 \\ 7 & 8 & 9 \end{pmatrix} \begin{pmatrix} 0 \\ 1 \\ 0 \end{pmatrix}$.

(b) $\begin{pmatrix} 0 & 1 & 1 \\ 1 & 0 & 1 \\ 1 & 1 & 0 \end{pmatrix} \begin{pmatrix} 2 \\ 1 \\ 3 \end{pmatrix}$.

(c) $(2 \quad 1 \quad 4) \begin{pmatrix} 3 \\ 5 \\ 7 \end{pmatrix}$.

(d) $\begin{pmatrix} 3 \\ 5 \\ 7 \end{pmatrix} (2 \quad 1 \quad 4)$.

(e) $\begin{pmatrix} 2 & 1 \\ 5 & 6 \\ 3 & 4 \end{pmatrix} \begin{pmatrix} 1 & -1 & 1 \\ -1 & 1 & -1 \end{pmatrix}$.

(f) $\begin{pmatrix} 2 & 0 & 0 \\ 0 & 4 & 0 \\ 0 & 0 & 5 \end{pmatrix} \begin{pmatrix} -1 \\ 1 \\ -1 \end{pmatrix}$.

10. Prove Equations 1, 2, 3, 5 of Theorem 1.1.

11. If A is a square matrix, it can be multiplied by itself, and we can define $A^2 = AA$, $A^3 = AAA = A^2A$, $A^n = AA \ldots A$ (n factors). These powers of A all have the same dimension as A. Find A^2 and A^3 if

(a) $A = \begin{pmatrix} 2 & 1 \\ 0 & 1 \end{pmatrix}$.

(b) $A = \begin{pmatrix} 1 & 0 & -1 \\ -1 & 0 & 1 \\ 2 & 1 & -1 \end{pmatrix}$.

(c) $A = \begin{pmatrix} 2 & 0 \\ 1 & 1 \end{pmatrix}$.

(d) $A = \begin{pmatrix} 2 & 0 \\ 0 & 3 \end{pmatrix}$.

[Note that 0 is the only number whose cube is 0. Part (b) of this problem thus illustrates another difference between the arithmetic of numbers and of matrices.]

12. The numerical equation $a^2 = 1$ has $a = 1$ and $a = -1$ as its only solutions.

(a) Show that if $A = I$ or $-I$, then $A^2 = I$, where I is an identity matrix of any dimension.

(b) Show that $\begin{pmatrix} a & b \\ c & -a \end{pmatrix}^2 = \begin{pmatrix} 1 & 0 \\ 0 & 1 \end{pmatrix}$ if $a^2 + bc = 1$; so the equation $A^2 = I$ has infinitely many different solutions in the set of 2-by-2 matrices.

(c) Show that every 2-by-2 matrix A for which $A^2 = I$ is either I, $-I$, or one of the matrices described in part (b).

13. (a) Express the matrix $\begin{pmatrix} 1 & -1 \\ 1 & 0 \end{pmatrix}$ as a linear combination of the matrices $\begin{pmatrix} 2 & -1 \\ 0 & 1 \end{pmatrix}$ and $\begin{pmatrix} 5 & -3 \\ 1 & 2 \end{pmatrix}$.

(b) Can an arbitrary 2-by-2 matrix be expressed as a linear combination of the last two matrices in part (a)?

14. Verify the matrix multiplications in Example 9 of the text.

15. Show that if A is an n-by-n matrix and I is the n-by-n identity matrix, then

(a) $(A - I)(A + I) = A^2 - I$.

(b) $(A + I)^2 = A^2 + 2A + I$.

(c) Show that, if B is also n-by-n, it is not true in general that $(A - B)(A + B) = A^2 - B^2$ or that $(A + B)^2 = A^2 + 2AB + B^2$.

16. Find inverses for those of the following 2-by-2 matrices that have inverses.

(a) $\begin{pmatrix} 1 & 1 \\ 1 & 2 \end{pmatrix}$.

(b) $\begin{pmatrix} 3 & 6 \\ 2 & 4 \end{pmatrix}$.

(c) $\begin{pmatrix} \frac{1}{2} & \frac{1}{4} \\ \frac{1}{4} & \frac{1}{5} \end{pmatrix}.$

(d) $\begin{pmatrix} -7 & -5 \\ 12 & 9 \end{pmatrix}.$

17. Solve the matrix equation $A\mathbf{x} = \mathbf{b}$ by multiplying both sides by A^{-1} with

(a) $A = \begin{pmatrix} 2 & -1 \\ 3 & 4 \end{pmatrix};\ \mathbf{b} = \begin{pmatrix} 1 \\ 1 \end{pmatrix}.$

(b) $A = \begin{pmatrix} 7 & 2 \\ 1 & 1 \end{pmatrix};\ \mathbf{b} = \begin{pmatrix} -2 \\ 4 \end{pmatrix}.$

18. (a) Show that if A and B are invertible matrices of the same dimension, then AB is invertible and $(AB)^{-1} = B^{-1}A^{-1}$.

(b) Show that, if $A_1, \ldots, A_n$ are invertible matrices with the same dimension, then the matrix product $A_1A_2 \ldots A_n$ is invertible and $(A_1A_2 \ldots A_n)^{-1} = A_n^{-1} \ldots A_2^{-1}A_1^{-1}$.

(c) Show that if $AB = BA = I$ and $AC = CA = I$, then $B = C$.

1F Determinants

For a square matrix A, the determinant is a numerical-valued function written det A. However, for a displayed matrix it is also customary just to replace the parentheses by vertical bars. Thus the notations

$$\begin{vmatrix} 1 & 4 & 5 \\ 6 & 7 & -3 \\ -2 & 1 & 0 \end{vmatrix}, \qquad \begin{vmatrix} a & b \\ c & d \end{vmatrix}$$

mean the same as

$$\det\begin{pmatrix} 1 & 4 & 5 \\ 6 & 7 & -3 \\ -2 & 1 & 0 \end{pmatrix}, \qquad \det\begin{pmatrix} a & b \\ c & d \end{pmatrix}.$$

Our definition of determinant will be inductive; that is, we will define det A first for 1-by-1 matrices, and then for each n define the determinant of an n-by-n matrix in terms of determinants of certain $(n-1)$-by-$(n-1)$ matrices called minors. For any matrix A, the matrix obtained by deleting the ith row and jth column of A is called the ijth **minor** of A and is denoted by A_{ij}. Recall that we use the small letter a_{ij} to denote the ijth entry of a matrix A. Thus the ijth minor A_{ij} corresponds to the entry a_{ij} in a natural way, because the minor is obtained by deleting the row and column containing a_{ij}.

EXAMPLE 13   Let

$$A = \begin{pmatrix} -5 & -6 & 7 \\ 8 & -9 & 0 \\ -3 & 4 & 2 \end{pmatrix}, \qquad B = \begin{pmatrix} 1 & 2 \\ 3 & 4 \end{pmatrix}.$$

Then some examples of entries and corresponding minors are

$$a_{11} = -5, \qquad A_{11} = \begin{pmatrix} -9 & 0 \\ 4 & 2 \end{pmatrix},$$

$$a_{23} = 0, \qquad A_{23} = \begin{pmatrix} -5 & -6 \\ -3 & 4 \end{pmatrix},$$

$$b_{11} = 1, \qquad B_{11} = (4),$$

$$b_{12} = 2, \qquad B_{12} = (3).$$

We can now give the definition of **determinant**. For a 1-by-1 matrix, $A = (a)$, we define

$$\det A = a.$$

For an n-by-n matrix, $A = (a_{ij})$, $i, j = 1, \ldots, n$, we define

1.3 $\qquad \det A = a_{11} \det A_{11} - a_{12} \det A_{12} + \cdots - (-1)^n a_{1n} \det A_{1n}.$

The definition is *inductive* in the sense that defining the determinant of an n-by-n matrix A requires us to know the determinants of the $(n - 1)$-by-$(n - 1)$ minors A_{ij}. But the simple definition for the 1-by-1 case allows us to go on to 2-by-2, then 3-by-3, and so on. In words, the formula says that $\det A$ is the sum, with alternating signs, of the elements of the first row of A, each multiplied by the determinant of its corresponding minor. For this reason, the numbers

$$\det A_{11}, \; -\det A_{12}, \; \ldots, \; (-1)^{n+1} \det A_{1n}$$

are called the *cofactors* of the corresponding elements of the first row of A. In general, the **cofactor** of the entry a_{ij} in A is defined to be $(-1)^{i+j} \det A_{ij}$. Thus, in Example 13 the entry $a_{23} = 0$ in the matrix A has cofactor

$$(-1)^{2+3} \det \begin{pmatrix} -5 & -6 \\ -3 & 4 \end{pmatrix} = 38.$$

The factor $(-1)^{i+j}$ associates plus and minus signs with $\det A_{ij}$ according to the pattern.

$$\begin{pmatrix} + & - & + & - & \cdots \\ - & + & - & + & \cdots \\ + & - & + & - & \cdots \\ - & + & - & + & \cdots \\ \cdot & \cdot & \cdot & \cdot & \cdot \\ \cdot & \cdot & \cdot & \cdot & \cdot \\ \cdot & \cdot & \cdot & \cdot & \cdot \end{pmatrix}.$$

EXAMPLE 14

(a) $\det \begin{pmatrix} 1 & 2 \\ 3 & 4 \end{pmatrix} = (1)(4) - (2)(3) = 4 - 6 = -2.$

(b) $\det \begin{pmatrix} -5 & -6 & 7 \\ 8 & -9 & 0 \\ -3 & 4 & 2 \end{pmatrix} = -5 \det \begin{pmatrix} -9 & 0 \\ 4 & 2 \end{pmatrix} - (-6) \det \begin{pmatrix} 8 & 0 \\ -3 & 2 \end{pmatrix}$

$$+ 7 \det\begin{pmatrix} 8 & -9 \\ -3 & 4 \end{pmatrix}$$
$$= (-5)(-18-0) + (6)(16 - 0)$$
$$+ 7(32 - 27)$$
$$= 90 + 96 + 35 = 221.$$

(c) $\det\begin{pmatrix} a & b \\ c & d \end{pmatrix} = ad - bc.$

The result of the last example is worth remembering as a rule of calculation. *The determinant of a 2-by-2 matrix is the product of the entries on the main diagonal minus the product of the other two entries.* Thus 2-by-2 determinants can often be computed mentally, and 3-by-3 determinants in one or two lines.

It is an important fact, the proof of which we omit, that if in Equation 1.3 the elements and cofactors of the first row are replaced by the elements and cofactors of any other row, or of any column, then the expansion is still valid. Formally, the statement can be expressed as follows.

1.4 Theorem.   If A is a square matrix, then

$$\det A = \sum_{j=1}^{n} (-1)^{i+j} a_{ij} \det A_{ij} \qquad\qquad \textbf{expansion by }i\textbf{th row}$$

and

$$\det A = \sum_{i=1}^{n} (-1)^{i+j} a_{ij} \det A_{ij} \qquad\qquad \textbf{expansion by }j\textbf{th column}$$

Notice that Equation 1.3, which we used to define determinant, appears as a special case of the first equation in the theorem when we let $i = 1$. The alternating pattern of cofactor signs applies to all expansions by row or column.

EXAMPLE 15   The determinant computation in part (b) of Example 13 can be done also by using the elements and cofactors of the second row:

$$\det\begin{pmatrix} -5 & -6 & 7 \\ 8 & -9 & 0 \\ -3 & 4 & 2 \end{pmatrix}$$
$$= -(8) \det\begin{pmatrix} -6 & 7 \\ 4 & 2 \end{pmatrix} + (-9) \det\begin{pmatrix} -5 & 7 \\ -3 & 2 \end{pmatrix} - (0) \det\begin{pmatrix} -5 & -6 \\ -3 & 4 \end{pmatrix}$$
$$= -8(-12 - 28) - 9(-10 + 21)$$
$$= 221.$$

Computing the same determinant using the elements and cofactors of the third column gives

$$\det\begin{pmatrix} -5 & -6 & 7 \\ 8 & -9 & 0 \\ -3 & 4 & 2 \end{pmatrix}$$

$$= 7 \det\begin{pmatrix} 8 & -9 \\ -3 & 4 \end{pmatrix} - (0) \det\begin{pmatrix} -5 & -6 \\ -3 & 4 \end{pmatrix} + 2 \det\begin{pmatrix} -5 & -6 \\ 8 & -9 \end{pmatrix}$$

$$= 7(32 - 27) + 2(45 + 48)$$

$$= 221.$$

The following theorem shows the effect on det A of the row operations we use to solve linear systems. The proof is an easy consequence of the definition of det A (see Exercise 16).

1.5 Theorem.   Let A be a square matrix. Then

1. Multiplying a row, or a column, of A by a number r multiplies det A by r.
2. Adding a multiple of one row to another leaves det A unchanged; likewise for columns.
3. Interchanging two rows or two columns changes the sign of det A.

Thus, in putting a matrix in another form B, we just need to keep track of the row multipliers r and the sign changes that occur when rows are interchanged. Then k det $A = \det B$, where k is plus or minus the product of the row multipliers. A little thought shows that the row operations just listed can be used to alter a matrix so that the first nonzero entry in each row is 1 and so that a column containing a 1 has all its other entries equal to zero. Such a matrix R is said to be in **reduced form**, and det R is either 0, 1, or -1. An algorithm, suitable for machine computation, that puts a matrix in reduced form is described in Section 6 of this chapter.

EXAMPLE 16   Let

$$A = \begin{pmatrix} 1 & 3 & -2 \\ 2 & -4 & 1 \\ 3 & 5 & -2 \end{pmatrix}, \qquad C = \begin{pmatrix} 1 & 3 & 0 \\ 2 & -4 & 5 \\ 3 & 5 & 4 \end{pmatrix}.$$

The third column of C is equal to the third column of A plus 2 times the first column. We compute

$$\det C = (1)(-16 - 25) - (3)(8 - 15) + (0)(10 + 12) = -20.$$

It follows that det $A = -20$ also.

EXAMPLE 17   Let

$$A = \begin{pmatrix} 2 & 4 & -1 & 0 \\ 3 & 0 & 2 & 3 \\ -1 & 2 & 3 & 1 \\ 0 & 1 & -2 & -1 \end{pmatrix}.$$

By adding 2 times column 3 to column 1, and 4 times column 3 to column 2, we obtain

$$B = \begin{pmatrix} 0 & 0 & -1 & 0 \\ 7 & 8 & 2 & 3 \\ 5 & 14 & 3 & 1 \\ -4 & -7 & -2 & -1 \end{pmatrix},$$

and by Theorem 1.5(2), $\det A = \det B$. The expansion of $\det B$ by the first row has only one nonzero term, and we get

$$\det B = (-1) \det \begin{pmatrix} 7 & 8 & 3 \\ 5 & 14 & 1 \\ -4 & -7 & -1 \end{pmatrix}$$

$$= -\det \begin{pmatrix} 7 & 1 & 3 \\ 5 & 9 & 1 \\ -4 & -3 & -1 \end{pmatrix} \qquad \begin{array}{l}[\text{subtract column 1} \\ \text{from column 2}]\end{array}$$

$$= -\det \begin{pmatrix} 0 & 1 & 0 \\ -58 & 9 & -26 \\ 17 & -3 & 8 \end{pmatrix} \qquad \begin{array}{l}[\text{subtract 7 times} \\ \text{column 2 from} \\ \text{column 1 and 3} \\ \text{times column 2} \\ \text{from column 3}].\end{array}$$

Then $\det B = -(-1)((-58)(8) - (17)(-26)) = -22$.

The product of two square matrices is again a square matrix. It is a remarkable fact that the determinant of the product equals the product of the determinants of the individual matrices. The theorem is called the **product rule** for determinants.

The proof is fairly long, and we omit it.

1.6 Theorem.   If A and B are square matrices of the same size, then

$$\det (AB) = (\det A)(\det B).$$

A consequence of the product rule is that, for an invertible matrix A,

$$\det (A^{-1}) = \frac{1}{\det A}.$$

There is a determinant formula for the inverse of an invertible matrix that generalizes the very simple Equation 1.2 that holds for 2-by-2 matrices. The formula is reasonably efficient for simple 3-by-3 or even 4-by-4 matrices. For large matrices, it is usually preferable to use row operations to compute a determinant. However, the inversion formula allows us to see some general facts about determinants that are awkward to derive in other ways. We first prove a theorem from which the facts about inverses follow easily.

1.7 Theorem.   For an n-by-n matrix A,

$$\sum_{i=1}^{n} (-1)^{i+j} a_{ik} \det A_{ij} = \begin{cases} \det A & \text{if } k = j, \\ 0 & \text{if } k \neq j, \end{cases}$$

and

$$\sum_{j=1}^{n} (-1)^{i+j} a_{kj} \det A_{ij} = \begin{cases} \det A & \text{if } k = i, \\ 0 & \text{if } k \neq i. \end{cases}$$

Proof.   Consider the expression

$$\sum_{i=1}^{n} (-1)^{i+j} x_i \det A_{ij},$$

where $x_1, \ldots, x_n$ may be any n numbers. We see that this is a certain determinant; in fact, it is the expansion, using the jth column, of the matrix obtained from A by replacing the jth column by $x_1, \ldots, x_n$. Now consider what happens if we take $x_1 = a_{1k}, x_2 = a_{2k}, \ldots, x_n = a_{nk}$, in other words, if we enter the elements from the kth column in the jth column. If $k = j$, we just get $\det A$. If $k \neq j$, we have the determinant of a matrix with two columns (the jth and kth) identical, so the result is zero. (Exercise 14.) This proves the first equation; the second is proved similarly by reversing the roles of row and column.

At this point we need the idea of transposing a square matrix, that is, reflecting it across its main diagonal. The matrix A and its **transpose** A' are related as shown:

$$A = \begin{pmatrix} a_{11} & a_{12} & \cdots \\ a_{21} & a_{22} & \cdots \\ \vdots & \vdots & \end{pmatrix}; \qquad A' = \begin{pmatrix} a_{11} & a_{21} & \cdots \\ a_{12} & a_{22} & \cdots \\ \vdots & \vdots & \end{pmatrix}.$$

Thus forming the transpose interchanges rows and columns.

Now the number $(-1)^{i+j} \det A_{ij}$ is the ijth cofactor of A; we will abbreviate the cofactor $\tilde{a}_{ij}$ and write $\tilde{A}$ for the matrix with entries $\tilde{a}_{ij}$. Using this notation, we can interpret the previous theorem as a statement about the matrix products $\tilde{A}'A$ and $A\tilde{A}'$. In fact, we have

1.8 $$\tilde{A}'A = A\tilde{A}' = (\det A)I,$$

because the jkth entry in the product $\tilde{A}'A$ is the product of the jth row of $\tilde{A}'$ (i.e., the jth column of $\tilde{A}$) and the kth column of A, in other words, the sum $\sum_{i=1}^{n} \tilde{a}_{ij} a_{ik}$. But by the previous theorem this sum is equal to det A if $k = j$ and 0 otherwise. Hence $\tilde{A}^t A$ is equal to $(\det A)I$, a numerical multiple of the identity matrix. A similar calculation using the second equation in Theorem 1.7 shows that $A\tilde{A}^t = (\det A)I$ also. Thus we have proved Formula 1.8.

We can now easily write down a formula for A^{-1} when $\det A \neq 0$.

1.9 Theorem.   If $\det A \neq 0$, then A is invertible, and

$$A^{-1} = \frac{1}{\det A}\tilde{A}^t,$$

where $\tilde{A}^t$ is the transpose of the matrix of cofactors of A. Conversely, if A is invertible, then $\det A \neq 0$.

Proof. If $\det A \neq 0$, then Formula 1.6 can be written

$$\left(\frac{1}{\det A}\tilde{A}^t\right)A = A\left(\frac{1}{\det A}\tilde{A}^t\right) = I.$$

Hence A^{-1} exists and equals $(\det A)^{-1}\tilde{A}^t$. Conversely, if A is invertible, then there is a matrix B such that $AB = BA = I$. By the product rule for determinants,

$$(\det A)(\det B) = \det I = 1,$$

so neither $\det A$ nor $\det B$ can be zero.

EXAMPLE 18   We can use Theorem 1.9 to compute the inverse of the matrix

$$A = \begin{pmatrix} 2 & 3 & 4 \\ 5 & 6 & 7 \\ 8 & 9 & 0 \end{pmatrix}.$$

The matrix with entries $\det A_{ij}$ is easily computed to be

$$\begin{pmatrix} -63 & -56 & -3 \\ -36 & -32 & -6 \\ -3 & -6 & -3 \end{pmatrix}.$$

To get the matrix of cofactors, we insert the factor $(-1)^{i+j}$. This changes the sign of every second entry, giving

$$\tilde{A} = \begin{pmatrix} -63 & 56 & -3 \\ 36 & -32 & 6 \\ -3 & 6 & -3 \end{pmatrix}.$$

The transposed matrix is

$$\tilde{A}^t = \begin{pmatrix} -63 & 36 & -3 \\ 56 & -32 & 6 \\ -3 & 6 & -3 \end{pmatrix}.$$

The cofactors in $\tilde{A}$ can be used to expand det A using the last column. We expand det A by its last column to get

$$\det A = 4(-3) + 7(6) + 0(-3) = 30.$$

Finally,

$$A^{-1} = \frac{1}{\det A}\tilde{A}^t = \begin{pmatrix} -\frac{63}{30} & \frac{36}{30} & -\frac{3}{30} \\ \frac{56}{30} & -\frac{32}{30} & \frac{6}{30} \\ -\frac{3}{30} & \frac{6}{30} & -\frac{3}{30} \end{pmatrix}.$$

EXERCISES

1. Find AB, BA, and the determinants of A, B, AB, and BA when

   (a) $A = \begin{pmatrix} 1 & -2 \\ 3 & 1 \end{pmatrix}$, $B = \begin{pmatrix} 0 & 1 \\ 2 & -3 \end{pmatrix}$.

   (b) $A = \begin{pmatrix} 2 & 0 & 0 \\ 0 & 3 & 0 \\ 0 & 0 & 4 \end{pmatrix}$, $B = \begin{pmatrix} -1 & 0 & 1 \\ 2 & -1 & -3 \\ 0 & 3 & 5 \end{pmatrix}$.

2. For the matrices in Exercise 1(b), what are det $(2A)$ and det $(2B)$?

3. A **diagonal matrix** D is a square matrix such that only the main diagonal entries $d_i = d_{ii}$ are allowed to be nonzero. We sometimes write $D = \operatorname{diag}(d_1, \ldots , d_n)$. Show that det $D = d_1, d_2, \cdots d_n$. In particular, det $I = 1$.

4. What is the relation between det A and det $(-A)$?

5. Verify the product rule for the pairs of matrices (a) and (b) in Exercise 1.

6. Apply the product rule to show that, if A is invertible, then det $A \neq 0$ and $(\det A^{-1}) = (\det A)^{-1}$.

7. Let A be an m-by-m matrix and B an n-by-n matrix. Consider the $(m + n)$-by-$(m + n)$ matrix $\begin{pmatrix} A & 0 \\ 0 & B \end{pmatrix}$, which has A in the upper left corner, B in the lower right corner, and zeros elsewhere. Show that its determinant is equal to $(\det A)(\det B)$. (*Hint:* Consider the cases $A = I$ and $B = I$. Then use the product rule.)

8. Use the method of Example 17 of the text to evaluate

   (a) $\det\begin{pmatrix} -1 & 0 & 1 & 2 \\ 0 & 1 & 2 & -1 \\ 1 & 2 & -1 & 0 \\ 2 & -1 & 0 & 1 \end{pmatrix}$.   (b) $\det\begin{pmatrix} 1 & 1 & 1 & 1 \\ 1 & 2 & 4 & 8 \\ 1 & 3 & 9 & 27 \\ 1 & 4 & 16 & 64 \end{pmatrix}$.

9. (a) Compute

   $$\det\begin{pmatrix} 1 & 2 & 3 & 4 \\ 0 & -1 & 5 & 6 \\ 0 & 0 & 3 & -1 \\ 0 & 0 & 0 & 4 \end{pmatrix}.$$

(b) A matrix A, like the one in part (a), in which every element below the diagonal is 0, is said to be **upper triangular.** Show that if A is any triangular matrix then det A is equal to the product of the diagonal elements.

10. Let A be the 3-by-3 matrix

$$\begin{pmatrix} 1 & 4 & 1 \\ 2 & 5 & 6 \\ -1 & 3 & 7 \end{pmatrix}.$$

(a) What is A'?

(b) Compute det A and det A'.

11. Show that for a square matrix A

$$\det (A\tilde{A}') = \det (\tilde{A}'A) = (\det A)^2.$$

12. (a) Show that if A, B, and C are n-by-n matrices such that

$$AB = CA = I$$

then $B = C$. (*Hint:* Multiply $CA = I$ by B on the right.)

(b) Show that if A and B are n-by-n matrices such that $AB = I$, then A and B are both invertible, with $A^{-1} = B$ and $B^{-1} = A$. [*Hint:* Use the product rule for determinants, together with part (a).]

13. Use Theorem 1.9 to determine which of the following matrices have inverses, and find the inverses of the ones that are invertible.

(a) $\begin{pmatrix} 1 & 0 & 0 \\ 3 & 1 & 5 \\ -2 & 0 & 1 \end{pmatrix}.$
   (b) $\begin{pmatrix} 1 & 2 & 3 \\ -1 & 1 & 0 \\ 0 & 3 & 3 \end{pmatrix}.$
   (c) $\begin{pmatrix} 2 & 4 & 8 \\ 1 & 0 & 0 \\ 1 & -3 & -7 \end{pmatrix}.$

(d) $\begin{pmatrix} t & 0 & 0 \\ 0 & 2 & 0 \\ 0 & 0 & 1 \end{pmatrix},$ t real.
   (e) $\begin{pmatrix} 1 & 2 & 1 \\ 0 & 0 & 1 \\ 0 & 0 & 3 \end{pmatrix}.$
   (f) $\begin{pmatrix} 1 & -1 & 1 \\ 0 & -1 & 1 \\ 0 & 0 & 1 \end{pmatrix}.$

(g) $\begin{pmatrix} 1 & 2 & -1 & 3 \\ 0 & 2 & 0 & 1 \\ 0 & 0 & 1 & 1 \\ 0 & 0 & 0 & 4 \end{pmatrix}.$
   (h) $\begin{pmatrix} 1 & 0 & 1 & 0 \\ 0 & 2 & 0 & 0 \\ 0 & 0 & 3 & 0 \\ 0 & 0 & 0 & 4 \end{pmatrix}.$
   (i) $\begin{pmatrix} 1 & 0 & 0 \\ 0 & e^t & te^t \\ 0 & 0 & e^t \end{pmatrix},$ t real.

14. Show that, if a square matrix A has two rows proportional, then det $A = 0$. Show that the same result holds for columns.

15. Let A and B be identical square matrices except in some one row or column. Let C also be the same as A and B except that in that one row or column its entries are the sums of the corresponding entries in A and B. Show that det $C = $ det A + det B.

16. Prove (1), (2), and (3) of Theorem 1.5. The first two follow immediately from expansion by minors of the distinguished row. The third is immediate for 2-by-2 matrices and can be proved by induction using expansion by minors of a row different from the two to be interchanged.

2 EIGENVALUES AND EIGENVECTORS

2A Exponential Solutions

The vector differential equation

$$\frac{d\mathbf{x}}{dt} = A\mathbf{x}$$

has been discussed for several examples in Chapter 4. The examples show that, if A is a constant n-by-n matrix, then exponential solutions can be expected. For example, in the case $n = 1$, we would have $\dot{x} = ax$, with solutions of the form $x(t) = ce^{at}$. Consequently, we try solutions

2.1 $$\mathbf{x}(t) = e^{\lambda t}\mathbf{u},$$

where $\mathbf{u}$ is a constant vector in $\mathfrak{R}^n$. Differentiation of $\mathbf{x}(t)$ gives

$$\frac{d\mathbf{x}}{dt} = \lambda e^{\lambda t}\mathbf{u}.$$

The justification for this differentiation is that differentiation of vector-valued functions can be carried out one entry at a time. For example, if

$$\mathbf{x}(t) = e^{\lambda t}\begin{pmatrix} u_1 \\ u_2 \end{pmatrix} = \begin{pmatrix} e^{\lambda t}u_1 \\ e^{\lambda t}u_2 \end{pmatrix},$$

then

$$\frac{d\mathbf{x}(t)}{dt} = \begin{pmatrix} \lambda e^{\lambda t}u_1 \\ \lambda e^{\lambda t}u_2 \end{pmatrix} = \lambda e^{\lambda t}\begin{pmatrix} u_1 \\ u_2 \end{pmatrix}.$$

Since $e^{\lambda t}$ is a numerical factor, it can be taken past the matrix A, and

$$A\mathbf{x} = A(e^{\lambda t}\mathbf{u}) = e^{\lambda t}A\mathbf{u}.$$

To solve the differential equation, we then have to equate these two vectors to get

2.2 $$A\mathbf{u} = \lambda\mathbf{u}.$$

The choice $\mathbf{u} = 0$ is too trivial to be interesting. We ignore that possibility and define a nonzero vector $\mathbf{u}$ to be an **eigenvector** of A with **eigenvalue** λ if $\mathbf{u}$ and λ satisfy Equation 2.2. Then Equation 2.1 provides a nontrivial solution to the vector differential equation whenever $\mathbf{u}$ is an eigenvector of A with corresponding eigenvalue λ. The problem has been reduced to an algebraic problem to which we can apply the ideas and techniques of the previous section. More specifically, we want to find all real, or complex, numbers λ such that the homogeneous equation

2.3 $$(A - \lambda I)\mathbf{u} = 0$$

has a nontrivial solution $\mathbf{u}$. While Equation 2.2 is equivalent to Equation 2.3, the latter equation has the technical advantage that it displays the matrix of the vector equation in one piece. If $A - \lambda I$ should happen to be invertible, we could apply its inverse to both sides of Equation 2.3 and conclude that $\mathbf{u}$ must be 0. Since this is just what we do not want, we have to choose λ so that $(A - \lambda I)^{-1}$ fails to exist. But Theorem 1.9 of Section 1 tells us that what we want happens precisely when

2.4
$$\det (A - \lambda I) = 0.$$

For example,

$$\det \left[\begin{pmatrix} 2 & 1 \\ 3 & 2 \end{pmatrix} - \lambda \begin{pmatrix} 1 & 0 \\ 0 & 1 \end{pmatrix} \right] = \det \begin{pmatrix} 2 - \lambda & 1 \\ 3 & 2 - \lambda \end{pmatrix} = 0,$$

or

$$(2 - \lambda)^2 - 3 = \lambda^2 - 4\lambda + 1 = 0.$$

This last equation is a condition on λ alone. Having found all values of λ that satisfy it, we can then go ahead to find corresponding eigenvectors $\mathbf{u}$. Note that if $\mathbf{u}$ is a nonzero solution of Equation 2.2 or 2.3 then so is $c\mathbf{u}$ for any nonzero constant c. If we have solutions of the form $e^{\lambda_k t}\mathbf{u}_k$ it is easy to verify that linear combinations

$$\mathbf{x}(t) = c_1 e^{\lambda_1 t}\mathbf{u}_1 + \cdots + c_n e^{\lambda_n t}\mathbf{u}_n$$

are solutions also.

EXAMPLE 1   To solve the system

$$\begin{pmatrix} dx_1/dt \\ dx_2/dt \end{pmatrix} = \begin{pmatrix} x_1 + x_2 \\ 4x_1 + x_2 \end{pmatrix}$$

$$= \begin{pmatrix} 1 & 1 \\ 4 & 1 \end{pmatrix} \begin{pmatrix} x_1 \\ x_2 \end{pmatrix},$$

try to find nonzero vectors $\mathbf{u} = (u_1, u_2)$ that satisfy the eigenvector equation

$$\begin{pmatrix} 1 & 1 \\ 4 & 1 \end{pmatrix} \begin{pmatrix} u_1 \\ u_2 \end{pmatrix} = \lambda \begin{pmatrix} u_1 \\ u_2 \end{pmatrix}$$

for some number λ. In other words, we find numbers λ such that the equation

$$\begin{pmatrix} 1 - \lambda & 1 \\ 4 & 1 - \lambda \end{pmatrix} \begin{pmatrix} u_1 \\ u_2 \end{pmatrix} = \begin{pmatrix} 0 \\ 0 \end{pmatrix}$$

has nonzero solutions. If the 2-by-2 matrix is invertible, then only the solution $(u_1, u_2) = (0, 0)$ exists, so we must have

$$\det \begin{pmatrix} 1 - \lambda & 1 \\ 4 & 1 - \lambda \end{pmatrix} = 0.$$

In other words,

$$(1 - \lambda)^2 - 4 = \lambda^2 - 2\lambda - 3$$
$$= (\lambda - 3)(\lambda + 1) = 0.$$

The only solutions are $\lambda = 3$ and $\lambda = -1$.

 Case (a): $\lambda = 3$. We want nonzero vectors (u_1, u_2) such that

$$\begin{pmatrix} -2 & 1 \\ 4 & -2 \end{pmatrix}\begin{pmatrix} u_1 \\ u_2 \end{pmatrix} = \begin{pmatrix} 0 \\ 0 \end{pmatrix}.$$

The two numerical equations

$$-2u_1 + u_2 = 0$$

$$4u_1 - 2u_2 = 0$$

are equivalent to $u_2 = 2u_1$, so we can choose $u_1 = 1$, $u_2 = 2$. Thus, since $\lambda = 3$,

$$\mathbf{x}_1(t) = e^{3t}\begin{pmatrix} 1 \\ 2 \end{pmatrix}$$

is a solution, along with any numerical multiple.

 Case (b): $\lambda = -1$. We want nonzero vectors (v_1, v_2) such that

$$\begin{pmatrix} 2 & 1 \\ 4 & 2 \end{pmatrix}\begin{pmatrix} v_1 \\ v_2 \end{pmatrix} = \begin{pmatrix} 0 \\ 0 \end{pmatrix}.$$

Clearly, $v_1 = 1$, $v_2 = -2$ will do, so

$$\mathbf{x}_2(t) = e^{-t}\begin{pmatrix} 1 \\ -2 \end{pmatrix}$$

is a solution, as well as any numerical multiple. We will see that the general solution of the vector differential equation is

$$\mathbf{x}(t) = c_1 e^{3t}\begin{pmatrix} 1 \\ 2 \end{pmatrix} + c_2 e^{-t}\begin{pmatrix} 1 \\ -2 \end{pmatrix}$$

$$= \begin{pmatrix} c_1 e^{3t} + c_2 e^{-t} \\ 2c_1 e^{3t} - 2c_2 e^{-t} \end{pmatrix}.$$

 To see the geometric significance of this computation, observe that, as shown in Figure 1, neither of the two vectors

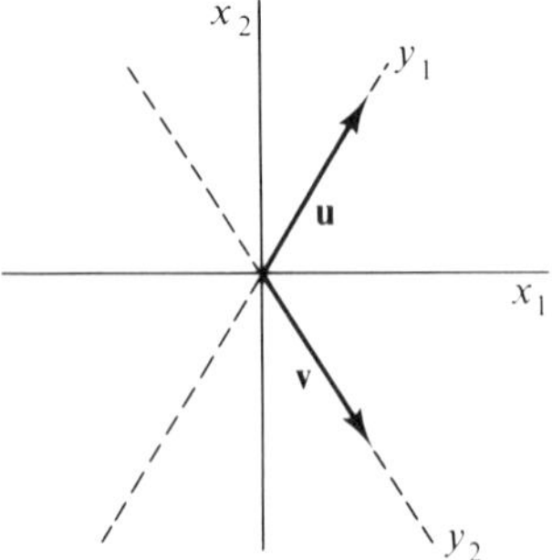

Figure 1

$$\mathbf{u} = \begin{pmatrix} 1 \\ 2 \end{pmatrix}, \qquad \mathbf{v} = \begin{pmatrix} 1 \\ -2 \end{pmatrix}$$

is a multiple of the other. Thus any vector $\mathbf{x}$ can be written in the form

$$\mathbf{x} = y_1\mathbf{u} + y_2\mathbf{v}.$$

By the linearity of multiplication by A,

$$A\mathbf{x} = Ay_1\begin{pmatrix} 1 \\ 2 \end{pmatrix} + Ay_2\begin{pmatrix} 1 \\ -2 \end{pmatrix}$$

$$= y_1 A\begin{pmatrix} 1 \\ 2 \end{pmatrix} + y_2 A\begin{pmatrix} 1 \\ -2 \end{pmatrix}.$$

Since $(1, 2)$ and $(1, -2)$ are eigenvectors, with eigenvalues 3 and -1, respectively,

$$A\mathbf{x} = 3y_1\begin{pmatrix} 1 \\ 2 \end{pmatrix} - y_2\begin{pmatrix} 1 \\ -2 \end{pmatrix}.$$

In other words, A has the effect of multiplying the first vector $\mathbf{u} = (1, 2)$ by 3 and the second vector $\mathbf{v} = (1, -2)$ by -1: $A\mathbf{x} = 3y_1\mathbf{u} - y_2\mathbf{v}$. Since also

$$\frac{d\mathbf{x}}{dt} = \frac{dy_1}{dt}\mathbf{u} + \frac{dy_2}{dt}\mathbf{v},$$

the equation $d\mathbf{x}/dt = A\mathbf{x}$ allows us to equate coefficients of the independent vectors $\mathbf{u}$ and $\mathbf{v}$ to get the two equations

$$\frac{dy_1}{dt} = 3y_1,$$

$$\frac{dy_2}{dt} = -y_2.$$

These two differential equations are particularly simple to solve, because each involves only one unknown function. In fact, we have

$$y_1(t) = c_1 e^{3t},$$
$$y_2(t) = c_2 e^{-t}.$$

This explains why the general solution can be written

$$\mathbf{x}(t) = c_1 e^{3t}\begin{pmatrix} 1 \\ 2 \end{pmatrix} + c_2 e^{-t}\begin{pmatrix} 1 \\ -2 \end{pmatrix}.$$

The procedure in the previous example can be tried in any number of dimensions. To solve the n-dimensional, constant-coefficient equation

$$\frac{d\mathbf{x}}{dt} = A\mathbf{x},$$

proceed as follows:

1. Find the eigenvalues of A by solving the polynomial equation $\det(A - \lambda I) = 0$.
2. For each eigenvalue λ_k, find an eigenvector $\mathbf{u}_k$ by solving

$$(A - \lambda_k I)\mathbf{u} = 0.$$

3. If the eigenvectors $\mathbf{u}_1, \ldots, \mathbf{u}_n$ are linearly independent, form the linear combination

$$\mathbf{x}(t) = c_1 e^{\lambda_1 t}\mathbf{u}_1 + \cdots + c_n e^{\lambda_n t}\mathbf{u}_n.$$

If the eigenvectors of A are *not* linearly independent, the procedure outlined produces solutions but not the most general one. In this case, we can use the elimination method explained in Chapter 4, or the general method of the following Section 5. However, it can be shown that, if the eigenvalues of A are all different, then the corresponding eigenvectors will always be linearly independent, and Step 3 produces the most general solution.

If the eigenvalues of A are complex, then the same method still works, with the complex exponential replacing the real exponential.

EXAMPLE 2   The system

$$\frac{dx_1}{dt} = x_1 - x_2,$$

$$\frac{dx_2}{dt} = x_1 + x_2$$

has matrix

$$A = \begin{pmatrix} 1 & -1 \\ 1 & 1 \end{pmatrix}.$$

The eigenvalues are solutions of

$$\det\begin{pmatrix} 1 - \lambda & -1 \\ 1 & 1 - \lambda \end{pmatrix} = 0;$$

this is the same as

$$(1 - \lambda)^2 + 1 = \lambda^2 - 2\lambda + 2 = 0,$$

having solutions $\lambda_1 = 1 + i$ and $\lambda_2 = 1 - i$.

Case (a). $\lambda = 1 + i$. The eigenvectors are the nonzero solutions of

$$\begin{pmatrix} -i & -1 \\ 1 & -i \end{pmatrix}\begin{pmatrix} u_1 \\ u_2 \end{pmatrix} = \begin{pmatrix} 0 \\ 0 \end{pmatrix},$$

that is, of

$$-iu_1 - u_2 = 0,$$

$$u_1 - iu_2 = 0.$$

One solution is

$$\begin{pmatrix} u_1 \\ u_2 \end{pmatrix} = \begin{pmatrix} 1 \\ -i \end{pmatrix}.$$

Case (*b*). $\lambda = 1 - i$. The eigenvectors are the nonzero solutions of

$$\begin{pmatrix} i & -1 \\ 1 & i \end{pmatrix}\begin{pmatrix} v_1 \\ v_2 \end{pmatrix} = \begin{pmatrix} 0 \\ 0 \end{pmatrix},$$

that is, of

$$iv_1 - v_2 = 0,$$

$$v_1 + iv_2 = 0.$$

One solution is

$$\begin{pmatrix} v_1 \\ v_2 \end{pmatrix} = \begin{pmatrix} 1 \\ i \end{pmatrix}.$$

The general solution of the differential equation is

$$\mathbf{x}(t) = c_1 e^{(1+i)t}\begin{pmatrix} 1 \\ -i \end{pmatrix} + c_2 e^{(1-i)t}\begin{pmatrix} 1 \\ i \end{pmatrix}$$

$$= c_1 e^t\begin{pmatrix} \cos t + i \sin t \\ -i \cos t + \sin t \end{pmatrix} + c_2 e^t\begin{pmatrix} \cos t - i \sin t \\ i \cos t + \sin t \end{pmatrix}$$

$$= \begin{pmatrix} (c_1 + c_2)e^t \cos t + i(c_1 - c_2)e^t \sin t \\ -i(c_1 - c_2)e^t \cos t + (c_1 + c_2)e^t \sin t \end{pmatrix}.$$

If we rename the constants so that $c_1 + c_2 = d_1$ and $i(c_1 - c_2) = d_2$, then

$$\mathbf{x}(t) = \begin{pmatrix} d_1 e^t \cos t + d_2 e^t \sin t \\ -d_2 e^t \cos t + d_1 e^t \sin t \end{pmatrix}$$

$$= d_1\begin{pmatrix} e^t \cos t \\ e^t \sin t \end{pmatrix} + d_2\begin{pmatrix} e^t \sin t \\ -e^t \cos t \end{pmatrix}.$$

Because $c_1 = (d_1 - id_2)/2$ and $c_2 = (d_1 + id_2)/2$, the constants c_1 and c_2 can always be chosen so that d_1 and d_2 have any preassigned values; in particular, they can be chosen to be real numbers.

2B Eigenvector Matrices

As usual, we may want to choose arbitrary constants in general solutions such as those preceding so as to satisfy prescribed initial conditions. This calculation can be reduced to a routine as follows, if we assume that the eigenvectors $\mathbf{u}_1, \ldots, \mathbf{u}_n$ of the n-by-n matrix A are linearly independent. We denote by U the matrix whose columns are the column vectors $\mathbf{u}_1, \ldots, \mathbf{u}_n$ in some fixed order. The matrix $U = (\mathbf{u}_1, \ldots, \mathbf{u}_n)$ is called the **eigenvector matrix** of the system. We denote by Λ_t the diagonal matrix

$$
\Lambda_t = \begin{pmatrix} e^{\lambda_1 t} & 0 & \cdots & 0 \\ 0 & e^{\lambda_2 t} & \cdots & 0 \\ \vdots & \vdots & & \vdots \\ 0 & 0 & \cdots & e^{\lambda_n t} \end{pmatrix},
$$

with corresponding eigenvalues λ_k in the *same order* as the eigenvectors. If $\mathbf{c}$ is any constant column vector with entries $c_1, \ldots, c_n$, we can form the vector-valued function

$$
\mathbf{x}(t) = U \Lambda_t \mathbf{c}.
$$

Clearly, $\mathbf{x}(t)$ has the form

$$
\mathbf{x}(t) = U \begin{pmatrix} c_1 e^{\lambda_1 t} \\ \vdots \\ c_n e^{\lambda_n t} \end{pmatrix}
$$

$$
= c_1 e^{\lambda_1 t} \mathbf{u}_1 + \cdots + c_n e^{\lambda_n t} \mathbf{u}_n,
$$

and so is a solution of $dx/dt = A\mathbf{x}$. But if the columns $\mathbf{u}_1, \ldots, \mathbf{u}_n$ of U are linearly independent, then, by Theorem 6.6, U is invertible and we can let $\mathbf{c} = U^{-1}\mathbf{x}_0$ for some $\mathbf{x}_0$ in $\mathcal{R}^n$. Thus

2.5
$$
\mathbf{x}(t) = U \Lambda_t U^{-1} \mathbf{x}_0.
$$

Since $\Lambda_0 = I$, we have $\mathbf{x}(0) = UU^{-1}\mathbf{x}_0$; so $\mathbf{x}(0) = \mathbf{x}_0$. Thus $\mathbf{x}(t)$ is the solution of the vector differential equation that satisfies $\mathbf{x}(0) = \mathbf{x}_0$. To satisfy an initial condition at $t = t_0$, we let

$$
\mathbf{c} = \Lambda_{-t_0} U^{-1} \mathbf{x}_0,
$$

so that

$$
\mathbf{x}(t) = U \Lambda_{(t - t_0)} U^{-1} \mathbf{x}_0
$$

is the desired solution.

EXAMPLE 3   In Example 1, we found for

$$
A = \begin{pmatrix} 1 & 1 \\ 4 & 1 \end{pmatrix}
$$

the eigenvalues $\lambda_1 = 3$, $\lambda_2 = -1$ with eigenvectors

$$
\mathbf{u}_1 = \begin{pmatrix} 1 \\ 2 \end{pmatrix}, \qquad \mathbf{u}_2 = \begin{pmatrix} 1 \\ -2 \end{pmatrix}.
$$

We have

$$
\Lambda_t = \begin{pmatrix} e^{3t} & 0 \\ 0 & e^{-t} \end{pmatrix},
$$

$$U = \begin{pmatrix} 1 & 1 \\ 2 & -2 \end{pmatrix},$$

$$U^{-1} = \begin{pmatrix} \frac{1}{2} & \frac{1}{4} \\ \frac{1}{2} & -\frac{1}{4} \end{pmatrix}.$$

Thus, if $\mathbf{x}_0 = (2, 3)$, the solution

$$\mathbf{x}(t) = \begin{pmatrix} 1 & 1 \\ 2 & -2 \end{pmatrix}\begin{pmatrix} e^{3t} & 0 \\ 0 & e^{-t} \end{pmatrix}\begin{pmatrix} \frac{1}{2} & \frac{1}{4} \\ \frac{1}{2} & -\frac{1}{4} \end{pmatrix}\begin{pmatrix} 2 \\ 3 \end{pmatrix}$$

$$= \begin{pmatrix} \frac{7}{4}e^{3t} + \frac{1}{4}e^{-t} \\ \frac{7}{2}e^{3t} - \frac{1}{2}e^{-t} \end{pmatrix}$$

satisfies the differential equation $d\mathbf{x}/dt = A\mathbf{x}$ and also the initial condition $\mathbf{x}(0) = (2, 3)$.

To satisfy instead the initial condition $\mathbf{x}(1) = (3, 4)$, we form

$$\mathbf{x}(t) = \begin{pmatrix} 1 & 1 \\ 2 & -2 \end{pmatrix}\begin{pmatrix} e^{3(t-1)} & 0 \\ 0 & e^{-(t-1)} \end{pmatrix}\begin{pmatrix} \frac{1}{2} & \frac{1}{4} \\ \frac{1}{2} & -\frac{1}{4} \end{pmatrix}\begin{pmatrix} 3 \\ 4 \end{pmatrix}$$

$$= \begin{pmatrix} \frac{5}{2}e^{3(t-1)} + \frac{1}{2}e^{-(t-1)} \\ 5e^{3(t-1)} - e^{-(t-1)} \end{pmatrix}.$$

EXERCISES

1. The matrix

$$A = \begin{pmatrix} 1 & 12 \\ 3 & 1 \end{pmatrix}$$

has eigenvalues 7 and -5. Which of the following vectors is an eigenvector of A? For those that are, what is the corresponding eigenvalue?

$$\begin{pmatrix} 2 \\ 1 \end{pmatrix}, \quad \begin{pmatrix} -2 \\ 1 \end{pmatrix}, \quad \begin{pmatrix} -4 \\ -2 \end{pmatrix}, \quad \begin{pmatrix} -2 \\ 2 \end{pmatrix}, \quad \begin{pmatrix} 1 \\ 1 \end{pmatrix}.$$

2. Find all the eigenvalues of each of the following matrices.

(a) $\begin{pmatrix} 1 & 4 \\ 1 & 1 \end{pmatrix}$. **(b)** $\begin{pmatrix} 0 & 4 \\ 1 & 0 \end{pmatrix}$. **(c)** $\begin{pmatrix} 2 & 4 \\ 1 & 2 \end{pmatrix}$. **(d)** $\begin{pmatrix} 1 & 0 & 0 \\ 2 & 1 & 0 \\ 0 & 1 & 2 \end{pmatrix}$.

3. For each matrix in Exercise 2, find an eigenvector corresponding to each eigenvalue.

4. Each of the following two-dimensional systems of differential equations has the form

$$\frac{d\mathbf{x}}{dt} = A\mathbf{x}$$

in which the matrix A has constant entries. In each example, find the eigenvalues of A, and for each eigenvalue find a corresponding eigenvector.

(a) $\begin{pmatrix} \dfrac{dx}{dt} \\ \dfrac{dy}{dt} \end{pmatrix} = \begin{pmatrix} -3 & 2 \\ -4 & 3 \end{pmatrix}\begin{pmatrix} x \\ y \end{pmatrix}.$

$\qquad\qquad$ [*Ans.* $\lambda_1 = 1,\ \mathbf{u}_1 = (1,\ 2);$ $\lambda_2 = -1,\ \mathbf{u}_2 = (1,\ 1).$]

(b) $\dfrac{dx}{dt} = 3x,$

$\qquad \dfrac{dy}{dt} = 2y.$ $\qquad\qquad$ [*Ans.* $\lambda_1 = 3,\ \mathbf{u}_1 = (1,\ 0);\ \lambda_2 = 2,\ \mathbf{u}_2 = (0,\ 1).$]

(c) $\dfrac{dx}{dt} = x + 4y,$

$\qquad \dfrac{dy}{dt} = 5y.$ $\qquad\qquad$ [*Ans.* $\lambda_1 = 5,\ \mathbf{u}_1 = (1,\ 1);\ \lambda_2 = 1,\ \mathbf{u}_2 = (1,\ 0).$]

(d) $\begin{pmatrix} \dfrac{dx}{dt} \\ \dfrac{dy}{dt} \end{pmatrix} = \begin{pmatrix} 2 & -1 \\ 1 & 2 \end{pmatrix}\begin{pmatrix} x \\ y \end{pmatrix}.$

$\qquad\qquad$ [*Ans.* $\lambda_1 = 2 + i,\ \mathbf{u}_1 = (i,\ 1);$ $\lambda_2 = 2 - i,\ \mathbf{u}_2 = (i,\ -1).$]

5. Use the eigenvalues and eigenvectors of each of the systems in Exercise 4 to write the general solution of the system in the form

$$\mathbf{x}(t) = c_1 e^{\lambda_1 t}\mathbf{u}_1 + c_2 e^{\lambda_2 t}\mathbf{u}_2.$$

In the case of complex eigenvalues, convert the solution to real form.

6. For the systems of differential equations in Exercise 4, find the particular solution that satisfies the corresponding condition listed.
(a) $(x(0),\ y(0)) = (1,\ 3).$ $\qquad$ **(b)** $(x(0),\ y(0)) = (-1,\ 0).$
(c) $(x(0),\ y(0)) = (1,\ 1).$ $\qquad$ **(d)** $(x(0),\ y(0)) = (0,\ 0).$

7. (a) Find the general homogeneous solution of the system

$$\frac{d\mathbf{x}}{dt} = \begin{pmatrix} 1 & -1 & 4 \\ 3 & 2 & -1 \\ 2 & 1 & -1 \end{pmatrix}\mathbf{x} + \begin{pmatrix} 1 \\ 0 \\ 2 \end{pmatrix}.$$

(b) Find a particular solution of the system in part (a).
(c) Find the particular solution $\mathbf{x}$ of the system in part (a) that satisfies the condition $\mathbf{x}(0) = (1,\ 1,\ 2).$

8. (a) Find the general solution of the system

$$\frac{dx}{dt} = y,$$

$$\frac{dy}{dt} = -x,$$

$$\frac{dz}{dt} = -z$$

by any method.
(b) What are the eigenvalues of the following matrix?

$$\begin{pmatrix} 0 & 1 & 0 \\ -1 & 0 & 0 \\ 0 & 0 & -1 \end{pmatrix}$$

9. (a) Find the general solution of the system

$$\frac{dx}{dt} = -x + y,$$

$$\frac{dy}{dt} = -y.$$

(b) What are the eigenvectors of the following matrix?

$$\begin{pmatrix} -1 & 1 \\ 0 & -1 \end{pmatrix}$$

10. Show that, if A is n-by-n with n independent eigenvectors, and the eigenvalues of A have negative real parts, then all solutions of $d\mathbf{x}/dt = A\mathbf{x}$ tend to zero as t tends to $+\infty$.

11. The second-order, constant-coefficient, differential equation

$$\frac{d^2 y}{dt^2} + a\frac{dy}{dt} + by = 0$$

is equivalent to the first-order system

$$\frac{dx}{dt} = -ax - by,$$

$$\frac{dy}{dt} = x.$$

Show that the eigenvalues of the matrix

$$\begin{pmatrix} -a & -b \\ 1 & 0 \end{pmatrix}$$

are the same as the characteristic roots of the second-order differential equation.

12. Let U be an invertible n-by-n matrix with columns $\mathbf{u}_1, \ldots, \mathbf{u}_n$. Let D be the diagonal matrix with diagonal entries $\lambda_1, \ldots, \lambda_n$, and define the n-by-n matrix A by $A = UDU^{-1}$.
(a) Show that $A\mathbf{u}_k = \lambda_k \mathbf{u}_k$, for $k = 1, \ldots, n$.
(b) Find the 2-by-2 matrix that has eigenvectors $\mathbf{u}_1 = (2, 3)$, $\mathbf{u}_2 = (1, 1)$, and corresponding eigenvalues $\lambda_1 = 3$, $\lambda_2 = 1$.
(c) Show that the system $d\mathbf{x}/dt = A\mathbf{x}$ has solutions $e^{\lambda_k t}\mathbf{u}_k$, for $k = 1, \ldots, n$.
(d) Find a two-dimensional system having

$$\mathbf{x}(t) = c_1 e^{3t}\begin{pmatrix} 2 \\ 3 \end{pmatrix} + c_2 e^{t}\begin{pmatrix} 1 \\ 1 \end{pmatrix}$$

as its general solution.

13. In applications of linear systems, we're sometimes concerned mainly with the general behavior of solutions. By finding the eigenvalues $\lambda = a + ib$ of the matrix A, find out whether the solutions of $d\mathbf{x}/dt = A\mathbf{x}$ oscillate ($b \neq 0$), tend to 0 as t tends to $+\infty$ ($a < 0$), or become unbounded ($a > 0$). Of course a single equation may have solutions exhibiting more than one kind of behavior.

(a) $\dfrac{d}{dt}\begin{pmatrix} x \\ y \end{pmatrix} = \begin{pmatrix} -1 & -2 \\ 1 & 0 \end{pmatrix}\begin{pmatrix} x \\ y \end{pmatrix}.$ 　　　　**(b)** $\dfrac{d}{dt}\begin{pmatrix} x \\ y \end{pmatrix} = \begin{pmatrix} 0 & -1 \\ 1 & 0 \end{pmatrix}\begin{pmatrix} x \\ y \end{pmatrix}.$

(c) $\dfrac{d}{dt}\begin{pmatrix} x \\ y \end{pmatrix} = \begin{pmatrix} 2x - y \\ 3x - y \end{pmatrix}.$ **(d)** $dx/dt = 2x + y,$
$dy/dt = 3x + y.$

14. Verify that if $x_1(t)$ and $x_2(t)$ are solutions of $x = Ax$, then so is $c_1 x_1(t) + c_2 x_2(t)$ for arbitrary constant c_1 and c_2.

3 MATRIX EXPONENTIALS

The exponential e^A of a square matrix A is fundamental for describing solutions of linear differential systems with constant coefficients. The ordinary numerical exponential e^x can be defined in several ways, and the simplest one on which to model our matrix definition is the infinite sum

$$e^x = 1 + \frac{x}{1!} + \frac{x^2}{2!} + \frac{x^3}{3!} + \cdots + \frac{x^k}{k!} \cdots$$

This formula is valid for all real or complex values of x, and while it is not usual to do so, all the properties of e^x can be derived directly from it. For example, notice that the formal derivative of the infinite sum is equal to the sum itself, which suggests that $de^x/dx = e^x$, as we know. Also, letting $x = 0$ shows that $e^0 = 1$. From these two properties alone it follows that e^x has all the familiar properties of the exponential function. Using similar arguments, we will be able to establish analogous properties for e^A, where A is a square matrix. Our initial approach will use an infinite series of matrices, because this allows us to see most directly the general properties of the solutions we are looking for. For the purpose of computing explicit solution formulas, however, the exponential series is effective only in special cases. Thus other methods will be introduced to find the desired formulas. The state of affairs is somewhat like that associated with the definition and computation of values for the ordinary exponential function e^x. The payoff for the work we do, in both the matrix and numerical case, is that we have a notation that strongly suggests the properties of the mathematical quantity that it represents.

If A has dimension n-by-n and I is the n-by-n identity matrix, we consider the sum

$$I + A + \frac{A^2}{2!} + \cdots + \frac{A^k}{k!} = \sum_{j=0}^{k} \frac{A^j}{j!}.$$

This sum of n-by-n matrices is, of course, also an n-by-n matrix. We define the **exponential** of A by

$$e^A = \lim_{k \to \infty} \sum_{j=0}^{k} \frac{A^j}{j!} = \sum_{j=0}^{\infty} \frac{A^j}{j!},$$

where the existence of the matrix limit is understood to mean that the limit exists in each of the n^2 entries in the matrix. It is sometimes convenient to use the notation of $\exp A$ for e^A. For example

$$A = \begin{pmatrix} 2 & 0 \\ 0 & 3 \end{pmatrix}, \; A^2 = \begin{pmatrix} 2^2 & 0 \\ 0 & 3^2 \end{pmatrix}, \; \ldots, \; A^j = \begin{pmatrix} 2^j & 0 \\ 0 & 3^j \end{pmatrix},$$

then

$$\exp \begin{pmatrix} 2 & 0 \\ 0 & 3 \end{pmatrix} = \lim_{k \to \infty} \sum_{j=0}^{k} \frac{1}{j!} \begin{pmatrix} 2^j & 0 \\ 0 & 3^j \end{pmatrix}$$

$$= \begin{pmatrix} \sum_{j=0}^{\infty} \dfrac{2^j}{j!} & 0 \\ 0 & \sum_{j=0}^{\infty} \dfrac{3^j}{j!} \end{pmatrix} = \begin{pmatrix} e^2 & 0 \\ 0 & e^3 \end{pmatrix}.$$

It is a remarkable and useful fact that the exponential of a square matrix always exists and has many of the properties of the ordinary real or complex exponential function. The justification is provided by the following fundamental theorem, in which the **derivative of a matrix** is defined to be simply the matrix obtained by differentiating the individual entries; thus

$$\frac{d}{dt} \begin{pmatrix} 1 & t \\ 0 & e^t \end{pmatrix} = \begin{pmatrix} 0 & 1 \\ 0 & e^t \end{pmatrix}.$$

3.1 Theorem.   If A is an n-by-n real or complex matrix, the matrix series

$$\sum_{j=0}^{\infty} \frac{t^j A^j}{j!}$$

converges to an n-by-n matrix e^{tA} having properties

(a). $e^{(t+s)A} = e^{tA} e^{sA} = e^{sA} e^{tA}$, for any numbers t and s.

(b). e^{tA} is invertible, and $e^{-tA} e^{tA} = e^{tA} e^{-tA} = I$.

(c). $\dfrac{d}{dt} e^{tA} = A e^{tA} = e^{tA} A$.

Proof. If $A = (a_{ij})$, choose a positive number b such that $|a_{ij}| \leq b$ for i, $j = 1, \ldots, n$. Since the entries in A^2 are of the form

$$a_{i1} a_{1j} + \cdots + a_{in} a_{nj},$$

it follows that they are all at most nb^2 in absolute value. Proceeding inductively, the entries in A^k are at most $n^{k-1} b^k$ in absolute value. It follows that each entry in e^A is defined by an absolutely convergent infinite series dominated by the convergent series

$$1 + b + \frac{nb^2}{2!} + \cdots + \frac{1}{j!} n^{j-1} b^j + \cdots .$$

Hence all entries exist and e^A is defined. These estimates show that, if the entries a_{ij} in A are replaced by the entries ta_{ij} in tA, then the convergence is uniform on every bounded interval $c \leq t \leq d$.

To prove property (a), we apply the binomial theorem to $(t + s)^j$ to get

$$e^{(t+s)A} = \sum_{j=0}^{\infty} \frac{(t + s)^j A^j}{j!} = \sum_{j=0}^{\infty} \frac{1}{j!} \left[\sum_{l=0}^{j} \binom{j}{l} t^l s^{j-l} A^j \right].$$

Since $\binom{j}{l} = j!/l!(j - l)!$, we can cancel $j!$ to get

$$e^{(t + s)A} = \sum_{j=0}^{\infty} \left[\sum_{l=0}^{j} \frac{t^l A^l}{l!} \frac{s^{j-l} A^{j-l}}{(j - 1)!} \right].$$

On the other hand, this last sum is just the product of the two absolutely convergent series that represent e^{tA} and e^{sA}, respectively. Note that the product can be formed in either order.

Property (b) follows from (a) on taking $t = 1$ and $s = -1$. Clearly, $e^0 = I$, so $I = e^A e^{-A} = e^{-A} e^A$.

Formally, the derivative of e^{tA} can be computed from the definition by

$$\frac{d}{dt} e^{tA} = \frac{d}{dt} \sum_{j=0}^{\infty} \frac{t^j A^j}{j!}$$

$$= \sum_{j=1}^{\infty} \frac{jt^{j-1} A^j}{j!}$$

$$= A \sum_{j=1}^{\infty} \frac{t^{j-1} A^{j-1}}{(j - 1)!} = Ae^{tA}.$$

Note that the factor A could be taken out on the right just as well on the left. This computation can be justified using term-by-term differentiation of series. The reason is that the differentiated series in each entry is uniformly convergent because of the estimate made in the first part of the proof.

3A Solving Systems

The quickest way to see the importance of e^{tA} is to show that

$$\mathbf{x}(t) = e^{tA} \mathbf{x_0}$$

defines the solution of

$$\frac{d\mathbf{x}}{dt} = A\mathbf{x}, \qquad \mathbf{x}(0) = \mathbf{x_0}.$$

First, to differentiate the vector $e^{tA} \mathbf{x_0}$, we need only differentiate each entry in the matrix e^{tA}, because $\mathbf{x_0}$ has constant entries. Hence

$$\frac{d}{dt} e^{tA} \mathbf{x_0} = Ae^{tA} \mathbf{x_0}$$

by part (c) of the previous theorem. Thus the differential equation is satisfied. Second, to show that the initial condition is satisfied, note that

$$\mathbf{x}(0) = I\mathbf{x}_0 = \mathbf{x}_0.$$

More generally,

$$\mathbf{x}(t) = e^{(t-t_0)A}\mathbf{x}_0$$

satisfies the initial condition $\mathbf{x}(t_0) = \mathbf{x}_0$.

3.2. Theorem   The unique solution of

$$\frac{d\mathbf{x}}{dt} = A\mathbf{x}, \ \mathbf{x}(0) = \mathbf{x}_0,$$

is given by $x(t) = e^{tA}\mathbf{x}_0$.

Proof. Multiply the differential equation by e^{-tA} to get

$$e^{-tA}\frac{d\mathbf{x}}{dt} - e^{-tA}A\mathbf{x} = 0.$$

This is the same as $d/dt\,(e^{-tA}\mathbf{x}) = 0$, so $e^{-tA}\mathbf{x} = \mathbf{c}$, for some constant vector $\mathbf{c}$. Since $e^{tA}e^{-tA} = I$, multiplying by e^{tA} gives $\mathbf{x}(t) = e^{tA}\mathbf{c}$. Finally $\mathbf{c} = e^{0A}\mathbf{c} = \mathbf{x}(0) = \mathbf{x}_0$.

EXAMPLE 1   The system

$$\frac{dx}{dt} = x$$

$$\frac{dy}{dt} = -y$$

can be written in matrix form as $d\mathbf{x}/dt = A\mathbf{x}$, where

$$A = \begin{pmatrix} 1 & 0 \\ 0 & -1 \end{pmatrix}.$$

Hence the system has general solution

$$\mathbf{x}(t) = e^{tA}\mathbf{c},$$

where $\mathbf{c}$ is the two-dimensional column vector with constant entries c_1, c_2. To compute the exponential matrix, write

$$\exp t\begin{pmatrix} 1 & 0 \\ 0 & -1 \end{pmatrix} = \exp \begin{pmatrix} t & 0 \\ 0 & -t \end{pmatrix}$$

$$= \sum_{j=0}^{\infty} \frac{1}{j!}\begin{pmatrix} t & 0 \\ 0 & -t \end{pmatrix}^{j}$$

$$= \sum_{j=0}^{\infty} \frac{1}{j!} \begin{pmatrix} t^j & 0 \\ 0 & (-t)^j \end{pmatrix}$$

$$= \begin{bmatrix} \sum_{j=0}^{\infty} \frac{1}{j} t^j & 0 \\ 0 & \sum_{j=0}^{\infty} \frac{1}{j!}(-t)^j \end{bmatrix} = \begin{pmatrix} e^t & 0 \\ 0 & e^{-t} \end{pmatrix}.$$

Hence the general solution is

$$\mathbf{x}(t) = \begin{pmatrix} e^t & 0 \\ 0 & e^{-t} \end{pmatrix} \begin{pmatrix} c_1 \\ c_2 \end{pmatrix}$$

$$= \begin{pmatrix} c_1 & e^t \\ c_2 & e^{-t} \end{pmatrix}.$$

Of course, this solution can easily be found by a glance at the original system; we derive it using the exponential matrix because it provides a very simple example of the definition and use of the exponential.

EXAMPLE 2   The system

$$\frac{dx}{dt} = x + y$$

$$\frac{dy}{dt} = y$$

can be written in matrix form $d\mathbf{x}/dt = A\mathbf{x}$ as

$$\begin{pmatrix} dx/dt \\ dy/dt \end{pmatrix} = \begin{pmatrix} 1 & 1 \\ 0 & 1 \end{pmatrix} \begin{pmatrix} x \\ y \end{pmatrix}.$$

We compute

$$A = \begin{pmatrix} 1 & 1 \\ 0 & 1 \end{pmatrix}, \; A^2 = \begin{pmatrix} 1 & 2 \\ 0 & 1 \end{pmatrix}, \ldots, A^k = \begin{pmatrix} 1 & k \\ 0 & 1 \end{pmatrix}, \ldots$$

Then

$$e^{tA} = \sum_{k=0}^{\infty} \frac{t^k}{k!} \begin{pmatrix} 1 & k \\ 0 & 1 \end{pmatrix} = \begin{pmatrix} \sum_{k=0}^{\infty} \frac{t^k}{k!} & \sum_{k=1}^{\infty} \frac{t^k}{(k-1)!} \\ 0 & \sum_{k=0}^{\infty} \frac{t^k}{k!} \end{pmatrix}$$

$$= \begin{pmatrix} e^t & te^t \\ 0 & e^t \end{pmatrix}.$$

Hence the solution with initial conditions $x(0) = 2$, $y(0) = -3$ is

$$\begin{pmatrix} x(t) \\ y(t) \end{pmatrix} = \begin{pmatrix} e^t & te^t \\ 0 & e^t \end{pmatrix} \begin{pmatrix} 2 \\ -3 \end{pmatrix}$$

$$= \begin{pmatrix} 2e^t - 3\,te^t \\ -3e^t \end{pmatrix}.$$

This system could also be solved by the elimination method.

3B Relationship to Eigenvectors

The connection between matrix exponential solutions and the eigenvector method is as follows. Assume that the eigenvectors of the square matrix A are such that the eigenvector matrix

$$U = (\mathbf{u}_1, \mathbf{u}_2, \ldots, \mathbf{u}_n)$$

with these vectors as columns is invertible. We also form the diagonal matrix

$$\Lambda_t = \begin{pmatrix} e^{\lambda_1 t} & \cdots & & & 0 \\ & \cdot & & & \\ & & \cdot & & \\ & & & \cdot & \\ 0 & & \cdots & & e^{\lambda_n t} \end{pmatrix},$$

where λ_k is the eigenvalue of $\mathbf{u}_k$. Then we have seen that

$$\mathbf{x}(t) = U\Lambda_t U^{-1}\mathbf{x}_0$$

solves the initial-value problem for the equation $d\mathbf{x}/dt = A\mathbf{x}$. Since

$$\mathbf{x}(t) = e^{tA}\mathbf{x}_0$$

solves the same problem, we are faced with the question of whether the two solutions are the same. It turns out by Theorem 3.2 that there is in fact only one solution satisfying $\mathbf{x}(0) = \mathbf{x}_0$. Hence $e^{tA}\mathbf{x}_0 = U\Lambda_t U^{-1}\mathbf{x}_0$ for all t. It follows, since $\mathbf{x}_0$ is arbitrary, that

3.3 $$e^{tA} = U\Lambda_t U^{-1}.$$

EXAMPLE 3 In Example 1 of Section 2, we solved the system

$$\begin{pmatrix} dx_1/dt \\ dx_2/dt \end{pmatrix} = \begin{pmatrix} 1 & 1 \\ 4 & 1 \end{pmatrix} \begin{pmatrix} x_1 \\ x_2 \end{pmatrix}$$

by finding the eigenvalues $\lambda_1 = 3$, $\lambda_2 = -1$ and corresponding eigenvectors $(1, 2)$, $(1, -2)$ of the 2-by-2 matrix A of the system. Thus

$$U = \begin{pmatrix} 1 & 1 \\ 2 & -2 \end{pmatrix}, \qquad \Lambda_t = \begin{pmatrix} e^{3t} & 0 \\ 0 & e^{-t} \end{pmatrix}, \qquad U^{-1} = \begin{pmatrix} \frac{1}{2} & \frac{1}{4} \\ \frac{1}{2} & -\frac{1}{4} \end{pmatrix}$$

and

$$e^{tA} = U\Lambda_t U^{-1} = \begin{pmatrix} \frac{1}{2}e^{3t} + \frac{1}{2}e^{-t} & \frac{1}{4}e^{3t} - \frac{1}{4}e^{-t} \\ e^{3t} - e^{-t} & \frac{1}{2}e^{3t} + \frac{1}{2}e^{-t} \end{pmatrix}.$$

As a check on the computation, notice that $e^{0A} = I$. This example shows that, if the eigenvectors of A are linearly independent, it may well be easier to use them to compute e^{tA} than to use the matrix power series definition.

By equation 3.3 and Theorem 3.2 the general solution of $dx/dt = Ax$ can be written

$$\mathbf{x}(t) = U\Lambda_t U^{-1}\mathbf{x}_0$$

$$= U\Lambda_t \mathbf{c},$$

where $\mathbf{c}$ is the column vector with entries $c_1, \ldots, c_n$. Multiplying this last expression out gives the usual form

$$\mathbf{x}(t) = c_1 e^{\lambda_1 t}\mathbf{u}_1 + \cdots + c_n e^{\lambda_n t}\mathbf{u}_n$$

for the solution when the eigenvector matrix U is invertible.

EXERCISES

1. Find the exponential e^{tA} of each of the following matrices A by first computing the successive terms $I, tA, t^2 A^2/2! \ldots$ in the series definition.

(a) $A = \begin{pmatrix} -1 & 0 \\ 0 & 1 \end{pmatrix}.$ **(b)** $A = \begin{pmatrix} 1 & 0 \\ 1 & 0 \end{pmatrix}.$

(c) $A = \begin{pmatrix} i & 0 \\ 0 & -i \end{pmatrix}.$ **(d)** $A = \begin{pmatrix} 1 & 0 & 0 \\ 0 & 2 & 0 \\ 0 & 0 & 4 \end{pmatrix}.$

2. In Example 2 of the text, it was shown that

$$e^{t\left(\begin{smallmatrix} 1 & 1 \\ 0 & 1 \end{smallmatrix}\right)} = \begin{pmatrix} e^t & te^t \\ 0 & e^t \end{pmatrix}.$$

Verify directly for this example that

(a) $e^{t\left(\begin{smallmatrix} 1 & 1 \\ 0 & 1 \end{smallmatrix}\right)}$ and $e^{-t\left(\begin{smallmatrix} 1 & 1 \\ 0 & 1 \end{smallmatrix}\right)}$ are inverse to one another.

(b) $e^{t\left(\begin{smallmatrix} 1 & 1 \\ 0 & 1 \end{smallmatrix}\right)}e^{s\left(\begin{smallmatrix} 1 & 1 \\ 0 & 1 \end{smallmatrix}\right)} = e^{(t + s)\left(\begin{smallmatrix} 1 & 1 \\ 0 & 1 \end{smallmatrix}\right)}.$

(c) $\dfrac{d}{dt}e^{t\left(\begin{smallmatrix} 1 & 1 \\ 0 & 1 \end{smallmatrix}\right)} = \begin{pmatrix} 1 & 1 \\ 0 & 1 \end{pmatrix}e^{t\left(\begin{smallmatrix} 1 & 1 \\ 0 & 1 \end{smallmatrix}\right)}.$

(d) What it the solution of the system

$$\begin{pmatrix} \dfrac{dx}{dt} \\ \dfrac{dy}{dt} \end{pmatrix} = \begin{pmatrix} 1 & 1 \\ 0 & 1 \end{pmatrix}\begin{pmatrix} x \\ y \end{pmatrix},$$

satisfying

$$\begin{pmatrix} x(0) \\ y(0) \end{pmatrix} = \begin{pmatrix} -1 \\ 2 \end{pmatrix}?$$

3. If A is a 1-by-1 matrix with entry a, what is e^{tA}?

4. Let

$$A = \begin{pmatrix} -1 & -2 & -2 \\ 0 & 1 & -1 \\ 0 & 0 & 2 \end{pmatrix}.$$

Using methods derived in Section 5, we will be able to show quite easily that

$$e^{tA} = \begin{pmatrix} e^{-t} & e^{-t} - e^{t} & e^{-t} - e^{t} \\ 0 & e^{t} & e^{t} - e^{2t} \\ 0 & 0 & e^{2t} \end{pmatrix}.$$

(a) Replace t by $-t$ in e^{tA} and verify directly that $e^{-tA}e^{tA} = I$ and that $e^{0A} = I$.

(b) If $B(t) = (d/dt)e^{tA}$, verify that $B(0) = A$.

5. (a) Find the general solution of the system

$$\begin{pmatrix} \dfrac{dx}{dt} \\ \dfrac{dy}{dt} \end{pmatrix} = \begin{pmatrix} 9 & -4 \\ 4 & 1 \end{pmatrix}\begin{pmatrix} x \\ y \end{pmatrix}$$

by the method of elimination.

(b) Use the result of part (a) to compute the matrix e^{tA}, where

$$A = \begin{pmatrix} 9 & -4 \\ 4 & 1 \end{pmatrix}.$$

[*Hint:* Find solutions such that $\mathbf{x}_1(0) = \begin{pmatrix} 1 \\ 0 \end{pmatrix}$ and $\mathbf{x}_2(0) = \begin{pmatrix} 0 \\ 1 \end{pmatrix}$.]

6. Let A be an n-by-n matrix with real entries. The matrix e^{itA} is defined by

$$e^{itA} = \sum_{k=0}^{\infty} \frac{1}{k!}(i)^{k}t^{k}A^{k}.$$

Define $\cos tA$ to be the real part of the series and $\sin tA$ to be the imaginary part, so that

$$e^{itA} = \cos tA + i \sin tA.$$

Show that the matrices $\cos tA$ and $\sin tA$ satisfy

(a) $\cos(-t)A = \cos tA$, $\sin(-t)A = -\sin tA$.

(b) $(d/dt)\cos tA = -A \sin tA$, $(d/dt)\sin tA = A \cos tA$.

   (*Hint:* Express $\cos tA$ and $\sin tA$ in terms of e^{itA}.)

(c) $(\cos tA)^2 + (\sin tA)^2 = I$, where I is the n-by-n identity matrix.

7. Let A be the 2-by-2 matrix

$$\begin{pmatrix} 1 & 1 \\ 0 & 1 \end{pmatrix}.$$

Define $\cos tA$ and $\sin tA$ as in Exercise 6, and verify the formulas given in (a), (b), and (c).

8. Show that if A is an n-by-n matrix then a system of the form

$$\frac{d^2\mathbf{x}}{dt^2} + A^2\mathbf{x} = 0$$

has solutions of the form

$$\mathbf{x}(t) = (\cos tA)\mathbf{c}_1 + (\sin tA)\mathbf{c}_2,$$

where $\mathbf{c}_1$ and $\mathbf{c}_2$ are constant n-dimensional vectors, whereas $\cos tA$ and $\sin tA$ are the n-by-n matrices defined in Exercise 6. Is this always the most general solution?

9. For each of the following matrices A, compute e^{tA} by using Formula 3.3. Then find the inverse matrix e^{-tA}, and check your original computation by showing that the derivative of e^{tA} at $t = 0$ is equal to A.

(a) $\begin{pmatrix} 0 & 1 \\ -6 & 5 \end{pmatrix}.$ (b) $\begin{pmatrix} -4 & 4 \\ -6 & 6 \end{pmatrix}.$ (c) $\begin{pmatrix} -1 & 0 & 0 \\ 0 & \frac{3}{2} & -\frac{1}{2} \\ 0 & -\frac{1}{2} & \frac{3}{2} \end{pmatrix}.$

10. Use the answers provided along with Exercise 4 of Section 2 to find e^{tA} for each of the following matrices A.

(a) $\begin{pmatrix} -3 & 2 \\ -4 & 3 \end{pmatrix}.$ (b) $\begin{pmatrix} 3 & 0 \\ 0 & 2 \end{pmatrix}.$

(c) $\begin{pmatrix} 1 & 4 \\ 0 & 5 \end{pmatrix}.$ (d) $\begin{pmatrix} 2 & -1 \\ 1 & 2 \end{pmatrix}.$

11. Let A be the n-by-n matrix with all entries equal to 1.
(a) Show that $A^2 = nA$.

(b) Show that $e^{tA} = I - \frac{1}{n}(e^{nt} - 1)A$.

(c) Find the four entries in e^{tA} when $n = 2$.

3C Independent Solutions

The discussion leading up to the previous Example 2 shows that, for every equation $d\mathbf{x}/dt = A\mathbf{x}$ with A a constant square matrix, there is an exponential solution formula $\mathbf{x}(t) = e^{tA}\mathbf{c}$. In the example we had

$$e^{tA}\mathbf{c} = \begin{pmatrix} e^t & te^t \\ 0 & e^t \end{pmatrix}\begin{pmatrix} c_1 \\ c_2 \end{pmatrix}$$

$$= c_1\begin{pmatrix} e^t \\ 0 \end{pmatrix} + c_2\begin{pmatrix} te^t \\ e^t \end{pmatrix}.$$

Since c_1, c_2 are arbitrary constants, it is good to avoid the redundancy that would occur in the formula in case one of the two columns in the matrix is a constant multiple of the other. We will see that this cannot happen in general, but to state the general result we need to know what is meant by linear independence of vector functions. Let $\mathbf{x}_1(t)$, $\mathbf{x}_2(t)$, . . . ,$\mathbf{x}_m(t)$ be n-dimensional column vectors whose entries are functions on some common interval $a < t < b$. (It is not ruled out of course that some or all of the entries may happen to be constant.) Then the $\mathbf{x}_k(t)$, $k = 1, \ldots, m$ are **linearly independent** if the only choices of c_k for which we always have

$$c_1\mathbf{x}_1(t) + c_2\mathbf{x}_2(t) + \cdots + c_m\mathbf{x}_m(t) = 0$$

are $c_1 = c_2 = \cdots = c_m = 0$. When we have only two vectors ($m = 2$), linear independence is the same as saying that neither vector is a constant multiple of the other. The reason is that if either c_1 or c_2 is not zero we could divide by it and express one vector as a multiple of the other. Similarly, if we have $m > 2$ vectors, linear independence means that none of the vectors is a linear combination of the others. The negation of linear independence of a set of vectors is called **linear dependence**, and it means simply that at least one of the vectors is a linear combination of the others.

EXAMPLE 4   The vector functions

$$\mathbf{x}_1(t) = \begin{pmatrix} e^t \\ 0 \end{pmatrix}, \qquad \mathbf{x}_2 = \begin{pmatrix} te^t \\ e^t \end{pmatrix}$$

that form the exponential matrix in Example 1 are linearly independent. For

$$c_1\begin{pmatrix} e^t \\ 0 \end{pmatrix} + c_2\begin{pmatrix} te^t \\ e^t \end{pmatrix} = \begin{pmatrix} 0 \\ 0 \end{pmatrix}, \qquad -\infty < t < \infty$$

is the same as

$$c_1 e^t + c_2 te^t = 0$$
$$c_2 e^t = 0.$$

It follows that $c_2 = 0$. Hence $c_1 = 0$. This conclusion holds for any fixed value of t, so in particular the constant vectors

$$\begin{pmatrix} 1 \\ 0 \end{pmatrix}, \qquad \begin{pmatrix} 0 \\ 1 \end{pmatrix}$$

are linearly independent. Just set $t = 0$.

EXAMPLE 5   Consider the vector functions

$$\mathbf{x}_1(t) = \begin{pmatrix} e^t \\ 0 \\ 0 \end{pmatrix}, \qquad \mathbf{x}_2(t) = \begin{pmatrix} 0 \\ e^t \\ te^t \end{pmatrix}, \qquad \mathbf{x}_3(t) = \begin{pmatrix} 0 \\ e^t \\ t^2 e^t \end{pmatrix}.$$

The check for independence for $-\infty < t < \infty$ is to solve

$$c_1 \begin{pmatrix} e^t \\ 0 \\ 0 \end{pmatrix} + c_2 \begin{pmatrix} 0 \\ e^t \\ te^t \end{pmatrix} + c_3 \begin{pmatrix} 0 \\ e^t \\ t^2 e^t \end{pmatrix} = \begin{pmatrix} 0 \\ 0 \\ 0 \end{pmatrix}$$

for c_1, c_2, c_3. This is the same as

$$c_1 e^t = 0$$
$$c_2 e^t + c_3 e^t = 0$$
$$c_2 t e^t + c_3 t^2 e^t = 0.$$

The first equation shows that $c_1 = 0$. The middle equation implies $c_3 = -c_2$, so the last equation says $c_2 t - c_2 t^2 = 0$ for all t. Thus $c_2 = c_3 = 0$, so the vector functions are independent as defined on an interval $a < t < b$. Note, however, that when $t = 0$ we get

$$\mathbf{x}_1(0) = \begin{pmatrix} 1 \\ 0 \\ 0 \end{pmatrix}, \qquad \mathbf{x}_2(0) = \begin{pmatrix} 0 \\ 1 \\ 0 \end{pmatrix}, \qquad \mathbf{x}_3(0) = \begin{pmatrix} 0 \\ 1 \\ 0 \end{pmatrix},$$

and these constant vectors are linearly dependent. This shows that functions may be linearly independent while their restrictions to some smaller domain (in this example a single point) may be linearly dependent.

Here is the theorem that guarantees independence of the columns of an exponential matrix for any and all values of t.

3.4 Theorem.   Let A be an n-by-n matrix with constant entries and let $\mathbf{x}_k(t)$ be the kth column of the exponential matrix e^{tA}. Then the vector functions $\mathbf{x}_1(t), \ldots, \mathbf{x}_n(t)$ are linearly independent over any set of t-values.

Proof. Apply the matrix e^{tA} to both sides of the vector equation

$$c_1 \mathbf{x}_1(t) + \cdots + c_n \mathbf{x}_n(t) = 0.$$

Using the properties 3 and then 2 of Theorem 1.1, we get

$$c_1 e^{-tA} \mathbf{x}_1(t) + \cdots + c_n e^{-tA} \mathbf{x}_n(t) = 0.$$

But e^{-tA} is the inverse of the matrix whose kth column is $\mathbf{x}_k(t)$. Hence $e^{-tA} \mathbf{x}_k(t)$ is the kth column of the identity matrix I. Thus our equation becomes

$$c_1 \mathbf{e}_1 + \cdots + c_k \mathbf{e}_k + \cdots + c_n \mathbf{e}_n = 0,$$

where $\mathbf{e}_k$ is the column vector with 1 in the kth entry and 0 elsewhere. Adding up the linear combination gives

$$\begin{pmatrix} c_1 \\ \cdot \\ \cdot \\ \cdot \\ c_k \\ \cdot \\ \cdot \\ \cdot \\ c_n \end{pmatrix} = \begin{pmatrix} 0 \\ \cdot \\ \cdot \\ \cdot \\ 0 \\ \cdot \\ \cdot \\ \cdot \\ 0 \end{pmatrix},$$

which shows that the vectors $\mathbf{x}_k(t)$ are linearly independent.

Theorem 3.3 is a special case of the following more general theorem, the proof of which is deferred to Section 6 of this chapter.

3.5 Theorem.   The columns of a square matrix A are linearly independent if, and only if, the matrix A is invertible.

The proof of Theorem 3.4 is just the easier part of the proof of Theorem 3.5 applied to the special case of an exponential matrix, which we already know is always invertible.

EXERCISES

1. Each of the given matrices is the exponential matrix of some constant matrix A. In each case, express the vector function $\mathbf{x}(t) = e^{tA}\mathbf{c}$ as a linear combination of the columns of e^{tA}.

   (a) $e^{tA} = \begin{pmatrix} 1 & 0 \\ 0 & e^t \end{pmatrix}.$

   (b) $e^{tA} = \begin{pmatrix} e^t & 0 \\ 0 & e^{2}t \end{pmatrix}.$

   (c) $e^{tA} = \begin{pmatrix} e^t & 0 & 0 \\ 0 & e^{2t} & 0 \\ 0 & 0 & e^{3t} \end{pmatrix}.$

   (d) $e^{tA} = \begin{pmatrix} e^t & te^t & 0 \\ 0 & e^t & 0 \\ 0 & 0 & e^{2t} \end{pmatrix}.$

2. For each of the exponential matrices in Exercise 1, find A by computing the derivative of e^{tA} at $t = 0$. Then verify that each column of e^{tA} is a solution of the differential equation $\dot{\mathbf{x}} = A\mathbf{x}$.

3. Not every square matrix with linearly independent columns is an exponential matrix. For example, an exponential matrix e^{tA} must equal I when $t = 0$. Show that each of the following matrices has linearly independent columns, but is not an exponential matrix.

   (a) $\begin{pmatrix} e^t & t \\ 0 & e^t \end{pmatrix}.$
   (b) $\begin{pmatrix} e^{2t} & t^2e^t \\ 0 & e^t \end{pmatrix}.$
   (c) $\begin{pmatrix} 0 & e^t \\ e^t & e^t \end{pmatrix}.$

4. Theorem 3.4 is a simple consequence of the following more general theorem: If $A(t)$ is an invertible square matrix for each t in some interval $a < t < b$, then the columns of $A(t)$ are linearly independent vector functions on this interval. Show how to prove this theorem using the ideas in the proof of Theorem 3.4.

4 NONHOMOGENEOUS SYSTEMS

4A Solution Formula

To develop efficient methods for solving nonhomogeneous systems, we need a formula for the derivative of a matrix product, or, more particularly, the product of a matrix and a vector. The rule is similar to the usual product rule for derivatives and appears in Theorem 1.1 of Section 1:

4.1
$$\frac{d}{dt}[A(t)B(t)] = \left[\frac{d}{dt}A(t)\right]B(t) + A(t)\left[\frac{d}{dt}B(t)\right].$$

Recall that the derivative of a matrix is just the matrix obtained by differentiating the entries. However, the order of the factors on the right is important because it involves matrix multiplication, which is not in general commutative. To prove the formula, we differentiate one entry at a time on the left side; the derivative of the ijth entry is

$$\frac{d}{dt}\sum_{k=0}^{n} a_{ik}(t)b_{kj}(t) = \sum_{k=0}^{n}\frac{da_{ik}}{dt}(t)b_{kj}(t) + \sum_{k=0}^{n}a_{ik}(t)\frac{db_{kj}}{dt}(t).$$

But this is just the ijth entry in the sum of products of matrices on the right, so the formula is proved.

To solve the nonhomogeneous vector equation

$$\frac{d\mathbf{x}}{dt} = A\mathbf{x} + \mathbf{b}(t),$$

where A is a constant matrix, we write

$$\frac{d\mathbf{x}}{dt} - A\mathbf{x} = \mathbf{b}(t)$$

and multiply by the matrix e^{-tA} to get

$$e^{-tA}\frac{d\mathbf{x}}{dt} - e^{-tA}A\mathbf{x} = e^{-tA}\mathbf{b}(t).$$

But by the product rule for differentiation of matrices, this is the same as

$$\frac{d}{dt}(e^{-tA}\mathbf{x}) = e^{-tA}\mathbf{b}(t).$$

Integration of both sides gives

$$e^{-tA}\mathbf{x} = \int e^{-tA}\mathbf{b}(t)\,dt + \mathbf{c}.$$

Since e^{tA} is the inverse of e^{-tA}, we multiply by it to get

4.2
$$\mathbf{x}(t) = e^{tA}\int e^{-tA}\mathbf{b}(t)\,dt + e^{tA}\mathbf{c}$$

for the **general solution**, depending on the constant vector **c**. A review of the steps used in deriving Equation 4.2 shows that the term "general solution" is justified: if $\mathbf{x}(t)$ is some solution of the given equation, then it must have the form of Equation 4.2 for some choice of the vector **c**.

EXAMPLE 1   In the second example of the previous section, we saw that the system

$$\begin{pmatrix} dx/dt \\ dy/dt \end{pmatrix} = \begin{pmatrix} 1 & 1 \\ 0 & 1 \end{pmatrix}\begin{pmatrix} x \\ y \end{pmatrix}$$

had associated with it the fundamental matrix exponential

$$e^{tA} = \begin{pmatrix} e^t & te^t \\ 0 & e^t \end{pmatrix}.$$

Hence, to find a particular solution of

$$\begin{pmatrix} dx/dt \\ dy/dt \end{pmatrix} = \begin{pmatrix} 1 & 1 \\ 0 & 1 \end{pmatrix}\begin{pmatrix} x \\ y \end{pmatrix} + \begin{pmatrix} e^t \\ e^{-t} \end{pmatrix},$$

we compute the particular solution

$$\begin{aligned}
e^{tA} \int e^{-tA} \begin{pmatrix} e^t \\ e^{-t} \end{pmatrix} dt &= \begin{pmatrix} e^t & te^t \\ 0 & e^t \end{pmatrix} \int \begin{pmatrix} e^{-t} & -te^{-t} \\ 0 & e^{-t} \end{pmatrix}\begin{pmatrix} e^t \\ e^{-t} \end{pmatrix} dt \\
&= \begin{pmatrix} e^t & te^t \\ 0 & e^t \end{pmatrix} \int \begin{pmatrix} 1 - te^{-2t} \\ e^{-2t} \end{pmatrix} dt \\
&= \begin{pmatrix} e^t & te^t \\ 0 & e^t \end{pmatrix}\begin{pmatrix} t + \frac{1}{2}te^{-2t} + \frac{1}{4}e^{-2t} \\ -\frac{1}{2}e^{-2t} \end{pmatrix} \\
&= \begin{pmatrix} te^t + \frac{1}{4}e^{-t} \\ -\frac{1}{2}e^{-t} \end{pmatrix}.
\end{aligned}$$

Adding the particular solution just found to the general homogeneous solution we already had gives

$$\begin{aligned}
\mathbf{x}(t) &= \begin{pmatrix} e^t & te^t \\ 0 & e^t \end{pmatrix}\begin{pmatrix} c_1 \\ c_2 \end{pmatrix} + \begin{pmatrix} te^t + \frac{1}{4}e^{-t} \\ -\frac{1}{2}e^{-t} \end{pmatrix} \\
&= \begin{pmatrix} c_1 e^t + c_2 te^t + te^t + \frac{1}{4}e^{-t} \\ c_2 e^t - \frac{1}{2}e^{-t} \end{pmatrix}
\end{aligned}$$

for the general solution.

4B Linearity

Matrix multiplication acts linearly on n-dimensional columns **x** in the usual sense that

$$A(c_1\mathbf{x}_1 + c_2\mathbf{x}_2) = c_1 A\mathbf{x}_1 + c_2 A\mathbf{x}_2.$$

Similarly, d/dt acts linearly:

$$\frac{d}{dt}(c_1\mathbf{x}_1 + c_2\mathbf{x}_2) = c_1\frac{d}{dt}\mathbf{x}_1 + c_2\frac{d}{dt}\mathbf{x}_2.$$

Now suppose $A = A(t)$ has functions of t for entries and define the operator $D - A(t)$ by

$$[D - A(t)]\mathbf{x} = \frac{d\mathbf{x}}{dt} - A(t)\mathbf{x}.$$

It is an easy exercise to show that $D - A(t)$ is a linear operator. This is helpful to know here because of what Theorem 3.1 of Chapter 2 says in this instance about the general solution $\mathbf{x}(t)$ of

$$\frac{d\mathbf{x}}{dt} - A(t)\mathbf{x} = \mathbf{b}(t):$$

find a particular solution $\mathbf{x}_p(t)$ and add it to the general solution $\mathbf{x}_h(t)$ of the **homogenous equation**

$$\frac{d\mathbf{x}}{dt} - A(t)\mathbf{x} = 0$$

to get $\mathbf{x}(t) = \mathbf{x}_p(t) + \mathbf{x}_h(t)$. Equation 4.2 of this section has just this form for the case of constant entries in A. Section 4C shows how to find $\mathbf{x}_p(t)$ if we have $\mathbf{x}_h(t)$. General methods for finding $\mathbf{x}_h(t)$ when $A = A(t)$ has nonconstant entries are not available. We can return to the elimination method, in which case we are faced with the type of problem discussed in Chapter 4 for two-dimensional examples. The other alternative is to resort to the numerical methods of Chapter 4.

4C Variation of Parameters

It is an important fact that even in the case of a nonconstant, continuous n-by-n matrix $A(t)$ there is a formula for a particular solution of

$$\frac{d\mathbf{x}}{dt} = A(t)\mathbf{x} + \mathbf{b}(t)$$

in terms of solutions of the associated homogeneous equation. Suppose $\mathbf{x}_1(t), \ldots,$ $\mathbf{x}_n(t)$ is a set of n linearly independent solutions of

$$\frac{d\mathbf{x}}{dt} = A(t)\mathbf{x}.$$

Solutions would be independent, for example, if their values for some $t = t_0$ were independent vectors. We now form the n-by-n **fundamental matrix**

$$X(t) = (\mathbf{x}_1(t) \ldots \mathbf{x}_n(t))$$

whose columns are these independent vector solutions. If A has constant entries, the columns of e^{tA} will do: $\mathbf{x}_k(t) = e^{tA}\mathbf{e}_k$.

EXAMPLE 2   If A is a constant matrix, then the matrix $X(t) = e^{tA}$ is an example of a fundamental matrix, because its columns are linearly independent solutions of $d\mathbf{x}/dt = A\mathbf{x}$. In Example 1, a fundamental matrix is

$$X_1(t) = e^{tA} = \begin{pmatrix} e^t & te^t \\ 0 & e^t \end{pmatrix}.$$

Another fundamental matrix for the same system is

$$X_2(t) = \begin{pmatrix} te^t & e^t \\ e^t & 0 \end{pmatrix},$$

although $X_2(t)$ is not an exponential matrix, because $X_2(0) \neq I$.

Next we try to find a vector-valued function $\mathbf{v}(t)$ such that

$$\mathbf{x}_p(t) = X(t)\mathbf{v}(t)$$

is a solution of the nonhomogeneous equation. It turns out that this can always be done as follows. Using the product rule for differentiation, we substitute $X(t)\mathbf{v}(t)$ into the nonhomogeneous equation to get

$$\frac{dX(t)}{dt}\mathbf{v}(t) + X(t)\frac{d\mathbf{v}(t)}{dt} = A(t)X(t)\mathbf{v}(t) + \mathbf{b}(t).$$

Since each column of $X(t)$ is a solution of the homogeneous equation, we have

$$\frac{dX(t)}{dt} = A(t)X(t).$$

Therefore, the first term cancels on each side, leaving

$$X(t)\frac{d\mathbf{v}(t)}{dt} = \mathbf{b}(t).$$

If $X(t)$ is invertible, we can multiply both sides of this last equation by $X^{-1}(t)$ to get

$$\frac{d\mathbf{v}(t)}{dt} = X^{-1}(t)\mathbf{b}(t).$$

Integration gives a formula for $\mathbf{v}(t)$. (We need only one solution, so we skip the constant of integration.)

$$\mathbf{v}(t) = \int X^{-1}(t)\mathbf{b}(t)\, dt.$$

Finally,

4.3
$$\mathbf{x}_p(t) = X(t)\mathbf{v}(t)$$

$$= X(t)\int X^{-1}(t)\mathbf{b}(t)\,dt.$$

Notice that this formula is the same as the one previously derived in the constant-coefficient case, with e^{tA} now replaced by the more general $X(t)$. This process for finding $\mathbf{x}_p$ is called **variation of parameters**, because to find it we allow the constant vector $\mathbf{v}_0$ in the homogeneous vector solution $X(t)\mathbf{v}_0$ to vary as a function of t.

It will follow from Theorem 3.1 of Chapter 7 that a fundamental matrix $X(t)$ is always invertible; in specific examples, we will always verify this simply by finding the inverse matrix.

EXAMPLE 3   It is easy to verify that the homogeneous system associated with

$$\frac{d\mathbf{x}}{dt} = \begin{pmatrix} 1 & e^t \\ 0 & 1 \end{pmatrix}\mathbf{x} + \begin{pmatrix} e^t \\ e^{2t} \end{pmatrix}$$

has independent solutions

$$\mathbf{x}_1(t) = \begin{pmatrix} e^t \\ 0 \end{pmatrix}, \qquad \mathbf{x}_2(t) = \begin{pmatrix} e^{2t} \\ e^t \end{pmatrix}.$$

We form a fundamental matrix $X(t)$ and its inverse:

$$X(t) = \begin{pmatrix} e^t & e^{2t} \\ 0 & e^t \end{pmatrix}, \qquad X^{-1}(t) = \begin{pmatrix} e^{-t} & -1 \\ 0 & e^{-t} \end{pmatrix}.$$

Formula 4.3 gives the particular solution

$$\mathbf{x}_p(t) = \begin{pmatrix} e^t & e^{2t} \\ 0 & e^t \end{pmatrix}\int \begin{pmatrix} e^{-t} & -1 \\ 0 & e^{-t} \end{pmatrix}\begin{pmatrix} e^t \\ e^{2t} \end{pmatrix}\,dt$$

$$= \begin{pmatrix} e^t & e^{2t} \\ 0 & e^t \end{pmatrix}\int \begin{pmatrix} 1 - e^{2t} \\ e^t \end{pmatrix}\,dt$$

$$= \begin{pmatrix} e^t & e^{2t} \\ 0 & e^t \end{pmatrix}\begin{pmatrix} t - \tfrac{1}{2}e^{2t} \\ e^t \end{pmatrix} = \begin{pmatrix} te^t + \tfrac{1}{2}e^{3t} \\ e^{2t} \end{pmatrix}.$$

The general solution is then

$$\mathbf{x}(t) = c_1\mathbf{x}_1(t) + c_2\mathbf{x}_2(t) + \mathbf{x}_p(t).$$

If the matrix A in $\dot{\mathbf{x}} = A\mathbf{x} + \mathbf{b}$ is constant, the practical significance of the fundamental matrix $X(t)$ as compared with the more special exponential matrix e^{tA} is that the exponential matrix must equal the identity matrix when $t = 0$, whereas the columns of $X(t)$ are required only to be independent solutions of the homogeneous equation $\dot{\mathbf{x}} = A\mathbf{x}$. The general theory of Chapter 7 shows that a fundamental matrix can also be described as an *invertible* matrix whose columns are solutions of the homogeneous equation, a property that is of course shared by the exponential matrix.

4D Summary of Methods

For linear systems in the standard form $d\mathbf{x}/dt = A\mathbf{x} + \mathbf{b}$, and hence for the systems and equations reducible to this form, we usually proceed as follows:

1. Find the general solution of the homogeneous equation $d\mathbf{x}/dt = A\mathbf{x}$, either by elimination, by the eigenvector method in Section 2, or by finding e^{tA} directly. In the constant-coefficient case the homogeneous solution is always of the form $\mathbf{x}_h(t) = e^{tA}\mathbf{c}$, where $\mathbf{c}$ is a constant vector. If A is not constant, there is no general method for finding $\mathbf{x}_h(t)$, and we will very likely have to use numerical methods.
2. Find a particular solution to the nonhomogeneous equation, either as a by-product of the elimination method, by undetermined coefficient, if applicable, or by Formula 4.2 or 4.3.
3. Write the general solution as $\mathbf{x}(t) = \mathbf{x}_h(t) + \mathbf{x}_p(t)$.

EXERCISES

1. Use Formula 4.2 to solve the following equations of the form $d\mathbf{x}/dt = A\mathbf{x} + \mathbf{b}(t)$. The associated homogeneous equations $d\mathbf{x}/dt = A\mathbf{x}$ were found in Exercise 10 of the previous section 3B to have exponential matrices e^{tA} as shown.

   (a) $\begin{pmatrix} \dfrac{dx}{dt} \\ \dfrac{dy}{dt} \end{pmatrix} = \begin{pmatrix} -3 & 2 \\ -4 & 3 \end{pmatrix}\begin{pmatrix} x \\ y \end{pmatrix} + \begin{pmatrix} e^{2t} \\ 1 \end{pmatrix}$; $e^{tA} = \begin{pmatrix} -e^{t} + 2e^{-t} & e^{t} - e^{-t} \\ -2e^{t} + 2e^{-t} & 2e^{t} - e^{-t} \end{pmatrix}$.

   (b) $\begin{pmatrix} \dfrac{dx}{dt} \\ \dfrac{dy}{dt} \end{pmatrix} = \begin{pmatrix} 3 & 0 \\ 0 & 2 \end{pmatrix}\begin{pmatrix} x \\ y \end{pmatrix} + \begin{pmatrix} e^{t} + 1 \\ e^{-t} \end{pmatrix}$; $e^{tA} = \begin{pmatrix} e^{3t} & 0 \\ 0 & e^{2t} \end{pmatrix}$.

   (c) $\begin{pmatrix} \dfrac{dx}{dt} \\ \dfrac{dy}{dt} \end{pmatrix} = \begin{pmatrix} 1 & 4 \\ 0 & 5 \end{pmatrix}\begin{pmatrix} x \\ y \end{pmatrix} + \begin{pmatrix} 1 \\ e^{t} \end{pmatrix}$; $e^{tA} = \begin{pmatrix} e^{t} & e^{5t} - e^{t} \\ 0 & e^{5t} \end{pmatrix}$.

   (d) $\begin{pmatrix} \dfrac{dx}{dt} \\ \dfrac{dy}{dt} \end{pmatrix} = \begin{pmatrix} 2 & -1 \\ 1 & 2 \end{pmatrix}\begin{pmatrix} x \\ y \end{pmatrix} + \begin{pmatrix} e^{2t} \\ 2e^{2t} \end{pmatrix}$; $e^{tA} = \begin{pmatrix} e^{2t}\cos t & -e^{2t}\sin t \\ e^{2t}\sin t & e^{2t}\cos t \end{pmatrix}$.

2. Find a particular solution of each of the differential equations in Exercise 1 that satisfies the following corresponding initial condition:

(a) $\mathbf{x}(0) = (-1, -1)$.   **(b)** $\mathbf{x}(0) = (0, 0)$.
(c) $\mathbf{x}(0) = (0, 1)$.   **(d)** $\mathbf{x}(0) = (-1, -1)$.

3. (a) Show that, for a solution of the form

$$\mathbf{x}(t) = e^{tA}\mathbf{c}, \qquad \mathbf{c} \text{ constant}$$

to satisfy the condition $\mathbf{x}(t_0) = \mathbf{x}_0$, we must have $\mathbf{c} = e^{-t_0 A}\mathbf{x}_0$.

(b) Show that if $X(t)$ is an n-by-n matrix with linearly independent columns, in particular if $X(t)$ is a fundamental matrix, then for

$$\mathbf{x}(t) = X(t)\mathbf{c}, \qquad \mathbf{c} \text{ constant}$$

to satisfy $\mathbf{x}(t_0) = \mathbf{x}_0$ we must have $\mathbf{c} = X^{-1}(t_0)\mathbf{x}_0$.

4. Let $X(t)$ be a fundamental matrix whose columns span the set of solutions of the homogeneous equation

$$\frac{d\mathbf{x}}{dt} = A(t)x.$$

(a) Show that if $\mathbf{x}_p(t)$ is a particular solution of the nonhomogeneous system

$$\frac{d\mathbf{x}}{dt} = A(t)\mathbf{x} + \mathbf{b}(t),$$

then the general solution of the nonhomogeneous system can be written

$$\mathbf{x}(t) = \mathbf{x}_p(t) + X(t)\mathbf{c},$$

where $\mathbf{c}$ is a constant.

(b) Show that, if the general solution in part (a) is to satisfy an initial condition $\mathbf{x}(t_0) = \mathbf{x}_0$, then $\mathbf{c}$ should be chosen so that

$$\mathbf{c} = X^{-1}(t_0)(\mathbf{x}_0 - \mathbf{x}_p(t_0)).$$

5. The systems in Exercise 1 can be solved by the method of undetermined coefficients: Form linear combinations of the terms, and their derivatives, that occur in each entry of the nonhomogeneous part of the differential equation, taking care to include appropriate multiples by powers of t for terms that are also homogeneous solutions. Then substitute into the equation to determine the coefficients of combination. Use this method on each of the parts of Exercise 1.

6. Each of the following systems has the *homogeneous* solutions shown. Verify that these are linearly independent solutions. Find a particular solution of the nonhomogeneous equation, using Equation 4.3.

(a)
$$\begin{pmatrix} \dfrac{dx}{dt} \\[2mm] \dfrac{dy}{dt} \end{pmatrix} = \begin{pmatrix} \dfrac{3}{2t} & -\dfrac{1}{2} \\[2mm] -\dfrac{1}{2t^2} & \dfrac{1}{2t} \end{pmatrix}\begin{pmatrix} x \\ y \end{pmatrix} + \begin{pmatrix} t^3 \\ 3t^2 \end{pmatrix}; \quad \begin{pmatrix} t \\ 1 \end{pmatrix}, \begin{pmatrix} -t^2 \\ t \end{pmatrix}.$$

(b)
$$\begin{pmatrix} \dfrac{dx}{dt} \\[2mm] \dfrac{dy}{dt} \end{pmatrix} = \begin{pmatrix} \dfrac{t}{t-1} & -\dfrac{1}{t-1} \\[2mm] 1 & 0 \end{pmatrix}\begin{pmatrix} x \\ y \end{pmatrix} + \begin{pmatrix} 1-t \\ 1-t^2 \end{pmatrix}; \quad \begin{pmatrix} 1 \\ t \end{pmatrix}, \begin{pmatrix} e^t \\ e^t \end{pmatrix}.$$

7. (a) Let $\mathbf{x}_1(t), \ldots, \mathbf{x}_n(t)$ be linearly independent, continuously differentiable functions taking values in $\mathcal{R}^n$. Let $X(t)$ be the n-by-n matrix with columns $\mathbf{x}_1(t), \ldots, \mathbf{x}_n(t)$. Show that if we define

$$A(t) = X'(t)X^{-1}(t)$$

then the system $d\mathbf{x}/dt = A(t)\mathbf{x}$ has $\mathbf{x}_1(t), \ldots, \mathbf{x}_n(t)$ as solutions and thus has $X(t)$ as a fundamental matrix.

   (b) Find a first-order homogeneous linear system of the form $d\mathbf{x}/dt = A(t)\mathbf{x}$ having

$$\mathbf{x}_1(t) = \begin{pmatrix} e^t \\ 2e^{2t} \end{pmatrix}, \qquad \mathbf{x}_2(t) = \begin{pmatrix} 1 \\ e^t \end{pmatrix}$$

   as solutions. Are these two solutions linearly independent?

8. Let A be an n-by-n invertible matrix of constants, and let $\mathbf{b}$ be a fixed vector in $\mathcal{R}^n$. Show that the equation

$$\frac{d\mathbf{x}}{dt} = A\mathbf{x} + \mathbf{b}$$

   always has $\mathbf{x}_p = -A^{-1}\mathbf{b}$ for a particular solution.

9. Show that if A is a constant n-by-n matrix then the equation

$$\frac{d\mathbf{x}}{dt} = A\mathbf{x}$$

   has $X(t) = e^{tA}$ for its fundamental matrix of independent column solutions with $X(0) = I$.

10. Let $A(t)$ be a square matrix with entries that are differentiable on some interval.
   (a) Show that

$$\frac{dA^2}{dt} = 2A\frac{dA}{dt}$$

   when A and dA/dt commute.
   (b) Show that

$$\frac{de^A}{dt} = e^A\frac{dA}{dt}$$

   when A and dA/dt commute.

11. Modify the derivation of Formula 4.2 to show that, for A constant and $\mathbf{b}(t)$ continuous, the initial-value problem

$$\frac{d\mathbf{x}}{dt} = A\mathbf{x} + \mathbf{b}(t), \qquad \mathbf{x}(t_0) = \mathbf{x}_0$$

   has a unique solution of the form

$$\mathbf{x}(t) = e^{tA}\int_{t_0}^{t} e^{-uA}\mathbf{b}(u)\, du + e^{(t - t_0)A}\mathbf{x}_0.$$

   (*Hint:* Integrate from t_0 to t instead of using an indefinite integral.)

5 COMPUTING e^{tA} IN GENERAL

So far we have seen two different ways to find e^{tA}. One way is to use the elimination method to find the general solution of the system $\dot{\mathbf{x}} = A\mathbf{x}$ and then to use the general solution to find the particular solutions $\mathbf{x}_k(t)$ that satisfy $\mathbf{x}_k(0) = \mathbf{e}_k$. Then $\mathbf{x}_k(t)$ is the kth column of e^{tA}. This method has the advantage that you can be opportunistic and use any special characteristics that you can spot in the system to simplify the computation. The disadvantage is the other side of that coin: choices have to be made at each step, so the process is hard to reduce to a single routine that works efficiently for all systems. The method of eigenvalues and eigenvectors is somewhat more routine, even for large systems, but has the disadvantage that it does not work, without substantial extension, for all n-by-n matrices A.

The method described next is completely routine, in fact so much so that it is quite simple to program for a computing system capable of symbolic computation. In addition, the method compares favorably with other methods when carried out by hand for systems of only moderately large dimension.

The idea behind the method is the following remarkable theorem:

5.1   Cayley–Hamilton Theorem.   If A is an n-by-n matrix with characteristic polynomial

$$P(\lambda) = \det(A - \lambda I),$$

then the matrix polynomial obtained by substituting A^k for λ^k in $P(\lambda)$ satisfies $P(A) = 0$, with the understanding that $A^0 = I$ replaces $\lambda^0 = 1$.

This theorem has several interesting proofs, but they would take us rather far afield at this point, so we will illustrate its use by example.

EXAMPLE 1   Let

$$A = \begin{pmatrix} 1 & 1 \\ 0 & 1 \end{pmatrix}.$$

Then $P(\lambda) = \det(A - \lambda I)$ is the same as

$$P(\lambda) = \det\begin{pmatrix} 1 - \lambda & 1 \\ 0 & 1 - \lambda \end{pmatrix}$$

$$= (1 - \lambda)^2 = 1 - 2\lambda + \lambda^2.$$

The corresponding matrix polynomial is $P(A) = I - 2A + A^2$. It is simple to verify that $P(A)$ is the zero matrix:

$$I - 2A + A^2 = 0,$$

or

$$\begin{pmatrix} 1 & 0 \\ 0 & 1 \end{pmatrix} - 2\begin{pmatrix} 1 & 1 \\ 0 & 1 \end{pmatrix} + \begin{pmatrix} 1 & 2 \\ 0 & 1 \end{pmatrix} = \begin{pmatrix} 0 & 0 \\ 0 & 0 \end{pmatrix}.$$

It follows not only that

$$A^2 = -I + 2A,$$

but that

$$\begin{aligned} A^3 = AA^2 &= A(-I + 2A) \\ &= -A + 2A^2 \\ &= -A + 2(-I + 2A) \\ &= -2I + 3A. \end{aligned}$$

Since $A^4 = AA^3$, we find similarly that

$$\begin{aligned} A^4 &= A(-2I + 3A) \\ &= -2A + 3A^2 \\ &= -2A + 3(-I + 2A) \\ &= -3I + 4A. \end{aligned}$$

In this way, a power A^k can be expressed in the form $c_k I + d_k A$ using the Cayley–Hamilton theorem to reduce the exponents.

For any n-by-n matrix, the Cayley–Hamilton theorem allows us to express A^n, and all higher powers of A, as polynomials of degree at most $n - 1$ in A. Since e^{tA} is defined to be a power series in A, it may seem that a similar reduction of the series might fail, because of the infinitely many terms in the expansion. However, the reduction of the series of e^{tA} always works, although it is most convenient to express it in terms of matrices of the form $(A - \lambda_k I)$; the numbers λ_k are the **characteristic roots,** or **eigenvalues** of A, that is, the roots of the polynomial

$$P(\lambda) = \det(A - \lambda I).$$

If A is an n-by-n matrix, the equation

$$\det(A - \lambda I) = 0$$

has degree n, and has n roots $\lambda_1, \ldots, \lambda_n$ if we allow for possible multiple roots.

EXAMPLE 2   If

$$A = \begin{pmatrix} 1 & 1 \\ 0 & 1 \end{pmatrix},$$

then $\det(A - \lambda I) = (1 - \lambda)^2$, as we saw in Example 1. Thus the characteristic roots are $\lambda_1 = \lambda_2 = 1$, so there is a double root.

5.2   Theorem.   Let A be an n-by-n matrix and let $\lambda_1, \ldots, \lambda_n$ be the roots of the characteristic equation with some λ's possibly equal. Then $e^{tA} = E(t)$, where

$$E(t) = a_1(t)B_1 + a_2(t)B_2 + \cdots + a_n(t)B_n,$$

with $a_k(t)$ determined recursively by

$$a_1(t) = e^{\lambda_1 t}, \qquad a_k(t) = \int_0^t e^{\lambda_k(t-u)} a_{k-1}(u)\, du, \qquad k = 2, \ldots, n,$$

and the constant matrices B_k determined by

$$B_1 = I, \qquad B_k = (A - \lambda_1 I) \ldots (A - \lambda_{k-1} I), \qquad k = 2, \ldots, n.$$

Proof. We will show that $E'(t) = AE(t)$. Differentiating the formula for $a_k(t)$ shows that

$$a_1'(t) = \lambda_1 a_1(t), \qquad a_k'(t) = \lambda_k a_k(t) + a_{k-1}(t), \qquad k = 2, \ldots, n.$$

Then

$$E'(t) = a_1'(t)B_1 + \cdots + a_k'(t)B_k + \cdots + a_n'(t)B_n$$

$$= \lambda_1 a_1(t)I + \cdots + [\lambda_k a_k(t) + a_{k-1}(t)]B_k + \cdots + [\lambda_n a_n(t) + a_{n-1}(t)]B_n$$

$$= \sum_{k=1}^{n} \lambda_k a_k(t)B_k + \sum_{k=2}^{n} a_{k-1}(t)B_k.$$

But $\lambda_k B_k = AB_k - B_{k+1}$, since $B_{k+1} = (A - \lambda_k I)B_k$. Hence

$$E'(t) = \sum_{k=1}^{n} a_k(t)(AB_k - B_{k+1}) + \sum_{k=2}^{n} a_{k-1}(t)B_k$$

$$= A \sum_{k=1}^{n} a_k(t)B_k - \sum_{k=1}^{n} a_k(t)B_{k+1} + \sum_{k=1}^{n-1} a_k(t)B_{k+1}$$

$$= AE(t) - a_n(t)B_{n+1}.$$

But $B_{n+1} = (A - \lambda_1 I) \ldots (A - \lambda_n I)$ is the zero matrix by the Cayley–Hamilton theorem. Hence $E'(t) - AE(t) = 0$. Multiplying by e^{-tA} gives

$$e^{-tA} E'(t) - Ae^{-tA} E(t) = [e^{-tA} E(t)]' = 0.$$

Thus $e^{-tA} E(t) = C$ for some constant matrix C. Since $E(0) = I$, then $C = I$ and $E(t) = e^{tA}$.

EXAMPLE 3   Suppose A and its characteristic equation are

$$A = \begin{pmatrix} 1 & 1 \\ 4 & 1 \end{pmatrix}, \qquad \det\begin{pmatrix} 1 - \lambda & 1 \\ 4 & 1 - \lambda \end{pmatrix} = (1 - \lambda)^2 - 4 = 0.$$

Thus the characteristic roots, or eigenvalues, are the roots of

$$\lambda^2 - 2\lambda - 3 = (\lambda - 3)(\lambda + 1) = 0.$$

We set $\lambda_1 = -1$, $\lambda_2 = 3$. Since $B_1 = I$, we compute only

$$B_2 = \begin{pmatrix} 1 - \lambda_1 & 1 \\ 4 & 1 - \lambda_1 \end{pmatrix} = \begin{pmatrix} 2 & 1 \\ 4 & 2 \end{pmatrix}.$$

With $a_1(t) = e^{-t}$, we then compute

$$a_2(t) = \int_0^t e^{3(t-u)} a_1(u) \, du$$

$$= \int_0^t e^{3(t-u)} e^{-u} \, du$$

$$= e^{3t} \int_0^t e^{-4u} \, du = -\tfrac{1}{4} e^{-t} + \tfrac{1}{4} e^{3t}.$$

Then by Theorem 5.2, $e^{tA} = a_1(t)I + a_2(t)B_2$, so

$$e^{tA} = e^{-t}\begin{pmatrix} 1 & 0 \\ 0 & 1 \end{pmatrix} + \left(-\tfrac{1}{4} e^{-t} + \tfrac{1}{4} e^{3t} \right)\begin{pmatrix} 2 & 1 \\ 4 & 2 \end{pmatrix}$$

$$= \begin{pmatrix} \tfrac{1}{2} e^{-t} + \tfrac{1}{2} e^{3t} & -\tfrac{1}{4} e^{-t} + \tfrac{1}{4} e^{3t} \\ -e^{-t} + e^{3t} & \tfrac{1}{2} e^{-t} + \tfrac{1}{2} e^{3t} \end{pmatrix}.$$

EXAMPLE 4   Suppose A and its characteristic equation are

$$A = \begin{pmatrix} 1 & -1 \\ 1 & 1 \end{pmatrix}, \qquad \det\begin{pmatrix} 1 - \lambda & -1 \\ 1 & 1 - \lambda \end{pmatrix} = (1 - \lambda)^2 + 1.$$

The eigenvalues are the roots of

$$\lambda^2 - 2\lambda + 2 = 0.$$

The quadratic formula shows that the roots are

$$\frac{2 \pm \sqrt{-4}}{2} \quad \text{or} \quad \lambda_1 = 1 - i, \qquad \lambda_2 = 1 + i.$$

Then

$$B_2 = \begin{pmatrix} 1 - \lambda_1 & -1 \\ 1 & 1 - \lambda_1 \end{pmatrix} = \begin{pmatrix} i & -1 \\ 1 & i \end{pmatrix}.$$

Since $a_1(t) = e^{(1-i)t}$, we compute

$$a_2(t) = \int_0^t e^{(1+i)(t-u)} a_1(u) \, du = \int_0^t e^{(1+i)(t-u)} e^{(1-i)u} \, du$$

$$= e^{(1+i)t} \int_0^t e^{-2iu} \, du = e^{(1+i)t}\left[\frac{1}{-2i}(e^{-2it} - 1) \right]$$

$$= \frac{1}{-2i}(e^{(1-i)t} - e^{(1+i)t}) = e^t \sin t.$$

By Theorem 5.2, $e^{tA} = a_1(t)I + a_2(t)B_2$, so

$$e^{tA} = e^{(1-i)t}\begin{pmatrix} 1 & 0 \\ 0 & 1 \end{pmatrix} + e^t \sin t \begin{pmatrix} i & -1 \\ 1 & i \end{pmatrix}$$

$$= \begin{pmatrix} e^t \cos t & -e^t \sin t \\ e^t \sin t & e^t \cos t \end{pmatrix}.$$

EXAMPLE 5   Consider the system

$$\frac{d\mathbf{x}}{dt} = \begin{pmatrix} 1 & 0 & 1 & -1 \\ -1 & 2 & 0 & 1 \\ -1 & 1 & 0 & 2 \\ 0 & 0 & 0 & 1 \end{pmatrix}\mathbf{x}.$$

Expansion of the determinant by the last row gives the eigenvalue equation

$$\det\begin{pmatrix} 1-\lambda & 0 & 1 & -1 \\ -1 & 2-\lambda & 0 & 1 \\ -1 & 1 & -2 & 2 \\ 0 & 0 & 0 & 1-\lambda \end{pmatrix} = (\lambda - 1)^4 = 0.$$

Hence $\lambda_1 = \lambda_2 = \lambda_3 = \lambda_4 = 1$ is an eigenvalue of multiplicity 4. We have, as always, $B_1 = I$. In this example,

$$B_2 = (A - I) = \begin{pmatrix} 0 & 0 & 1 & -1 \\ -1 & 1 & 0 & 1 \\ -1 & 1 & -1 & 2 \\ 0 & 0 & 0 & 0 \end{pmatrix}$$

and

$$B_3 = (A - I)^2 = \begin{pmatrix} -1 & 1 & -1 & 2 \\ -1 & 1 & -1 & 2 \\ 0 & 0 & 0 & 0 \\ 0 & 0 & 0 & 0 \end{pmatrix}.$$

Finally, we have $B_4 = (A - I)^3$, which just happens to turn out to be the zero matrix. Thus we need to use only the coefficients $a_1(t)$, $a_2(t)$, and $a_3(t)$, since $a_4(t)$ would not affect the expression for e^{tA}. Then

$$a_1(t) = e^t, \qquad a_2(t) = \int_0^t e^{t-u}e^u \, du$$

$$= e^t \int_0^t du = te^t,$$

$$a_3(t) = \int_0^t e^{t-u}ue^u$$

$$= e^t \int_0^t u \, du = \frac{1}{2}t^2 e^t.$$

The end result, since $B_4 = 0$, is that

$$e^{tA} = e^t I + te^t B_2 + \tfrac{1}{2}t^2 e^t B_3$$

$$= \begin{pmatrix} (1 - \tfrac{1}{2}t^2)e^t & \tfrac{1}{2}t^2 e^t & (t - \tfrac{1}{2}t^2)e^t & (-t + t^2)e^t \\ -(t + \tfrac{1}{2}t^2)e^t & (1 + t + \tfrac{1}{2}t^2)e^t & -\tfrac{1}{2}t^2 e^t & (t + t^2)e^t \\ -te^t & te^t & (1 - t)e^t & 2te^t \\ 0 & 0 & 0 & e^t \end{pmatrix}.$$

Note that e^t can be factored out of the entire matrix, leaving $e^t F(t)$, where $F(t)$ has only polynomial entries.

EXERCISES

1. Solve $\dfrac{d\mathbf{x}}{dt} = A\mathbf{x}$ for the matrices A by finding e^{tA}; the general solution is then a linear combination of the columns of e^{tA}.

(a) $\begin{pmatrix} 8 & -3 \\ 10 & -3 \end{pmatrix}.$

(b) $\begin{pmatrix} 8 & 9 \\ -4 & -4 \end{pmatrix}.$

(c) $\begin{pmatrix} 1 & 1 & 2 \\ 0 & 1 & -1 \\ 0 & 0 & 2 \end{pmatrix}.$

(d) $\begin{pmatrix} 1 & 1 & 1 \\ 0 & 1 & 0 \\ 0 & 0 & 2 \end{pmatrix}.$

(e) $\begin{pmatrix} 1 & 1 & 1 & 0 \\ 0 & 0 & -1 & 0 \\ 1 & 1 & 2 & 0 \\ 1 & 0 & -1 & 1 \end{pmatrix}.$

(f) $\begin{pmatrix} 1 & 0 & 0.5 & -0.5 \\ 1 & 0 & -1 & 0 \\ 0 & 2 & 2.5 & 0.5 \\ -1 & 1 & 0.5 & 1.5 \end{pmatrix}.$

2. Solve the system

$$\frac{dx}{dt} = 2x \qquad\quad + z,$$

$$\frac{dy}{dt} = -x + 3y + z,$$

$$\frac{dz}{dt} = -x \qquad\quad + 4z.$$

(*Hint:* $\lambda = 3$ is a triple eigenvalue.)

3. Solve the system

$$\frac{dx}{dt} = 2x \qquad + z,$$

$$\frac{dy}{dt} = \qquad\; y \qquad + w,$$

$$\frac{dz}{dt} = \qquad 2z + w,$$

$$\frac{dw}{dt} = \quad -y \qquad + w.$$

4. Find e^{-tA} for each of the matrices A in Exercise 1(a) and (b).

5. Find the solution to the equation in Exercise 1 that satisfies the corresponding initial condition given here.

(a) $\mathbf{x}(0) = \begin{pmatrix} 2 \\ -3 \end{pmatrix}.$ **(b)** $\mathbf{x}(1) = \begin{pmatrix} 1 \\ 1 \end{pmatrix}.$

(c) $\mathbf{x}(0) = \begin{pmatrix} 0 \\ 1 \\ 0 \end{pmatrix}.$ **(d)** $\mathbf{x}(0) = \begin{pmatrix} 2 \\ 1 \\ 1 \end{pmatrix}.$

6. Let $A = \begin{pmatrix} \alpha & 1 \\ 0 & \beta \end{pmatrix}$, where α, β are real numbers.

(a) Show that if $\alpha \neq \beta$ then A has two linearly independent eigenvectors.

(b) Show that if $\alpha = \beta$ then the only eigenvectors of A are of the form $\mathbf{u} = \begin{pmatrix} c \\ 0 \end{pmatrix}, c \neq 0.$

(c) Compute e^{tA} in each of the two cases.

7. Show by using the definition of e^{tA} as a matrix power series that $e^{tI} = e^{t}I$.

8. An n-by-n matrix of the form

$$J = \begin{pmatrix} a & 1 & 0 & . & . & . & 0 \\ 0 & a & 1 & . & . & . & 0 \\ 0 & 0 & a & . & . & . & 0 \\ . & . & . & & & & . \\ . & . & . & & & & . \\ . & . & . & & & & . \\ 0 & 0 & 0 & . & . & . & 1 \\ 0 & 0 & 0 & . & . & . & a \end{pmatrix}$$

is called a **Jordan matrix.**

(a) Show that the only eigenvalue of J is $\lambda = a$, and that its only eigenvector is of the form $c\mathbf{e}_1, c \neq 0.$

(b) Show that $(J - aI)\mathbf{e}_j = \mathbf{e}_{j-1}, j = 2, \ldots, n,$ and conclude that J has $\mathbf{e}_2, \ldots, \mathbf{e}_n$, as generalized eigenvectors.

(c) Show that $(J - aI)^n = 0$ and use Theorem 5.2 to show that the entries in e^{tJ} are of the form $p(t)e^{at}$, where $p(t)$ is a polynomial of degree at most $n - 1$.

9. For each n-by-n in matrix A in Exercise 1, use the Cayley-Hamilton theorem to express A^n as a linear combination of lower powers of A.

6 ALGEBRAIC EQUATIONS

To solve a linear system of constant-coefficient differential equations, we can use the elimination method of Chapter 4 or, alternatively, we can try to find the appropriate exponential matrix. However, computing an exponential matrix directly is sometimes too difficult. It happens that computational methods for large systems often boil down to using some kind of elimination method for certain systems of linear algebraic equations. As a special case of this method, the computation of inverse matrices can be effectively done by elimination.

6A Row Operations

Using the product of an m-by-n matrix A and an n-dimensional column vector $\mathbf{x}$, we can write any system of linear algebraic equations in the form $A\mathbf{x} = \mathbf{b}$, where $\mathbf{b}$ is an m-dimensional column vector.

> **EXAMPLE 1**   Each system on the left is equivalent to the matrix equation on the right:

$$
\begin{aligned}
4x + 3y &= 1 \\
-x + 2y &= 2
\end{aligned}
\qquad
\begin{pmatrix} 4 & 3 \\ -1 & 2 \end{pmatrix}
\begin{pmatrix} x \\ y \end{pmatrix}
=
\begin{pmatrix} 1 \\ 2 \end{pmatrix}.
$$

$$
\begin{aligned}
2x + y + 2z &= -1 \\
x + 2y + z &= 0
\end{aligned}
\qquad
\begin{pmatrix} 2 & 1 & 2 \\ 1 & 2 & 1 \end{pmatrix}
\begin{pmatrix} x \\ y \\ z \end{pmatrix}
=
\begin{pmatrix} -1 \\ 0 \end{pmatrix}.
$$

$$
\begin{aligned}
x_1 + x_2 &= 1 \\
x_1 - x_2 &= 0 \\
x_1 + 2x_2 &= 1
\end{aligned}
\qquad
\begin{pmatrix} 1 & 1 \\ 1 & -1 \\ 1 & 2 \end{pmatrix}
\begin{pmatrix} x_1 \\ x_2 \end{pmatrix}
=
\begin{pmatrix} 1 \\ 0 \\ 1 \end{pmatrix}.
$$

In solving systems of linear equations, the operations we use are "multiplying an equation by a number" and "adding a multiple of one equation to another." It is easy to verify afterward by substitution that we have found a solution. In fact, the final system will turn out to be **equivalent** to the system we started with in the sense that both systems will have exactly the same solutions. The general method then is to reduce a system to an equivalent one that is easy to solve by inspection, or at least with very little arithmetic. Before proving that this can always be done, we define precisely what operations will be used. The definitions will be given in terms of matrices and will refer to the general matrix equation $A\mathbf{x} = \mathbf{b}.$

The elimination method consists of performing on a set of equations simple operations that can just as well be performed on the rows of the matrix A and vector $\mathbf{b}$ if the equations are written in vector form $A\mathbf{x} = \mathbf{b}.$ These **row operations** on a matrix or column vector are as follows:

1. Multiplication of a row by a nonzero number.
2. Adding a multiple of one row to another row.
3. Changing the order of the rows.

These are just the operations that we perform on a system of equations to find its solutions. Two remarks are in order. First, a row operation leaves the set of solutions of a system of equations unchanged. In fact, there is always a reverse row operation that will take us back to where we started, so a solution of one system is always a solution of the other. Second, the purpose in performing the operations is to arrive at a system of equations for which the solutions are easy to describe. An example will make this clear.

EXAMPLE 2   We will solve a system in matrix form, writing the system of linear equations next to it for comparison. In either form the operations must be applied to *both sides* of the equalities.

$$\begin{array}{rcl} x - y & = & 1 \\ 2x - 3y + z & = & 0 \\ x + y - 5z & = & 2 \end{array} \qquad \begin{pmatrix} 1 & -1 & 0 \\ 2 & -3 & 1 \\ 1 & 1 & -5 \end{pmatrix}\begin{pmatrix} x \\ y \\ z \end{pmatrix} = \begin{pmatrix} 1 \\ 0 \\ 2 \end{pmatrix}.$$

Add (-2) times the first row to the second and (-1) times the first row to the third:

$$\begin{array}{rcl} x - y & = & 1 \\ -y + z & = & -2 \\ 2y - 5z & = & 1 \end{array} \qquad \begin{pmatrix} 1 & -1 & 0 \\ 0 & -1 & 1 \\ 0 & 2 & -5 \end{pmatrix}\begin{pmatrix} x \\ y \\ z \end{pmatrix} = \begin{pmatrix} 1 \\ -2 \\ 1 \end{pmatrix}.$$

Add twice the second row to the third:

$$\begin{array}{rcl} x - y & = & 1 \\ -y + z & = & -2 \\ -3z & = & -3 \end{array} \qquad \begin{pmatrix} 1 & -1 & 0 \\ 0 & -1 & 1 \\ 0 & 0 & -3 \end{pmatrix}\begin{pmatrix} x \\ y \\ z \end{pmatrix} = \begin{pmatrix} 1 \\ -2 \\ -3 \end{pmatrix}.$$

We see right away from the last row that $z = 1$, from the second row that then $y = 3$, and finally from the first row that $x = 4$. Then $(x, y, z) = (4, 3, 1)$ is the solution.

EXAMPLE 3   We now exhibit a system of equations with infinitely many solutions. Consider the matrix equation, with entries x, y, and z in $\mathbf{x}$,

$$\begin{pmatrix} 1 & -2 & -3 \\ \frac{1}{2} & -2 & -\frac{13}{2} \\ -3 & 5 & 4 \end{pmatrix}\mathbf{x} = \begin{pmatrix} 2 \\ 7 \\ 0 \end{pmatrix}.$$

Add $(-\frac{1}{2})$ times the first row to the second row, and then add 3 times the first row to the third row to produce zeros in the second and third entries of the first column, and obtain

$$\begin{pmatrix} 1 & -2 & -3 \\ 0 & -1 & -5 \\ 0 & -1 & -5 \end{pmatrix}\mathbf{x} = \begin{pmatrix} 2 \\ 6 \\ 6 \end{pmatrix}.$$

Multiply the second row by (-1) to obtain

$$\begin{pmatrix} 1 & -2 & -3 \\ 0 & 1 & 5 \\ 0 & -1 & -5 \end{pmatrix} \mathbf{x} = \begin{pmatrix} 2 \\ -6 \\ 6 \end{pmatrix}.$$

Add 2 times the second row to the first and then add 1 times the second row to the third to obtain

$$\begin{pmatrix} 1 & 0 & 7 \\ 0 & 1 & 5 \\ 0 & 0 & 0 \end{pmatrix} \mathbf{x} = \begin{pmatrix} -10 \\ -6 \\ 0 \end{pmatrix}.$$

Let us put $\mathbf{x} = \begin{pmatrix} x \\ y \\ z \end{pmatrix}$, and translate the matrix equation back into a system of linear equations. The result is

$$\begin{aligned} x \qquad\quad + 7z &= -10, \\ y + 5z &= -\ 6, \\ 0x + 0y + 0z &= \quad 0. \end{aligned}$$

The third equation is satisfied for any values of x, y, z. The first two equations may be rewritten as $x = -10 - 7z$ and $y = -6 - 5z$. Thus, for any value of z,

$$\begin{pmatrix} -7z - 10 \\ -5z -\ 6 \\ z \end{pmatrix} = z \begin{pmatrix} -7 \\ -5 \\ 1 \end{pmatrix} + \begin{pmatrix} -10 \\ -6 \\ 0 \end{pmatrix}$$

is a solution, and every solution has this form for some value of z. We have now described the set of solutions of

$$\begin{pmatrix} 1 & 0 & 7 \\ 0 & 1 & 5 \\ 0 & 0 & 0 \end{pmatrix} \mathbf{x} = \begin{pmatrix} -10 \\ -6 \\ 0 \end{pmatrix},$$

and by reversing steps we know that this is the same as the set of solutions of the matrix equation with which we started.

EXAMPLE 4   Consider the matrix equation

$$\begin{pmatrix} 1 & -2 & -3 \\ \frac{1}{2} & -2 & -\frac{13}{2} \\ -3 & 5 & 4 \end{pmatrix} \mathbf{x} = \begin{pmatrix} 2 \\ 7 \\ 2 \end{pmatrix}.$$

The matrix on the left is the same as the matrix in Example 3. Application of the same sequence of elementary operations yields

$$\begin{pmatrix} 1 & 0 & 7 \\ 0 & 1 & 5 \\ 0 & 0 & 0 \end{pmatrix} \mathbf{x} = \begin{pmatrix} -10 \\ -6 \\ 2 \end{pmatrix}.$$

Whatever $\mathbf{x}$ is, the third row in the product $\begin{pmatrix} 1 & 0 & 7 \\ 0 & 1 & 5 \\ 0 & 0 & 0 \end{pmatrix}\mathbf{x}$ will be zero because the third row of the left factor is zero. Thus no value of $\mathbf{x}$ can give a column vector with 2 in the third row, and we conclude that the equation has *no* solution. If we put $\mathbf{x} = (x, y, z)$ and write the matrix equation out as a system of equations, we obtain

$$x \qquad + 7z = -10,$$

$$y + 5z = -\ 6,$$

$$0x + 0y + 0z = \quad 2.$$

The last equation obviously cannot be satisfied by any values of x, y, z.

6B Reduced Matrices

We are going to look here at what it means for a matrix equation to be "easy to solve by inspection." A **zero row** in a matrix is a row of nothing but zeros. A **leading entry** in a matrix is the first nonzero entry in a row. In these terms a matrix is **reduced** if the following three conditions hold:

  (i) Every *column* containing a leading entry is zero except for the leading entry.
 (ii) Every leading entry is 1.
(iii) Any zero rows are at the bottom of the matrix.

Here is the general procedure for putting a matrix in reduced form.

1. If a leading entry r in some row is not 1, multiply by r^{-1} to make it 1.
2. If some entries in a column with a leading entry are not zero, make them zero by adding the proper multiples of the row with the leading entry to the other rows. Another column that already has a leading entry s, and zeros elsewhere, will not be affected by this, because we will be operating with a different row from the row that contains s.
3. Put zero rows at the bottom.

We can state the final result of all this as a theorem.

6.1 Theorem.   Any matrix A can be converted to reduced form by a sequence of row operations.

From Theorem 6.1 we can get an important fact about **homogeneous systems** of equations, that is, systems that look like $A\mathbf{x} = 0$ in matrix form.

6.2 Theorem.   A homogeneous system $A\mathbf{x} = 0$ with more unknowns than equations always has infinitely many nonzero solutions.

Proof. To see why the theorem is true, put A in reduced form. Any zero rows are consistent, because the right side is zero also. Since there are more unknowns than equations, there must be at least one unknown that corresponds to a nonleading entry. Specifying the infinitely many arbitrary values for the unknowns that correspond to nonleading entries then determines the values of the ones corresponding to the leading entries.

The simple example

$$x + y = 0$$

$$x + y = 1$$

shows that the theorem is false for $A\mathbf{x} = \mathbf{b}$ without the assumption that $\mathbf{b} = 0$.

EXAMPLE 5   If

$$A = \begin{pmatrix} 1 & 2 & 0 \\ 0 & 0 & 1 \\ 0 & 0 & 0 \end{pmatrix} \quad \text{and} \quad B = \begin{pmatrix} 0 & 0 & 0 \\ 1 & 1 & 1 \\ 0 & 2 & 0 \end{pmatrix},$$

the matrix A is in reduced form because the only zero row is at the bottom and the other two rows have leading entry 1 with only zeros in the columns containing the leading entries. The matrix B is not in reduced form because all three conditions are violated. The reduced form of A gives an easy solution to the matrix equation $A\mathbf{x} = \mathbf{b}$, with $\mathbf{b} = \begin{pmatrix} 1 \\ 2 \\ 0 \end{pmatrix}$. The matrix equation $A\mathbf{x} = \mathbf{b}$ represents the system

$$x + 2y = 1,$$

$$z = 2,$$

$$0 = 0.$$

So the general solution is, with $y = t$, the set of vectors

$$(x, y, z) = (1 - 2t, t, 2) = t(-2, 1, 0) + (1, 0, 2).$$

On the other hand, with $\mathbf{b} = \begin{pmatrix} 1 \\ 2 \\ 1 \end{pmatrix}$, the last row becomes $0 = 1$, so that system is inconsistent, that is, has no solutions at all.

EXAMPLE 6   The matrix equation

$$\begin{pmatrix} 1 & -2 & 1 \\ 2 & 1 & -3 \end{pmatrix}\begin{pmatrix} x \\ y \\ z \end{pmatrix} = \begin{pmatrix} 0 \\ 0 \end{pmatrix}$$

is an example of Theorem 6.2, because it is homogeneous and has more unknowns than equations. The solutions can be thought of as the intersection of two planes. But since the system is homogeneous, it has the trivial zero solution $(x, y, z) = (0, 0, 0)$. Hence the planes intersect in at least an entire line of points. In fact, it is easy to see that all solutions are of the form $(x, y, z) = t(1, 1, 1)$, where t ranges over all real numbers; just look at the reduced form with $z = t$.

$$\begin{pmatrix} 1 & 0 & -1 \\ 0 & 1 & -1 \end{pmatrix}\begin{pmatrix} x \\ y \\ z \end{pmatrix} = \begin{pmatrix} 0 \\ 0 \end{pmatrix}.$$

EXERCISES

1. Write each of the following systems in matrix form; that is, find a matrix A and a vector $\mathbf{b}$ such that the system is equivalent to the equation $A\mathbf{x} = \mathbf{b}$.

(a) $\begin{aligned} 3x - 2y &= 1, \\ x - 3y &= 2. \end{aligned}$

(b) $\begin{aligned} 3x + y + z &= 1, \\ x - y - z &= 0. \end{aligned}$

(c) $\begin{aligned} x + y \quad\ \ &= 1, \\ y - z &= 1, \\ x \quad\ + z &= 0. \end{aligned}$

(d) $\begin{aligned} x + 2y &= 0, \\ x -\ y &= 0, \\ -x +\ y &= 1. \end{aligned}$

2. Write a system of equations equivalent to each of the following matrix equations.

(a) $\begin{pmatrix} 1 & 2 \\ 3 & 1 \end{pmatrix}\begin{pmatrix} x \\ y \end{pmatrix} = \begin{pmatrix} 1 \\ 0 \end{pmatrix}.$

(b) $\begin{pmatrix} -1 & 2 \\ 0 & 1 \end{pmatrix}\mathbf{x} = \begin{pmatrix} 0 \\ 0 \end{pmatrix}.$

(c) $\begin{pmatrix} 1 & 0 & 1 \\ 0 & 1 & 0 \\ 1 & 1 & 0 \end{pmatrix}\mathbf{x} = \begin{pmatrix} 0 \\ 1 \\ 0 \end{pmatrix}.$

(d) $\begin{pmatrix} 2 & 1 \\ 1 & 2 \\ 3 & 3 \end{pmatrix}\mathbf{x} = \begin{pmatrix} 1 \\ 0 \\ 2 \end{pmatrix}.$

3. Solve each of the systems in Exercise 1 by applying a sequence of row operations to both sides.

4. Solve each of the equations in Exercise 2 by applying a sequence of row operations to both sides so as to put the matrix in reduced form.

5. Solve the following real or complex systems. (*Note:* $i^2 = -1$.)

(a) $\begin{pmatrix} 1 & 1 & 1 \\ -1 & 2 & -4 \\ 1 & 3 & 9 \end{pmatrix}\begin{pmatrix} x \\ y \\ z \end{pmatrix} = \begin{pmatrix} 2 \\ 2 \\ 0 \end{pmatrix}.$

(b) $\begin{pmatrix} 1 & i & 0 \\ -2i & 1 & 1 \\ 3 & -1 & 2 \end{pmatrix}\begin{pmatrix} u \\ v \\ w \end{pmatrix} = \begin{pmatrix} 0 \\ 1 \\ -1 \end{pmatrix}.$

6. Find a reduced form of each of the following matrices.

(a) $\begin{pmatrix} 1 & -2 & 1 \\ 2 & 1 & -3 \end{pmatrix}.$

(b) $\begin{pmatrix} 0 & 1 & 1 \\ 1 & 1 & 1 \end{pmatrix}.$

$$\text{(c)} \begin{pmatrix} 1 & 1 & 1 \\ 1 & 1 & 0 \\ 1 & 0 & 0 \end{pmatrix}. \qquad\qquad \text{(d)} \begin{pmatrix} 0 & 0 & 2 \\ 2 & 0 & 0 \\ 0 & 0 & 3 \end{pmatrix}.$$

7. Solve the system $A\mathbf{x} = 0$, where A is each of the matrices in Exercise 6.

8. (a) Express the vectors $\mathbf{e}_1$, $\mathbf{e}_2$ in $\mathcal{R}^2$ as a linear combination of $(1, 2)$ and $(2, 3)$ by solving an appropriate system of equations for the coefficients of combination.

  (b) Express the vectors $\mathbf{e}_1$, $\mathbf{e}_2$, $\mathbf{e}_3$ in $\mathcal{R}^3$ as a linear combination of $(1, 1, 1)$, $(1, 1, 0)$, and $(1, 0, 0)$.

  (c) Express the vector $(5, 0, 1, 2)$ as a linear combination of $(1, 2, 1, 0)$ and $(2, -1, 0, 1)$.

  (d) Can an arbitrary vector in $\mathcal{R}^4$ be expressed as a linear combination of the last two vectors in part (c)? Explain.

6C Computing A^{-1}

While we have a simple formula for the inverse of a 2-by-2 matrix, and similar formulas exist for n-by-n matrices, efficient computation of A^{-1} when A is large requires another approach. What it comes down to is finding an n-by-n matrix X such that $AX = I$, where I is the n-by-n identity. Now the jth column of I is just $\mathbf{e}_j$ with a 1 in the jth entry and 0 elsewhere. If we denote the jth column of X by $\mathbf{x}_j$, then finding X amounts to solving all the equations

$$A\mathbf{x}_j = \mathbf{e}_j, \qquad j = 1, 2, \ldots, n.$$

The appropriate row operations are the same for every equation, and what we try to do is reduce A to I. If $\mathbf{c}_j$ is the result of applying these operations to $\mathbf{e}_j$, we would get

$$I\mathbf{x}_j = \mathbf{c}_j, \qquad j = 1, 2, \ldots, n.$$

In other words, $\mathbf{x}_j = \mathbf{c}_j$ are the desired columns of X. Hence the matrix C with jth column $\mathbf{c}_j$ satisfies $AC = I$. It turns out that, if we are successful in reducing A to I, then C really is the inverse of A. Here is the complete theorem.

6.3 Theorem.   A square matrix A is invertible if and only if it can be reduced to I by a sequence of row operations. If C is the result of applying the same sequence of row operations to I, then $A^{-1} = C$.

Proof. The discussion leading up to the statement proves that $AC = I$. We need to verify that $CA = I$. Since C was found by applying row operations to I, we can get I back again from C by applying the reverse row operations in the opposite order. Applying to C the argument we applied earlier to A, we find a matrix D such that $CD = I$. But then

$$A = AI = A(CD) = (AC)D = ID = D.$$

Thus $A = D$, so $CA = CD = I$. By definition, A is invertible and C is its inverse.

Conversely, assume A is invertible. Then $A\mathbf{x} = 0$ has only the solution $\mathbf{x} = 0$. If R is a reduced form for A, then $R\mathbf{x} = 0$ also has only the solution $\mathbf{x} = 0$. But then

R can have neither a row with a nonleading entry different from zero, nor a zero row, either of which would imply infinitely many solutions. Hence R must be a row permutation of I.

EXAMPLE 7   To find the inverse of

$$A = \begin{pmatrix} 2 & 4 & 8 \\ 1 & 0 & 0 \\ 1 & -3 & -7 \end{pmatrix},$$

we perform the same sequence of row operations on both A and I: We start with

$$\begin{pmatrix} 2 & 4 & 8 \\ 1 & 0 & 0 \\ 1 & -3 & -7 \end{pmatrix}, \qquad \begin{pmatrix} 1 & 0 & 0 \\ 0 & 1 & 0 \\ 0 & 0 & 1 \end{pmatrix}.$$

Add -2 times the second row to the first and -1 times the second row to the third to get

$$\begin{pmatrix} 0 & 4 & 8 \\ 1 & 0 & 0 \\ 0 & -3 & -7 \end{pmatrix}, \qquad \begin{pmatrix} 1 & -2 & 0 \\ 0 & 1 & 0 \\ 0 & -1 & 1 \end{pmatrix}.$$

Multiply the first row by $\frac{1}{4}$ and then add 3 times the first row to the third to get

$$\begin{pmatrix} 0 & 1 & 2 \\ 1 & 0 & 0 \\ 0 & 0 & -1 \end{pmatrix}, \qquad \begin{pmatrix} \frac{1}{4} & -\frac{1}{2} & 0 \\ 0 & 1 & 0 \\ \frac{3}{4} & -\frac{5}{2} & 1 \end{pmatrix}.$$

Multiply the third row by -1 and then add -2 times the third row to the first to get

$$\begin{pmatrix} 0 & 1 & 0 \\ 1 & 0 & 0 \\ 0 & 0 & 1 \end{pmatrix}, \qquad \begin{pmatrix} \frac{7}{4} & -\frac{11}{2} & 2 \\ 0 & 1 & 0 \\ -\frac{3}{4} & \frac{5}{2} & -1 \end{pmatrix}.$$

Transpose the first and second rows to get

$$\begin{pmatrix} 1 & 0 & 0 \\ 0 & 1 & 0 \\ 0 & 0 & 1 \end{pmatrix}, \qquad \begin{pmatrix} 0 & 1 & 0 \\ \frac{7}{4} & -\frac{11}{2} & 2 \\ -\frac{3}{4} & \frac{5}{2} & -1 \end{pmatrix}.$$

The last matrix on the right is A^{-1}, as may be verified by multiplying by A.

Since many different sequences of row operations could be used to reduce a matrix to the identity, it is important to have the following theorem:

6.4 Theorem.   A matrix A has at most one inverse.

Proof. Suppose there are two matrices, B and C, such that both

$$AB = BA = I \quad \text{and} \quad AC = CA = I.$$

Then $AB = AC$, because both products equal I. Multiplying by B on the left gives

$$BAB = BAC.$$

But since $BA = I$, substitution gives $IB = IC$, from which it follows that $B = C$, as was to be proved.

6D Linear Independence

Checking m row or column vectors $\mathbf{a}_j$ with n entries for linear independence amounts to showing that the vector equation

6.5
$$c_1\mathbf{a}_1 + c_2\mathbf{a}_2 + \cdots + c_m\mathbf{a}_m = \mathbf{b}$$

has the unique solution $\mathbf{c} = (c_1, c_2, \ldots, c_m) = 0$ when $\mathbf{b} = 0$. An alternative way to put this is to consider the matrix A whose jth column is $\mathbf{a}_j$, write $\mathbf{c}$ as a column, and observe that Equation 6.5 is equivalent to $A\mathbf{c} = \mathbf{b}$. Thus linear independence of the columns of A amounts to the equation $A\mathbf{c} = 0$ having only the solution $\mathbf{c} = 0$. The following theorems relate these ideas more closely.

6.6 Theorem.   Let A be an n-by-n matrix with jth column $\mathbf{a}_j$. Then A is invertible if and only if the vectors $\mathbf{a}_j$ are linearly independent.

6.7 Theorem.   If $\mathbf{a}_1, \ldots, \mathbf{a}_m$ are n-dimensional row or column vectors, with $m > n$, then the $\mathbf{a}_j$ are linearly dependent.

6.8 Theorem.   If $\mathbf{a}_1, \ldots, \mathbf{a}_n$ are linearly independent n-dimensional row (or column) vectors, then any n-dimensional row (or column) vector $\mathbf{b}$ can be written as a linear combination of the $\mathbf{a}_j$. The coordinates $c_1, \ldots, c_n$ in the representation 6.5 are the entries in the vector $\mathbf{c} = A^{-1}\mathbf{b}$.

Proof of Theorems 6.6, 6.7, and 6.8. Let R be a reduced form for A. Thus the equation $A\mathbf{c} = 0$ is equivalent to $R\mathbf{c} = 0$ for some vector $\mathbf{d}$. But if A^{-1} exists, then R must be a row permutation of I; so $\mathbf{c} = 0$ and the columns $\mathbf{a}_j$ of A are independent. If the $\mathbf{a}_j$ are independent, then $A\mathbf{c} = 0$ has the unique solution $\mathbf{c} = 0$, so R must be a row permutation of I. Thus A^{-1} exists and $A\mathbf{x} = \mathbf{b}$ has the unique solution $\mathbf{x} = A^{-1}\mathbf{b}$. This much proves Theorems 6.6 and 6.8. If A is m-by-n with $m > n$, then $A\mathbf{c} = 0$ has infinitely many solutions $\mathbf{c}$, all but one different from the zero vector. This proves Theorem 6.7.

The next example shows how Theorems 6.6, 6.7, and 6.8 are often applied.

EXAMPLE 8   Consider the three-dimensional column vectors

$$\begin{pmatrix} 2 \\ 1 \\ 1 \end{pmatrix}, \quad \begin{pmatrix} 4 \\ 0 \\ -3 \end{pmatrix}, \quad \begin{pmatrix} 8 \\ 0 \\ -7 \end{pmatrix}, \quad \begin{pmatrix} 2 \\ -1 \\ -4 \end{pmatrix}.$$

The matrix A formed by the first three columns can be reduced to the identity I by row operations. This was done in Example 7. It follows that the first three vectors are linearly independent by Theorem 6.6. By Theorem 6.7, all four vectors are necessarily linearly dependent, just because there are more than three of them. In the process of reducing A in Example 7, we found the matrix A^{-1}. By Theorem 6.8, the coordinates we need to represent the fourth vector $\mathbf{b}$ as a linear combination of the first three are just the entries in $A^{-1}\mathbf{b}$:

$$\begin{pmatrix} 0 & 1 & 0 \\ 7/4 & -11/2 & 2 \\ -3/4 & 5/2 & -1 \end{pmatrix} \begin{pmatrix} 2 \\ -1 \\ -4 \end{pmatrix} = \begin{pmatrix} -1 \\ 1 \\ 0 \end{pmatrix}.$$

Thus the fourth vector is -1 times the first plus 1 times the second.

EXERCISES

1. Use row operations to find A^{-1} if A equals

(a) $\begin{pmatrix} 1 & 0 & 0 \\ 3 & 1 & 5 \\ -2 & 0 & 1 \end{pmatrix}.$ (b) $\begin{pmatrix} 1 & 2 & 3 \\ -1 & 1 & 0 \\ 0 & 3 & 3 \end{pmatrix}.$ (c) $\begin{pmatrix} 2 & 4 & 8 \\ 1 & 0 & 0 \\ 1 & -3 & -7 \end{pmatrix}.$

2. Solve the matrix equation for X.

$$\begin{pmatrix} 1 & 2 \\ 5 & 6 \end{pmatrix} X = \begin{pmatrix} 0 & -3 & 4 \\ 1 & 2 & 0 \end{pmatrix}.$$

3. Find inverses for those of the following matrices that have inverses.

(a) $\begin{pmatrix} 1 & 0 & 0 \\ 0 & 2 & 0 \\ 0 & 0 & 1 \end{pmatrix}.$ (b) $\begin{pmatrix} 1 & 2 & 1 \\ 0 & 0 & 1 \\ 0 & 0 & 3 \end{pmatrix}.$ (c) $\begin{pmatrix} 1 & -1 & 1 \\ 0 & -1 & 1 \\ 0 & 0 & 1 \end{pmatrix}.$

(d) $\begin{pmatrix} 1 & 2 & -1 & 3 \\ 0 & 2 & 0 & 1 \\ 0 & 0 & 1 & 1 \\ 0 & 0 & 0 & 4 \end{pmatrix}.$ (e) $\begin{pmatrix} 1 & 1 & 0 & 0 \\ 0 & 2 & 0 & 0 \\ 0 & 0 & 3 & 0 \\ 0 & 0 & 0 & 4 \end{pmatrix}.$

4. Show that if A is an invertible matrix then the only solution of $A\mathbf{x} = 0$ is $\mathbf{x} = 0$.

5. Determine whether the following sets of vectors are linearly independent or not.

(a) $\begin{pmatrix} 1 \\ -1 \\ 1 \end{pmatrix},$ $\begin{pmatrix} 2 \\ 1 \\ 2 \end{pmatrix},$ $\begin{pmatrix} 3 \\ 1 \\ 4 \end{pmatrix}.$

(b) $\begin{pmatrix} 1 \\ -1 \\ 2 \end{pmatrix},$ $\begin{pmatrix} 2 \\ 1 \\ 2 \end{pmatrix},$ $\begin{pmatrix} 3 \\ 0 \\ 4 \end{pmatrix}.$

(c) $\begin{pmatrix} 2 \\ 2 \\ 3 \\ 2 \end{pmatrix},$ $\begin{pmatrix} 1 \\ -1 \\ 1 \\ 0 \end{pmatrix},$ $\begin{pmatrix} 0 \\ 0 \\ 2 \\ 5 \end{pmatrix},$ $\begin{pmatrix} 3 \\ 2 \\ 5 \\ 6 \end{pmatrix}.$

(d) $\begin{pmatrix} 1 \\ 2 \end{pmatrix}$, $\qquad \begin{pmatrix} 2 \\ 1 \end{pmatrix}$, $\qquad \begin{pmatrix} 4 \\ -1 \end{pmatrix}$.

(e) $\begin{pmatrix} 1 \\ 2 \\ -1 \\ 0 \end{pmatrix}$, $\qquad \begin{pmatrix} 2 \\ 2 \\ 2 \\ 2 \end{pmatrix}$, $\qquad \begin{pmatrix} 3 \\ 4 \\ 2 \\ 2 \end{pmatrix}$.

(f) $\begin{pmatrix} 2 \\ 1 \\ 2 \end{pmatrix}$, $\qquad \begin{pmatrix} 2 \\ -1 \\ 1 \end{pmatrix}$, $\qquad \begin{pmatrix} -1 \\ 1 \\ 3 \end{pmatrix}$.

6. For each set of vectors, determine whether or not the last vector is a linear combination of the others; if it is, find coefficients that make it so. Are the coefficients uniquely determined?

(a) $\begin{pmatrix} 1 \\ -1 \\ 2 \end{pmatrix}$, $\qquad \begin{pmatrix} 2 \\ 1 \\ 2 \end{pmatrix}$, $\qquad \begin{pmatrix} 3 \\ 1 \\ 4 \end{pmatrix}$; $\qquad \begin{pmatrix} 2 \\ 6 \\ 0 \end{pmatrix}$.

(b) $\begin{pmatrix} 1 \\ -1 \\ 2 \end{pmatrix}$, $\qquad \begin{pmatrix} 2 \\ 1 \\ 2 \end{pmatrix}$, $\qquad \begin{pmatrix} 3 \\ 0 \\ 4 \end{pmatrix}$; $\qquad \begin{pmatrix} 2 \\ 2 \\ 2 \end{pmatrix}$.

(c) $\begin{pmatrix} 1 \\ 2 \end{pmatrix}$, $\qquad \begin{pmatrix} 2 \\ 1 \end{pmatrix}$; $\qquad \begin{pmatrix} 3 \\ 4 \end{pmatrix}$.

(d) $\begin{pmatrix} 6 \\ 1 \end{pmatrix}$, $\qquad \begin{pmatrix} 2 \\ 2 \end{pmatrix}$, $\qquad \begin{pmatrix} 3 \\ 4 \end{pmatrix}$; $\qquad \begin{pmatrix} 0 \\ 1 \end{pmatrix}$.

7. Let $\mathbf{x}_1$ and $\mathbf{x}_2$ be linearly independent vectors in $\mathcal{R}^3$. Show that not every vector $\mathbf{v}$ in $\mathcal{R}^3$ can be written as a linear combination of $\mathbf{x}_1$ and $\mathbf{x}_2$. [*Hint*: Let $\mathbf{v}$ be perpendicular to $\mathbf{x}_1$ and $\mathbf{x}_2$.]

6

Variable Coefficient Methods

In Chapter 2, we saw that the differential equation

$$y'' + ay' + by = f(x)$$

has for its general solution

1.1
$$y = c_1 y_1(x) + c_2 y_2(x) + y_p(x),$$

where y_1 and y_2 are independent solutions of the associated homogeneous equation, and y_p is some solution of the nonhomogeneous equation. One condition under which this was proved was that $f(x)$ be continuous on finite intervals. An additional condition was that a and b be constant, and in fact with that assumption we could be quite explicit about the possibilities for y_1 and y_2: they can be chosen to be of the form $x^l e^{\alpha x} \cos \beta x$, and/or $x^l e^{\alpha x} \sin \beta x$. The purpose of this chapter is to see what is different if $a = a(x)$ and $b = b(x)$ are allowed to be continuous functions on some interval. It turns out that the general formula 1.1 for solutions is still valid, but that we cannot be nearly as explicit about the form of y_1, y_2 and y_p without knowing more about $a(x)$ and $b(x)$. In particular, we cannot get away with looking only at the elementary functions of calculus. For technical reasons, the proof that Formula 1.1 holds in general is most conveniently constructed using systems of differential equations (see Chapter 7).

The fact that $L = D^2 + a(x)D + b(x)$ acts as a linear operator on twice differentiable functions, that is,

$$L(c_1 y_1 + c_2 y_2) = c_1 L(y_1) + c_2 L(y_2),$$

underlies almost everything in this chapter. Checking the linearity of L is no more complicated than it is when a and b are constant, but it is an exercise that is worth doing again in the present context.

As for the reasons for wanting $a(x)$ and $b(x)$ to be functions of x, one reason is that in some applications these functions represent physical parameters that vary with x. An equally important reason in practice is that variable coefficients arise naturally from attempts to reduce the solution of the partial differential equations of Chapter 8 to the solution of ordinary differential equations.

1 INDEPENDENT SOLUTIONS

Recall that two functions y_1, y_2 defined on the same interval are **linearly independent** if neither is a constant multiple of the other; correspondingly, y_1 and y_2 are **linearly dependent** if one of them *is* a constant multiple of the other, say $y_2 = cy_1$. It is only when y_1 and y_2 are linearly independent that there is not some redundancy in the general linear combination $c_1y_1 + c_2y_2$. While the discussion of linear independence in the chapter on linear systems is in some ways more general, what we are concerned with here looks different enough that we will treat it separately and then unify the ideas later.

EXAMPLE 1   A glance at the graphs of $y_1(x) = e^x$ and $y_2(x) = e^{-x}$ in Figure 1(a) shows that neither of these functions is a multiple of the other on any interval, because one of them increases, the other decreases, and both are positive. Hence the two functions are linearly independent. The same is true of any two exponentials $e^{\alpha x}$, $e^{\beta x}$, where $\alpha \neq \beta$. For if $e^{\alpha x} = ce^{\beta x}$, then $e^{(\alpha - \beta)x} = c$. But the exponential is a constant only if $\alpha = \beta$. Hence $e^{\alpha x}$, $e^{\beta x}$ are independent unless they are equal.

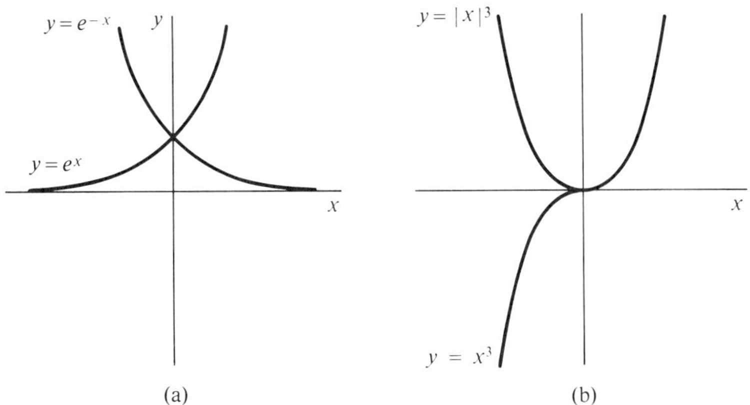

Figure 1

EXAMPLE 2   The formulas $y_1(x) = x^3$, $y_2(x) = |x|^3$ define independent functions on some intervals and dependent functions on other intervals. On any interval consisting of only positive numbers, the two functions are equal and so are dependent. On a negative interval, $y_2(x) = -y_1(x)$, as shown in Figure 1(b). This is again a dependence relation. However, over an entire interval $a < x < b$ with $a < 0 < b$, neither function is a unique constant multiple of the other, so the functions thus defined are linearly independent.

It often happens that it is fairly easy to find one nontrivial (i.e., not identically zero) solution y_1 of a homogeneous equation

$$y'' + a(x)y' + b(x)y = 0.$$

For example, power series, discussed in Chapter 9, may provide a solution, or we may be fortunate enough to guess one. But then to be able to write the general solution in the form

$$y = c_1 y_1 + c_2 y_2,$$

we need to be able to find a linearly independent solution y_2. It turns out that this can always be done just by finding a suitable factor $u(x)$ that makes

$$y_2(x) = y_1(x)u(x)$$

a solution also. The reason the method works is that having one factor as a solution of the differential equation makes several terms drop out when the product $y_1 u$ is substituted into the equation. To illustrate the method most simply, we start with a constant-coefficient equation for which we already know how to write down the general solution.

EXAMPLE 3   It is easy to check that

$$y'' - 2y' + y = 0$$

has $y_1(x) = e^x$ for a solution. To find another, try

$$y(x) = y_1(x)u(x)$$
$$= e^x u(x).$$

We find

$$y = e^x u,$$
$$y' = e^x u + e^x u',$$
$$y'' = e^x u + 2e^x u' + e^x u''.$$

When we substitute these expressions into the differential equation, we can collect the terms having u as a factor at the beginning to get

$$(e^x u - 2e^x u + e^x u) + (e^x u'' + 2e^x u' - 2e^x u') = 0.$$

The first three terms add up to zero, and the second three boil down to $e^x u''$. We are then left with $e^x u'' = 0$; so divide out e^x, leaving $u'' = 0$. The solution is

$$u(x) = c_1 x + c_2.$$

So, since $y = u y_1$ and $y_1 = e^x$, we get

$$y(x) = y_1(x)(c_1 x + c_2) = (c_1 x + c_2)e^x.$$

This is in fact the general solution, but we can obviously pick out a solution independent of $y_1(x) = e^x$ by setting $c_1 = 1$ and $c_2 = 0$: $y_2(x) = xe^x$. If this method for finding a second independent solution seems shorter than what we did in Chapter 2, it is only because this time we started with one solution. Furthermore, the present method does not, by itself, prove that we have found the most general solution.

The routine we applied in Example 3 works quite generally, and rather than go through it repeatedly we can condense most of it into a single formula that requires only a couple of integrations. If $y_1(x)$ is a solution of

$$y'' + a(x)y' + b(x)y = 0,$$

we try

$$y = y_1 u,$$

$$y' = y_1' u + y_1 u',$$

$$y'' = y_1'' u + 2y_1' u' + y_1 u''.$$

Substitution into the differential equation allows us to collect the terms containing u at the beginning:

$$(y_1'' + a(x)y_1' + b(x)y_1)u + y_1 u'' + (2y_1' + a(x)y_1)u' = 0.$$

The first three terms add up to zero, because y_1 is a solution. The remaining terms form a differential equation of first order for u',

$$(u')' + \left(\frac{2y_1'}{y_1} + a(x)\right)u' = 0,$$

valid on any interval where $y_1(x) \neq 0$. An exponential multiplier for this first-order linear equation is the exponential of

$$\int \left(\frac{2y_1'}{y_1} + a(x)\right) dx = 2 \log|y_1| + \int a(x)\, dx.$$

We find

$$u' = c_1 e^{-2\log|y_1| - \int a(x)\, dx}$$

$$= c_1 \frac{\exp(-\int a(x)\, dx)}{y_1^2}.$$

Then

$$u(x) = c \int \frac{\exp(-\int a(x)\, dx)}{y_1^2(x)}\, dx + c_2.$$

We can take $c_1 = 1$ and $c_2 = 0$ to get the solution $y_2 = y_1 u$ in the form

1.2
$$y_2(x) = y_1(x) \int \frac{\exp(-\int a(x)\, dx)}{y_1^2(x)}\, dx.$$

This formula defines a solution independent of y_1, because for the integral factor to be a constant, the integrand $y_1^{-2} \exp(-\int a\, dx)$ would have to be identically zero on the interval in question. But this is impossible, given the nature of the two factors: one is an exponential and the other is the reciprocal of a continuous function. Because the technique that leads to Equation 1.2 involves solving only a first-order equation, it is often described as an **order reduction method.**

To apply Equation 1.2, it is important, because of the way $a(x)$ comes in, that the differential equation to be solved should have the coefficient of y'' normalized to be 1. The next example illustrates how the apparent problem with y_1 being zero actually washes out in the computation.

EXAMPLE 4   The **Bessel equation** of index p can be written

$$x^2 y'' + xy' + (x^2 - p^2)y = 0.$$

Suitably normalized, the equation is

$$y'' + \frac{1}{x}y' + \left(1 - \frac{p^2}{x^2}\right)y = 0.$$

When $p = \frac{1}{2}$, substitution shows that the formula

$$y_1(x) = \frac{\sin x}{\sqrt{x}}, \qquad x > 0,$$

provides a solution. Since $a(x) = 1/x$, Formula 1.2 becomes

$$y_2(x) = \frac{\sin x}{\sqrt{x}} \int \frac{x}{\sin^2 x} e^{-\ln x}\, dx$$

$$= \frac{\sin x}{\sqrt{x}} \int \frac{dx}{\sin^2 x}$$

$$= -\frac{\sin x}{\sqrt{x}} \cot x = -\frac{\cos x}{\sqrt{x}}.$$

The solutions y_1 and y_2 are clearly independent.

The order-reduction method produces a formula for a second, independent solution of a second-order differential equation if we already have one solution. Finding that first solution may be fairly easy, involving only a little intelligent guessing. At the other extreme, we may begin with no particularly handy formula for a solution and have to rely on an existence theorem to tell us that there is a solution. From there we may proceed to study some particular solution, using a combination of numerical methods and the form of the differential equation to deduce properties of the solution. Once we are comfortable with this first solution, we can then proceed to derive an independent solution by the order-reduction method. The solutions arrived at in this way form the bulk of the so-called higher transcendental functions, and they lie outside the scope of this text.

EXERCISES

1. Decide whether the following parts of functions are linearly independent on the indicated interval. Sketch the relevant graphs.
   (a) $\cos x$, $\sin x$; $-\infty < x < \infty$.
   (b) x, x^2; $-\infty < x < \infty$.
   (c) $\tan x$, $\sin x$; $-\pi/2 < x < \pi/2$.
   (d) x^5, $|x|^5$; $-\infty < x < \infty$.
   (e) x^5, $|x|^5$, $-\infty < x < 0$.
   (f) e^x, xe^{2x}; $-\infty < x < \infty$.

2. Each of the following equations has the given function as a solution. Verify this, and find a linearly independent solution using Equation 1.2.
   (a) $y'' - 4y' + 4y = 0$; $y_1(x) = e^{2x}$.
   (b) $xy'' - (2x + 1)y' + (x + 1)y = 0$; $y_1(x) = e^x$.
   (c) $(x - 1)y'' - xy' + y = 0$; $y_1(x) = e^x$.
   (d) $x^2y'' - 2xy' + 2y = 0$; $y_1(x) = x$, $x > 0$.
   (e) $(x^2 - 1)y'' - 2xy' + 2y = 0$; $y_1(x) = x$.

3. Each of the following equations is of the kind that can be treated using the method of this section, but because the term in y is already absent, it is simpler to solve each equation directly by treating it as a first-order equation in the unknown function y', and then find y by integration. For each equation, find the general solution in the form $c_1 y_1 + c_2 y_2$. Use either method.

   (a) $y'' + xy' = 0$.   (b) $y'' + \dfrac{1}{x}y' = 0$.

4. The **Legendre equation** of index 1 is

$$(1 - x^2)y'' - 2xy' + 2y = 0$$

   and has a solution $y_1(x) = x$.
   (a) Use Equation 1.2 to find an independent solution in terms of elementary functions.
   (b) Find the solution $y(x)$ to the Legendre equation such that $y(0) = 1$ and $y'(0) = 2$.

5. The **Hermite equation** of index 1 is

$$y'' - 2xy' + 2y = 0$$

and has a solution $y_1(x) = 2x$. Use Equation 1.2 to find an independent solution in power series form.

6. Let $L(y) = x^2 y'' - 2xy' + 2y$. Verify that $y_1(x) = x$ is a solution of $L(y) = 0$. Then find a solution of the nonhomogeneous equation $L(y) = x^2$ in the form $y(x) = y_1(x)u(x) = xu(x)$. [*Hint:* Substitute $y = xu(x)$ into the equation to be solved and then solve the resulting equation for u.]

7. The derivation of Equation 1.2 entails the formula

$$u' = c\, \frac{\exp\left(-\int a(x)\, dx\right)}{y_1^2},$$

where $y_2 = y_1 u$.

(a) Replace u by y_2/y_1 to show that

$$y_1 y_2' - y_2 y_1' = c \exp\left(-\int a(x)\, dx\right).$$

This result is called **Abel's** formula.

(b) Show that Abel's formula can be written using determinants as

$$\begin{vmatrix} y_1(x) & y_2(x) \\ y_1'(x) & y_2'(x) \end{vmatrix} = c \exp\left(-\int a(x)\, dx\right).$$

The determinant is called the **Wronskian** of y_1, y_2.

8. Verify that, if $L = D^2 + a(x)D + b(x)$ on some interval $x_1 < x < x_2$, then L acts linearly on twice differentiable functions y_1, y_2 on that interval. That is, show that

$$L(c_1 y_1 + c_2 y_2) = c_1 L(y_1) + c_2 L(y_2), \qquad c_1, c_2 \text{ constant.}$$

9. (a) Show that the **Euler differential equation**

$$x^2 y'' + axy' + by = 0, \qquad a, b \text{ real constants,}$$

can be solved as follows, assuming $x > 0$. Let $y = x^r$, $y' = rx^{r-1}$, and $y'' = r(r-1)x^{r-2}$, and show that r satisfies the **characteristic equation.**

$$r^2 + (a-1)r + b = 0.$$

(b) Show that if the characteristic equation has real roots $r_1 \neq r_2$ then $y_1 = x^{r_1}$, $y_2 = x^{r_2}$ are solutions.

(c) Show that if the characteristic equation has complex conjugate roots $r_1 = \alpha + i\beta$, $r_2 = \alpha - i\beta$ then $y_1 = x^\alpha \cos(\beta \ln x)$, $y_2 = x^\alpha \sin(\beta \ln x)$ are solutions. Note that, by definition, $x^{i\beta} = e^{i\beta \ln x}$.

(d) Show that if r_0 is a double root of the characteristic equation then $y_1 = x^{r_0}$, $y_2 = x^{r_0} \ln x$ are solutions. Use Formula 1.2 for this.

10. Use the results of Exercise 9 to solve the following Euler equations.

(a) $x^2 y'' + xy' - y = 0$.   **(b)** $x^2 y'' + 4xy' + y = 0$.

(c) $x^2 y'' + 3xy' + y = 0$.   **(d)** $x^2 y'' + xy' + y = 0$.

11. The following systems have solutions that can be found by first eliminating one of the unknown functions of t. Find the general solution.

(a) $\dot{x} = x - (t - 1)y,$
  $\dot{y} = 0.$
  [*Ans.* $x = c_1 t + c_2 e^t, y = c_1.$]

(b) $\dot{x} = (1/t^2)y, t > 0,$
  $\dot{y} = x.$
  [*Hint*: Use Exercise 9.]

(c) $\dot{x} = 2(t + 1)^{-1}x - 2(t + 1)^{-2}y,$
  $\dot{y} = x.$

(d) $\dot{x} = \left(\dfrac{t + 2}{t}\right)x - \left(\dfrac{2 + 2t - t^2}{2t^2}\right)y.$
  $\dot{y} = 2x - y.$

12. Let $f(x)$ and $g(x)$ be twice differentiable functions on an interval $a < x < b$ on which $f(x)g'(x) - f'(x)g(x) \neq 0$.

(a) Show that the 3-by-3 determinant equation

$$\begin{vmatrix} y & f & g \\ y' & f' & g' \\ y'' & f'' & g'' \end{vmatrix} = 0$$

is a second order homogeneous linear equation on the interval (a, b) having $f(x)$ and $g(x)$ as solutions.

(b) Find a second order homogeneous linear equation having $f(x) = \sin x$ and $g(x) = x \sin x$ as solutions for $0 < x < \pi$.

2 NONHOMOGENEOUS EQUATIONS

2A Variation of Parameters

The technique used in Section 1 to find an independent solution of a homogeneous differential equation can be applied just as well to solving a nonhomogeneous equation as long as we have in hand a single nontrivial solution y_1 of the associated homogeneous equation. Because the multiplicative parameter u in the trial solution

$$y(x) = y_1(x)u(x)$$

is allowed to be a function of x, the procedure is sometimes called *variation of parameters*, although this term is often reserved for a refinement to be described later. The substitution works for the same reason it works for the homogeneous equation: it leaves us with a first-order linear equation that can be solved for u'.

EXAMPLE 1   The equation

$$x^2y'' - 2xy' + 2y = x^4$$

has $y_1(x) = x$ for a *homogeneous* solution. Letting $y = y_1(x)u(x) = xu(x)$ we compute

$$y = xu, \qquad y' = xu' + u, \qquad y'' = xu'' + 2u'.$$

After substitution and simplification of the differential equation, we end up with

$$x^3u'' = x^4.$$

This equation is easily solved for u:

$$u = \tfrac{1}{6}x^3 + c_1 x + c_2.$$

So our solution is

$$y = y_1(x)u(x)$$

$$= \tfrac{1}{6}x^4 + c_1 x^2 + c_2 x.$$

Finding the homogeneous solution $y_1(x) = x$ was done by guessing, but lacking a correct guess we could have tried to find a solution in power series form using methods described in Chapter 9. (See also Exercise 9 of the previous section.)

When the previous routine is applied, the number of integrations required can be reduced if we already know two independent solutions y_1, y_2 of the associated homogeneous equation. Thus suppose $a(x)$, $b(x)$ are continuous on some interval and that $f(x)$ is continuous. Then consider the **normalized equation**

$$y'' + a(x)y' + b(x)y = f(x),$$

with coefficient of y'' equal to 1. If y_1, y_2 are homogeneous solutions, we form the linear combination

$$y(x) = y_1(x)u_1(x) + y_2(x)u_2(x),$$

where $u_1(x)$ and $u_2(x)$ are to be determined so that $y(x)$ is a solution of the non-homogeneous equation. [Note that if u_1 and u_2 were constant, $y(x)$ would be a solution of the associated homogeneous equation.] What we do now is much like what we did in the previous example: compute the derivatives y', y'' and substitute into the non-homogeneous equation. Then rearrange the terms as follows:

$$(y_1'' + ay_1' + by_1)u_1 + (y_2'' + ay_2' + by_2)u_2 + (y_1u_1' + y_2u_2')'$$
$$+ a(y_1u_1' + y_2u_2') + (y_1'u_1' + y_2'u_2') = f.$$

The first two collections of terms are zero because y_1 and y_2 are homogeneous solutions. What remains of the equations will be satisfied if we can choose u_1 and u_2 so that the two equations

$$y_1u_1' + y_2u_2' = 0,$$

2.1

$$y_1'u_1' + y_2'u_2' = f$$

hold indentically for all x on the interval in question. This little system of equations can be solved for u_1' and u_2' by elimination, with the result that

$$u_1'(x) = \frac{-y_2(x)f(x)}{y_1(x)y_2'(x) - y_2(x)y_1'(x)},$$

$$u_2'(x) = \frac{y_1(x)f(x)}{y_1(x)y_2'(x) - y_2(x)y_1'(x)}.$$

The expression in the denominators is the same in both formulas and can be written as the 2-by-2 determinant

$$w(x) = \begin{vmatrix} y_1(x) & y_2(x) \\ y_1'(x) & y_2'(x) \end{vmatrix} = y_1(x)y_2'(x) - y_2(x)y_1'(x),$$

called the **Wronskian determinant** of y_1, y_2. [In Chapter 7 it is proved that $w(x)$ is never zero if y_1, y_2 are linearly independent solutions, so the examples we consider here will have that property.] To complete the solution, integrate the formula for u_1', u_2' to find u_1 and u_2, and then combine with y_1, y_2 to get a particular solution

2.2
$$y_p(x) = y_1(x)u_1(x) + y_2(x)u_2(x),$$

$$y_p(x) = y_1(x) \int \frac{-y_2(x)f(x)}{w(x)}\, dx + y_2(x) \int \frac{y_1(x)f(x)}{w(x)}\, dx.$$

Because Equations 2.1 are easier to remember than Equation 2.2, people often prefer to start with them in each problem and carry out the rest of the computation to arrive at Equation 2.2. The next example will be done that way. Because of the way that f enters Equations 2.1 and 2.2, the equation to be solved must be in normalized form to make these formulas valid.

EXAMPLE 2   The equation $x^2y'' - 2xy' + 2y = x^3$ is normalized, for example for positive x, to

$$y'' - \frac{2}{x}y' + \frac{2}{x^2}y = x, \qquad x > 0,$$

and has solutions $y_1(x) = x$, $y_2(x) = x^2$. The latter can be discovered by the order-reduction method of Section 1. Equations 2.1 are in this example

$$xu_1' + x^2u_2' = 0,$$

$$u_1' + 2xu_2' = x.$$

Multiplying the second equations by x and then subtracting the first equation from it gives $x^2u_2' = x^2$, or $u_2' = 1$. It then follows from the first equation that $u_1' = -x$. Integrating to find u_1, u_2 gives

$$u_1(x) = -\frac{x^2}{2}, \qquad u_2(x) = x.$$

A particular solution is

$$y_p = y_1u_1 + y_2u_2$$

$$= x \cdot \left(-\frac{x^2}{2}\right) + x^2 \cdot x = \frac{1}{2}x^3.$$

Adding constants of integration to u_1 and u_2 would only add a linear combination of homogeneous solutions to y_p. In any case, we have the general solution

$$y = c_1 x + c_2 x^2 + \frac{1}{2}x^3.$$

We will now solve a generalization of Example 1, but using Equation 2.2 instead of Equation 2.1.

EXAMPLE 3   In normalized form, we consider

$$y'' - \frac{2}{x}y' + \frac{2}{x^2}y = f(x), \qquad x > 0.$$

Homogeneous solutions are $y_1(x) = x$, $y_2(x) = x^2$. The Wronskian determinant of y_1, y_2 is

$$w(x) = \begin{vmatrix} x & x^2 \\ 1 & 2x \end{vmatrix} = 2x^2 - x^2 = x^2.$$

Equation 2.2 reduces to

$$y_p(x) = x \int \frac{-x^2 f(x)}{x^2}\, dx + x^2 \int \frac{xf(x)}{x^2}\, dx$$

$$= -x \int f(x)\, dx + x^2 \int \frac{f(x)}{x}\, dx.$$

To make the integration fairly easy, we can use the example $f(x) = x \cos x$. For this choice, we get

$$y_p(x) = -x \int x \cos x\, dx + x^2 \int \cos x\, dx$$

$$= -x\left(x \sin x - \int \sin x\, dx\right) + x^2 \int \cos x\, dx$$

$$= -x(x \sin x + \cos x) + x^2 \sin x$$

$$= -x \cos x.$$

2B Green's Functions

A slight modification of Equation 2.2 gives us the analogue of the Green's function formula that we found in Chapter 2 for the initial-value problem $y(x_0) = y'(x_0) = 0$ associated with a constant-coefficient operator. All we need to do to get this representation is to convert the integrals in Equation 2.2 into definite integrals over the interval from x_0 to x:

$$y_p(x) = y_1(x) \int_{x_0}^{x} \frac{-y_2(t)f(t)}{w(t)}\, dt + y_2(x) \int_{x_0}^{x} \frac{y_1(t)f(t)}{w(t)}\, dt,$$

2.3

$$y_p(x) = \int_{x_0}^{x} \frac{y_1(t)y_2(x) - y_2(t)y_1(x)}{w(t)} f(t)\, dt$$

where $w(t) = y_1(t)y_2'(t) - y_2(t)y_1'(t)$ is the Wronskian determinant of $y_1(t)$, $y_2(t)$.

Setting $x = x_0$ in Equation 2.3 gives $y_p(x_0) = 0$. It is left as an exercise to show that $y_p'(x_0) = 0$ also. Thus the **Green's function**

$$G(x,\ t) = \frac{y_1(t)y_2(x) - y_2(t)y_1(x)}{w(t)}$$

generates the solution

$$y_p(x) = \int_{x_0}^{x} G(x,\ t)f(t)\, dt$$

to the initial-value problem

$$y'' + a(x)y' + b(x)y = f(x), \qquad y(x_0) = y'(x_0) = 0.$$

Quite apart from the neatness with which Equation 2.3 displays a solution, it is convenient for calculating solutions when the forcing function $f(x)$ is discontinuous.

EXAMPLE 4   The equation

$$y'' - \frac{2x}{x^2 - 1}y' + \frac{2}{x^2 - 1}y = f(x)$$

has homogeneous solutions $y_1(x) = x$, $y_2(x) = x^2 + 1$. Their Wronskian is

$$w(x) = \begin{vmatrix} x & x^2 + 1 \\ 1 & 2x \end{vmatrix} = x^2 - 1.$$

The Green's function for an initial-value problem is then

$$G(x,\ t) = \frac{t(x^2 + 1) - (t^2 + 1)x}{t^2 - 1}.$$

Suppose we want the solution with forcing function

$$f(x) = \begin{cases} 0, & -1 < x < 0, \\ 1, & 0 \le x \le \tfrac{1}{2}, \\ 0, & \tfrac{1}{2} < x < 1, \end{cases}$$

and satisfying $y(0) = y'(0) = 0$. This is

$$y_p(x) = \int_{0}^{x} G(x,\ t)f(t)\, dt$$

$$= \begin{cases} 0, & -1 < x < 0, \\ \displaystyle\int_0^x G(x,\, t)\, dt, & 0 \le x \le \dfrac{1}{2}, \\ \displaystyle\int_0^{1/2} G(x,\, t)\, dt, & \dfrac{1}{2} < x < 1. \end{cases}$$

We compute, for $0 \le x < 1$,

$$\int_0^x G(x,\, t)\, dt = (x^2 + 1)\int_0^x \frac{t}{t^2 - 1}\, dt - x \int_0^x \frac{t^2 + 1}{t^2 - 1}\, dt$$

$$= \frac{1}{2}(x^2 + 1) \ln(1 - x^2) - x\left[x + \ln\left(\frac{1 - x}{1 + x}\right)\right].$$

The complete solution is then

$$y_p(x) = \begin{cases} 0, & -1 < x < 0, \\ \dfrac{1}{2}(x^2 + 1)\ln(1 - x^2) - x \ln\dfrac{1 - x}{1 + x} - x^2, & 0 \le x \le 1/2 \\ \dfrac{1}{2}\ln\!\left(\dfrac{3}{4}\right)(x^2 + 1) - \left(\dfrac{1}{2} - \ln 3\right)x, & \dfrac{1}{2} < x < 1. \end{cases}$$

The first and third lines are of course solutions of the homogeneous equation on their respective intervals, because $f(x) = 0$ there. Finding the solution that satisfies more general initial conditions, $y(0) = y_0$, $y'(0) = z_0$, is just a matter of solving for c_1 and c_2 from

$$y(x) = c_1 x + c_2(x^2 + 1) + y_p(x).$$

We have

$$y'(x) = c_1 + 2c_2 x + y_p'(x).$$

But $y_p(0) = y_p'(0) = 0$, so we find c_1, c_2 from equations $c_2 = y_0$, $c_1 = z_0$.

2C Summary

What we have seen in this section and the previous sections is a collection of methods for finding explicit solution formulas of the form

2.4 $$y(x) = c_1 y_1(x) + c_2 y_2(x) + y_p(x)$$

for differential equations of the form

2.5 $$y'' + a(x)y' + b(x)y = f(x).$$

We already proved in Chapter 2 that if $a(x)$ and $b(x)$ are constant then Equation 2.5 has its most general solution of the form 2.4. Furthermore, we saw there that by choosing c_1 and c_2 properly we could satisfy arbitrary initial conditions of the form $y(x_0) = y_0$, $y'(x_0) = z_0$. This last possibility followed simply from the meaning of c_1

and c_2 as arbitrary constants of integration. If $a(x)$ and $b(x)$ are continuous functions, proving the existence of solutions is a more theoretical undertaking and we defer it to Chapter 7, after the necessary machinery has been developed. The results to be proved can be summarized as follows.

2.6 Theorem.   Let $a(x)$ and $b(x)$ be continuous and $f(x)$ be continuous on an interval $a < x < b$. Then the general solution of Equation 2.5 can be written in the form of Equation 2.4, where y_1 and y_2 are independent solutions of the associated homogeneous equation. Every pair of initial conditions $y(x_0) = y_0$, $y'(x_0) = z_0$ can be satisfied by a unique choice of c_1 and c_2 in Equation 2.4.

EXERCISES

1. For each of the following differential equations, try to find or guess a solution y_1 of the associated homogeneous equation. Then determine $u(x)$ so that $y(x) = y_1(x)u(x)$ is the general solution of the given equation.

   (a) $y'' - 4y' + 4y = e^x$.

   (b) $y'' + \dfrac{1}{x}y' = x$, $x > 0$.

   (c) $x^2 y'' - 3xy' + 3y = x^4$, $x > 0$. (*Hint:* Try $y_1 = x^n$, some n).

   $$\left[Ans. \ \frac{1}{3}x^4 + c_1 x^3 + c_2 x. \right]$$

   (d) $xy'' - (2x + 1)y' + (x + 1)y = 3x^2 e^x$, $x > 0$. [*Hint:* Try $y_1(x) = e^{rx}$.]
   $$[Ans. \ x^3 e^x + c_1 x^2 e^x + c_2 e^x.]$$

2. Find a particular solution y_p by solving Equation 2.1 for $u_1(x)$, $u_2(x)$ to get $y_p(x) = y_1(x)u_1(x) + y_2(x)u_2(x)$. If suitable y_1 and y_2 are not given, find them first. Then write down the general solution.

   (a) $y'' + y' - 2y = e^{2x}$.

   (b) $y'' + y = \tan x$, $-\pi/2 < x < \pi/2$.

   (c) $y'' + y = \sec x$, $-\pi/2 < x < \pi/2$.

   (d) $y'' - y = xe^x$.

   (e) $x^2 y'' - 2xy' + 2y = 1$; $y_1(x) = x$, $y_2(x) = x^2$, $x > 0$.

3. Use Equation 2.2 to find a formula for a solution $y_p(x)$ to each of the following equations, with homogeneous solutions $y_1(x)$, $y_2(x)$ given in some cases. Complete the required integration if you can. Don't forget to make sure that the equation is normalized.

   (a) $y'' - 2y' + y = e^x$.   (b) $y'' + 3y' + 2y = 1 + e^x$.

   (c) $y'' + 3y' + 2y = (1 + e^x)^{-1}$.   (d) $2y'' + 8y = e^x$.

   (e) $(\cos x - \sin x)y'' + 2 \sin xy' - (\cos x + \sin x)y = \cos x - \sin x$; $y_1(x) = e^x$, $y_2(x) = \sin x$.

4. Find the Green's function $G(x, t)$ for an initial-value problem for each of the following equations. Then express the solution y_p satisfying $y_p(x_0) = y_p'(x_0) = 0$ in terms of G.

   (a) $y'' + 3y' + 2y = f(x)$.   (b) $2y'' + 4y = f(x)$.

   (c) $y'' + \dfrac{1}{x}y' = f(x)$, $x > 0$.   (d) $x^2 y'' - 2xy' + 2y = f(x)$.

5. Using the same equations as in Exercise 4 and the solutions of the initial-value problems found there, solve the following corresponding initial-value problems with the given choice for $f(x)$.

(a) $f(x) = \begin{cases} 0, & x < 1, \\ 1, & 1 \le x; \end{cases}$ $y(0) = 1, y'(0) = 2.$

(b) $f(x) = \begin{cases} 1, & x < 1, \\ 0, & 1 \le x; \end{cases}$ $y(0) = -1, y'(0) = 1.$

(c) $f(x) = \dfrac{1}{x}; y(1) = 0, y'(1) = 2.$

(d) $f(x) = \begin{cases} 0, & 0 < x < 2, \\ 3, & 2 \le x \le 4, \\ 1, & 4 < x; \end{cases}$ $y(3) = 0, y'(3) = 0.$

6. Show that the derivative of the Green's function Formula 2.3 can be written

$$y_p'(x) = \int_{x_0}^{x} \frac{y_1(t)y_2'(x) - y_2(t)y_1'(x)}{w(t)} f(t)\, dt.$$

Show then that $y_p'(x_0) = 0$. (*Hint:* Separate the integral in Equation 2.3 into two integrals, and then apply the product rule for differentiation.)

7. The **Leibnitz rule** for differentiating an integral states that

$$\frac{d}{dx}\int_{a(x)}^{b(x)} F(x, t)\, dt = \int_{a(x)}^{b(x)} \frac{\partial F}{\partial x}(x, t)\, dt + b'(x)F(x, b(x)) - a'(x)F(x, a(x)),$$

if $\partial F/\partial x$ is continuous. Use this result to establish the formula for $y_p'(x)$ in Exercise 6.

8. The constant coefficient equation $y'' + ay' + by = f(x)$ has homogeneous solutions $y_1(x) = e^{r_1 x}$, $y_2(x) = e^{r_2 x}$. These solutions are independent if $r_1 \ne r_2$; otherwise we consider $y_1(x) = e^{r_1 x}$, $y_2(x) = xe^{r_1 x}$. Find the Green's function for the equation in the following cases by using the approach of the present section.

(a) $r_1 \ne r_2.$ (b) $r_1 = r_2.$ (c) $r_1 = \alpha + i\beta, r_2 = \alpha - i\beta, \beta \ne 0.$

3 BOUNDARY PROBLEMS

3A Two-Point Boundaries

Initial-value problems for second-order linear differential equations are those in which we specify for a solution $y = y(x)$ the values $y(a)$ and $y'(a)$ at a single point $x = a$. We have seen that, under reasonable hypotheses, such problems always have a unique solution. If instead we try to specify the solution values $y(a)$ and $y(b)$ at two distinct points $x = a$ and $x = b$, then the variety of possibilities for a solution is much wider: there may be a unique solution, there may be infinitely many solutions, or there may be no solution at all. It is customary to call the search for a solution with values given at two points a **boundary-value problem**. The problem we're considering has an important physical interpretation for the oscillator equation

$$\ddot{y} = -k(t)\dot{y} - h(t)y$$

with specified boundary values $y(t_0) = \alpha$ and $y(t_1) = \beta$. The solutions $y(t)$ of such an equation often represent the displacements of an oscillating mechanism as a function of time t. Our question then becomes the following: over the time interval from t_0 to t_1 is it possible to go from displacement $y = \alpha$ to displacement $y = \beta$, and if so, is there a unique way to do this? The next example illustrates all of these possibilities.

EXAMPLE 1   The differential equation $y'' + y = 0$ has the general solution

$$y(x) = c_1 \cos x + c_2 \sin x.$$

To satisfy the boundary conditions

$$y(0) = 1, \qquad y(\pi/2) = 2,$$

we must have

$$y(0) = c_1 = 1, \qquad y(\pi/2) = c_2 = 2.$$

Thus c_1 and c_2 are uniquely determined, and the resulting solution is

$$y(x) = \cos x + 2 \sin x$$

$$= \sqrt{3} \sin\left(x + \frac{\pi}{6}\right).$$

The graph of the solution is shown in Figure 2(a).

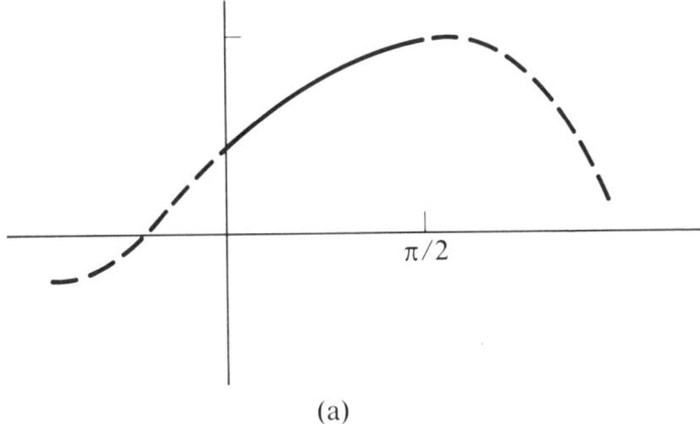

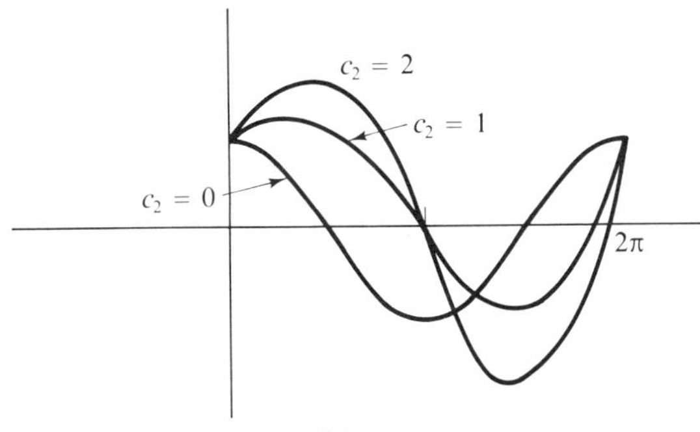

Figure 2

If we keep the same differential equation, but change the boundary conditions to

$$y(0) = 1, \qquad y(2\pi) = 1$$

then the constants c_1 and c_2 in the general solution are subject to the requirement

$$y(0) = c_1 = 1, \qquad y(2\pi) = c_1 = 1,$$

while c_2 can evidently be chosen arbitrarily. The resulting infinite collection of solutions is made up of all functions of the form

$$y(x) = \cos x + c_2 \sin x.$$

The graphs of three of these solutions are shown in Figure 2(b).

Still keeping the same differential equation, we can change the boundary conditions to

$$y(0) = 1, \qquad y(\pi) = 1.$$

Then we require of c_1 and c_2 that

$$y(0) = c_1 = 1, \qquad y(\pi) = -c_1 = 1.$$

Since we cannot have both $c_1 = 1$ and $c_1 = -1$, it follows that there is no solution satisfying this third set of conditions.

It turns out that the case of nonexistence and the case of multiple existence can together be characterized in terms of the differential equation and the boundary points, the boundary values themselves being irrelevant. The statement is as follows:

3.1 Theorem.   Let the differential equation

$$y'' + p(x)y' + q(x)y = 0$$

have continuous coefficients on the interval $a < x < b$, and let α and β be arbitrary numbers. Then the differential equation has a unique solution satisfying

$$y(a) = \alpha, \qquad y(b) = \beta,$$

if and only if *either* of the following equivalent statements holds:

(i). The only solution satisfying $y(a) = y(b) = 0$ is the identically zero solution.

(ii). For two independent solutions, we have the 2-by-2 determinant condition

$$y_1(a)y_2(b) - y_2(a)y_1(b) = \begin{vmatrix} y_1(a) & y_2(a) \\ y_1(b) & y_2(b) \end{vmatrix} \neq 0.$$

Proof. Let the general solution of the differential equation be

$$y(x) = c_1 y_1(x) + c_2 y_2(x).$$

The boundary conditions $y(a) = \alpha$, $y(b) = \beta$ amount to

$$c_1 y_1(a) + c_2 y_2(a) = \alpha,$$

$$c_1 y_1(b) + c_2 y_2(b) = \beta.$$

It is easy to show (see Exercise 11) that these equations have a unique solution for c_1 and c_2 if and only if

$$y_1(a)y_2(b) - y_2(a)y_1(b) \neq 0.$$

But the condition that the determinant is not zero is precisely the condition that the equations

$$c_1 y_1(a) + c_2 y_2(a) = 0,$$

$$c_1 y_1(b) + c_2 y_2(b) = 0$$

have only the trivial solution $c_1 = c_2 = 0$. This proves the theorem.

Theorem 3.1 shows that, given the differential equation, the existence of a unique boundary-value solution depends on the location of the boundary points. The term **conjugate points** is used to denote a pair of boundary points for which there *fails* to exist a unique solution. Thus a pair of points is conjugate relative to a homogeneous differential equation if there is a nontrivial solution that is zero at the two points. From the statement of the theorem, we see that, equivalently, a pair of points a and b is conjugate if and only if, for two linearly independent solutions y_1 and y_2, the relation

$$\begin{vmatrix} y_1(a) & y_2(a) \\ y_1(b) & y_2(b) \end{vmatrix} = 0$$

holds.

EXAMPLE 2   Referring to Example 1, the pairs $0, 2\pi$ and $0, \pi$ are conjugate for $y'' + y = 0$. More generally, (a, b) is a conjugate pair if

$$\begin{vmatrix} \cos a & \sin a \\ \cos b & \sin b \end{vmatrix} = 0,$$

that is, if $\sin b \cos a - \cos b \sin a = \sin(b - a) = 0$. Thus (a, b) is a conjugate pair if $b - a = k\pi$ for some integer k. Then the boundary-value problem

$$y'' + y = 0, \qquad y(a) = \alpha, \qquad y(b) = \beta$$

has a unique solution precisely when a and b do *not* differ by an integer multiple of π.

EXAMPLE 3   For the equation $y'' - y = 0$, the general solution is expressible as

$$y(x) = c_1 e^x + c_2 e^{-x}.$$

The arbitrary pair a, b of boundary points is conjugate precisely when the equations

$$y(a) = c_1 e^a + c_2 e^{-a} = 0,$$

$$y(b) = c_1 e^b + c_2 e^{-b} = 0$$

are satisifed by some c_1 and c_2 not both zero. But for real a and b,

$$\begin{vmatrix} e^a & e^{-a} \\ e^b & e^{-b} \end{vmatrix} = 0$$

only when $a = b$. So there is always a unique solution to the boundary-value problem

$$y'' - y = 0, \qquad y(a) = \alpha, \qquad y(b) = \beta.$$

3B Nonhomogeneous Equations

We consider boundary-value problems for the nonhomogeneous equation with continuous coefficients $p(x)$, $q(x)$, and $f(x)$,

$$y'' + p(x)y' + q(x)y = f(x).$$

The key is first to find a solution satisfying the special **homogeneous** boundary conditions

$$y(a) = 0, \qquad y(b) = 0.$$

It turns out that this can always be done provided a and b are not conjugate points for the differential equation. For recall the variation of parameters formula,

$$y(x) = -y_1(x) \int \frac{y_2(x)f(x)}{w(x)} \, dx + y_2(x) \int \frac{y_1(x)f(x)}{w(x)} \, dx,$$

where y_1 and y_2 are independent solutions, and w is their Wronskian. By choosing the constants of integration properly in the two integrals, we can get the most general solution of the nonhomogeneous differential equation. In particular, we can specify these constants by replacing the indefinite integrals by definite integrals as follows:

$$y_p(x) = -y_1(x) \int_b^x \frac{y_2(t)f(t)}{w(t)} \, dt + y_2(x) \int_a^x \frac{y_1(t)f(t)}{w(t)} \, dt$$

$$= \int_a^x \frac{y_2(x)y_1(t)}{w(t)} f(t) \, dt + \int_x^b \frac{y_1(x)y_2(t)}{w(t)} f(t) \, dt.$$

Now suppose that the linearly independent solutions y_1 and y_2 have been chosen so that $y_1(a) = 0$ and $y_2(b) = 0$. It then follows by setting $x = a$ and $x = b$ respectively, that

$$y_p(a) = 0 \qquad \text{and} \qquad y_p(b) = 0;$$

thus we have found a particular solution satisfying the preceding boundary condition.

The fact that a and b are assumed not to be conjugate points ensures that the desired independent solutions y_1 and y_2 can always be found. We have, in fact, the following theorem:

3.2 Theorem. If the equation $y'' + p(x)y' + q(x)y = 0$ has continuous coefficients, and $x = a$, $x = b$ is a pair of nonconjugate boundary points for the equation, then there are linearly independent solutions y_1 and y_2 satisfying $y_1(a) = 0$, $y_2(b) = 0$.

Proof. All we have to do for y_1 is pick a nontrivial initial-value solution, with $y_1(a) = 0$, $y_1'(a) \neq 0$. Then automatically $y_1(b) \neq 0$, because otherwise the fact that a and b are not conjugate would imply that y_1 is necessarily identically zero. Similarly, choose y_2 so that $y_2(b) = 0$, $y_2'(b) \neq 0$, with the result that $y_2(a) \neq 0$. Then y_1 and y_2 are linearly independent because, if one were a multiple of the other, each would have to be zero at both boundary points a and b. But $y_1(a)y_2(b) \neq y_1(b)y_2(a)$, so this is impossible.

We denote by G the function of two variables defined for $a \leq x \leq b$ and $a \leq t \leq b$ by

$$G(x, t) = \begin{cases} \dfrac{y_2(x)y_1(t)}{w(t)}, & a \leq t \leq x, \\[2mm] \dfrac{y_1(x)y_2(t)}{w(t)}, & x \leq t \leq b, \end{cases}$$

where $y_1(a) = y_2(b) = 0$. Then G is called a **Green's function** for the boundary-value problem, and we can write

$$y_p(x) = \int_a^b G(x, t)f(t)\, dt$$

for the solution of

$$y'' + p(x)y' + q(x)y = f(x), \qquad y(a) = y(b) = 0.$$

It is important to understand that while the independent homogeneous solutions y_1 and y_2 must satisfy the conditions $y_1(a) = 0$ and $y_2(b) = 0$, respectively, there is still some arbitrariness in the choice that we make of them. We can illustrate this point with a very simple example.

EXAMPLE 4 Suppose we want to solve the equation $y'' = f(x)$, subject to $y(0) = 0$, $y(1) = 0$. (This could, of course, be done by repeated integration.)

The most general solution of the associated homogeneous equation is

$$y(x) = c_1 + c_2 x,$$

so to satisfy $y_1(0) = 0$ we need to have $c_1 = 0$. Hence we can choose for simplicity $c_2 = 1$. At $x = 1$ we must have $c_1 + c_2 = 0$, so we can choose this time $c_1 = 1$, $c_2 = -1$. Thus we have selected

$$y_1(x) = x, \qquad y_2(x) = 1 - x$$

for our independent homogeneous solutions. The Wronskian $w[y_1, y_2](x)$ is

$$\begin{vmatrix} x & 1 - x \\ 1 & -1 \end{vmatrix} = -1,$$

which happens to be a constant. Thus the Green's function is given by

$$G(x, t) = \begin{cases} \dfrac{(1 - x)t}{-1}, & 0 \le t \le x, \\[2ex] \dfrac{x(1 - t)}{-1}, & x \le t \le 1. \end{cases}$$

The formula for the particular solution of $y'' = 1$ is obtained by taking $f(x) = 1$ in the general Green's formula:

$$y_p(x) = -(1 - x) \int_0^x t \, dt - x \int_x^1 (1 - t) \, dt$$

$$= -(1 - x)\frac{x^2}{2} - x\left(\frac{1}{2} - x + \frac{x^2}{2}\right) = \frac{x^2}{2} - \frac{x}{2}.$$

Clearly, $y_p(0) = y_p(1) = 0$ and $y_p''(x) = 1$, so we have found the correct solution, although by a more elaborate method than necessary.

3C Nonhomogeneous Boundary Conditions

To solve the nonhomogeneous equation

$$y'' + p(x)y' + q(x)y = f(x)$$

with general **nonhomogeneous** boundary conditions

$$y(a) = \alpha, \qquad y(b) = \beta,$$

we continue to assume that a and b are not conjugate points of the associated homogeneous equation. Then first solve the equation with zero boundary conditions

$$y(a) = 0, \qquad y(b) = 0,$$

for example by the method just outlined, calling this solution y_p. Now the general solution of the nonhomogeneous equation is

$$y(x) = c_1 y_1(x) + c_2 y_2(x) + y_p(x),$$

where y_1 and y_2 are independent homogeneous solutions and c_1 and c_2 are constants. Since $y_p(a) = y_p(b) = 0$, all we have to do to satisfy $y(a) = \alpha$, $y(b) = \beta$ is choose c_1 and c_2 so that

$$c_1 y_1(a) + c_2 y_2(a) = \alpha,$$

$$c_1 y_1(b) + c_2 y_2(b) = \beta.$$

This can always be done if a and b are not conjugate, because in that case the determinant of the system of linear equations for c_1 and c_2 is not zero. See Exercise 11(a).

EXAMPLE 5   The differential equation

$$y'' + y = 1$$

has linearly independent homogeneous solutions $\cos x$ and $\sin x$. To satisfy boundary conditions of the form

$$y(0) = \alpha, \qquad y\!\left(\frac{\pi}{2}\right) = \beta,$$

we can check that the points $x = 0$ and $x = \pi/2$ are not conjugate by observing that

$$\begin{vmatrix} \cos 0 & \cos \pi/2 \\ \sin 0 & \sin \pi/2 \end{vmatrix} = 1.$$

Rather than construct a Green's function, we can observe by inspection that $y(x) = 1$ is a particular solution, so

$$y(x) = c_1 \cos x + c_2 \sin x + 1$$

is the most general solution of the differential equation. To satisfy $y(0) = \alpha$, $y(\pi/2) = \beta$, we solve the equations

$$c_1 \cos 0 + c_2 \sin 0 + 1 = \alpha,$$

$$c_1 \cos \frac{\pi}{2} + c_2 \sin \frac{\pi}{2} + 1 = \beta$$

for c_1 and c_2 to get $c_1 = \alpha - 1$, $c_2 = \beta - 1$. Thus

$$y(x) = (\alpha - 1) \cos x + (\beta - 1) \sin x + 1$$

is the unique solution satisfying the boundary conditions.

On the other hand, the points $x = 0$, $x = \pi$ are conjugate for $y'' + y = 0$, because $\sin x$ is a nontrivial solution that is zero at both points. Thus an attempt to solve

$$c_1 \cos 0 + c_2 \sin 0 + 1 = \alpha$$

$$c_1 \cos \pi + c_2 \sin \pi + 1 = \beta$$

leads to $c_1 = \alpha - 1$ *and* $-c_1 = \beta - 1$. Thus there is no solution to the boundary-value problem with boundary conditions

$$y(0) = \alpha, \qquad y(\pi) = \beta$$

unless $\alpha - 1 = -\beta + 1$ or $\alpha + \beta = 2$. In that case, there are infinitely many solutions, all of the form

$$y(x) = (\alpha - 1) \cos x + c_2 \sin x + 1,$$

where c_2 is arbitrary. Clearly, these solutions all satisfy

$$y(0) = \alpha, \qquad y(\pi) = 2 - \alpha.$$

EXERCISES

1. For each of the following differential equations, determine whether (i) there is a unique solution, (ii) there are infinitely many solutions, or (iii) there is no solution satisfying the given boundary conditions.
   (a) $y'' + y = 0$, $y(0) = 1$, $y(\pi/4) = 0$.
   (b) $y'' + y = 0$, $y(0) = 1$, $y(2\pi) = 0$.
   (c) $y'' + y = 0$, $y(0) = 1$, $y(2\pi) = 1$.
   (d) $y'' + 2y = 0$, $y(1) = 1$, $y(2) = 0$.

2. Find all pairs of conjugate points for the following differential equations of the form $L[y] = 0$.
   (a) $y'' + 2y = 0$.   [*Ans.* (a, b), where $b - a = k\pi/\sqrt{2}$.]
   (b) $y'' - 2y = 0$.   [*Ans.* No conjugate pairs.]
   (c) $y'' + y' + y = 0$.   [*Ans.* (a, b), where $b - a = 2k\pi/\sqrt{3}$.]
   (d) $y'' + xy' + y = 0$. [*Hint:* $e^{-x^2/2}$ is a solution; show that $e^{-x^2/2} \int_0^x e^{t^2/2}\, dt$ is an
   independent one.]   [*Ans.* No conjugate pairs.]

3. A pair of nonconjugate boundary points is given for each of the differential equations in Exercise 2. Find in each case a pair of linearly independent solutions y_1 and y_2 that satisfy $y_1(a) = 0$ and $y_2(b) = 0$. Then use these solutions to construct a Green's function for the equation on the interval $[a, b]$.
   (a) $a = 0$, $b = \pi/(2\sqrt{2})$.
   (b) $a = 1$, $b = 2$.
   (c) $a = 0$, $b = \pi/\sqrt{3}$.
   (d) $a = 0$, $b = 1$.

4. Use the Green's function found in each part of Exercise 3 to represent the solution $y(x)$ of the corresponding differential equation $L[y] = f$, where f is defined as follows, and where $y(a) = y(b) = 0$. Compute the solution by any method.
   (a) $f(x) = x$, $0 \le x \le \pi/(2\sqrt{2})$.

(b) $f(x) = \begin{cases} -1, & 0 \le x < 1, \\ 1, & 1 \le x \le 2. \end{cases}$

(c) $f(x) = \cos x, \quad 0 \le x \le \pi/\sqrt{3}.$

(d) $f(x) = e^{-x}, \quad 0 \le x \le 1.$

5. Modify the solutions found in Exercise 4 so that, instead of satisfying the homogeneous conditions $y(a) = y(b) = 0$, they satisfy the following corresponding nonhomogeneous conditions.

   (a) $y(0) = 1, \, y(\pi/(2\sqrt{2})) = 0.$

   (b) $y(1) = 2, \, y(2) = -1.$

   (c) $y(0) = -1, \, y(\pi/\sqrt{3}) = 0.$

   (d) $y(0) = 1, \, y(1) = 1.$

6. Show that the points $a = 0$, $b = \pi\sqrt{2}$ are conjugate for the differential equation $y'' + 2y = 0$, and find conditions on α and β under which there are solutions of this equation that satisfy $y(0) = \alpha$, $y(\pi\sqrt{2}) = \beta$. Find all such solutions. Does one of these solutions also satisfy $y'(0) = 1$?

7. For a homogeneous linear boundary problem, the only way to get a nontrivial (i.e., not identically zero) solution is to have the solution nonunique. In particular, consider the problem of finding a number λ and a nontrivial solution $y(x)$ to the equation

$$L[y] = \lambda y,$$

   where L is a second-order linear operator. (Such a number λ is called an **eigenvalue**, and a corresponding nontrivial solution is called an **eigenfunction**.) Show that if $a < b$, the problem

$$y'' = \lambda y, \qquad y(a) = y(b) = 0,$$

   has nontrivial solutions if and only if $\lambda = -k^2\pi^2/(b - a)^2$, where $k = 0, 1, \ldots$. What are the corresponding solutions $y_k(x)$?

8. Consider the oscillator equation

$$my'' + ky' + hy = 0,$$

   where m, k, and h are constant and $m > 0$.

   (a) Show that if $k^2 \ge 4mh$, then there are no conjugate pairs a, b. (The case $k^2 = 4mh$ requires special attention.)

   (b) Explain the physical significance of the conclusion in part (a).

   (c) Find all conjugate pairs a, b in case $k^2 < 4mh$.

9. **(a)** Show that the boundary conditions

$$A[y] = a_1 y(0) + a_2 y'(0) = 0,$$

$$B[y] = b_1 y(\pi) + b_2 y'(\pi) = 0,$$

   when applied to the differential equation $y'' + y = 0$, can be satisfied by a solution $y(x) \ne 0$ precisely when $a_1 b_2 = a_2 b_1$.

   (b) Show that $y'' + y = f(x)$ has a unique solution for each continuous f on $0 \le x \le \pi$ and satisfying $A[y] = \alpha$, $B[y] = \beta$, precisely when $a_1 b_2 \ne a_2 b_1$.

10. The downward deflection $y = y(x)$ of a uniform horizontal beam of length L satisfies the fourth-order differential equation

$$y'''' = R, \qquad 0 \le x \le L,$$

where R is an appropriate constant and x represents distance from one end of the beam. If the beam has a simple support at an end $x = x_0$, then $y(x_0) = y''(x_0) = 0$. If the beam is held rigidly at x_0, say by imbedding it in a concrete wall, then $y(x_0) = y'(x_0) = 0$. If the beam is unsupported at $x = x_0$, then $y''(x_0) = y'''(x_0) = 0$. Solve the differential equation under the following sets of conditions and arbitrary $R > 0$.

(a) $y(0) = y''(0) = 0$, $y(L) = y''(L) = 0$.
(b) $y(0) = y'(0) = 0$, $y(L) = y''(L) = 0$.
(c) $y(0) = y'(0) = 0$, $y''(L) = y'''(L) = 0$.
(d) $y(0) = y''(0) = 0$, $y''(L) = y'''(L) = 0$. Give a physical explanation for what goes wrong under these conditions.

11. Consider the algebraic equations

$$Az + Bw = K$$

$$Cz + Dw = L.$$

(a) Show that these equations have a unique solution for z, w if $AD - BC \neq 0$.
(b) Show that if $AD - BC = 0$, then these equations have either infinitely many solutions or else no solutions at all. [*Hint*: if A, B, C, D are all zero, the conclusion is obvious. If for example $B \neq 0$, then $C = AD/B$.]

12. In certain one-dimensional diffusion processes the concentration $y(x)$ of the diffusing substance depends only on the distance x from one end of the medium. Then $y(x)$ can be shown to satisfy the equation

$$y'' = ky \exp[\gamma(1 - y)/(1 + \delta(1 - y))],$$

with boundary conditions $y(0) = 0$, $y(1) = 1$. Here k, γ, and δ are nonnegative constants.
(a) Solve the equation under the assumption $\gamma = 0$ and arbitrary $k \geq 0$.
(b) Sketch the graph of the solution found in part (a) for $k = 1$.
(c) For general constants γ and δ see Exercise 8 of Section 4.

13. Under circumstances different from those assumed in Exercise 12, a diffusion concentration may satisfy an equation of the form

$$y'' = py' - qy^2,$$

where p and q are nonnegative constants, and there are boundary conditions of the form $y(0) = 0$, $y(1/p) = 1$.
(a) Solve the equation with $q = 0$ and arbitrary $p \geq 0$.
(b) Sketch the graph of the solution found in part (a).
(c) For general constants p and q see Exercise 9 of Section 4.

4 SHOOTING

Shooting is a method for applying numerical solution of initial-value problems to numerical solution of boundary-value problems. Assuming there is a solution to

$$y'' + a(x)y' + b(x)y = F(x), \qquad y(a) = \alpha, \qquad y(b) = \beta,$$

we can try to approximate it as follows. Replace the condition $y(b) = \beta$ by a condition $y'(a) = z_0$. Suppose we are very lucky in our choice of z_0. It might happen that the solution of the problem with initial value $y(a) = \alpha$, $y'(a) = z_0$ turns out to satisfy $y(b) = \beta$ nearly enough that we are satisfied with the result, and so we accept this for our approximate solution. If the discrepancy $y(b) - \beta$ is too great, we can then use this difference to guide us in adjusting z so as to reduce $|\,y(b) - \beta\,|$ to an acceptable size. The method applies just as well to the general second-order problem

$$y'' = f(x, y, y'), \qquad y(a) = \alpha, \qquad y(b) = \beta.$$

The term "shooting" suggests the idea that from $y(a)$ we "aim" at $y(b)$ by adjusting $y'(a)$ and making a succession of "shots" after each adjustment. The values of each approximate solution should be recorded in some way in case they turn out to be acceptable. See Figure 3.

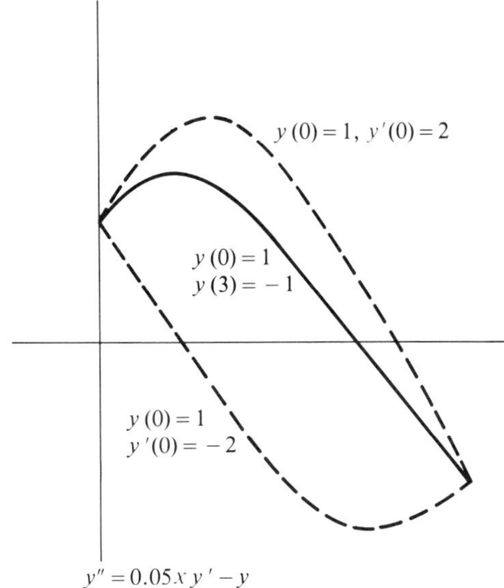

Figure 3

To start the method, we make two trial shots with $y'(a) = z_1$, z_2 and note the corresponding approximate values y_1, y_2 that we get at $x = b$. Of course, $y = \beta$ is what we want, so we choose the next value for $z = y'(a)$ to satisfy the linear interpolation equation

4.1
$$z_3 = z_2 + \frac{z_1 - z_2}{y_1(b) - y_2(b)}\,(\beta - y_2(b)).$$

This is just the equation in the y, z-plane of the line through (y_1, z_1) and (y_2, z_2), but evaluated at $y = \beta$. If the shot aimed with the choice $z = z_3$ is unacceptable, then try again, replacing z_1 by z_2 and z_2 by z_3 in Equation 4.1 to get z_4. It is a good idea to try to choose z_1, z_2 so that the corresponding y-values $y_1(b)$, $y_2(b)$ straddle the desired value $y = \beta$ and are not too close together. Then at each shot after that use z-values that lead to a straddle.

The central part of a computer program to implement the method is a subprogram to approximate the solution of an initial-value problem with $y(a) = \alpha$, $y'(a) = z$. The subprogram is called repeatedly until $|\,y(b) - \beta\,|$ is small enough.

EXAMPLE 1   The equation

$$\frac{d^2y}{dt^2} - 0.1t\,\frac{dy}{dt} + y = 0$$

is linear with a single nonconstant coefficient. If t is restricted to be fairly small, the solutions of the equation can be expected to behave something like the solutions of

$$\frac{d^2y}{dt^2} + y = 0;$$

they are $y = c_1 \cos t + c_2 \sin t$. For the given equation, consider the boundary-value problem of finding a solution satisfying

$$y(0) = 1, \qquad y(3) = -1.$$

To estimate a trial value for $y'(0)$, we solve the boundary problem for the constant-coefficient equation, with the result

$$y_c(t) = \cos t - 0.07 \sin t,$$

$$y_c'(t) = \sin t - 0.07 \cos t.$$

Thus we could try $z_0 = -0.07$ for a first shot. A more conservative approach would be to try $z_0 = 2$ and $z_1 = -2$ for the first two shots. With this choice, and with 500 steps of the modified Euler method, the result for the third shot turns out to give the value $y(3) = -1$. The graphs of the three solutions are shown in Figure 4.

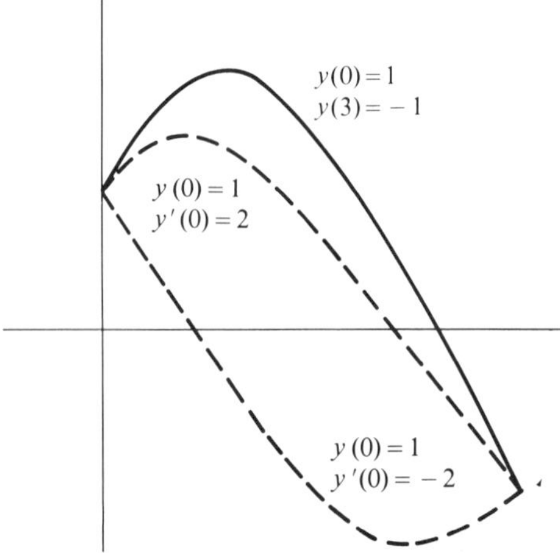

Figure 4

It is not hard to prove that the method described, when applied to a *linear* equation, will in principle lead to the desired solution at the third shot. The proof is left as an exercise. As an alternative, here is another method that applies only to linear equations and which, if applied correctly, is guaranteed to take only three shots. The approximate values from the first two shots have to be filed away and then combined with the results of the third shot. The three approximate solutions are designed to fit the following conditions:

$$y_p{:}y'' + a(x)y' + b(x)y = F(x), \qquad y(a) = 0, \qquad y'(a) = 0,$$

$$y_1{:}y'' + a(x)y' + b(x)y = 0, \qquad y(a) = \alpha, \qquad y'(a) = 0,$$

$$y_2{:}y'' + a(x)y' + b(x)y = 0, \qquad y(a) = 0, \qquad y'(a) = c.$$

The arbitrary constant c should be chosen to make $y_2(b) \neq 0$, and in fact to keep $y_2(b)$ from being near zero. It is easy to verify that the function

4.2 $$y(x) = y_p(x) + y_1(x) + \frac{\beta - y_p(b) - y_1(b)}{y_2(b)} y_2(x)$$

satisfies $y(a) = \alpha$, $y(b) = \beta$. In principle, $y(x)$ also satisfies the nonhomogeneous equation. The accuracy of the final approximation depends only on the accuracy of y_p, y_1, and y_2 and on the small amount of arithmetic required to combine them. If $y_2(b)$ happens to be too close to zero, a different choice for $y'(a) = c$ may be necessary. Of course, if you have a homogeneous equation, $F(x) = 0$ and $y_p(x) = 0$, so y_p need not be computed.

EXAMPLE 2   The problem

$$y'' + \frac{2x}{1 + x^2}y' + \frac{1}{20}xy = \frac{1}{5}, \qquad y(0) = -1, \qquad y(10) = 1.5,$$

has a solution that has been approximated as described in the preceding. The solution, together with y_p, y_1, and y_2 are plotted in Figure 5.

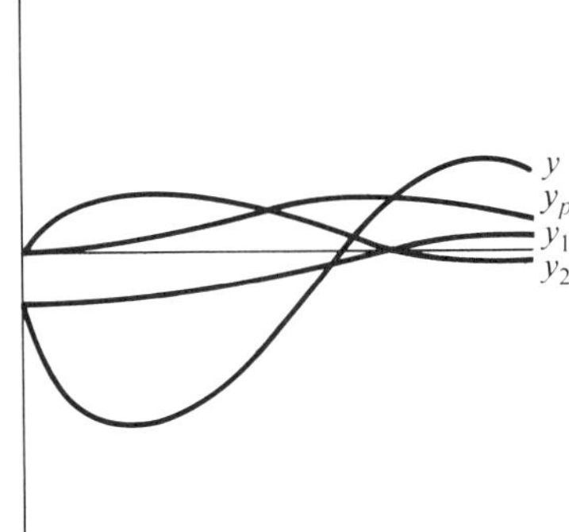

Figure 5

EXERCISES

1. **(a)** The differential equation $y'' - y = 1$ has the solution

$$y = (e + 1)^{-1}(e^x + e^{-x+1}) - 1$$

   satisfying $y(0) = y(1) = 0$. Find solutions y_1, y_2 satisfying

$$y_1(0) = 0, \qquad y_1'(0) = z_1 = 1,$$

$$y_2(0) = 0, \qquad y_2'(0) = z_2 = -1.$$

   Then find a third solution y_3 such that $y_3(0) = 0$, $y_3'(0) = z_3$ where z_3 is determined from Equation 4.1. Compare the result with the preceding solution.
   (b) Derive the solution given at the beginning of part (a) by using a Green's function.

2. **(a)** The differential equation $y'' + y = 1$ has the solution

$$y = -\cos x + \sin x + 1$$

   satisfying $y(0) = y(2\pi) = 0$. Find solutions y_1, y_2 satisfying

$$y_1(0) = 0, \qquad y_1'(0) = 1,$$

$$y_2(0) = 0, \qquad y_2'(0) = -1.$$

   Then explain why you cannot use Equation 4.1 to find a third solution.
   (b) Find all solutions of the given boundary-value problem.

3. Write a computer program to implement the shooting method for the approximate solution of

$$y'' = f(x, y, y'), \qquad y(a) = \alpha, \, y(b) = \beta.$$

   This program should contain some method at least as accurate as modified Euler for solution of the initial-value problem $y'' = f(x, y, y')$, $y(a) = \alpha, y'(a) = z$.

4. **(a)** Find graphical or numerical data for the solution of the boundary-value problem

$$y'' + \frac{2x}{1 + x^2}y' = F(x), \qquad y(0) = -1, \qquad y(10) = 1.5,$$

   for the case $F(x) = \frac{1}{5}$.
   (b) Show that the general solution of the differential equation in part (a) has the form

$$y = c_1 + c_2 \arctan x + \int_0^x (1 + t^2)(\tan^{-1}x - \tan^{-1}t)F(t)\, dt.$$

   (c) Explain how to determine c_1 and c_2 in part (b) in order to find a solution to the boundary-value problem in part (a).

5. Verify that, if y_1, y_2, and y_p are determined as they are for Equation 4.2, then the formula given there satisfies $y(a) = \alpha$, $y(b) = \beta$, provided that $y_2(b) \neq 0$.

6. Consider the nonlinear pendulum problem

$$\frac{d^2y}{dt^2} = -\sin y, \qquad y(0) = -1, \qquad y(3) = 1.$$

(a) Find a numerical solution for this problem and estimate the value of $y'(0)$ for your solution.
(b) Give a physical interpretation of your results in part (a).

7. Consider the boundary-value problem

$$y'' + a(x)y' + b(x)y = F(x),$$

$$A_1 y(a) + A_2 y'(a) = \alpha,$$

$$B_1 y(b) + B_2 y'(b) = \beta.$$

Let $y_p(x)$ be a solution of the differential equation satisfying $y_p(a) = y_p'(a) = 0$, and let $y_1(x)$ and $y_2(x)$ be linearly independent solutions of the associated **homogeneous** equation. The problem is to determine c_1, c_2 so that

$$y(x) = y_p(x) + c_1 y_1(x) + c_2 y_2(x)$$

satisfies the boundary conditions with A's, B's, α, and β given.
(a) Show that c_1, c_2 must satisfy

$$[A_1 y_1(a) + A_2 y_1'(a)]c_1 + [A_1 y_2(a) + A_2 y_2'(a)]c_2 = \alpha$$

$$[B_1 y_1(b) + B_2 y_1'(b)]c_1 + [B_1 y_2(b) + B_2 y_2'(b)]c_2 = \beta - B_1 y_p(b) - B_2 y_p'(b).$$

(b) Solve the equations in part (a) for c_1 and c_2 for the particular case of boundary conditions

$$y(0) = 0, \qquad y'(1) = 1.$$

Then show that the resulting solution of the boundary-value problem has the form

$$y(x) = y_p(x) + \frac{1 - y_p'(1)}{y_2'(1)} y_2(x), \qquad \text{if } y_2'(1) \neq 0.$$

(c) Using the boundary conditions given in part (b), find a numerical solution of the nonhomogeneous Airy equation

$$y'' - xy = 1.$$

8. Apply the shooting method to the boundary problem described in Exercise 12 of the previous section under the assumptions $k = 1$, $\gamma = .3$, $\delta = 5$. Compare your solution numerically with the exact solution for the case $k = 1$, $\gamma = 0$.

9. Apply the shooting method to the boundary problem described in Exercise 13 of the previous section under the assumptions $p = 1$, $q = 2$. Compare your solution numerically with the exact solution for the case $p = 1$, $q = 0$.

7

Existence and Uniqueness

1 INTRODUCTION

Quite a bit can be learned from the chapter without even looking at the statements, let alone the proofs, of the theorems. In particular, the Picard method described at the beginning of Section 1A is based on a valuable idea, the iterative solution of equations, that has important applications in many branches of applied mathematics. Beyond that, each section contains specific examples that have appeared earlier in the book and that can now be seen in a new way.

1A The Picard Method

In this section we take up an iterative method for finding approximate solutions to the vector differential equation with initial condition:

$$\frac{d\mathbf{x}}{dt} = F(t, \mathbf{x}), \qquad \mathbf{x}(t_0) = \mathbf{x}_0.$$

The method has the advantage that it can be used to prove the existence of solutions under certain hypotheses, although we will not complete the proof in all its details.

The first step in applying the Picard method to the preceding vector differential equation is to replace it by an equivalent vector **integral equation:**

$$\mathbf{x}(t) = \mathbf{x}_0 + \int_{t_0}^{t} F(u, \mathbf{x}(u)) \, du.$$

It is easy to see that, if $\mathbf{x} = \mathbf{x}(t)$ is a solution of either equation, then it is a solution of the other, in particular because the initial condition $\mathbf{x}(t_0) = \mathbf{x}_0$ is satisfied by the integral equation. The vector integral is, of course, computed by integrating the individual coordinate functions of the integrand. Next we compute a sequence of approximations to the desired solution by substituting each approximation into the next integrand:

$$\mathbf{x}_1(t) = \mathbf{x}_0 + \int_{t_0}^{t} F(u, \mathbf{x}_0)\, du,$$

$$\mathbf{x}_2(t) = \mathbf{x}_0 + \int_{t_0}^{t} F(u, \mathbf{x}_1(u))\, du,$$

$$\vdots$$

$$\mathbf{x}_{k+1}(t) = \mathbf{x}_0 + \int_{t_0}^{t} F(u, \mathbf{x}_k(u))\, du.$$

To get some feeling for the kind of approximation that the method produces, we take up an example for which a solution formula can be found by ordinary methods.

EXAMPLE 1   The real-valued differential equation

$$\frac{dx}{dt} = x + 1$$

is easily seen to have the solution

$$x(t) = -1 + e^t,$$

satisfying the initial condition $x(0) = 0$. Since $t_0 = 0$, $x_0 = 0$, and $F(t, x) = 1 + x$, we find

$$x_1(t) = \int_0^t 1\, du = t,$$

$$x_2(t) = \int_0^t (1 + u)\, du = t + \frac{t^2}{2},$$

$$x_3(t) = \int_0^t \left(1 + u + \frac{u^2}{2}\right) du = t + \frac{t^2}{2} + \frac{t^3}{3 \cdot 2},$$

$$\vdots$$

$$x_k(t) = \int_0^t \left(1 + u + \frac{u^2}{2} + \cdots + \frac{u^{k-1}}{(k-1)!}\right) du = t + \frac{t^2}{2} + \cdots + \frac{t^k}{k!}.$$

If we let $k \to \infty$, we get

$$\lim_{k \to \infty} x_k(t) = \lim_{k \to \infty} \left(t + \frac{t^2}{2} + \cdots + \frac{t^k}{k!} \right)$$

$$= \sum_{k=1}^{\infty} \frac{t^k}{k!} = -1 + e^t.$$

Thus, the kth Picard approximation happens in this example to be just the first k terms in the power series expansion of the solution. Figure 1 shows the graphs of the first few approximations, as well as the solution itself.

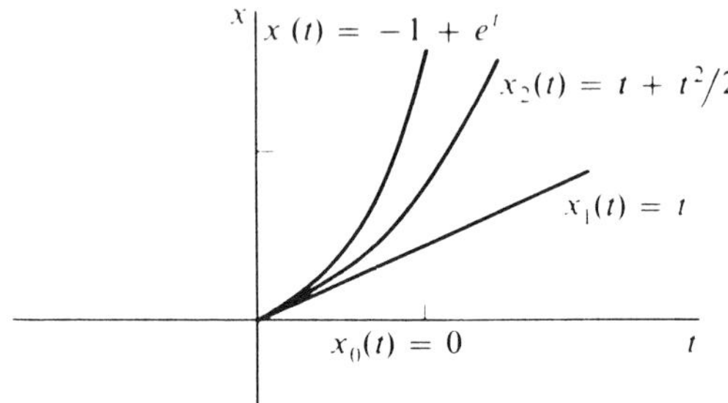

Figure 1

EXAMPLE 2   The nonlinear system

$$\frac{dx}{dt} = xy, \qquad x(0) = -1,$$

$$\frac{dy}{dt} = x + y, \qquad y(0) = 1$$

has Picard approximations given by the formulas

$$x_{k+1}(t) = -1 + \int_0^t x_k(u)\, y_k(u)\, du,$$

$$y_{k+1}(t) = 1 + \int_0^t (x_k(u) + y_k(u))\, du.$$

Then the first approximation, made using $(x_0, y_0) = (-1, 1)$, is

$$x_1(t) = -1 + \int_0^t (-1)\, du = -1 - t,$$

$$y_1(t) = 1 + \int_0^t 0\, du = 1.$$

The next step gives

$$x_2(t) = -1 + \int_0^t (-1 - u)\, du = -1 - t - \frac{t^2}{2},$$

$$y_2(t) = 1 + \int_0^t -u \; du = 1 - \frac{t^2}{2}.$$

In this example, we can sketch the *trajectories* of the two approximations, which are shown in Figure 2.

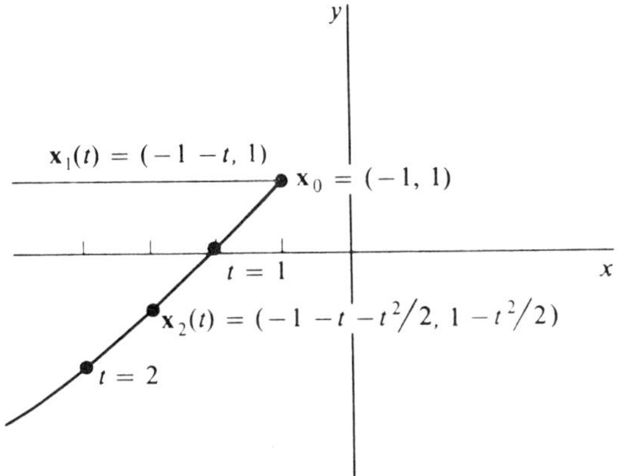

Figure 2

EXAMPLE 3   The Picard method can be applied to differential equations of order higher than one. For example, the initial-value problem

$$\ddot{x} + tx = 1, \qquad x(0) = 0, \qquad \dot{x}(0) = 1,$$

is equivalent to the system

$$\dot{x} = y, \qquad x(0) = 0,$$

$$\dot{y} = -tx + 1, \qquad y(0) = 1.$$

Thus $F(t, x, y) = (y, -tx + 1)$, so the Picard iterates start with

$$x_1(t) = \int_0^t 1 \; du = t,$$

$$y_1(t) = 1 + \int_0^t 1 \; du = 1 + t.$$

Next comes

$$x_2(t) = \int_0^t (1 + u) \; du = t + \frac{t^2}{2},$$

$$y_2(t) = 1 + \int_0^t (-u^2 + 1) \; du = 1 + t - \frac{t^3}{3}.$$

Figure 3 shows the graph of $x_1(t)$ and $x_2(t)$, since it is this rather than the system trajectory that we are interested in. Notice that the general solution of the equation $\ddot{x} + tx = 1$ is not available in a simple formula. However, we will be

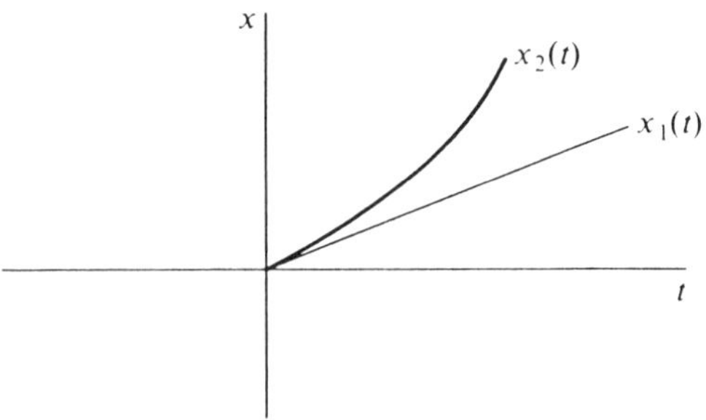

Figure 3

able to prove, under reasonable assumptions, that the Picard approximations converge to the true solution $x(t)$ satisfying $x(0) = 0$, $\dot{x}(0) = 1$.

Without some restriction on the nature of the function $F(t, \mathbf{x})$, we cannot prove that the Picard iterates $\mathbf{x}_k(t)$ converge to a solution $\mathbf{x}(t)$ of $d\mathbf{x}/dt = F(t, \mathbf{x})$. The most widely used kind of restriction is a **Lipschitz condition:** there is a number L such that

$$|F(t, \mathbf{x}) - F(t, \mathbf{y})| \leq L|\mathbf{x} - \mathbf{y}|$$

for all $\mathbf{x}$ and $\mathbf{y}$ in some subset S of $\mathcal{R}^n$ and for all t in some interval. If the function $F(t, \mathbf{x})$ satisfies a Lipschitz condition as a function of the vector $\mathbf{x}$, then the Picard iteration converges to a solution of the differential equation. The precise statement follows.

1B Existence Theorem

1.1 Theorem.   Suppose that $F(t, \mathbf{x})$ is continuous and satisfies a Lipschitz condition

$$|F(t, \mathbf{x}) - F(t, \mathbf{y})| \leq L|\mathbf{x} - \mathbf{y}|$$

for $t_0 \leq t \leq m$, and for all $\mathbf{x}$ and $\mathbf{y}$ in $\mathcal{R}^n$. Then for any $\mathbf{x}_0$ in $\mathcal{R}^n$, the differential equation

$$\frac{d\mathbf{x}}{dt} = F(t, \mathbf{x})$$

has a solution defined for $t_0 \leq t \leq m$ and satisfying the initial condition $\mathbf{x}(t_0) = \mathbf{x}_0$. The error

$$E_n = \max_{t_0 \leq t \leq m} |\mathbf{x}(t) - \mathbf{x}_n(t)|$$

made by using $\mathbf{x}_n$, the nth Picard approximation, can be estimated by the inequality

$$E_n \leq [\max_{t_0 \leq t \leq m} |F(t, \mathbf{x}_0)|] \sum_{k=n}^{\infty} \frac{L^k(m - t_0)^{k+1}}{(k + 1)!}.$$

Proof. To prove convergence, we use repeatedly the norm inequality for vector integrals

$$\left| \int_a^b f(u)\, du \right| \le \int_a^b |f(u)|\, du.$$

This inequality is a generalization of a familiar fact about real-valued integrals, and its proof is outlined in Exercise 6. As usual, we let $\mathbf{x}_k(t)$ stand for the kth Picard approximation. Then the norm inequality implies for $\mathbf{x}_1(t) - \mathbf{x}_0$ that

$$|\mathbf{x}_1(t) - \mathbf{x}_0| = \left| \int_{t_0}^t F(u, \mathbf{x}_0)\, du \right| \le \int_{t_0}^t |F(u, \mathbf{x}_0)|\, du \le M \int_{t_0}^t du$$

$$= (t - t_0)M.$$

Here, M is the maximum of $|F(u, \mathbf{x}_0)|$ for $t_0 \le u \le m$. We next form the difference $\mathbf{x}_2(t) - \mathbf{x}_1(t)$ and apply first the norm inequality, then the Lipschitz condition, to get

$$|\mathbf{x}_2(t) - \mathbf{x}_1(t)| \le \int_{t_0}^t |F(u, \mathbf{x}_1(u)) - F(u, \mathbf{x}_0)|\, du$$

$$\le \int_{t_0}^t L |\mathbf{x}_1(u) - \mathbf{x}_0|\, du \le LM \int_{t_0}^t (u - t_0)\, du$$

$$= \frac{LM(t - t_0)^2}{2!}.$$

Proceeding by induction, we find that

$$|\mathbf{x}_{k+1}(t) - \mathbf{x}_k(t)| \le \frac{ML^k(t - t_0)^{k+1}}{(k + 1)!}$$

$$\le \frac{ML^k(m - t_0)^{k+1}}{(k + 1)!}.$$

Hence

$$\sum_{k=0}^{\infty} |\mathbf{x}_{k+1}(t) - \mathbf{x}_k(t)| \le \frac{M}{L} \sum_{k=0}^{\infty} L^{k+1} \frac{(m - t_0)^{k+1}}{(k + 1)!}$$

$$= \frac{M}{L} \left[e^{L(m - t_0)} - 1 \right].$$

It follows that the series on the left converges, and so the series with partial sums

$$\sum_{k=0}^{n-1} (\mathbf{x}_{k+1}(t) - \mathbf{x}_k(t)) = \mathbf{x}_n(t) - \mathbf{x}_0$$

converges absolutely in each coordinate. Since $\mathbf{x}_0$ is a constant vector, the limit

$$\lim_{n \to \infty} \mathbf{x}_n(t) = \mathbf{x}(t)$$

defines $\mathbf{x} = \mathbf{x}(t)$ as a function of t for $t_0 \le t \le m$. If we let $k \to \infty$ in the equation

$$\mathbf{x}_{k+1}(t) = \mathbf{x}_0 + \int_{t_0}^{t} F(u, \mathbf{x}_k(u))\, du,$$

we get

$$\mathbf{x}(t) = \mathbf{x}_0 + \lim_{k \to \infty} \int_{t_0}^{t} F(u, \mathbf{x}_k(u))\, du.$$

To calculate the remaining limit, we observe that

$$\left| \mathbf{x}(t) - \mathbf{x}_n(t) \right| = \left| \sum_{k=n}^{\infty} (\mathbf{x}_{k+1}(t) - \mathbf{x}_k(t)) \right|$$

$$\leq \frac{M}{L} \sum_{k=n}^{\infty} \frac{L^{k+1}(m - t_0)^{k+1}}{(k + 1)!} = R_n,$$

where $\lim_{n \to \infty} R_n = 0$ because R_n is the tail end of a convergent series of numbers. By the Lipschitz condition,

$$\left| F(t, \mathbf{x}(t)) - F(t, \mathbf{x}_n(t)) \right| \leq L \left| \mathbf{x}(t) - \mathbf{x}_n(t) \right| \leq L R_n.$$

Hence, for t such that $t_0 \leq t \leq m$, we have

$$\left| \int_{t_0}^{t} F(u, \mathbf{x}(u))\, du - \int_{t_0}^{t} F(u, \mathbf{x}_n(u))\, du \right| \leq L R_n (m - t_0).$$

Since R_n tends to zero, the right-hand integral tends to the one to its left; therefore,

$$\mathbf{x}(t) = \mathbf{x}_0 + \int_{t_0}^{t} F(u, \mathbf{x}(u))\, du.$$

Differentiation with respect to t shows that $d\mathbf{x}/dt = F(t, \mathbf{x})$. Finally, it is clear that $\mathbf{x}(t_0) = \mathbf{x}_0$ by setting $t = t_0$.

It turns out that if $F(t, \mathbf{x})$ is linear as a function of $\mathbf{x}$, or, more generally, has the form $F(t, \mathbf{x}) = A(t)\mathbf{x} + \mathbf{b}(t)$, then it automatically satisfies a Lipschitz condition in all of $\mathfrak{R}^n$, provided that the coefficients are continuous functions of t on some interval. These and most other cases of practical interest are covered by the following criterion.

1.2 Theorem.   If $F(t, \mathbf{x}) = (F_i(t, x_1, \ldots, x_n))$ has partial derivatives satisfying

$$\left| \frac{\partial F_i(t, \mathbf{x})}{\partial x_k} \right| \leq M$$

for all $\mathbf{x}$ in some convex set S of $\mathfrak{R}^n$ and for t in some interval, then $F(t, \mathbf{x})$ has a Lipschitz constant $L = nM$ in the same set.

Proof. The idea is to apply the mean-value theorem to the segment joining $\mathbf{x}$ and $\mathbf{x} + \mathbf{y}$. The chain rule for derivatives shows that for each F_i

$$\frac{d}{du} F_i(t, \mathbf{x} + u\mathbf{y}) = \sum_{k=1}^{n} \frac{\partial F_i}{\partial x_k} (t, \mathbf{x} + u\mathbf{y}) y_k.$$

If $\mathbf{x}$ and $\mathbf{x} + \mathbf{y}$ are in S, then the convexity of S allows us to apply the mean-value theorem to F_i as a function of u, giving for some v, with $0 < v < 1$,

$$F_i(t, \mathbf{x} + \mathbf{y}) - F_i(t, \mathbf{x}) = \sum_{k=1}^{n} \frac{\partial F_i}{\partial x_k}(t, \mathbf{x} + v\mathbf{y})y_k.$$

The expression on the right is a dot product of two vectors, so we can apply the Cauchy–Schwarz inequality to get

$$(F_i(t, \mathbf{x} + \mathbf{y}) - F_i(t, \mathbf{x}))^2 \leq \left[\sum_{k=1}^{n}\left(\frac{\partial F_i}{\partial x_k}\right)^2\right]\left(\sum_{k=1}^{n} y_k^2\right).$$

Summing over i gives the norm squared

$$|F(t, \mathbf{x} + \mathbf{y}) - F(t, \mathbf{x})|^2 \leq n\left(\sum_{i=1}^{n} M^2\right)|\mathbf{y}|^2 = n^2 M^2 |\mathbf{y}|^2.$$

Taking square roots shows that

$$|F(t, \mathbf{x} + \mathbf{y}) - F(t, \mathbf{x})| \leq L|\mathbf{y}|,$$

with $L = nM$.

EXAMPLE 4   If F has the form

$$F(t, \mathbf{x}) = A(t)\mathbf{x} + \mathbf{b}(t),$$

then

$$F_i(t, x_1, \ldots, x_n) = \sum_{k=1}^{n} a_{ik}(t)x_k + b_i(t).$$

The partial derivatives in question are

$$\frac{\partial F_i}{\partial x_k}(t, \mathbf{x}) = a_{ik}(t),$$

so we can take M to be the maximum of all of $|a_{ik}(t)|$, for t in whatever interval we happen to be interested. A Lipschitz constant is then nM.

EXAMPLE 5   If F and x are real valued and F is given by $F(t, x) = 1 + x^2$, then

$$\frac{\partial F}{\partial x}(t, x) = 2x,$$

and for $|x| \leq b$ we find

$$\left|\frac{\partial F}{\partial x}(t, x)\right| = |2x| \leq 2b.$$

Hence we can take $L = M = 2b$ on the interval $-b \leq x \leq b$. However, F does not satisfy a Lipschitz condition unrestrictedly in x, because

$$|F(t, x) - F(t, y)| = |x^2 - y^2| = |x + y||x - y|.$$

Since the factor $|x + y|$ can become arbitrarily large without some restriction on x and y, it is impossible to find a Lipschitz constant that will do for all x in $\mathcal{R}^1$. Indeed, the conclusion of Theorem 1.1 fails for the differential equation

$$\frac{dx}{dt} = 1 + x^2,$$

without some restriction on x. For if we take $x(t) = \tan t$, we see that we get a solution for $-\pi/2 < t < \pi/2$, but in no larger interval containing $t = 0$.

EXERCISES

1. For each differential equation, write an equivalent integral equation.

   (a) $\dfrac{dx}{dt} = tx + t^2,\ x(0) = 1.$

   (b) $\dfrac{dx}{dt} = 1,\ x(0) = 2.$

   (c) $\dfrac{dx}{dt} = x - t,\ x(0) = 0.$

2. For each equation given in Exercise 1, find the Picard approximate solutions $x_1(t)$ and $x_2(t)$. Start with $x_0(t) \equiv x(0)$ in each case.

$$\left[\text{(a) } x_1(t) = 1 + \frac{t^2}{2} + \frac{t^2}{3}.\right]$$

3. (a) Find a vector integral equation equivalent to the system

$$\dot{x} = ty, \qquad x(0) = 1,$$
$$\dot{y} = tx, \qquad y(0) = 2.$$

   (b) Find the Picard approximate solution $(x_1(t), y_1(t))$ by starting with $(x(0), y(0)) = (1, 2)$.

   (c) Find the Picard approximate solution $(x_2(t), y_2(t))$ from the approximation found in part (b).

$$\left[Ans.\ 1 + t^2 + \frac{t^4}{8},\ 2 + \frac{t^2}{2} + \frac{t^4}{4}.\right]$$

   (d) Sketch the trajectory of the approximate solution found in part (c).

4. (a) The second-order differential equation

$$\frac{d^2y}{dt^2} + ty = 0, \qquad y(0) = 0, \qquad \dot{y}(0) = 1$$

is equivalent to a first-order system of differential equations. What is that system?

(b) Find a pair of integral equations equivalent to the system found in part (a).

(c) Starting with the constant initial solution, find the next two Picard approximate solutions and pick out the relevant part as an approximate solution to the second-order differential equation in part (a).

5. Find Lipschitz constants for each of the following functions on the specified set.

   (a) $F(t, x) = t^2 + x^2,\ 0 \le t \le 1,\ 0 \le x \le 1.$

   (b) $F(t, x) = x,\ -\infty < t < \infty,\ -\infty < x < \infty.$

   (c) $F(t, x, y) = tx + y^2,\ 0 \le t \le 1,\ 0 \le x,\ 0 \le y \le 1.$

6. If f is continuous on $[a, b]$ and takes its values in $\mathcal{R}^n$, show that

   (a)
   $$\int_a^b \mathbf{k} \cdot f(u)\ du = \mathbf{k} \cdot \int_a^b f(u)\ du,$$

   for any constant vector $\mathbf{k}$ in $\mathcal{R}^n$.

   (b) Show that

   $$\left| \int_a^b f(u)\ du \right| \le \int_a^b |f(u)|\ du.$$

   [*Hint.* By the Cauchy–Schwarz inequality,

   $$f(t) \cdot \int_a^b f(u)\ du \le |f(t)| \left| \int_a^b f(u)\ du \right|$$

   for each t in $[a, b]$. Integrate with respect to t and apply the result of part (a).]

7. Let A be an n-by-n constant matrix. Prove that the Picard iterates $\mathbf{x}_n(t)$ for the differential equation $d\mathbf{x}/dt = A\mathbf{x}$ are just the partial sums of the power series expansion of the matrix exponential e^{tA}.

2 UNIQUENESS

For many of the differential equations we have looked at so far in this book, the question of uniqueness of solutions satisfying prescribed initial conditions has been settled by a method of solution that proves there is only one solution satisfying the initial conditions. Thus there is no need to look for another solution. Such questions often become more difficult to answer in the case of differential equations for which no solution formulas exist; but at the same time it may be even more important to know the answer. For example, having found a numerical approximation by some method, it is possible that there may be another solution satisfying the same initial conditions, and that the other solution is the one we are really interested in. The main theoretical statements given here are made in terms of first-order systems, because they contain the other important systems as special cases, but the examples will just as often be about single differential equations.

EXAMPLE 1   The first-order linear equation

$$\frac{dx}{dt} + f(t)x = g(t), \qquad x(t_0) = x_0$$

has a unique solution satisfying the initial condition, on any interval containing t_0 for which f and g are continuous. The usual solution formula proves the existence of a solution; to see that it is unique, we need only review the way in which the formula is derived, that is, at each step observing that, if a solution exists, then it must be of a specific form. We first find an equivalent equation by using an exponential multiplier:

$$\frac{d}{dt}\left[\exp\left(\int_{t_0}^{t} f(u)\,du \right)x \right] = g(t)\exp\left(\int_{t_0}^{t} f(u)\,du \right).$$

Then

$$x(t) = \exp\left(-\int_{t_0}^{t} f(u)\,du \right) \int_{t_0}^{t} g(t)\exp\left(\int_{t_0}^{t} f(u)\,du \right) dt$$

$$+ x_0 \exp\left(-\int_{t_0}^{t} f(u)\,du \right).$$

The requirement that $x(t_0) = x_0$ *forces* us to choose the constant of integration to be x_0, thus ensuring that the solution is unique.

An nth-order differential equation of the form

$$\frac{d^n y}{dt^n} = F\left(t, y, \frac{dy}{dt}, \ldots, \frac{d^{n-1}y}{dt^{n-1}} \right)$$

has initial conditions of the form

$$y(t_0) = c_0, \frac{dy}{dt}(t_0) = c_1, \ldots, \frac{d^{n-1}y}{dt^{n-1}}(t_0) = c_{n-1}.$$

Such a differential equation can be written as a first-order system of dimension n by letting

$$\frac{dy}{dt} = y_1,$$

$$\frac{dy_1}{dt} = y_2,$$

$$\vdots$$

$$\frac{dy_{n-2}}{dt} = y_{n-1},$$

$$\frac{dy_{n-1}}{dt} = F(t, y, y_1, \ldots, y_{n-1}).$$

Thus, the kth derivative of y is equal to y_k, so the initial conditions can be written

$$y(t_0) = c_0,$$

$$y_1(t_0) = c_1,$$

$$\cdot$$
$$\cdot$$
$$\cdot$$

$$y_{n-1}(t_0) = c_{n-1}.$$

We have used this way of writing a differential equation as a system in earlier sections of this book. Here its importance lies in the fact that questions about existence and uniqueness of solutions are much more natural to treat for first-order systems; then it is an easy matter to translate the result into a statement about nth-order equations.

Simple examples show that, if we assume only continuity of F in the vector differential equation $dx/dt = F(t, \mathbf{x})$, then there may be more than one solution satisfying $\mathbf{x}(t_0) = \mathbf{c}$ for some vector $\mathbf{c}$ in $\mathcal{R}^n$. An additional assumption about F that will do to guarantee uniqueness is a **Lipschitz condition:** there is a constant $L > 0$ such that

$$|F(t, \mathbf{x}) - F(t, \mathbf{y})| \le L|\mathbf{x} - \mathbf{y}|$$

for all $\mathbf{x}$ and $\mathbf{y}$ in some subset S of $\mathcal{R}^n$ and for all t in some interval. When S is a convex set, Theorem 1.2 of the previous section gives a simple criterion for a Lipschitz condition, that the partial derivatives $\partial F_i/\partial x_k$ be bounded on S for all t in the interval. We first state the relevant theorem and then show how it works.

2.1 Theorem.   If F satisfies a Lipschitz condition on a set S in $\mathcal{R}^n$ for all t in an interval $a < t < b$, then there is at most one solution to the vector differential equation

$$\frac{d\mathbf{x}}{dt} = F(t, \mathbf{x})$$

satisfying $\mathbf{x}(t_0) = \mathbf{c}$, where $\mathbf{c}$ is in S and $a < t_0 < b$.

Proof. Suppose there are two solutions $\mathbf{x}(t)$ and $\mathbf{y}(t)$ satisfying $\mathbf{x}(t_0) = \mathbf{y}(t_0) = \mathbf{c}$. Let

$$\rho(t) = |\mathbf{x}(t) - \mathbf{y}(t)|^2 = (\mathbf{x}(t) - \mathbf{y}(t)) \cdot (\mathbf{x}(t) - \mathbf{y}(t)).$$

Then routine calculation shows that

$$\frac{d\rho}{dt}(t) = 2(\mathbf{x}(t) - \mathbf{y}(t)) \cdot \left(\frac{d\mathbf{x}}{dt}(t) - \frac{d\mathbf{y}}{dt}(t)\right).$$

Because $\mathbf{x}(t)$ and $\mathbf{y}(t)$ satisfy the differential equation,

$$\frac{d\rho}{dt}(t) = 2(\mathbf{x}(t) - \mathbf{y}(t)) \cdot (F(t, \mathbf{x}(t)) - F(t, \mathbf{y}(t))).$$

By the Cauchy-Schwarz inequality, $|\mathbf{a} \cdot \mathbf{b}| \le |\mathbf{a}||\mathbf{b}|$,

$$\frac{d\rho}{dt}(t) \le 2|\mathbf{x}(t) - \mathbf{y}(t)| \, \|F(t, \mathbf{x}(t)) - F(t, \mathbf{y}(t))|.$$

Then, by the Lipschitz condition, there is a constant L such that

$$\frac{d\rho}{dt}(t) \le 2L|\mathbf{x}(t) - \mathbf{y}(t)|^2 = 2L\rho(t).$$

We now apply the exponential multiplier technique to the inequality, multiplying both sides by e^{-2Lt} to get

$$\frac{d}{dt}(e^{-2Lt}\rho) = e^{-2Lt}\left(\frac{d\rho}{dt} - 2L\rho\right) \le 0.$$

Since $e^{-2Lt}\rho$ has a nonpositive derivative, it is nonincreasing for increasing t. But $\rho(t_0) = 0$ and $e^{-2Lt}\rho(t) \ge 0$, so $\rho(t) = 0$ when $t_0 \le t \le b$. It follows that $\mathbf{x}(t) = \mathbf{y}(t)$ on the same interval. A similar argument, making the change of variable $t = a - u$, and applied to the interval $a - t_0 \le u < 0$, shows that $\mathbf{x}(t) = \mathbf{y}(t)$ when $a < t \le t_0$.

EXAMPLE 2   Consider the differential equation

$$\frac{dx}{dt} = \begin{cases} \sqrt{x}, & x \ge 0, \\ 0, & x < 0. \end{cases}$$

The equation has two solutions satisfying $x(0) = 0$:

$$x_1(t) = \begin{cases} \dfrac{t^2}{4}, & t \ge 0, \\ 0, & t < 0, \end{cases}$$

and
$$x_2(t) = 0 \qquad \text{for all } t.$$

However, there is no contradiction to the previous theorem here, because the function $F(x) = \sqrt{x}$ fails to satisfy a Lipschitz condition for a fixed constant $L > 0$. We would have to have

$$|\sqrt{x} - \sqrt{y}| \le L|x - y|$$

or
$$|\sqrt{x} - \sqrt{y}| \le L|(\sqrt{x} - \sqrt{y})(\sqrt{x} + \sqrt{y})|.$$

In other words, $0 < 1/L \le \sqrt{x} + \sqrt{y}$. But as x and y get close to 0, the right side gets close to 0, which is impossible if $L > 0$.

For an autonomous system $d\mathbf{x}/dt = F(\mathbf{x})$ satisfying the conditions of the uniqueness theorem, the solutions have a very natural dependence on the initial conditions; to express it, we denote the solution starting at $\mathbf{x}_0$ when $t = 0$ by $\mathbf{x} = \mathbf{x}(t, \mathbf{x}_0)$. Thus, at time t the solution is at $\mathbf{x}(t, \mathbf{x}_0)$, whereas $\mathbf{x}(0, \mathbf{x}_0) = \mathbf{x}_0$. If we now start a solution at the point $\mathbf{x}(t, \mathbf{x}_0)$, then after time s we wind up at $\mathbf{x}(s, \mathbf{x}(t, \mathbf{x}_0))$. But by the uniqueness theorem we would have arrived at the same point by starting at $\mathbf{x}_0$ and allowing time $t + s$ to elapse. Hence

$$\mathbf{x}(s, \mathbf{x}(t, \mathbf{x}_0)) = \mathbf{x}(t + s, \mathbf{x}_0).$$

If we use the notation $T_t(\mathbf{c}) = \mathbf{x}(t, \mathbf{c})$, the previous equation can be written

$$T_s(T_t(\mathbf{x}_0)) = T_{t+s}(\mathbf{x}_0);$$

this generalizes the equation $e^{sA}e^{tA} = e^{(t+s)A}$, that holds if $F(\mathbf{x}) = A\mathbf{x}$.

EXAMPLE 3   The autonomous differential equation

$$\frac{dx}{dt} = x - 1$$

has the general solution $x(t) = ce^t + 1$. The solution satisfying the initial condition $x(0) = x_0$ is

$$T_t(x_0) = (x_0 - 1)e^t + 1.$$

Then

$$T_s(T_t(x_0)) = [((x_0 - 1)e^t + 1) - 1]e^s + 1$$
$$= (x_0 - 1)e^{t+s} + 1 = T_{t+s}(x_0).$$

On the other hand, the nonautonomous equation

$$\frac{dx}{dt} = tx$$

has the general solution $x(t) = ce^{t^2/2}$. Starting at x_0 when $t = 0$, we get

$$x(t, x_0) = x_0 e^{t^2/2}.$$

Starting over again at $x(t, x_0)$, we get

$$x(s, x(t, x_0)) = x(t, x_0)e^{s^2/2}$$
$$= x_0 e^{t^2/2}e^{s^2/2} = x_0 e^{(t^2+s^2)/2}.$$

But

$$x(t + s, x_0) = x_0 e^{(t+s)^2/2} \neq x_0 e^{(t^2+s^2)/2}.$$

EXERCISES

1. Show by a logical argument based on a specific method of solution that each of the following differential equations has, for all x, a unique solution satisfying the given initial condition.

   (a) $\dfrac{dy}{dx} = \sin x$, $y(0) = 1$.   [*Ans.* $2 - \cos x$.]

   (b) $\dfrac{dy}{dx} = xy$, $y(0) = 1$, for $-1 \leq x \leq 1$.

   (c) $\dfrac{d^2y}{dx^2} + 2\dfrac{dy}{dx} + y = 0$, $y(0) = 0$, $y'(0) = 1$.

2. Show that Theorem 2.1 can be used to guarantee the existence of a unique solution for each differential equation in Exercise 1.

3. (a) Show that the function $F(t, x) = t\sqrt{x}$ does not satisfy a Lipschitz condition for $0 \leq t \leq m$, $0 \leq x$.

   (b) Show that the function $F(t, x) = t\sqrt{x}$ does satisfy a Lipschitz condition for $0 \leq t \leq m$, $a < x$, provided that $a > 0$.

   (c) Find all solutions of $(dx/dt) = t\sqrt{x}$.

4. Show that the differential equation

$$\frac{dy}{dx} = 3y^{2/3}$$

   has the solutions

$$y = \begin{cases} (x - a)^3, & x < a, \\ 0, & a \leq x \leq b, \\ (x - b)^3 & b \leq x. \end{cases}$$

   Sketch some of these solutions for various values of the constants a and b. Are there any other solutions?

5. Show that the differential equation in Exercise 4 fails to satisfy a Lipschitz condition on an interval of the form $-a < y < a$, $a > 0$.

3 INITIAL CONDITIONS

The ideas in this section all have to do with linear differential equations and the way in which their solutions depend on initial conditions. It is clear from the earlier chapters that the set S of all solutions of a homogeneous linear equation can be expected to be a linear combination of certain basic solutions; this is often expressed by saying that the basic solutions **span** S. Here we can see why this is true even for equations with variable coefficients, and also see why it is true that there are, under fairly general conditions, solutions satisfying arbitrary initial conditions.

For linear systems with continuous coefficients, the convergence of the Picard approximations guarantees the existence of solutions with prescribed initial conditions. For systems of the special form $d\mathbf{x}/dt = A(t)\mathbf{x}$, the possibility of prescribing the initial conditions is equivalent to the existence of linearly independent solutions. The following theorem describes the equivalence precisely.

3.1 Theorem.   Let $d\mathbf{x}/dt = A(t)\mathbf{x}$ be a linear homogeneous vector differential equation with n-by-n coefficient matrix having continuous entries for $a < t < b$. If

$$\mathbf{x}_1(t), \ \mathbf{x}_2(t), \ \ldots, \ \mathbf{x}_k(t)$$

are vector solutions of the differential equation, then these solutions are linearly independent as functions of t if and only if for an arbitrary t_0, with $a < t_0 < b$, the constant initial vectors

$$\mathbf{x}_1(t_0), \ \mathbf{x}_2(t_0), \ \ldots, \ \mathbf{x}_k(t_0)$$

are independent vectors in $\mathcal{R}^n$. In the case of independence and in case $k = n$, the initial vectors span $\mathcal{R}^n$, and the corresponding solutions span the n-dimensional space of all solutions of the differential equation.

Proof. Independence of the constant vectors $\mathbf{x}_j(t_0)$ implies the independence of the functions $\mathbf{x}_j(t)$ even if these functions are not solutions of a differential equation. For if

$$\sum_{j=1}^{k} c_j\mathbf{x}_j(t) = 0, \qquad a < t < b,$$

then the equation holds in particular for $t = t_0$. But since the $\mathbf{x}_j(t_0)$ are assumed to be independent, all the c_j must be zero. This proves the functions linearly independent.

Conversely, if the solutions $\mathbf{x}_j(t)$ are linearly independent, we can show that $\mathbf{x}_j(t_0), j = 1, \ldots, k$, form an independent set of vectors in $\mathcal{R}^n$. For if

$$\sum_{j=1}^{k} c_j\mathbf{x}_j(t_0) = 0,$$

then the linear combination

$$\mathbf{x}(t) = \sum_{j=1}^{k} c_j\mathbf{x}_j(t)$$

defines a solution $\mathbf{x}$ of $d\mathbf{x}/dt = A(t)\mathbf{x}$, satisfying the initial condition $\mathbf{x}(t_0) = 0$. But the linear vector differential equation has the *unique* solution identically equal to zero that satisfies $\mathbf{x}(t_0) = 0$. Hence

$$\sum_{j=1}^{k} c_j\mathbf{x}_j(t) = 0, \qquad a < t < b,$$

which proves that all c_i are zero, because the functions $\mathbf{x}_j(t)$ are assumed independent. Thus the vectors $\mathbf{x}_j(t_0)$ are independent in $\mathcal{R}^n$.

The last statement in the theorem now follows from Theorem 6.8 of Chapter 5, because if we form a linear combination of n initial vectors to get an arbitrary vector $\mathbf{c}$ in $\mathcal{R}^n$ then the same linear combination of the corresponding solutions will be the unique solution with initial vector $\mathbf{c}$.

EXAMPLE 1   Consider the system

$$\begin{pmatrix} \dfrac{dx}{dt} \\ \dfrac{dy}{dt} \end{pmatrix} = \begin{pmatrix} 2 & 4 \\ 1 & -1 \end{pmatrix}\begin{pmatrix} x \\ y \end{pmatrix} + \begin{pmatrix} 2 \\ 4 \end{pmatrix}.$$

In Example 2, Section 2, Chapter 4, it was shown that the associated homogeneous system had solutions

$$x_1(t) = \begin{pmatrix} -e^{-2t} \\ e^{-2t} \end{pmatrix}, \qquad x_2(t) = \begin{pmatrix} 4e^{3t} \\ e^{3t} \end{pmatrix}.$$

We can check that these two vector functions are linearly independent by observing that when $t = 0$ we get the independent vectors

$$\begin{pmatrix} -1 \\ 1 \end{pmatrix}, \qquad \begin{pmatrix} 4 \\ 1 \end{pmatrix}.$$

Going the other way, since the functions are independent, the vectors

$$\mathbf{x}_1(t_0) = \begin{pmatrix} -e^{-2t_0} \\ e^{-2t_0} \end{pmatrix}, \qquad \mathbf{x}_2(t_0) = \begin{pmatrix} 4e^{3t_0} \\ e^{3t_0} \end{pmatrix}$$

are independent for every choice of t_0. It follows that a given vector $\mathbf{x}_0$ in $\mathfrak{R}^2$ can be expressed as a linear combinaton

$$\mathbf{x}_0 = c_1\mathbf{x}_1(t_0) + c_2\mathbf{x}_2(t_0),$$

so that the solution $\mathbf{x}$ defined by

$$\mathbf{x}(t) = c_1\begin{pmatrix} -e^{-2t} \\ e^{-2t} \end{pmatrix} + c_2\begin{pmatrix} 4e^{3t} \\ e^{3t} \end{pmatrix}$$

satisfies the initial condition $\mathbf{x}(t_0) = \mathbf{x}_0$.

To solve the nonhomogeneous system, all we have to do is observe that the vector

$$\mathbf{x}_p = \begin{pmatrix} -3 \\ 1 \end{pmatrix}$$

is a solution of the differential equation. From the general theory of linear operators, it follows that *all* solutions are obtained simply by adding $\mathbf{x}_p$ to the general homogeneous solution (see Chapter 2, Section 3).

One of the most important applications of the previous theorem is to nth-order linear differential equations of the form

$$L(y) = y^{(n)} + a_{n-1}(t)y^{(n-1)} + \cdots + a_0(t)y = f(t),$$

with initial conditions

$$y(t_0) = c_0, \ldots, y^{(n-1)}(t_0) = c_{n-1}.$$

Every such differential equation can be written as a linear system,

$$
\begin{pmatrix} \dfrac{dy}{dt} \\ \cdot \\ \cdot \\ \cdot \\ \dfrac{dy_{n-1}}{dt} \end{pmatrix} = \begin{pmatrix} 0 & 1 & \ldots & 0 \\ 0 & 0 & \ldots & 0 \\ \cdot & \cdot & & \cdot \\ \cdot & \cdot & & \cdot \\ \cdot & \cdot & & \cdot \\ 0 & 0 & \ldots & 1 \\ -a_0(t) & -a_1(t) & \ldots & -a_{n-1}(t) \end{pmatrix} \begin{pmatrix} y \\ y_1 \\ \cdot \\ \cdot \\ \cdot \\ y_{n-1} \end{pmatrix} + \begin{pmatrix} 0 \\ 0 \\ \cdot \\ \cdot \\ \cdot \\ f(t) \end{pmatrix},
$$

in which $y_k = d^k y / dt^k$. The initial condition $y(t_0) = c_0, \ldots , y^{(n-1)}(t_0) = c_{n-1}$ gets translated into

$$
\begin{pmatrix} y(t_0) \\ y_1(t_0) \\ \cdot \\ \cdot \\ \cdot \\ y_{n-1}(t_0) \end{pmatrix} = \begin{pmatrix} c_0 \\ c_1 \\ \cdot \\ \cdot \\ \cdot \\ c_{n-1} \end{pmatrix}.
$$

For this reason, any statement about initial conditions for a solution of the nth-order equation can be treated as a statement about a first-order system. The following theorem is the translated version of Theorem 3.1.

3.2   Theorem.   Let $u_1(t), \ldots , u_k(t)$ be solutions of a homogeneous nth-order linear differential equation $L(y) = 0$ with continuous coefficients on an interval $a < t < b$. These solutions are linearly independent as functions of t if and only if, for an arbitrary t_0, with $a < t_0 < b$, the vectors

$$
\mathbf{u}_j(t_0) = \begin{pmatrix} u_j(t_0) \\ u_j'(t_0) \\ \cdot \\ \cdot \\ \cdot \\ u_j^{(n-1)}(t_0) \end{pmatrix}, \qquad j = 1, \ldots , k,
$$

are independent vectors in $\mathfrak{R}^n$. In the case of independence, and in case $k = n$, the vectors $\mathbf{u}_j(t_0)$ span $\mathfrak{R}^n$, and the corresponding solutions span the n-dimensional space of all solutions of the differential equation.

Proof. By Theorem 3.1, we see that the vectors $\mathbf{u}_j(t_0)$ are independent if and only if the vector solutions $\mathbf{u}_j(t)$ are independent as functions. But if the real-valued solutions $u_j(t)$ are independent real-valued functions, then the vector functions $\mathbf{u}_j(t)$ are also independent, just because their first entries are independent. Conversely, if the $u_j(t)$ are dependent, then so are the vector functions $\mathbf{u}_j(t)$; the reason is that if there are constants c_j, not all zero, such that

$$
\sum_{j=1}^{k} c_j u_j(t) = 0, \qquad a < t < b,
$$

then successive differentiation gives

$$\sum_{j=1}^{k} c_j u_j^{(l)}(t) = 0, \qquad l = 1, 2, \ldots, n - 1.$$

In other words,

$$\sum_{j=1}^{k} c_j \mathbf{u}_j(t) = 0, \qquad a < t < b.$$

This establishes the equivalence of the two independence statements.

In case we have n independent solutions, the n independent initial vectors span $\mathcal{R}^n$ so the corresponding linear combinations of solutions give the unique solutions satisfying those initial conditions.

Theorem 3.2 applies to a wider class of linear differential equations than those of Chapter 2, which is concerned only with constant-coefficient operators. Note, however, that the discussion in Chapter 2 is quite explicit about the exponential character of the solutions, while the theorem for operators with continuous coefficients gives no explicit description of solutions.

EXAMPLE 2   The differential equation

$$y'' + y = 0$$

has solutions $u_1(t) = \cos t$, $u_2(t) = \sin t$, for $-\infty < t < \infty$. The vectors

$$\begin{pmatrix} u_1(t_0) \\ u_1'(t_0) \end{pmatrix} = \begin{pmatrix} \cos t_0 \\ -\sin t_0 \end{pmatrix}, \qquad \begin{pmatrix} u_2(t_0) \\ u_2'(t_0) \end{pmatrix} = \begin{pmatrix} \sin t_0 \\ \cos t_0 \end{pmatrix}$$

are linearly independent for $t_0 = 0$, because then we get the vectors

$$\begin{pmatrix} 1 \\ 0 \end{pmatrix}, \qquad \begin{pmatrix} 0 \\ 1 \end{pmatrix}.$$

It follows that the solutions $u_1(t)$ and $u_2(t)$ are independent, and indeed that the two vectors $(u_1(t_0), u_1'(t_0))$, and $(u_2(t_0), u_2'(t_0))$ are independent for every choice of t_0.

EXAMPLE 3   The general second-order linear equation

$$L(x) = \ddot{x} + p(t)\dot{x} + q(t)x = f(t)$$

with continuous coefficients has initial conditions

$$x(t_0) = c_1, \qquad \dot{x}(t_0) = c_2.$$

The equivalent system is found by letting $\dot{x} = y$:

$$\begin{pmatrix} \dot{x} \\ \dot{y} \end{pmatrix} = \begin{pmatrix} y \\ -q(t)x - p(t)y + f(t) \end{pmatrix} = \begin{pmatrix} 0 & 1 \\ -q(t) & -p(t) \end{pmatrix}\begin{pmatrix} x \\ y \end{pmatrix} + \begin{pmatrix} 0 \\ f(t) \end{pmatrix}$$

with

$$\begin{pmatrix} x(t_0) \\ y(t_0) \end{pmatrix} = \begin{pmatrix} c_1 \\ c_2 \end{pmatrix}.$$

By Theorem 1.1, on existence of solutions satisfying prescribed initial conditions, there is a solution satisfying these conditions. By the uniqueness theorem, Theorem 2.1, there is only one such solution. It follows that there is a one-to-one correspondence between pairs of initial values $(x(t_0), \dot{x}(t_0)) = (c_1, c_2)$ and solutions to the second-order equation.

For the homogeneous equation $L(x) = 0$ and the pairs of initial values $(1, 0)$ and $(0, 1)$, we get two solutions $u_1(t)$ and $u_2(t)$, such that

$$u_1(t_0) = 1, \qquad \dot{u}_1(t_0) = 0,$$

and

$$u_2(t_0) = 0, \qquad \dot{u}_2(t_0) = 1.$$

Thus the formula

$$x(t) = c_1 u_1(t) + c_2 u_2(t)$$

gives the homogeneous solution with initial values $x(t_0) = c_1$, $\dot{x}(t_0) = c_2$. The solutions u_1 and u_2 form a basis for the homogeneous solutions and the formula

$$x(t) = c_1 u_1(t) + c_2 u_2(t) + x_p(t)$$

describes all solutions to the nonhomogeneous equation, provided x_p satisfies $L(x_p) = f$.

A set of n vectors in $\mathfrak{R}^n$ is linearly independent if and only if the matrix A with those vectors as columns has nonzero determinant. (See Theorem 6.6 of Chapter 5, Section 6.) When the vectors in question are the values of n vector-valued functions at some point t, then the determinant is a real-valued function of t called the **Wronskian determinant** of the functions:

$$W[\mathbf{u}_1, \mathbf{u}_2, \ldots, \mathbf{u}_n](t) = \det \begin{pmatrix} u_{11}(t) & \cdots & u_{n1}(t) \\ \cdot & & \cdot \\ \cdot & & \cdot \\ \cdot & & \cdot \\ u_{1n}(t) & \cdots & u_{nn}(t) \end{pmatrix},$$

where

$$\mathbf{u}_k(t) = \begin{pmatrix} u_{k1}(t) \\ \cdot \\ \cdot \\ \cdot \\ u_{kn}(t) \end{pmatrix}, \qquad k = 1, \ldots, n.$$

The Wronskian determinant comes up most often when the vector functions $\mathbf{u}_k$ have as entries the successive derivatives of some real-valued functions. Thus, if $f_1(t)$, $\ldots, f_n(t)$ have $n - 1$ derivatives, we can form

$$\mathbf{u}_k(t) = \begin{pmatrix} f_k(t) \\ f_k'(t) \\ \cdot \\ \cdot \\ \cdot \\ f_k^{(n-1)}(t) \end{pmatrix},$$

and the Wronskian determinant is written

$$W[f_1, \ldots, f_n](t) = \det \begin{pmatrix} f_1(t) & f_2(t) & \cdots & f_n(t) \\ f_1'(t) & f_2'(t) & \cdots & \\ \cdot & & & \cdot \\ \cdot & & & \cdot \\ \cdot & & & \cdot \\ f_1^{(n-1)}(t) & & \cdots & f_n^{(n-1)}(t) \end{pmatrix}.$$

Because of our earlier theorems, statements about linear independence of solutions of linear homogeneous differential equations can be phrased in terms of the Wronskian: briefly, *solutions are independent if and only if the Wronskian is never zero*. For arbitrary differentiable functions, the implication goes only one way, as is shown in Exercise 8.

EXAMPLE 4   In Example 1, we considered the vector solutions

$$\begin{pmatrix} -e^{-2t} \\ e^{-2t} \end{pmatrix}, \qquad \begin{pmatrix} 4e^{3t} \\ e^{3t} \end{pmatrix}.$$

Their Wronskian is

$$W(t) = \det \begin{pmatrix} -e^{-2t} & 4e^{3t} \\ e^{-2t} & e^{3t} \end{pmatrix}$$
$$= -5e^{t} \neq 0,$$

which again verifies that the solutions are independent.

EXAMPLE 5   In Example 2, we considered the solutions $y_1(t) = \cos t$, $y_2(t) = \sin t$ of $y'' + y = 0$. The Wronskian is

$$W[y_1, y_2](t) = \det \begin{pmatrix} y_1(t) & y_2(t) \\ y_1'(t) & y_2'(t) \end{pmatrix}$$
$$= \det \begin{pmatrix} \cos t & \sin t \\ -\sin t & \cos t \end{pmatrix}$$
$$= \cos^2 t + \sin^2 t = 1.$$

Since the determinant is not zero, the solutions are independent.

EXERCISES

1. Suppose that a two-dimensional system $d\mathbf{x}/dt = A(t)\mathbf{x}$ has vector solutions $\mathbf{x}_1(t)$ and $\mathbf{x}_2(t)$ on an interval and satisfying each of the following pairs of conditions. Under which pairs of conditions can you conclude that the solutions must be linearly independent as vector-valued functions?

   (a) $\mathbf{x}_1(0) = \begin{pmatrix} 1 \\ 2 \end{pmatrix}$, $\mathbf{x}_2(0) = \begin{pmatrix} 1 \\ 1 \end{pmatrix}$.

   (b) $\mathbf{x}_1(0) = \begin{pmatrix} 1 \\ 1 \end{pmatrix}$, $\mathbf{x}_2(0) = \begin{pmatrix} -1 \\ -1 \end{pmatrix}$.

   (c) $\mathbf{x}_1(\pi) = \begin{pmatrix} -1 \\ 1 \end{pmatrix}$, $\mathbf{x}_2(\pi) = \begin{pmatrix} 0 \\ 1 \end{pmatrix}$.

   (d) $\mathbf{x}_1(0) = \begin{pmatrix} 1 \\ 0 \end{pmatrix}$, $\mathbf{x}_2(1) = \begin{pmatrix} 0 \\ 1 \end{pmatrix}$.

2. Which of the pairs of solutions specified in Exercise 1 must span the vector space of all solutions of the differential equations on the interval?

3. Suppose that a three-dimensional system $d\mathbf{x}/dt = B(t)\mathbf{x}$ has vector solutions defined on an interval and satisfying each of the following sets of conditions. Under which conditions can you conclude that the solutions must be linearly independent as vector-valued functions?

   (a) $\mathbf{x}_1(0) = \begin{pmatrix} 1 \\ 0 \\ 1 \end{pmatrix}$, $\mathbf{x}_2(0) = \begin{pmatrix} 0 \\ 1 \\ 0 \end{pmatrix}$, $\mathbf{x}_3(0) = \begin{pmatrix} 0 \\ 0 \\ -1 \end{pmatrix}$.

   (b) $\mathbf{x}_1(0) = \begin{pmatrix} 1 \\ 0 \\ 0 \end{pmatrix}$, $\mathbf{x}_2(0) = \begin{pmatrix} 1 \\ 1 \\ 1 \end{pmatrix}$.

   (c) $\mathbf{x}_1(0) = \begin{pmatrix} 2 \\ 1 \\ 1 \end{pmatrix}$, $\mathbf{x}_2(0) = \begin{pmatrix} 1 \\ 0 \\ 1 \end{pmatrix}$, $\mathbf{x}_3(0) = \begin{pmatrix} 0 \\ -2 \\ -1 \end{pmatrix}$.

   (d) $\mathbf{x}_1(0) = \begin{pmatrix} 1 \\ 0 \\ 0 \end{pmatrix}$, $\mathbf{x}_2(\tfrac{1}{2}) = \begin{pmatrix} 1 \\ 1 \\ 0 \end{pmatrix}$, $\mathbf{x}_3(1) = \begin{pmatrix} 1 \\ 1 \\ 1 \end{pmatrix}$.

4. Which of the sets of solutions specified in Exercise 3 must span the set of all solutions of the differential equation on the interval?

5. A differential equation of the form

$$\ddot{y} + a(t)\dot{y} + b(t)y = 0$$

   with coefficients continuous on an interval has solutions satisfying each of the following sets of conditions. Which sets of solutions must necessarily be linearly independent as functions?

   (a) $y_1(0) = 0$, $\dot{y}_1(0) = 1$, $y_2(0) = 1$, $\dot{y}_2(0) = 1$.
   (b) $y_1(0) = 1$, $\dot{y}_1(0) = 1$, $y_2(0) = 0$, $\dot{y}_2(0) = 0$.
   (c) $y_1(0) = 0$, $\dot{y}_1(0) = 1$, $y_2(0) = 1$, $\dot{y}_2(0) = 1$, $y_3(0) = -1$, $\dot{y}_3(0) = 1$.
   (d) $y_1(0) = 0$, $\dot{y}_1(0) = 1$, $y_2(1) = 1$, $\dot{y}_2(1) = 1$.

6. (a) Let $y_1(t)$ and $y_2(t)$ be twice differentiable functions on an interval $a < t < b$. Show that the equation

$$W[y, y_1, y_2] = \det \begin{pmatrix} y & y_1 & y_2 \\ y' & y_1' & y_2' \\ y'' & y_1'' & y_2'' \end{pmatrix} = 0$$

defines a second-order differential equation having y_1 and y_2 as solutions.

(b) Find a second-order differential equation that has $y_1(t) = e^t$ and $y_2(t) = \cos t$ as solutions for $-\pi < t < 4$. Are these solutions linearly independent?

(c) Find a third-order, constant-coefficient, homogeneous, linear differential equation that has $y_1(t) = e^t$, $y_2(t) = \cos t$, and $y_3(t) = \sin t$ as solutions.

7. (a) Compute the Wronskian determinant of the three real-valued functions $y_1(t) = e^t$, $y_2(t) = e^{2t}$, $y_3(t) = e^{3t}$.

(b) Explain why the computation in part (a) proves the functions y_1, y_2, and y_3 linearly independent over an arbitrary interval.

8. (a) Compute the Wronskian determinant of the two real-valued functions $y_1(t) = t^3$, $y_2(t) = |t|^3$ for $-\infty < t < \infty$.

(b) Are the functions y_1 and y_2 linearly independent?

(c) Are the functions y_1 and y_2 linearly independent when restricted so that $0 < t$? When restricted so that $t < 0$?

8

Linear Partial Differential Equations

1 INTRODUCTION

1A Equations and Solutions

Differential equations that involve partial derivatives of an unknown function $u(x_1, x_2, \ldots)$ of several variables arise just as often in applied mathematics as do ordinary differential equations for functions of one variable. For example, the height $u(x, t)$ of a water wave moving in a narrow channel is usually treated as a function of position x in the channel and time t, and the differential equations that $u(x, t)$ satisfies under various physical conditions will contain partial derivatives

$$\frac{\partial u}{\partial x}, \quad \frac{\partial u}{\partial t}, \quad \frac{\partial^2 u}{\partial x^2}, \quad \frac{\partial^2 u}{\partial t\,\partial x}, \quad \frac{\partial^2 u}{\partial t^2}, \ldots,$$

both with respect to x and t. (To make the formulas easier to write, we will sometimes use the alternative notations

$$u_x, \quad u_t, \quad u_{xx}, \quad u_{xt}, \quad u_{tt}, \ldots$$

for these partials.) Our examples will be restricted to linear equations of at most second order in two independent variables, say x and y; such an equation has the form

$$\textbf{1.1} \qquad\qquad a u_{xx} + b u_{xy} + c u_{yy} + d u_x + e u_y + f u = g,$$

where the coefficients a, b, c, d, e, f and the function g are assumed to be continuous functions, perhaps constant, of x and y in some common region of the (x, y)-plane. In addition, we will look only for solutions $u(x, y)$ all of whose first- and second-order partial derivatives are continuous, so that in particular the relation $u_{xy} = u_{yx}$ will hold for these solutions.

EXAMPLE 1   In the region of the (x, y)-plane for which $y > 0$, the **upper half-plane,** we will solve the equation

$$yu_{xy} = x.$$

Since $y > 0$, we can write the equation as $u_{xy} = x/y$ and integrate both sides with respect to y, for each *fixed* x, getting

$$u_x(x, y) = x \ln y + g(x).$$

The integration constant $g(x)$, will in general be a differentiable function of x. Now integrate both sides with respect to x:

$$u(x, y) = \tfrac{1}{2}x^2 \ln y + G(x) + H(y).$$

Here $G'(x) = g(x)$, and $H(y)$ is a constant (with respect to x) of integration that will be a twice-differentiable function of y. Clearly, we have found the most general twice-differentiable solution of the equation for $y > 0$.

For a second-order linear differential equation of ordinary type (i.e., with a single independent variable), we are used to having two arbitrary constants in the general solution formula. Notice, however, that in Example 1 the general solution contains two arbitrary twice-differentiable *functions,* allowing for a wide variety in the choice of a particular solution. This wide freedom of choice is an indication that we can expect to find particular solutions satisfying fairly complicated combinations of boundary and initial conditions. We will refer to boundary and initial conditions together as **side conditions,** and they will be explored later in the chapter in the context of specific physical problems.

EXAMPLE 2   The partial differential equation

$$u_{xx} - y^2 u = 0. \qquad y > 0,$$

is peculiar in that it contains a derivative only with respect to x, even though the solution will be of the form $u = u(x, y)$. Solving the equation depends only on techniques for ordinary differential equations, and we find

$$u(x, y) = c_1(y)e^{xy} + c_2(y)e^{-xy}.$$

A natural set of conditions to single out a particular solution is

$$u(0, y) = G(y), \qquad u_x(0, y) = H(y),$$

where G and H are arbitrary functions of y. (Note that in this example the solution need not be even once differentiable as a function of y.) The particular solution satisfying the preceding conditions is

$$u(x, y) = \frac{1}{2}\left[G(y) + \frac{1}{y}H(y)\right]e^{xy} + \frac{1}{2}\left[G(y) - \frac{1}{y}H(y)\right]e^{-xy}.$$

In particular, if $G(y) = u(0, y) = 1$ and $H(y) = u_x(0, y) = -y$, we get

$$u(x, y) = e^{-xy},$$

the graph of which is shown in Figure 1 for $0 \le x \le 5, 0 \le y \le 5$.

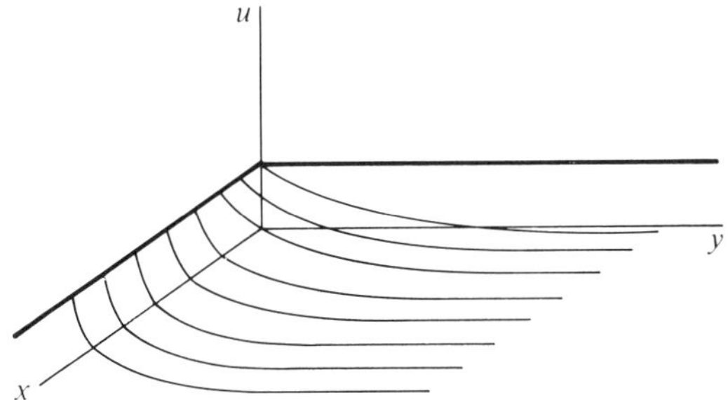

Figure 1   Graph of solution to $u_{xx} - y^2u = 0$, $u(0, y) = 1$, $u_x(0, y) = -y$.

The partial differential operator

$$L = a\frac{\partial^2}{\partial x^2} + b\frac{\partial^2}{\partial y\,\partial x} + c\frac{\partial^2}{\partial y^2} + d\frac{\partial}{\partial x} + e\frac{\partial}{\partial y} + f$$

is easily verified to be linear in its action on twice-differential functions u, v:

$$L(\alpha u + \beta v) = \alpha Lu + \beta Lv, \qquad \alpha,\ \beta \text{ constant.}$$

Just as for ordinary differential operators, it follows that we have a **superposition principle** for homogeneous equations

$$Lu = 0:$$

if u, v are homogeneous solutions, then so is a linear combination $\alpha u + \beta v$. For partial differential operators, some of the most important solution techniques use the extension to an arbitrary number of homogeneous solutions $u_1, \ldots, u_n$. Repeated application of the two-term formula gives

$$L(\alpha_1 u_1 + \cdots + \alpha_n u_n) = \alpha_1 Lu_1 + \cdots + \alpha_n Lu_n.$$

In this way, we start to construct solutions to general boundary problems from more elementary solutions.

1B Exponential Solutions

For a constant-coefficient homogeneous equation

1.2 $$au_{xx} + bu_{xy} + cu_{yy} + du_x + eu_y + fu = 0$$

we can always find solutions of exponential form, just as in the case of a linear ordinary differential equation with constant coefficients. Here we assume a solution of the form $u(x, y) = e^{rx+sy}$. We see that $u_x = ru$, $u_y = su$, $u_{xx} = r^2u$, $u_{xy} = rsu$, $u_{yy} = s^2u$. Thus a solution of the simple exponential form will have to have its parameters r, s chosen so that its **characteristic equation**

$$\textbf{1.3} \qquad\qquad ar^2 + brs + cs^2 + dr + es + f = 0$$

is satisfied.

EXAMPLE 3   The equation

$$u_{xx} - u_{yy} = 0$$

has characteristic equation $r^2 - s^2 = 0$ or $r^2 = s^2$. Thus exponential solutions have the form

$$u(x, y) = e^{rx+ry}, \quad \text{or} \quad u(x,y) = e^{rx-ry},$$

since $s = \pm r$. Corresponding to each complex number r, there is a pair of solutions

$$u(x, y) = \begin{cases} e^{r(x+y)}, \\ e^{r(x-y)}. \end{cases}$$

By forming linear combinations of these basic exponential solutions, it is possible to obtain representations for solutions satisfying many different boundary conditions. For example, the conditions

$$u(x, 0) = \cos \beta x,$$

$$u_y(x, 0) = 0,$$

for $-\infty < x < \infty$, are satisfied by $u(x, y) = \cos \beta x \cos \beta y$. But

$$\cos \beta x \cos \beta y = \tfrac{1}{2}(e^{i\beta x} + e^{-i\beta x}) \cdot \tfrac{1}{2}(e^{i\beta y} + e^{-i\beta y})$$

$$= \tfrac{1}{4}(e^{i\beta(x+y)} + e^{i\beta(x-y)} + e^{-i\beta(x+y)} + e^{-i\beta(x-y)}).$$

Hence $\cos \beta x \cos \beta y$ is a solution of the differential equation, as we see by letting $r = i\beta$ and $r = -i\beta$ in the exponential solutions. The technique of **Fourier expansion** will be introduced to make the computation of such representations fairly routine.

EXERCISES

1. Verify by direct substitution that each of the following differential equations is satisfied by the given function. Then verify that the function itself satisfies the given side condition.
   (a) $u_{xx} - u_y = 0$; $u(x, y) = e^{-y} \sin x$; $u(0, y) = u(\pi, y) = 0$.
   (b) $u_{xx} + u_{yy} = 0$; $u(x, y) = e^y \sin x$; $u(0, y) = u(\pi, y) = 0$.
   (c) $u_{xx} - u_{yy} = 0$; $u(x, y) = \sin x \sin y$; $u(x, 0) = u(x, \pi) = 0$.

(d) $u_{xx} - 4u_y = 0$; $u(x, y) = y^{-1/2}e^{-x^2/y}$, $y > 0$; $\lim_{x \to \infty} u(x, y) = 0$.

(e) $x^2 u_{xx} + xu_x + u_{yy} = 0$; $u(x, y) = x \sin y$; $u(x, 0) = u(x, \pi) = 0$.

2. **(a)** Solve the differential equation $u_{xy} = 0$ by integrating first with respect to y, then x.

   (b) Solve the differential equation $u_{xy} = 1$ by integrating first with respect to y, then x.

   (c) Verify that $u_p(x, y) = xy$ is a particular solution of $u_{xy} = 1$. Then use the linearity of the partial differential operator $L = \partial^2/\partial y\, \partial x$ to find the general twice-differentiable solution of $u_{xy} = 1$ by using the solution to part (a). (*Hint*: Use Theorem 3.1 of Chapter 2.)

3. Find the general twice-differentiable solution $u(x, y)$ of the partial differential equation

$$u_{xx} + 2u_x + u = y.$$

4. To find the general solution $u(x, y)$ of $u_{xx} - u_{yy} = 0$ for all x, y, define new variables z, w by

$$z = x + y,$$

$$w = x - y.$$

Define $\bar{u}$ as a function of two variables by

$$\bar{u}(z,w) = u\left(\frac{z + w}{2}, \frac{z - w}{2}\right).$$

   (a) Show that if (z, w) corresponds to (x, y) as above, then $\bar{u}(z, w) = u(x, y)$.

   (b) Assuming that $u_{xx} - u_{yy} = 0$, use the chain rule to show that $\bar{u}_{zw} = 0$. (*Hint*: Compute $u_{xx} - u_{yy}$ in terms of $\bar{u}_{zz}$, $\bar{u}_{zw}$ and $\bar{u}_{ww}$.)

   (c) Show that the general solution of $\bar{u}_{zw} = 0$ is $\bar{u}(z, w) = g(z) + h(w)$, where g, h are arbitrary twice-differentiable functions.

   (d) Conclude that the solution of the original equation is

$$u(x, y) = g(x + y) + h(x - y).$$

5. We consider the geometric properties of the set of points in the (r, s)-plane that satisfy the characteristic equation

$$ar^2 + brs + cs^2 + dr + es + f = 0$$

of the associated operator

$$L = a\,\frac{\partial^2}{\partial x^2} + b\,\frac{\partial^2}{\partial y\, \partial x} + c\,\frac{\partial^2}{\partial y^2} + d\,\frac{\partial}{\partial x} + e\,\frac{\partial}{\partial y} + f;$$

these properties provide an important classification of the operators L. If a, b, c, are not all zero, the characteristic equation is that of a (perhaps degenerate) conic section of elliptic, parabolic, or hyperbolic type. The corresponding operator L is then said to be respectively **elliptic, parabolic,** or **hyperbolic.** For example, $\partial^2/\partial x^2 + \partial^2/\partial y^2$ is elliptic, because $r^2 + s^2 = 0$ is a degenerate ellipse; $\partial^2/\partial x^2 + \partial/\partial y$ is parabolic, because $r^2 + s = 0$ is a parabola; $\partial^2/\partial x^2 - \partial^2/\partial y^2$ is hyperbolic, because $r^2 - s^2 = 0$ is a degenerate hyperbola. Classify each of the following operators as elliptic, parabolic, or hyperbolic.

   (a) $\dfrac{\partial^2}{\partial x^2} + \dfrac{\partial^2}{\partial y^2} - \dfrac{\partial}{\partial x}.$

(b) $\dfrac{\partial^2}{\partial x\, \partial y}$.

(c) $\dfrac{\partial^2}{\partial x^2} - 2\dfrac{\partial^2}{\partial y^2} + 4\dfrac{\partial}{\partial y}$.

(d) $\dfrac{\partial^2}{\partial y^2} - \dfrac{\partial}{\partial x} - 3$.

2 FOURIER SERIES

2A Introduction

Finite linear combinations of the exponential solutions discussed in the previous section are not enough to satisfy the requirements of the side conditions that come up in practice. However, if we allow infinite series whose partial sums are linear combinations of exponentials of the form $e^{i\beta x}$, or what amounts to the same thing, of $\cos \beta x$ and $\sin \beta x$, it turns out that we can produce solutions that satisfy many interesting side conditions. For the moment, disregarding any question of convergence, we define a **trigonometric series** to be a series of the form

2.1
$$\frac{a_0}{2} + \sum_{k=1}^{\infty} (a_k \cos kx + b_k \sin kx).$$

In the very special circumstance that the coefficients a_k, b_k are generated by an integrable function $f(x)$ through the **Euler formulas**

2.2
$$a_k = \frac{1}{\pi} \int_{-\pi}^{\pi} f(x) \cos kx\, dx, \qquad b_k = \frac{1}{\pi} \int_{-\pi}^{\pi} f(x) \sin kx\, dx,$$

then the trigonometric series is called the **Fourier series** of f, and the coefficients a_k, b_k as given by Equations 2.2 are called the **Fourier coefficients** of f. The most fundamental question about a Fourier series is the extent to which the series represents the function. The importance of such a representation stems from the fact that the individual terms of the series can be incorporated into a solution of some differential equation.

Our first examples are meant to illustrate the beautiful way in which the partial sums of a Fourier series attempts to mimic the function f that generates them. A partial sum

2.3
$$S_N(x) = \frac{a_0}{2} + \sum_{k=1}^{N} (a_k \cos kx + b_k \sin kx)$$

is called a **trigonometric polynomial.** Note that each term in S_N is periodic of period 2π. It follows that S_N is also periodic:

$$S_N(x + 2\pi) = S_N(x),$$

for all x. Note also that the Fourier coefficients a_k, b_k are determined by integral formulas that use the values of $f(x)$ only for $-\pi \leq x \leq \pi$. For these reasons, we will restrict attention to values of x in the interval of length 2π between $-\pi$ and π.

2B Orthogonality

The functions $\cos kx$, $\sin kx$ that occur in a Fourier series are the most important examples of **orthogonal** functions on the interval $-\pi \leq x \leq \pi$. For integer k and l:

$$\frac{1}{\pi} \int_{-\pi}^{\pi} \cos kx \sin lx \, dx = 0,$$

2.4
$$\frac{1}{\pi} \int_{-\pi}^{\pi} \cos kx \cos lx \, dx = \begin{cases} 0, & k \neq l, \\ 1, & k = l \neq 0. \end{cases}$$

$$\frac{1}{\pi} \int_{-\pi}^{\pi} \sin kx \sin lx \, dx = \begin{cases} 0, & k \neq l \text{ or } k = l = 0, \\ 1, & k = l \neq 0. \end{cases}$$

These formulas are easily proved using trigonometric identities, and are even more easily proved by first writing the sines and cosines in terms of e^{ikx} and e^{ilx} (see Exercise 5). As a sample application of the **orthogonality relations** 2.4, suppose that a trigonometric series satisfies some condition that allows us to integrate it term by term on the interval $-\pi \leq x \leq \pi$. For example, the series might converge uniformly on the interval, or it might just be a finite sum, with all a_k and b_k equal to zero from some point on. We can then denote the sum by $f(x)$ and write

$$f(x) = \frac{a_0}{2} + \sum_{k=1}^{\infty} (a_k \cos kx + b_k \sin kx).$$

Then for a fixed integer $l \geq 0$,

$$\frac{1}{\pi} \int_{-\pi}^{\pi} f(x) \cos lx \, dx = \frac{1}{\pi} \int_{-\pi}^{\pi} \left[\frac{a_0}{2} + \sum_{k=0}^{\infty} (a_k \cos kx + b_k \sin kx) \right] \cos lx \, dx$$

$$= \frac{a_0}{2\pi} \int_{-\pi}^{\pi} \cos lx \, dx + \sum_{k=1}^{\infty} \left[\frac{a_k}{\pi} \int_{-\pi}^{\pi} \cos kx \cos lx \, dx \right.$$

$$\left. + \frac{b_k}{\pi} \int_{-\pi}^{\pi} \sin kx \cos lx \, dx \right].$$

The first two relations in 2.4 show that all but one of these last terms is zero; the only term that survives is the term containing a_l, and we find, for $l \neq 0$,

$$\frac{1}{\pi} \int_{-\pi}^{\pi} f(x) \cos lx \, dx = \frac{a_l}{\pi} \int_{-\pi}^{\pi} \cos lx \cos lx \, dx$$

$$= a_l.$$

A similar computation shows that

$$\frac{1}{\pi} \int_{-\pi}^{\pi} f(x) \sin lx \, dx = b_l.$$

What the preceding argument shows is that, under fairly broad conditions on the coefficients of a trigonometric series, the coefficients must be given by the Euler Formulas 2.2. In other words, no way other than Formulas 2.2 of determining the a_k and b_k makes sense if we want to represent a reasonably large class of functions $f(x)$ by a trigonometric series of the form of Equation 2.1.

EXAMPLE 1   Let $f(x) = |x|$ for $-\pi \le x \le \pi$. Then

$$a_k = \frac{1}{\pi} \int_{-\pi}^{\pi} |x| \cos kx \, dx, \qquad b_k = \frac{1}{\pi} \int_{-\pi}^{\pi} |x| \sin kx \, dx.$$

Clearly, $|x| \sin kx$ has integral zero over $[-\pi, \pi]$, because the integrals over $[-\pi, 0]$ and $[0, \pi]$ are negatives of one another. Hence $b_k = 0$ for $k = 1, 2, \ldots$. On the other hand, the graph of $|x| \cos kx$ is symmetric about the y-axis. For $k \ne 0$ we integrate by parts, getting

$$\begin{aligned}
a_k &= \frac{2}{\pi} \int_0^{\pi} x \cos kx \, dx \\[2mm]
&= \frac{2}{\pi} \left[\frac{x \sin kx}{k} \right]_0^{\pi} - \frac{2}{\pi k} \int_0^{\pi} \sin kx \, dx \\[2mm]
&= \left[\frac{2}{\pi k^2} \cos kx \right]_0^{\pi} = \frac{2}{\pi k^2} (\cos k\pi - 1) \\[2mm]
&= \begin{cases} 0, & k = 2, 4, 6, \ldots, \\ -\dfrac{4}{\pi k^2}, & k = 1, 3, 5, \ldots. \end{cases}
\end{aligned}$$

When $k = 0$, we have

$$a_0 = \frac{2}{\pi} \int_0^{\pi} x \, dx = \pi.$$

To summarize,

$$a_0 = \pi, \qquad a_k = \begin{cases} 0, & k = 2, 4, 6, \ldots, \\ -\dfrac{4}{\pi k^2}, & k = 1, 3, 5, \ldots, \end{cases}$$

$$b_k = 0, \qquad k = 1, 2, 3, \ldots.$$

Hence, the Nth Fourier approximation is given for $N = 1, 3, 5, \ldots$ by the trigonometric polynomial

$$s_N(x) = \frac{\pi}{2} - \frac{4}{\pi} \cos x - \frac{4}{\pi} \frac{\cos 3x}{3^2} - \cdots - \frac{4}{\pi} \frac{\cos Nx}{N^2}.$$

If N is even, we have $s_N(x) = s_{N-1}(x)$. Figure 2 shows how the graphs of s_0, s_1, and s_3 approximate that of $|x|$ on $[-\pi, \pi]$.

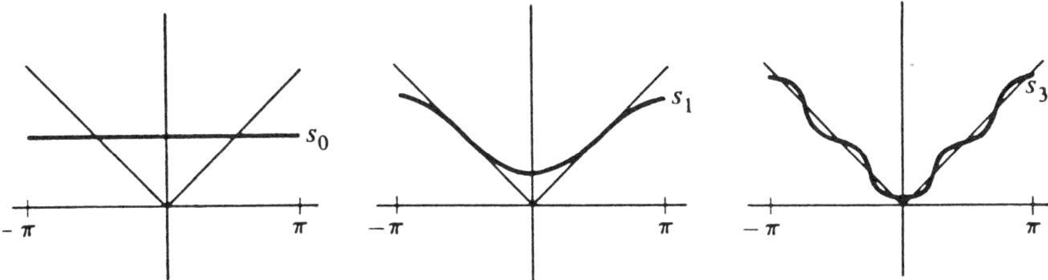

Figure 2

EXAMPLE 2   Let

$$g(x) = \begin{cases} 1, & 0 \le x \le \pi, \\ -1, & -\pi \le x < 0. \end{cases}$$

Then the Fourier coefficients of g are given by integration as

$$a_k = 0, \qquad k = 0, 1, 2, \ldots ,$$

$$b_k = \begin{cases} 0, & k = 2, 4, 6, \ldots , \\ \dfrac{4}{\pi k}, & k = 1, 3, 5, \ldots . \end{cases}$$

Hence, for N odd, the Nth Fourier approximation to g is given by

$$s_N(x) = \frac{4}{\pi} \sin x + \frac{4}{\pi} \frac{\sin 3x}{3} + \cdots + \frac{4}{\pi} \frac{\sin Nx}{N}.$$

The graphs of s_1, s_3, and s_5 are shown in Figure 3, together with that of $g(x)$.

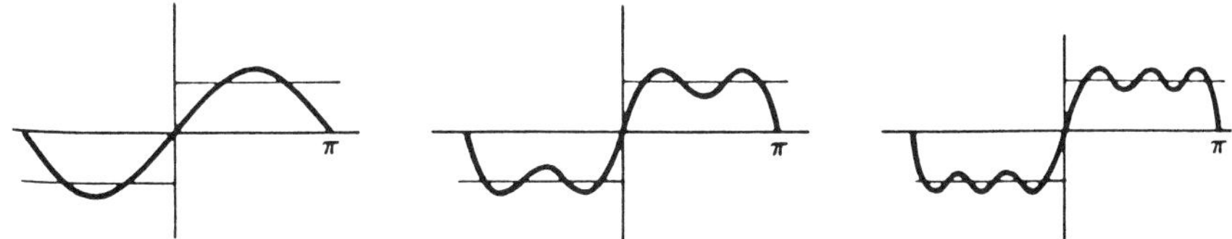

Figure 3

An important question is whether, for specific values of x, a Fourier approximation $s_N(x)$ converges as $N \to \infty$ to $f(x)$, where f is the function from which the Fourier coefficients are computed. We define the **Fourier series** of f to be the infinite series

$$\frac{a_0}{2} + \sum_{k=1}^{\infty} (a_k \cos kx + b_k \sin kx), \tag{1}$$

where a_k and b_k are given by Formula 2.2. Theorem 2.5, following, gives some conditions on f under which the Fourier series can be used to represent f. Indeed, suppose that the graph of f is **piecewise smooth.** This means that the interval $[-\pi, \pi]$ can be broken into finitely many subintervals, with endpoints $-\pi < x_1 < x_2 < \cdots < x_k < \pi$, such that f can be extended continuously from each open interval (x_k, x_{k+1}) to the closed interval $[x_k, x_{k+1}]$ so that f' is continuous on $[x_k, x_{k+1}]$. Then we can show that the Fourier series of f converges to $f(x)$ wherever f is continuous, and, at a possible discontinuity at x_k, will converge to the average value

$$(\tfrac{1}{2})[f(x_k-) + f(x_k+)].$$

Here $f(x-)$ stands for the left-hand limit of f at x, and $f(x+)$ represents the right-hand limit. The graph of a typical piecewise smooth function is shown in Figure 4, with the average value indicated by a dot at each jump.

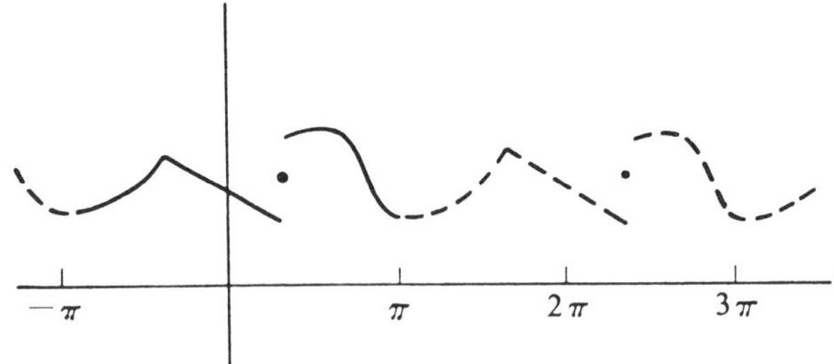

Figure 4

2.5 Theorem.   Let f be piecewise smooth on $(-\pi, \pi)$. Then the Fourier series of f converges at every point x of the interval to $(\tfrac{1}{2})[f(x-) + f(x+)]$. In particular, if f is continuous at x, then the series converges to $f(x)$. At $x = \pm\pi$, the series converges to $(\tfrac{1}{2})[f(\pi-) + f(-\pi+)]$.

The proof is omitted.

Examples 1 and 2 gave an indication of the way in which the partial sums of a Fourier series converge. In each of those examples, the function satisfies the condition of piecewise smoothness; hence the series converges to the appropriate value of the function for $-\pi < x < \pi$.

EXAMPLE 3   The function g defined in Example 2 is rather arbitrarily defined to have the value 1 at $x = 0$. In spite of this arbitrariness, Theorem 2.5 allows us to conclude that the Fourier series of g converges as follows:

$$\sum_{k=0}^{\infty} \frac{4}{\pi} \frac{\sin (2k + 1)x}{2k + 1} = \begin{cases} 1, & 0 < x < \pi, \\ 0, & x = 0, \\ -1, & -\pi < x < 0. \end{cases}$$

To be very specific, we can set $x = \pi/2$ and arrive at the alternating series expansion

$$\sum_{k=0}^{\infty} \frac{(-1)^k}{2k+1} = \frac{\pi}{4}.$$

Theorem 2.5 shows to some extent the reason for choosing the coefficients in a trigonometric polynomial according to Formula 2.2. The reason is that, under favorable circumstances, the resulting sequence of trigonometric polynomials converges, as the sequence of partial sums of a Fourier series, to the function f.

EXERCISES

1. Compute the Fourier coefficients of each of the following functions and write the corresponding Fourier series in the form of Equation 2.1. Sketch the graph of each function extended to have period 2π on the interval $-2\pi \le x \le 2\pi$. Finally, sketch the graphs of the first three partial sums $S_0(x)$, $S_1(x)$, $S_2(x)$ of the Fourier series.

   (a) $f(x) = x,\ -\pi \le x \le \pi.$

   (b) $f(x) = \begin{cases} -\pi - x, & -\pi < x < 0, \\ \pi - x, & 0 \le x \le \pi. \end{cases}$

   (c) $f(x) = x^2,\ -\pi \le x \le \pi.$

   (d) $f(x) = |x| + 1,\ -\pi \le x \le \pi.$

   (e) $f(x) = \begin{cases} 0, & -\pi < x \le 0, \\ 1, & 0 < x \le \pi. \end{cases}$

2. According to Theorem 2.5, the Fourier series of each function $f(x)$ in Exercise 1 converges to some function $F(x)$ whose graph may or may not differ at some points from the graph of $f(x)$. In each case, sketch $F(x)$ on the interval $-2\pi \le x \le 2\pi$.

3. Show that if $f(x)$ and $g(x)$ have the Fourier coefficients a_k, b_k and a_k', b_k' respectively, then $\alpha f(x) + \beta g(x)$ (α, β both constant) has Fourier coefficients $\alpha a_k + \beta a_k'$, $\alpha b_k + \beta b_k'$.

4. The Nth partial sum of a trigonometric series is called a **trigonometric polynomial** of degree N, and a trigonometric polynomial is necessarily the Fourier series of the function it represents. For example, the identity

$$\cos^2 x = \frac{1}{2} + \frac{1}{2}\cos 2x$$

   displays the Fourier series of $\cos^2 x$. Find the Fourier series of each of the following functions by finding appropriate identities.

   (a) $\sin^2 x.$ (b) $\cos^3 x.$ (c) $\sin 2x \cos x.$

5. Establish the orthogonality relations in Equations 2.4 of the text, as follows:

   (a) Use standard trigonometric identities to compute the integrals.

   (b) Use the identities

$$\cos nx = \frac{e^{inx} + e^{-inx}}{2}, \qquad \sin nx = \frac{e^{inx} - e^{-inx}}{2i},$$

   together with the standard identities for the exponential function.

3 ADAPTED FOURIER EXPANSIONS

3A General Intervals

The direct application of Fourier methods to practical problems usually requires some modification of the standard formulation presented in the previous section. In the present section we describe some of these modifications and calculate some examples.

While the interval $[-\pi, \pi]$ is a natural one for Fourier expansions because it is a period interval for the trigonometric functions, it may be that a function encountered in an application needs to be approximated on some other interval.

If the function f to be approximated is defined not on the interval $[-\pi, \pi]$ but on $[-p, p]$, a suitable change in the computation of the approximation can be made as follows. With f defined on $[-p, p]$, we define

$$f_p(x) = f\left(\frac{px}{\pi}\right), \qquad -\pi \le x \le \pi.$$

Then we can compute the Fourier coefficients of f_p by Formula 2.2. The resulting trigonometric polynomials s_N will approximate f_p on $[-\pi, \pi]$. To approximate f on $[-p, p]$, we consider

$$s_N\left(\frac{\pi x}{p}\right) = \frac{a_0}{2} + \sum_{k=1}^{N} \left(a_k \cos \frac{k\pi x}{p} + b_k \sin \frac{k\pi x}{p}\right), \qquad -p \le x \le p.$$

The coefficients a_k and b_k can be computed directly in terms of f by making a change of variable. We have

$$a_k = \frac{1}{\pi} \int_{-\pi}^{\pi} f_p(x) \cos kx \, dx = \frac{1}{\pi} \int_{-\pi}^{\pi} f\left(\frac{px}{\pi}\right) \cos kx \, dx$$

$$= \frac{1}{p} \int_{-p}^{p} f(x) \cos \left(\frac{k\pi x}{p}\right) dx.$$

A similar computation holds for b_k, and we have

3.1 $$a_k = \frac{1}{p} \int_{-p}^{p} f(x) \cos \frac{k\pi x}{p} \, dx, \qquad b_k = \frac{1}{p} \int_{-p}^{p} f(x) \sin \frac{k\pi x}{p} \, dx$$

for the coefficients in the Fourier approximation

$$\frac{a_0}{2} + \sum_{k=1}^{N} \left(a_k \cos \frac{k\pi x}{p} + b_k \sin \frac{k\pi x}{p}\right)$$

to the function f defined on $[-p, p]$.

EXAMPLE 1   If

$$h(x) = \begin{cases} 1, & 0 \le x \le p, \\ -1, & -p \le x < 0, \end{cases}$$

then

$$a_k = 0, \qquad k = 0, 1, 2, \ldots,$$

$$b_k = \frac{2}{p} \int_0^p \sin \frac{k\pi x}{p} \, dx$$

$$= \frac{2}{\pi} \int_0^\pi \sin kx \, dx = \begin{cases} 0, & k = 2, 4, 6, \ldots, \\ \dfrac{4}{\pi k}, & k = 1, 3, 5, \ldots. \end{cases}$$

Hence the Nth Fourier approximation to h is given, for odd N, by

$$s_N(x) = \frac{4}{\pi} \sin \frac{\pi x}{p} + \frac{4}{3\pi} \sin \frac{3\pi x}{p} + \cdots + \frac{4}{N\pi} \sin \frac{N\pi x}{p}, \qquad -p \leq x \leq p.$$

For a function f defined on an arbitrary interval $a \leq x \leq b$, it is helpful to think of a periodic extension f_E of f having period $b - a$ and defined for all real numbers x. Such an extension is illustrated in Figure 5. We set $2p = b - a$ so that $p = (b - a)/2$ and $-p = -(b - a)/2$. We then compute the Fourier coefficients of f_E over the interval $[-p, p]$ according to Formula 3.1. Furthermore, because the integrands in Formula 3.1 have period $2p$, we can use the fact, geometrically obvious, that the integration can be performed over any interval of length $2p = b - a$, in particular, over $[a, b]$. Thus Formula 3.1 can be rewritten:

3.2

$$a_k = \frac{2}{b - a} \int_a^b f(x) \cos \frac{2k\pi x}{b - a} \, dx,$$

$$b_k = \frac{2}{b - a} \int_a^b f(x) \sin \frac{2k\pi x}{b - a} \, dx.$$

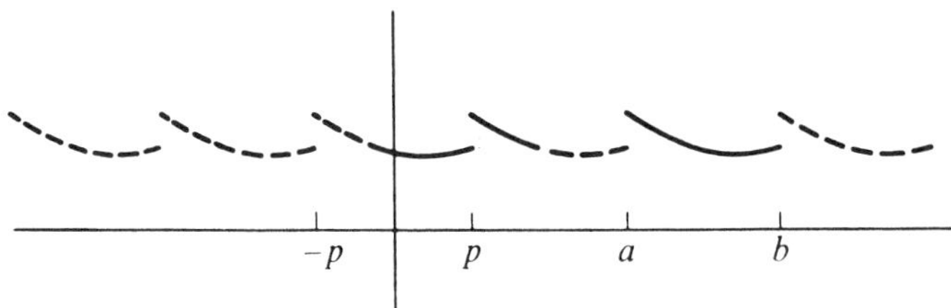

Figure 5

The associated trigonometric polynomials are

$$s_N(x) = \frac{a_0}{2} + \sum_{k=1}^{N} \left(a_k \cos \frac{2k\pi x}{b - a} + b_k \sin \frac{2k\pi x}{b - a} \right).$$

EXAMPLE 2   Let $f(x) = x$, for $0 \le x \le 1$. We find, integrating by parts,

$$a_k = 2 \int_0^1 x \cos 2k\pi x \, dx$$

$$= 2\left[x \frac{\sin 2k\pi x}{2k\pi} \right]_0^1 - \frac{2}{2k\pi} \int_0^1 \sin 2k\pi x \, dx = 0,$$

$$b_k = 2 \int_0^1 x \sin 2k\pi x \, dx$$

$$= 2\left[-x \frac{\cos 2k\pi x}{2k\pi} \right]_0^1 + \frac{2}{2k\pi} \int_0^1 \cos 2k\pi x \, dx$$

$$= -\frac{\cos 2k\pi}{k\pi} = -\frac{1}{k\pi}.$$

Then the Fourier series is

$$\frac{1}{2} - \frac{1}{\pi}\left(\sin 2\pi x + \frac{\sin 4\pi x}{2} + \frac{\sin 6\pi x}{3} + \cdots \right).$$

3B Sine and Cosine Expansions

It sometimes happens that an expansion in terms only of cosines or only of sines is more convenient to use than a general Fourier expansion. We begin with the observation that cosine is an even function [i.e., $\cos(-x) = \cos x$], and sine is an odd function [i.e., $\sin(-x) = -\sin x$]. In general, a function f is said to be **even** if $f(-x) = f(x)$ for all x in the domain of f and, alternatively, to be **odd** if $f(-x) = -f(x)$ always holds. The graphs of some even and odd functions are shown in Figure 6.

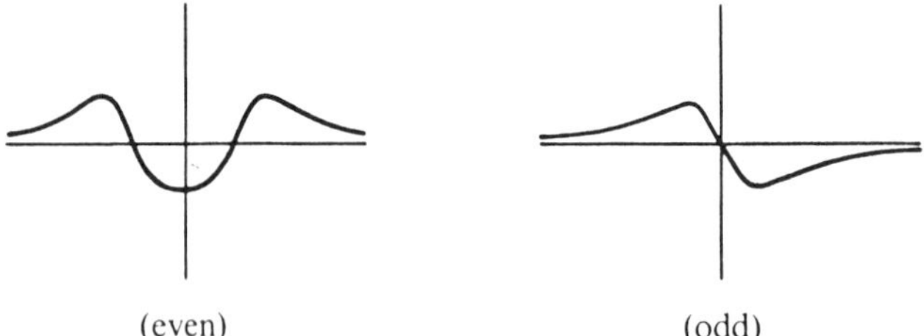

(even)   (odd)   **Figure 6**

Geometrically, a function is even if its graph is symmetric with respect to the y-axis and odd if its graph is symmetric with respect to the origin. It follows that

$$\int_{-p}^{p} f(x) \, dx = 2 \int_0^p f(x) \, dx, \qquad \text{for even } f \tag{1}$$

and

$$\int_{-p}^{p} f(x) \, dx = 0, \qquad \text{for odd } f. \tag{2}$$

Thus, if f is an even periodic function, the product

$$f(x) \sin \frac{k\pi x}{p}$$

is odd. (Why?) Therefore, for the Fourier sine coefficient b_k, we have, by Equation 3.1,

$$b_k = \frac{1}{p} \int_{-p}^{p} f(x) \sin \frac{k\pi x}{p} \, dx = 0. \tag{3}$$

It follows that *an even function has only cosine terms in its Fourier expansion.* Similarly, if f is an odd periodic function, the product

$$f(x) \cos \frac{k\pi x}{p}$$

is also odd; so for the Fourier cosine coefficient we have

$$a_k = \frac{1}{p} \int_{-p}^{p} f(x) \cos \frac{k\pi x}{p} \, dx = 0. \tag{4}$$

Thus *an odd function has only sine terms in its Fourier expansion.*

The facts in the preceding paragraph are the key to solving the following problem: given a function $f(x)$ defined just on the interval $0 \leq x \leq p$, find a trigonometric series expansion for f consisting only of cosine terms or, alternatively, only of sine terms. The trick is to extend the definition of f from the interval $0 \leq x \leq p$ to all real x in such a way that the extension is periodic of period $2p$ and either is even or is odd. We then compute the Fourier series of the extension. If f_e is an even periodic extension of f, then f_e will have only cosine terms in its Fourier series but will still agree with f on $0 \leq x \leq p$. Similarly, if f_o is an odd periodic extension of f, then f_o has only sine terms in its expansion but will agree with f for $0 \leq x \leq p$. For a **sine expansion** of $f(x)$ on $0 \leq x \leq p$ we have

3.3 $$a_k = 0, \quad b_k = \frac{2}{p} \int_{0}^{p} f(x) \sin \frac{k\pi x}{p} \, dx.$$

For a **cosine expansion** of $f(x)$ on $0 \leq x \leq p$ we have

3.4 $$a_k = \frac{2}{p} \int_{0}^{p} f(x) \cos \frac{k\pi x}{p} \, dx, \quad b_k = 0.$$

We illustrate the procedure with two examples.

EXAMPLE 3   We will compute the cosine expansion for the function defined by $f(x) = 1 - x$ for $0 \leq x \leq 2$. We consider the even periodic extension shown in Figure 7. To find the extension we define f_e by $f_e(x) = f(-x)$ for $-2 \leq x \leq 0$, and then extend periodically, with period 4, to the whole x-axis. We can use Formula 3.1 to compute the Fourier coefficients of f_e. Since f_e is

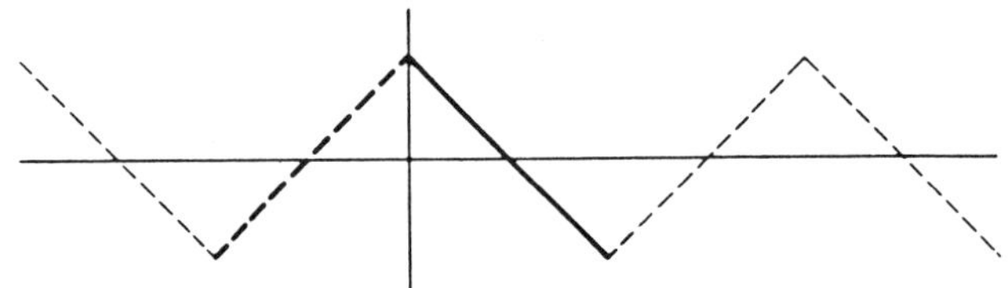

Figure 7

even, equation (3) shows that $b_k = 0$ for all k. Also, equation (1) allows us to write

$$a_k = \frac{1}{2} \int_{-2}^{2} f_e(x) \cos \frac{k\pi x}{2} \, dx$$

$$= \int_{0}^{2} f_e(x) \cos \frac{k\pi x}{2} \, dx.$$

Since, for $0 \le x \le 2$, the function f_e is the same as the given function $f(x) = 1 - x$, we have, for $k > 0$.

$$a_k = \int_{0}^{2} (1 - x) \cos \frac{k\pi x}{2} \, dx$$

$$= \left[\frac{2}{k\pi} (1 - x) \sin \frac{k\pi x}{2} \right]_{0}^{2} + \frac{2}{k\pi} \int_{0}^{2} \sin \frac{k\pi x}{2} \, dx$$

$$= \frac{4}{\pi^2 k^2} [1 - \cos k\pi]$$

$$= \begin{cases} 0, & k \text{ even,} \\ \dfrac{8}{\pi^2 k^2}, & k \text{ odd.} \end{cases}$$

Finally,

$$a_0 = \int_{0}^{2} (1 - x) \, dx = 0.$$

Thus the cosine expansion of f on $0 \le x \le 2$ has for its general nonzero term

$$\frac{8}{\pi^2 k^2} \cos \frac{k\pi x}{2}, \qquad k \text{ odd.}$$

Written out, the expansion looks like

$$\frac{8}{\pi^2} \left(\cos \frac{\pi x}{2} + \frac{\cos 3\pi x/2}{9} + \frac{\cos 5\pi x/2}{25} + \cdots \right).$$

EXAMPLE 4 Starting with the same function as in Example 3, $f(x) = 1 - x$ for $0 \le x \le 2$, we compute a sine expansion by considering the odd periodic extension shown in Figure 8. We first define $f_o(x) = -f(-x)$ for $-2 \le x < 0$, and then extend periodically with period 4. Using Formula 3.1 and equation (4), we find, as we intended, that $a_k = 0$ for all k. Also, by equation (1),

$$b_k = \frac{1}{2} \int_{-2}^{2} f_o(x) \sin \frac{k\pi x}{2} \, dx$$

$$= \int_{0}^{2} f_o(x) \sin \frac{k\pi x}{2} \, dx.$$

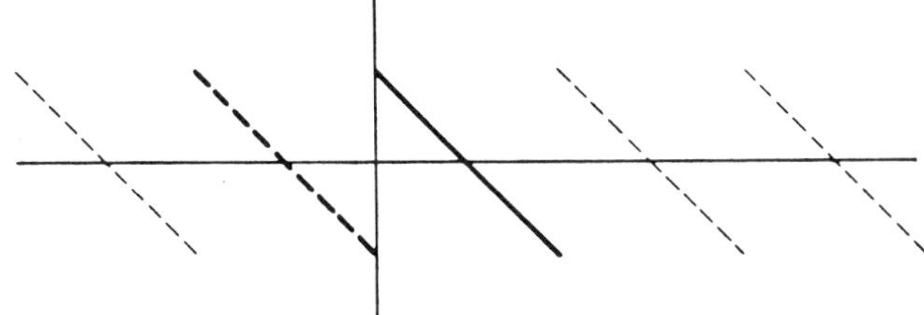

Figure 8

But $f_o(x) = 1 - x$ for $0 \le x \le 2$, so

$$b_k = \int_{0}^{2} (1 - x) \sin \frac{k\pi x}{2} \, dx$$

$$= \left[-\frac{2}{k\pi}(1 - x) \cos \frac{k\pi x}{2} \right]_{0}^{2} - \frac{2}{k\pi} \int_{0}^{2} \cos \frac{k\pi x}{2} \, dx$$

$$= \left[\frac{2}{k\pi}(-1)^k + \frac{2}{k\pi} \right] - \frac{2}{k\pi} \left[\frac{2}{k\pi} \sin \frac{k\pi x}{2} \right]_{0}^{2}$$

$$= \begin{cases} 0, & k \text{ odd}, \\ \dfrac{4}{k\pi}, & k \text{ even}. \end{cases}$$

Thus the general nonzero term in the sine expansion is

$$\frac{4}{k\pi} \sin \frac{k\pi x}{2}, \qquad \text{even } k.$$

The sine expansion is then

$$\frac{2}{\pi}\left(\sin \pi x + \frac{\sin 2\pi x}{2} + \frac{\sin 3\pi x}{3} + \cdots \right).$$

EXERCISES

1. Find the Fourier series for the function

$$f(x) = -x, \qquad -2 < x < 2.$$

To what values will the series converge at $x = 2$ and $x = -2$?

2. Find the Fourier series for the function

$$f(x) = 1 + x, \qquad 1 < x < 2.$$

To what values will the series converge at $x = 1$ and $x = 2$?

3. Find the sine expansion of each of the following functions. In each case sketch the graph of the odd periodic extension of f. The graph of the sum of the first two nonzero terms of the sine expansion should exhibit some reasonable relation to your sketch.

(a) $f(x) = 1,\ 0 < x < \pi$.

(b) $f(x) = x^2,\ 0 < x < \pi$.

(c) $f(x) = \cos x,\ 0 < x < \pi/2$.

(d) $f(x) = \begin{cases} 1, & 0 < x < 1, \\ 0, & 1 \le x < 2. \end{cases}$

(e) $f(x) = \begin{cases} 0, & 0 < x < 2, \\ x - 2, & 2 \le x < 3. \end{cases}$

4. Find the cosine expansion of each of the following functions. In each case sketch the graph of the even periodic extension of f. The graph of the sum of the first two nonzero terms of the cosine expansion should exhibit some reasonable relation to your sketch.

(a) $f(x) = 1,\ 0 < x < \pi$.

(b) $f(x) = x^2,\ 0 < x < \pi$.

(c) $f(x) = \sin x,\ 0 < x < \pi/2$.

(d) $f(x) = \begin{cases} 0, & 0 < x < 1, \\ 1, & 1 \le x < 2. \end{cases}$

(e) $f(x) = \begin{cases} x, & 0 < x < 1, \\ 0, & 1 \le x < 3. \end{cases}$

5. Find (a) the Fourier cosine expansion and (b) the Fourier sine expansion of the function

$$f(x) = x, \qquad 0 < x < \pi.$$

(c) Compare the results of (a) and (b) with the complete Fourier expansion of

$$g(x) = x, \qquad -\pi < x < \pi.$$

6. Show that every real-valued function f defined on a symmetric interval $[-a, a]$ can be written as the sum of an even function f_e and an odd function f_o. [*Hint:* Let $f_e(x) = (f(x) + f(-x))/2$.]

7. Show that, if the appropriate combinations are defined,

(a) a product of even functions is even.

(b) a product of odd functions is even.

(c) the product of an even function and an odd function is odd.

(d) a linear combination of even functions is even, and that of odd functions is odd.

8. Using elementary properties of integrals, prove equations (1) and (2) of the text.

9. Let f be an odd function on $[-\pi, \pi]$ [i.e., $f(-x) = -f(x)$], and let g be an even function [i.e., $g(-x) = g(x)$]. *Let a_k, b_k and a_k', b_k' be the Fourier coefficients of f and g, respectively. Show that*

$$a_k = 0, \qquad b_k = \frac{2}{\pi}\int_0^{\pi} f(x)\,\sin\,kx\,dx,$$

$$a_k' = \frac{2}{\pi}\int_0^{\pi} g(x)\,\cos\,kx\,dx, \qquad b_k' = 0.$$

4 HEAT AND WAVE EQUATIONS

In this section we show how a Fourier series can be used to solve some problems in heat conduction and wave motion. We first find a differential equation that is satisfied by the physical quantity being studied and then apply Fourier series to solve the equation.

4A One-dimensional Heat Equation

Suppose we are given a thin wire of uniform density and length p. Let $u(x, t)$ be the temperature, at time t, at a point x units from one end. Thus $0 \le x \le p$, and we assume $t \ge 0$. We will assume that the only heat transfer is along the direction of the wire and that the temperature at the two ends is held fixed. For this reason we can, without loss of generality, represent the wire as a straight segment along an x-axis and picture the temperature as the graph of a function $u = u(x, t)$, an example of which is shown in Figure 9.

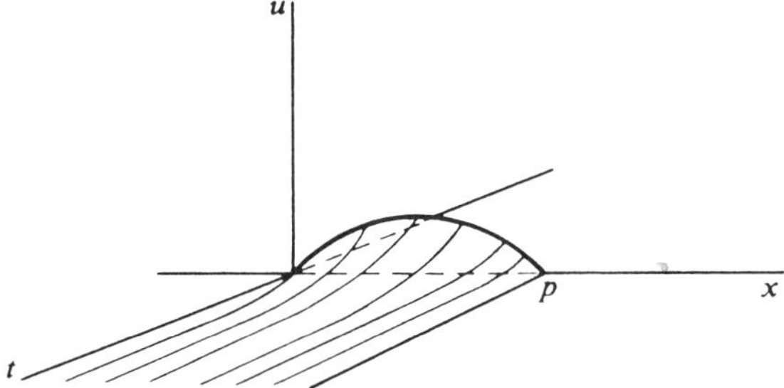

Figure 9

The basic physical principle of heat conduction is that heat flow is proportional to, and in the direction opposite to, the temperature gradient ∇u. Recall that ∇u is the direction in which the temperature is increasing most rapidly, so it is reasonable that heat should flow in the opposite direction, from hotter to colder. Since the medium is one-dimensional and is represented by a segment of the x-axis, the gradient is represented by $\partial u/\partial x$. Thus the rate of change of total heat in a segment $[x_1, x_2]$ is proportional to

$$- \frac{\partial u}{\partial x}(x_1, t) + \frac{\partial u}{\partial x}(x_2, t). \tag{1}$$

But it is also known that the rate of change of heat in the segment is also proportional to the average rate of change in temperature, times the length of the interval

$$\int_{x_1}^{x_2} \frac{\partial u}{\partial t}(x, t) \, dx, \tag{2}$$

the average being taken over the interval. By the fundamental theorem of calculus, expression (1) can be written as

$$\int_{x_1}^{x_2} \frac{\partial^2 u}{\partial x^2}(x, t) \, dx. \tag{3}$$

Hence the two proportional expressions (2) and (3) for rate of change of total heat can be combined to give

$$a^2 \int_{x_1}^{x_2} \frac{\partial^2 u}{\partial x^2}(x, t) \, dx = \int_{x_1}^{x_2} \frac{\partial u}{\partial t}(x, t) \, dx,$$

where a^2 is a positive proportionality constant. Allowing x_2 to vary, we can differentiate both sides of this last equation with respect to x_2, getting

4.1
$$a^2 \frac{\partial^2 u}{\partial x^2}(x, t) = \frac{\partial u}{\partial t}(x, t).$$

This is the one-dimensional **heat** or **diffusion equation.**

Equation 4.1 is linear in the sense that, if u_1 and u_2 are solutions, then so are linear combinations $b_1 u_1 + b_2 u_2$. To single out particular solutions, we impose linear boundary conditions of the form

$$u(0, t) = 0 \quad \text{and} \quad u(p, t) = 0, \tag{4}$$

together with an initial condition of the form

$$u(x, 0) = h(x).$$

The standard method of solution is by **separation of variables,** in which we start by trying to find product solutions of the form

$$u(x, t) = G(x)H(t)$$

with boundary condition $G(0) = G(p) = 0$. If such exist, substitution into $a^2 u_{xx} = u_t$ gives

$$a^2 G''(x)H(t) = G(x)H'(t),$$

for $0 \le x \le p, 0 < t$. Dividing through by $G(x)H(t)$ gives

$$a^2 \frac{G''(x)}{G(x)} = \frac{H'(t)}{H(t)}. \tag{5}$$

For this equation to be satisfied for varying x and t, both sides must be equal to a constant, which we denote by $-\lambda^2$. This procedure is the origin of the term separation of variables.

Setting both sides of equation (5) equal to $-\lambda^2$ gives two equations:

$$a^2 G'' + \lambda^2 G = 0, \tag{6}$$

$$H' + \lambda^2 H = 0. \tag{7}$$

Equation (6) has solutions

$$G(x) = c_1 \cos \left(\frac{\lambda}{a}\right) x + c_2 \sin \left(\frac{\lambda}{a}\right) x.$$

But $G(0) = 0$ requires $c_1 = 0$, and $G(p) = 0$ then requires $c_2 \sin (\lambda/a)p = 0$. This condition can be achieved, without making $c_2 = 0$, only by choosing λ so that $(\lambda/a)p = k\pi$, where k is an integer. That is, we must take $\lambda = (ka\pi)/p$, with the result that G has the form

$$G(x) = c_2 \sin \left(\frac{k\pi}{p}\right) x, \qquad k = 1, 2, \ldots .$$

Equation (7) has solution

$$H(t) = e^{-\lambda^2 t}$$

which, because $\lambda^2 = (k^2 a^2 \pi^2)/p^2$, becomes

$$H(t) = e^{-k^2(a^2 \pi^2/p^2)t}.$$

The product solution $u(x, t)$ is thus given, except for a constant factor, by

$$u_k(x, t) = e^{-k^2(a^2 \pi^2/p^2)t} \sin \left(\frac{k\pi}{p}\right) x, \qquad k = 1, 2, \ldots .$$

Thus a limit of linear combinations

$$\sum_{k=1}^{N} b_k u_k(x, t)$$

looks like

$$u(x, t) = \sum_{k=1}^{\infty} b_k e^{-k^2(a^2 \pi^2/p^2)t} \sin \left(\frac{k\pi}{p}\right) x. \tag{8}$$

To satisfy the initial condition $u(x, 0) = h(x)$, we require

$$\sum_{k=1}^{\infty} b_k \sin \left(\frac{k\pi}{p}\right) x = h(x), \tag{9}$$

for some b_k. If $h(x)$ can be expressed in this form, we can expect that a solution to the problem is given by equation (8). The boundary conditions, $u(0, t) = u(p, t) = 0$, require that the temperature remain zero at the ends, and the initial condition

$u(x, 0) = h(x)$ specifies the initial temperature at each point x of the wire between 0 and p. Thus we are naturally led to the problem of finding a Fourier sine series representation for $h(x)$. The coefficients in the expansion are given by

4.2
$$b_k = \frac{2}{p} \int_0^p h(x) \sin \frac{k\pi x}{p}\, dx.$$

The matter of conditions under which the infinite series (8) actually represents a solution of Equation 4.1 is taken up in Exercise 6.

EXAMPLE 1   To be more specific about solving the heat equation, we assume for simplicity that $p = \pi$ and recall that to solve $a^2 u_{xx} = u_t$ with boundary condition $u(0, t) = u(\pi, t) = 0$, and initial condition $u(x, 0) = h(x)$, we want in general to be able to represent h by an infinite series of the form

$$h(x) = \sum_{k=1}^{\infty} b_k \sin kx. \tag{10}$$

Suppose, for example, that h is given in $[0, \pi]$ by

$$h(x) = \begin{cases} x, & 0 \le x \le \pi/2, \\ \pi - x, & \pi/2 \le x \le \pi. \end{cases}$$

The graph of h is shown in Figure 10.

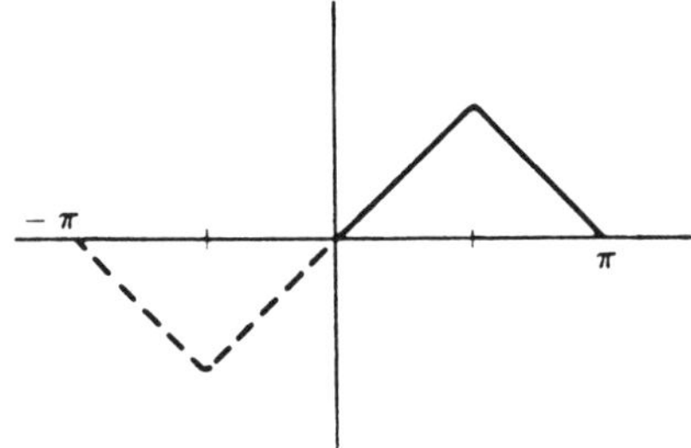

Figure 10

To make equation (10) represent the Fourier expansion of h on $[0, \pi]$, we extend h to the interval $[-\pi, \pi]$ in such a way that the cosine terms in the expansion of h will all be zero, leaving only the sine terms to be computed. We do this by extending the graph of h symmetrically about the origin, as shown in Figure 10. Then

$$a_k = \frac{1}{\pi} \int_{-\pi}^{\pi} h(x) \cos kx \, dx = 0,$$

because $h(-x) \cos k(-x) = -h(x) \cos kx$; therefore, the integrals over $[-\pi, 0]$ and $[0, \pi]$ are negatives of one another. To compute b_k, we use the fact that $h(-x) \sin k(-x) = h(x) \sin kx$, so the graph of this function is symmetric about the y-axis. Then

$$b_k = \frac{1}{\pi} \int_{-\pi}^{\pi} h(x) \sin kx \, dx$$

$$= \frac{2}{\pi} \int_{0}^{\pi} h(x) \sin kx \, dx$$

$$= \frac{2}{\pi} \int_{0}^{\pi/2} x \sin kx \, dx + \frac{2}{\pi} \int_{\pi/2}^{\pi} (\pi - x) \sin kx \, dx$$

$$= \frac{4}{\pi k^2} \sin \left(\frac{k\pi}{2} \right).$$

Hence

$$b_k = \begin{cases} 0, & k \text{ even}, \\[2mm] \dfrac{4}{\pi k^2}, & k = 1, 5, 9, \ldots, \\[2mm] \dfrac{-4}{\pi k^2}, & k = 3, 7, 11, \ldots. \end{cases}$$

Theorem 2.5 then implies that

$$h(x) = \frac{4}{\pi} \left(\frac{\sin x}{1^2} - \frac{\sin 3x}{3^2} + \frac{\sin 5x}{5^2} - \frac{\sin 7x}{7^2} + - \cdots \right)$$

for each x in $[0, \pi]$. Finally, from equation (8) we expect the solution to the equation $a^2 u_{xx} = u_t$ to be given by

$$u(x, t) = \frac{4}{\pi} \left(e^{-a^2 t} \sin x - \frac{1}{3^2} e^{-3^2 a^2 t} \sin 3x + \frac{1}{5^2} e^{-5^2 a^2 t} \sin 5x \right.$$

$$\left. - \frac{1}{7^2} e^{-7^2 a^2 t} \sin 7x + - \cdots \right).$$

Verification that $u(0, t) = u(\pi, t) = 0$ follows immediately from setting $x = 0$ and $x = \pi$. By setting $t = 0$, we get the representation of h by its Fourier series, which is guaranteed by Theorem 2.5. That $u(x, t)$ satisfies the equation $a^2 u_{xx} = u_t$ depends on term-by-term differentiation of the series for u.

4B Steady-State Solutions

A solution $u(x, t) = v(x)$ independent of t for the heat equation is called a **steady-state solution**. The heat equation for such functions becomes simply $v'' = 0$, and all solutions are necessarily of the form $u(x, t) = v(x) = \alpha + \beta x$, where α and β are constant. Solutions of this type are useful for solving the time-independent problem when we have **non-homogeneous boundary conditions** of the form

$$u(0, t) = u_0, \qquad u(p, t) = u_1, \qquad t > 0,$$

where at least one of u_0 and u_1 is a nonzero constant. The idea is to choose α, β in the steady-state solution so that

$$v(0) = u_0, \qquad v(p) = u_1.$$

Then

$$u(x, t) = w(x, t) + v(x)$$

will be the desired solution if $w(x, t)$ is a solution of the heat equation satisfying **homogeneous conditions**

$$w(0, t) = w(p, t) = 0, \qquad t > 0.$$

Note that the same function $w + v$ is indeed a solution of the heat equation, because both w and v are solutions and because the heat operator $a^2 D_{xx} - D_t$ acts linearly.

EXAMPLE 2   To solve the problems of the form

$$a^2 u_{xx} = u_t, \qquad u(0, t) = 10, \qquad u(5, t) = 30,$$

we first find a steady-state solution $v(x) = \alpha + \beta x$. We need $v(0) = 10 = \alpha$ and $v(5) = 30 = \alpha + 5\beta$. Solving for β, we find

$$v(x) = 10 + 4x.$$

The desired solutions

$$u(x, t) = w(x, t) + (10 + 4x)$$

come from finding solutions $w(x, t)$ that satisfy

$$w(0, t) = w(5, t) = 0, \qquad t \geq 0,$$

together with the requirement that an initial condition

$$u(x, 0) = w(x, 0) + v(x) = h(x)$$

imposes; this condition is just

$$w(x, 0) = h(x) - v(x).$$

Thus our solution $w(x, t)$ has the form of equation (8), where the b_k are Fourier sine coefficients computed by

$$b_k = \frac{2}{p} \int_0^p (h(x) - v(x)) \sin \frac{k\pi x}{p} \, dx.$$

EXERCISES

1. Solve the heat equation $a^2 u_{xx} = u_t$ subject to the following boundary and initial conditions.
   (a) $u(0, t) = u(p, t) = 0$; $u(x, 0) = \sin(\pi x/p)$, $0 < x < p$.

(b) $u(0, t) = u(\pi, t) = 0; u(x, 0) = x(\pi - x), 0 < x < \pi$.
(c) $u(0, t) = u(1, t) = 0; u(x, 0) = x, 0 < x < 1$.

2. Find steady-state solutions $u(x, t) = v(x)$ of the heat equation $a^2 u_{xx} = u_t$ that satisfy each of the following conditions:
   (a) $u(0, t) = -1, u(2, t) = 1$.
   (b) $u(0, t) = 0, u(100, t) = 100$.
   (c) $u_x(0, t) = 1, u(1, t) = 2$.

3. Solve the heat equation $a^2 u_{xx} = u_t$ subject to the following boundary and initial conditions.
   (a) $u(0, t) = 1, u(p, t) = 3; u(x, 0) = \sin(\pi x/p)\ 0 < x < p$.
   (b) $u(0, t) = -1, u(2, t) = 1; u(x, 0) = x, 0 < x < 2$.
   (c) $u(0, t) = 0, u(1, t) = 1; u(x, 0) = 1 - x, 0 < x < 1$.

4. Verify that the partial differential operator $L = a^2 D_{xx} - D_t$, defined by

$$L(u) = a^2 u_{xx} - u_t$$

   is linear in its action of suitably differential functions.

5. Find all solutions of the form $u(x, t) = X(x)T(t)$ for each of the following equations:
   (a) $u_{xx} + u_x = u_t$.   **(b)** $u_{xx} - u_x = u_t$.
   (c) $xu_x = 2u_t$.   **(d)** $u_{xx} = u_{tt}$.

6. **(a)** Verify that the series expansion for $u(x, t)$ given by equation (8) of the text is formally a solution of $a^2 u_{xx} = u_t$. Use term-by-term differentiation of the series.
   ***(b)** Show that the formal verification in part (a) is valid for $t > 0$, by showing that the differentiated series converge uniformly.
   Note: If f is smooth enough it can be shown that $\lim_{t \to 0+} u(x, t) = f(x)$.

7. Find the steady-state solution to each of the following problems.
   (a) $u_{xx} = u_t + 2; u(0, t) = 1, u(1, t) = 2$.
   (b) $u_{xx} = u_t + u; u(0, t) = 0, u(2, t) = 0$.
   (c) $u_{xx} = u_t + x; u(0, t) = 1, u(1, t) = 2$.

8. The one-dimensional heat flow problem

$$a^2 u_{xx} = u_t, \quad u_x(0, t) = u_x(p, t) = 0, \quad u(x, 0) = f(x)$$

   for $0 \le x \le p$, can be interpreted as requiring that the conducting medium have insulated ends at $x = 0$ and $x = p$. The reason is that the temperature gradient u_x is always 0 there, so there is no heat flow past those points.
   (a) Show that product solutions $u(x, t) = T(t)X(x)$ of the endpoint problem have the form

$$u_k(x, t) = a_k e^{-k^2 \pi^2 t/p} \cos \frac{k\pi}{p} x.$$

   (b) Use the Fourier cosine expansion for $f(x)$ on $0 \le x \le p$ to solve the boundary value problem for a general initial temperature $f(x)$.
   (c) What is the steady-state temperature function $u(x, \infty)$ for $0 \le x \le p$?

9. Suppose that $u(x, t)$ satisfies $a^2 u_{xx} = u_t$ and that $u(x, t_0)$ is concave up as a function of x for $a < x < b$. Show that for each x with $a < x < b$ there is a time $t(x)$ such that $u(x, t)$ increases as t increases from t_0 to $t(x)$. What if $u(x, t_0)$ is concave down?

4C One-dimensional Wave Equation

Consider for physical motivation a stretched elastic string of length p and uniform density ρ placed along the x-axis in $\mathcal{R}^3$. Suppose that the ends of the string are held fixed at $x = 0$ and $x = p$ by opposite forces of magnitude F. If the string is somehow made to vibrate, our problem is to determine in $\mathcal{R}^3$ the position $\mathbf{x}(s, t)$ at time t of a point on the string a distance s from the end fixed at $x = 0$. Figure 11 shows a possible configuration. We imagine the string subdivided into short pieces of length Δs and then derive two different expressions for the total force acting on a typical segment of the subdivision. If $\mathbf{t}(s)$ is the unit tangent vector to the string at $\mathbf{x}(s)$, then the opposing forces at $\mathbf{x}(s_0)$ and $\mathbf{x}(s_0 + \Delta s)$ are

$$F\mathbf{t}(x_0 + \Delta s) \quad \text{and} \quad -F\mathbf{t}(s_0).$$

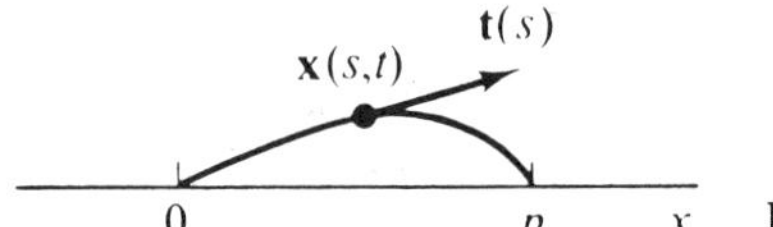

Figure 11

Hence the total force is

$$F[\mathbf{t}(s_0 + \Delta s) - \mathbf{t}(s_0)].$$

On the other hand, by Newton's law, the force equals mass, $\rho\,\Delta s$, times acceleration $\mathbf{a}(s_0)$. Hence

$$\rho\mathbf{a} = F\left[\frac{\mathbf{t}(s_0 + \Delta s) - \mathbf{t}(s_0)}{\Delta s}\right].$$

But

$$\mathbf{a} = \frac{\partial^2 \mathbf{x}}{\partial t^2}(s, t) \quad \text{and} \quad \mathbf{t}(s) = \frac{\partial \mathbf{x}}{\partial s}(s, t);$$

so, letting $\Delta s \to 0$, we get

$$\rho\frac{\partial^2 \mathbf{x}}{\partial t^2}(s, t) = F\frac{\partial^2 \mathbf{x}}{\partial s^2}(s, t). \tag{11}$$

This vector differential equation is of course equivalent to a system of three scalar equations

$$\frac{\partial^2 x}{\partial t^2} = a^2\frac{\partial^2 x}{\partial s^2},$$

$$\frac{\partial^2 y}{\partial t^2} = a^2\frac{\partial^2 y}{\partial s^2},$$

$$\frac{\partial^2 z}{\partial t^2} = a^2\frac{\partial^2 z}{\partial s^2},$$

where we have set $a^2 = F/\rho > 0$. In a formal sense, the problem of finding solution functions $x(s, t)$, $y(s, t)$, and $z(s, t)$ is the same for all three equations. In practice, however, the equation for $x(s, t)$ is usually set aside when the longitudinal motion (along the x-axis) is slight; we make this assumption. Between the other two equations there is little difference in physical significance unless some other special assumption is made. We suppose to be specific that the string has been plucked in such a way that its motion takes place entirely in the xy-plane, with $z(s, t) \equiv 0$. Finally, we assume that the displacements are small enough that we can replace $\partial^2 y/\partial s^2$ by $\partial^2 y/\partial x^2$. This last assumption is one that requires experimental justification in actual practice. Thus, with some loss of generality, we consider, instead of the vector equation (11), a single scalar equation for $y(x, t)$:

4.3
$$\frac{\partial^2 y}{\partial t^2} = a^2 \frac{\partial^2 y}{\partial x^2}.$$

This differential equation does not completely specify the vibration of a string unless we impose some initial conditions:

$$y(x, 0) = f(x), \tag{12}$$

$$\frac{\partial y}{\partial t}(x, 0) = g(x). \tag{13}$$

The first condition specifies the initial ($t = 0$) displacement in the y-direction as a function of x, rather than s, and the second equation specifies the initial velocity in the y-direction, also as a function of x. In addition, we also impose boundary conditions

$$y(0, t) = 0 \quad \text{and} \quad y(p, t) = 0. \tag{14}$$

As in the case of the heat equation, we use separation of variables and rely on the linearity of Equation 4.3 and the conditions (14) for constructing solutions that satisfy the conditions (12) and (13) also. We try

$$y(x, t) = G(x)H(t),$$

upon which Equation 4.2 becomes

$$G(x)H''(t) = a^2 G''(x)H(t)$$

or
$$\frac{H''(t)}{H(t)} = a^2 \frac{G''(x)}{G(x)}.$$

Since the left side is independent of t, both sides are constant; so we write

$$\frac{G''(x)}{G(x)} = \lambda \quad \text{and} \quad \frac{H''(t)}{H(t)} = a^2 \lambda.$$

The first equation, $G''(x) = \lambda G(x)$, has solutions

$$G(x) = c_1 e^{\sqrt{\lambda}x} + c_2 e^{-\sqrt{\lambda}x}.$$

The boundary conditions (14) require

$$G(0) = 0 \quad \text{and} \quad G(p) = 0;$$

so

$$c_1 + c_2 = 0 \quad \text{and} \quad c_1 e^{\sqrt{\lambda}p} + c_2 e^{-\sqrt{\lambda}p} = 0.$$

Solving for c_1 and c_2, we find that, for nonzero solutions to exist, we must have $c_1 = -c_2$ and $e^{-\sqrt{\lambda}p} = e^{\sqrt{\lambda}p}$, so $e^{2\sqrt{\lambda}p} = 1$. Thus, allowing complex exponents, we conclude that $2\sqrt{\lambda}p = 2\pi ki$ for some integer k, so that

$$\sqrt{\lambda} = \frac{\pi ki}{p}.$$

Solutions $G(x)$ are then of the form

$$G(x) = c_1 e^{(\pi ki/p)x} - c_1 e^{(-\pi ki/p)x}$$

$$= 2c_1 i \sin \frac{\pi k}{p} x.$$

We write

$$G_k(x) = b_k \sin \frac{\pi k}{p} x.$$

The differential equation for H, $H''(t) = a^2 \lambda H(t)$, now takes the form

$$H''(t) + \frac{\pi^2 k^2 a^2}{p^2} H(t) = 0,$$

since we have determined that $\lambda = -\pi^2 k^2 / p^2$. Solutions of this equation are of the form

$$H_k(t) = C_k \cos \frac{\pi ka}{p} t + D_k \sin \frac{\pi ka}{p} t;$$

therefore, product solutions $y_k = G_k H_k$ take the form

$$y_k(x, t) = \left[A_k \cos \frac{\pi ka}{p} t + B_k \sin \frac{\pi ka}{p} t \right] \sin \frac{\pi k}{p} x.$$

To satisfy the initial conditions (12) and (13), we form finite or infinite sums of the type

$$y(x, t) = \sum_{k=0}^{\infty} \left[A_k \cos \frac{\pi ka}{p} t + B_k \sin \frac{\pi ka}{p} t \right] \sin \frac{\pi k}{p} x.$$

Thus the initial conditions become formally

$$y(x, 0) = \sum_{k=0}^{\infty} A_k \sin \frac{\pi k}{p} x = f(x) \tag{15}$$

and

$$y_t(x,\, 0) = \sum_{k=0}^{\infty} \frac{\pi k a}{p}\, B_k \sin \frac{\pi k}{p}\, x = g(x). \tag{16}$$

The coefficients A_k and $(\pi k a / p)\, B_k$ are then determined so that they are the Fourier sine coefficients of f and g, respectively.

EXAMPLE 3 For a simple example, we consider a string that is initially stationary; hence, $y_t(x,\, 0) = 0$. We assume the initial displacement is given by

$$y(x,\, 0) = b \sin \frac{\pi x}{p}, \tag{17}$$

and $y(0,\, t) = y(p,\, t) = 0$. We choose $B_k = 0$ in equation (16) and $A_1 = b$ in equation (15), with $A_k = 0$ for $k \neq 1$. (It happens that f and g are so simple in this case that we can find their Fourier sine coefficients by inspection.) The solution to Equation 4.2 then takes the form

$$y(x,\, t) = b \cos \frac{\pi a}{p}\, t \sin \frac{\pi}{p}\, x.$$

Recall that for Equation 4.3 to be physically realistic the vibrations of the string should be fairly small; in other words, the coefficient b should be small. In Figure 12, the graph of our solution is shown with $a = 1$, $p = \pi$, and b chosen unrealistically large in order to bring out some qualitative features of the picture.

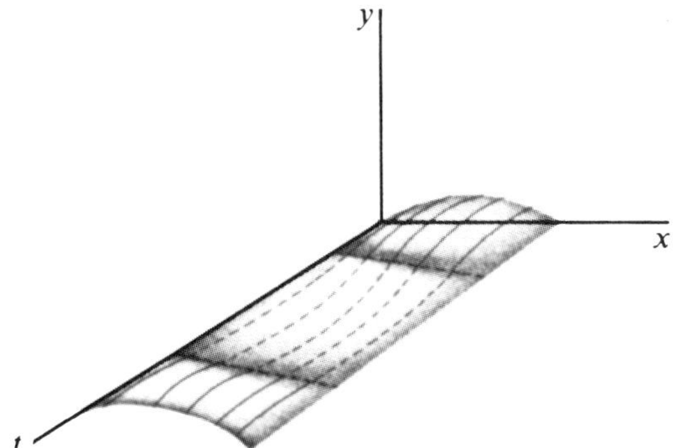

Figure 12

In general, the computation of the coefficients A_k, B_k in equations (16) is effected by the Fourier sine formulas

$$A_k = \frac{2}{p} \int_0^p f(x) \sin \frac{k \pi x}{p}\, dx,$$

$$B_k = \frac{2}{k \pi a} \int_0^p g(x) \sin \frac{k \pi x}{p}\, dx.$$

EXAMPLE 4 The solution of $a^2 u_{xx} = u_{tt}$ that satisfies

$$u(0, t) = u(1, t) = 0, \qquad u(x, 0) = x(1 - x), \qquad u_t(x, 0) = \sin \pi x$$

is found as follows. We compute, with $p = 1$ above,

$$A_k = 2 \int_0^1 x(1 - x) \sin k\pi x \, dx = \frac{4}{k^3 \pi^3} [1 - (-1)^k].$$

$$B_k = \frac{2}{k\pi a} \int_0^1 \sin \pi x \sin k\pi x \, dx = \begin{cases} \dfrac{1}{\pi a}, & k = 1 \\ 0, & k = 2, 3, \ldots . \end{cases}$$

The term corresponding to $k = 1$ in the general expansion is just

$$\left(\frac{8}{\pi^3} \cos \pi at + \frac{1}{\pi a} \sin \pi at \right) \sin \pi x.$$

All the other nonzero terms have odd index, so the solution formula is

$$u(x, t) = \left(\frac{8}{\pi^3} \cos \pi at + \frac{1}{\pi a} \sin \pi at \right) \sin \pi x$$

$$+ \sum_{j=1}^{\infty} \frac{8}{(2j + 1)^3 \pi^3} \cos (2j + 1)\pi at \sin (2j + 1)\pi x.$$

The method of Fourier series and separation of variables is important for the wave equation not only because it provides solution formulas. (In fact, the d'Alembert method discussed in Exercises 3 and 4 does this also, and in a way that is in many ways more direct.) What the Fourier method allows us to see is an analysis of vibratory motion into simpler vibrations, and it is this analysis, and its flexibility in applying to other problems, that makes the Fourier method so important.

EXAMPLE 5 Suppose $u(x, t)$ is the displacement at time t and position x of a taut string with ends fixed: $u(0, t) = u(p, t) = 0$. To simplify the discussion, suppose the initial velocity is zero: $u_t(x, 0) = 0$. Then the Fourier solution has the form

$$u(x, t) = \sum_{k=1}^{\infty} A_k \cos \frac{k\pi a}{p} t \sin \frac{k\pi}{p} x,$$

where the amplitudes A_k of the individual terms are determined as Fourier sine coefficients of the initial displacement $u(x, 0) = f(x)$. A typical term

$$u_k(x, t) = A_k \cos \frac{k\pi a}{p} t \sin \frac{k\pi}{p} x$$

is periodic in the time variable t with period T determined by $k\pi a T/p = 2\pi$, so that $T = 2p/ka$. The **frequency**, or number of vibrations per time unit, is thus

$v = ka/2p$. The **fundamental frequency** is the frequency that corresponds to the first nonzero A_k. If $A_1 \neq 0$, the fundamental vibration has frequency $a/2p$ and all others have frequencies that are integer mutiples of the fundamental frequency. If we think of each term $u_k(x, t)$ as generating an audible tone, it is usual to speak of the **fundamental** tone and its associated **overtones** or **harmonic tones** of higher frequency. It is the relative sizes of the amplitudes A_k that give a particular vibration in recognizable quality. Some typical displacements of a fundamental and its overtones are shown in Figure 13.

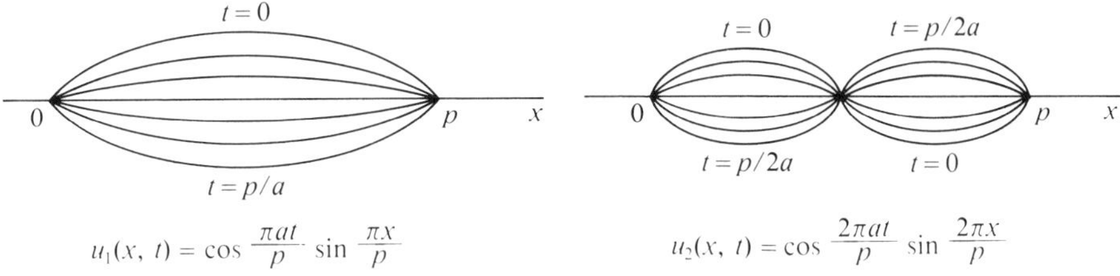

$$u_1(x, t) = \cos \frac{\pi a t}{p} \sin \frac{\pi x}{p}$$

$$u_2(x, t) = \cos \frac{2\pi a t}{p} \sin \frac{2\pi x}{p}$$

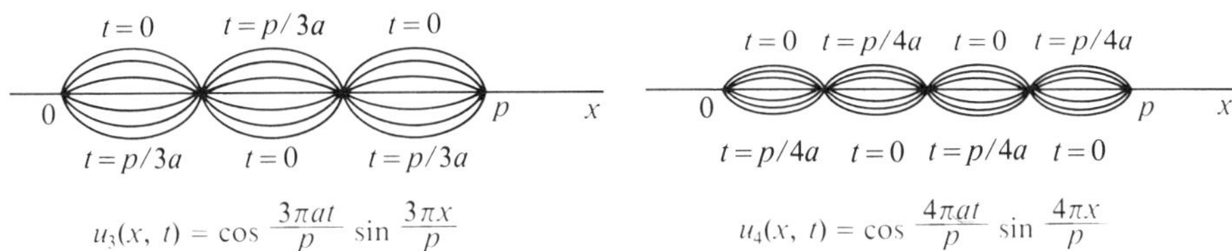

$$u_3(x, t) = \cos \frac{3\pi a t}{p} \sin \frac{3\pi x}{p}$$

$$u_4(x, t) = \cos \frac{4\pi a t}{p} \sin \frac{4\pi x}{p}$$

Figure 13

EXERCISES

1. Solve the wave equation $a^2 u_{xx} = u_{tt}$ subject to the following boundary and initial conditions.

(a) $u(0, t) = u(\pi, t) = 0$, $u(x, 0) = \sin x$, $u_t(x, 0) = 0$.

(b) $u(0, t) = u(\pi, t) = 0$, $u(x, 0) = \begin{cases} x, & 0 < x \leq \pi/2, \\ \pi - x, & \pi/2 < x < \pi, \end{cases}$ $u_t(x, 0) = 0$.

(c) $u(0, t) = u(\pi, t) = 0$, $u(x, 0) = 0$, $u_t(x, 0) = \begin{cases} 0, & 0 < x \leq \pi/2, \\ 1, & \pi/2 < x < \pi. \end{cases}$

(d) $u(0, t) = u(1, t) = 0$, $u(x, 0) = x(1 - x)$, $u_t(x, 0) = \sin \pi x$.

2. The **nonhomogeneous wave equation**

$$a^2 u_{xx} = u_{tt} - g$$

incorporates g, the accelaration of gravity, into the vibrating string problem.

(a) Find the steady-state solutions $u(x, t) = v(x)$ of the nonhomogeneous equation.

(b) Among the solutions found in part (a), select the solution that satisfies the boundary conditions $u(0, t) = u(p, t) = 0$.

(c) Explain how to modify the Fourier solution method for $a^2 u_{xx} = u_{tt}$ in order to cover the solution of the nonhomogeneous problem.

3. Let $U(x)$ and $V(x)$ be twice differentiable functions defined for all real x.

(a) Show that $u(x, t) = U(x + at) + V(x - at)$ defines a solution of $a^2 u_{xx} = u_{tt}$ that is valid for all real x and all real t.

(b) Assume that $f''(x)$ exists. Show that

$$u(x, t) = \frac{1}{2}[f(x + at) + f(x - at)]$$

is a solution to the wave equation having the form described in part (a) and also satisfies the initial conditions $u(x, 0) = f(x)$, $u_t(x, 0) = 0$.

(c) Assume that $g'(x)$ exists. Show that

$$u(x, t) = \frac{1}{2a} \int_{x-at}^{x+at} g(s)\, ds$$

is a solution to the wave equation having the form described in part (a) and also satisfies the initial $u(x, 0) = 0$, $u_t(x, 0) = g(x)$.

(d) Combine the results of parts (b) and (c) to find a solution formula for the wave equation subject to the initial conditions $u(x, 0) = f(x)$, $u_t(x, 0) = g(x)$.

4. **(a)** Let $f(x)$ be defined for $0 \leq x \leq p$. Extend $f(x)$ to the interval $-p < x < 0$ by defining

$$f(x) = -f(-x), \qquad \text{if } -p < x < 0.$$

Then extend $f(x)$ to $-\infty < x < \infty$ by defining f to have period $2p$. Show that $f(x)$ so extended is not only periodic, but odd: $f(-x) = -f(x)$ for $-\infty < x < \infty$.

(b) Sketch the odd periodic extension of $f(x) = x(1 - x)$, as described in part (a), from $0 \leq x \leq 1$ to $-\infty < x < \infty$.

(c) Show that if $f(x)$ is odd and has period $2p$ for $-\infty < x < \infty$ then the function

$$u(x, t) = \tfrac{1}{2}[f(x + at) + f(x - at)],$$

defined in Exercise 3(b), satisfies the boundary conditions $u(0, t) = u(p, t) = 0$ for all real t. The solution $u(x, t)$ to the wave equation with boundary and initial conditions described in Exercises 3 and 4 is called the **d'Alembert form** of the solution.

5. The wave equation with frictional drag accounted for has the form

$$a^2 u_{xx} = u_{tt} + Ku_t,$$

where K is a positive constant. Find all solutions of the form $u(x, t) = X(x)T(t)$.

5 THE LAPLACE EQUATION

The two-dimensional **Laplace equation**

5.1
$$u_{xx} + u_{yy} = 0,$$

arises naturally from the two-dimensional heat equation

$$u_{xx} + u_{yy} = \frac{1}{a^2}u_t,$$

The derivation of the above heat eqaution is similar to the one given in Section 4A. If R is a region in the (x, y)-plane whose boundary consists of finitely many smooth curves, we can let $u(x, y, t)$ be the temperature at (x, y), taken at time t. Let the temperature at the boundary be maintained so that it does not change as time goes on, although the temperature may be different at different points of the boundary. Now let enough time elapse so that there is no detectable change in the function u regarded as a function of t; in other words, wait long enough to be able to assume that $u_t(x, y, t) = 0$ throughout the interior of the region from that time on. In effect, we have assumed that $u(x, y, t) = u(x, y)$ is independent of t, in which case the heat equation reduces to the Laplace equation (5.1). It is often convenient to abbreviate the Laplace operator by

$$\Delta = \frac{\partial^2}{\partial x^2} + \frac{\partial^2}{\partial y^2}.$$

The operator Δ acts linearly on twice-differentiable functions $u(x, y)$.

Solving the Laplace equation in two independent variables x, y is somewhat different from solving the heat or wave equation in the variables x, t. One important difference is that the boundary and initial conditions on $u(x, t)$ that are appropriate for the heat and wave equations are replaced by a boundary condition on a solution $u(x, y)$ that is defined on some two-dimensional region. We will consider only rectangular regions, although techniques exist that are applicable to much more complicated regions.

The rectangle $0 \le x \le p$, $0 \le y \le q$ is shown in Figure 14. On each of the four sides we specify continuous boundary values for $u(x, y)$:

$$u(x, 0) = f_0(x), \qquad u(x, q) = f_q(x),$$

$$u(0, y) = g_0(y), \qquad u(p, y) = g_p(y).$$

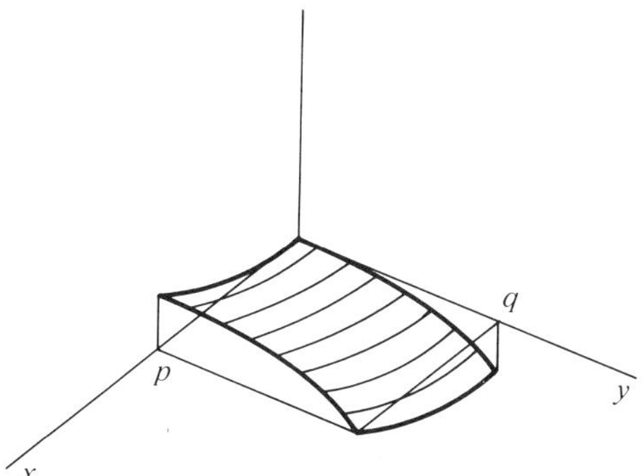

Figure 14

To simplify matters for the moment, suppose that $f_0(x)$, $g_0(y)$, and $g_p(y)$ are identically zero; we will see that the general case will follow from this special case in a simple way.

Using the **separation of variables** method, we look for product solutions in the form

$$u(x, y) = X(x)Y(y).$$

Substitution into $u_{xx} + u_{yy} = 0$ gives

$$X''Y + XY'' = 0$$

for $0 < x < p$, $0 < y < q$. Assuming $X \neq 0$, $Y \neq 0$, we divide by XY, separate the variables, and write

$$\frac{Y''}{Y} = -\frac{X''}{X} = \lambda,$$

where λ is constant with respect to x and y because Y''/Y does not depend on x and X''/X does not depend on y. This gives us the two ordinary differential equations

$$X'' + \lambda X = 0, \qquad Y'' - \lambda Y = 0.$$

If $X_\lambda(x)$ and $Y_\lambda(y)$ are the solutions of these equations for the same λ, then $u_\lambda(x, y) = X_\lambda(x)Y_\lambda(y)$ will be a solution of $u_{xx} + u_{yy} = 0$.

If our boundary conditions on $u(x, y)$ are assumed to be such that $u(x, 0) = 0$, $u(0, y) = 0$, $u(p, y) = 0$, the corresponding requirements on X and Y are

$$X(x)Y(0) = 0, \qquad 0 < x < p,$$

$$X(0)Y(y) = 0, \qquad 0 < y < q,$$

$$X(p)Y(y) = 0, \qquad 0 < y < q.$$

We achieve these conditions by making

$$X(0) = X(p) = 0, \qquad Y(0) = 0.$$

The resulting two-point boundary problem for X is then

$$X'' + \lambda X = 0, \qquad X(0) = X(p) = 0.$$

The nontrivial solutions of the problem are

$$X_k(x) = C_k \sin \frac{k\pi}{p} x, \qquad k = 1, 2, 3, \ldots$$

with $\lambda = k^2 \pi^2 / p^2$ being the only λ-values of interest from now on. The differential equation for Y is now

$$Y'' - \frac{k^2 \pi^2}{p^2} Y = 0, \qquad Y(0) = 0;$$

this has solutions that can be written either as

$$Y_k(y) = a_k e^{k\pi y/p} + b_k e^{-k\pi y/p}$$

or
$$Y_k(y) = A_k \cosh \frac{k\pi y}{p} + B_k \sinh \frac{k\pi y}{p}.$$

Recall that $\cosh x = (e^x + e^{-x})/2$ and $\sinh x = (e^x - e^{-x})/2$. Either form for $Y(y)$ will do, but writing the constants is a little simpler in terms of hyperbolic functions, so we will use them. We satisfy $Y(0) = 0$ by making $A_k = 0$. Writing $b_k = C_k B_k$, we have

$$X_k(x)Y_k(y) = b_k \sinh \frac{k\pi y}{p} \sin \frac{k\pi x}{p}.$$

The series representation

5.2
$$u(x, y) = \sum_{k=1}^{\infty} b_k \sinh \frac{k\pi y}{p} \sin \frac{k\pi x}{p}$$

will satisfy $u(x, q) = f_q(x)$, as well as $u(x, 0) = 0$, if we can choose the b_k so that

$$\sum_{k=1}^{\infty} b_k \sinh \frac{k\pi q}{p} \sin \frac{k\pi x}{p} = f_q(x).$$

So we take $b_k \sinh k\pi q/p$ to be the k-th Fourier sine coefficient of $f(x) = f_q(x)$:

5.3
$$b_k = \left(\sinh \frac{k\pi q}{p} \right)^{-1} \frac{2}{p} \int_0^p f(x) \sin \frac{k\pi x}{p}\, dx.$$

If f_q is smooth enough that its Fourier series converges to it, we will have satisfied the boundary condition at $y = q$. It can be shown that the second-order partial derivatives of $u(x, y)$ as defined by Equation 5.2 can be computed using term-by-term differentiation. Since each term satisfies the Laplace equation, the linearity of the equation guarantees that $u(x, y)$ will satisfy it also.

EXAMPLE 1   Suppose we want to solve the Laplace equation in the rectangle $0 < x < 2$, $0 < y < 3$ (i.e., $p = 2$, $q = 3$) with boundary conditions

$$u(x, 0) = 0, \qquad u(x, 3) = 4, \qquad 0 < x < 2,$$

$$u(0, y) = 0, \qquad u(2, y) = 0, \qquad 0 < y < 3.$$

Note that no condition is imposed at the corners of the rectangle. Equation 5.3 yields

$$b_k = \left(\sinh \frac{3k\pi}{2} \right)^{-1} \int_0^2 4 \sin \frac{k\pi x}{2}\, dx$$

$$b_k = \frac{8(1 - (-1)^k)}{k\pi \, \sinh 3k\pi/2}, \qquad k = 1, 2, 3, \ldots .$$

Notice that all even-order coefficients b_{2j} are zero. The solution formula 5.2 gives us the particular solution

$$u_3(x, y) = \sum_{j=0}^{\infty} \frac{16}{(2j + 1)\pi \, \sinh 3(2j + 1)\pi/2} \sinh \frac{(2j + 1)\pi y}{2} \sin \frac{(2j + 1)\pi x}{2}.$$

Replacing $\sinh k\pi y/p$ by $\sinh k\pi(q - y)/p$ in Equation 5.2 gives a different solution $u(x, y)$ of the Laplace equation, a solution that satisfies $u(x, q) = 0$. The corresponding coefficients b_k are then determined by Equation 5.3 with $f_0(x) = u(x, 0)$ plugged in for $f(x)$. This simple shift allows us to solve the Laplace equation with zero boundary values everywhere around the rectangle except the bottom edge.

EXAMPLE 2   Using the preceding remark, we can solve the Laplace equation with boundary conditions

$$u(x, 0) = x, \qquad u(x, 3) = 0, \qquad 0 < x < 2,$$

$$u(0, y) = 0, \qquad u(2, y) = 0, \qquad 0 < y < 3.$$

The coefficients b_k in Equation 5.2 are determined by the requirement $u(x, 0) = f_0(x) = x$ along the bottom edge of the rectangle. From Equation 5.3 we compute

$$b_k = \left(\sinh \frac{3k\pi}{2} \right)^{-1} \int_0^2 x \sin \frac{k\pi x}{2} \, dx$$

$$b_k = \left(\sinh \frac{3k\pi}{2} \right)^{-1} \frac{4(-1)^{k+1}}{k\pi} .$$

The particular solution we get is

$$u_0(x, y) = \sum_{k=1}^{\infty} \frac{4(-1)^{k+1}}{k\pi \, \sinh (3k\pi/2)} \sinh \frac{k\pi(3 - y)}{2} \sin \frac{k\pi x}{2} .$$

The factor $\sinh k\pi(3 - y)/2$ replaces $\sinh k\pi y/2$ as remarked earlier, so $u_0(x, 3) = 0$ will hold. But each term in the series is a solution of the Laplace equation, and it only requires justification of term-by-term differentiation of the series to show that u_0 itself satisfies the Laplace equation.

EXAMPLE 3   To solve the Laplace equation with boundary conditions

$$u(x, 0) = 5x, \qquad u(x, 3) = 4, \qquad 0 < x < 2$$

$$u(0, y) = 0, \qquad u(2, y) = 0, \qquad 0 < y < 3,$$

we just have to form a linear combination of the solutions u_3 and u_0 that we found in the two previous examples. The reason is, first, that $5u_0(x, y) + u_3(x, y)$ satisfies the Laplace equation because the Laplace operator Δ is linear:

$$\Delta(5u_0 + u_3) = 5\Delta u_0 + \Delta u_3 = 0 + 0.$$

Second,

$$5u_0(x, 3) + u_3(x, 3) = 5x + 0 = 5x,$$

and

$$5u_0(x, 0) + u_3(x, 0) = 0 + 4 = 4.$$

The reason this works is that each of u_3 and u_0 has a nonzero boundary value only where the other has a zero boundary value.

The previous examples have had nonzero boundary values imposed only on the top and bottom edges of the region in which a solution is sought. By reversing the roles of x and y in these examples, and in Equations 5.2 and 5.3, we can solve the Laplace equation with nonzero boundary values on the right and left edges of a rectangle.

EXERCISES

1. Each of the following functions $u(x, y)$ is a solution of the Laplace equation $u_{xx} + u_{yy} = 0$. Verify this, and verify that $u(x, y)$ satisfies the indicated boundary conditions. Then sketch the graph of $z = u(x, y)$ for (x, y) inside the indicated boundary.
   (a) $u(x, y) = (e^y - e^{-y}) \sin x$; $u(0, y) = u(\pi, y) = 0$ for $0 < y < \ln 3$,
   $u(x, 0) = 0$, $u(x, \ln 3) = (8/3) \sin x$ for $0 < x < \pi$.
   (b) $u(x, y) = x^2 - y^2$; $u(0, y) = -y^2$, $u(1, y) = -1 - y^2$ for $0 < y < 1$,
   $u(x, 0) = x^2$, $u(x, 1) = x^2 - 1$ for $0 < x < 1$.
   (c) $u(x, y) = e^x \cos y$; $u(0, y) = \cos y$, $u(1, y) = e \cos y$ for $0 < y < \pi/2$,
   $u(x, 0) = e^x$, $u(x, \pi/2) = 0$ for $0 < x < 1$.
   (d) $u(x, y) = (\sinh 2)^{-1} \sinh(2 - y) \sin x$; $u(0, y) = u(\pi, y) = 0$ for $0 < y < 2$,
   $u(x, 0) = \sin x$, $u(x, 2) = 0$ for $0 < x < \pi$.
2. As usual, define $\cosh y = (e^y + e^{-y})/2$, $\sinh y = (e^y - e^{-y})/2$. Prove the following:
   (a) $d \cosh y/dy = \sinh y$.   (b) $d \sinh y/dy = \cosh y$.
   (c) $\cosh^2 y - \sinh^2 y = 1$.   (d) $e^y = \cosh y + \sinh y$.
3. (a) Solve the Laplace equation $u_{xx} + u_{yy} = 0$ subject to the boundary conditions $u(0, y) = u(\pi, y) = 0$, $u(x, 0) = 0$, $u(x, \pi) = 0$.
   (b) Sketch the graph of the solution for $0 < x < \pi$, $0 < y < \pi$.
4. Find a series solution for the Laplace equation subject to each of the following sets of boundary conditions:
   (a) $u(x, 0) = u(x, 1) = 1$, $0 < x < 1$,
   $u(0, y) = 0$, $u(1, y) = 0$, $0 < y < 1$.
   (b) $u(x, 0) = 0$, $u(x, 1) = x(1 - x)$, $0 < x < 1$,
   $u(0, y) = u(1, y) = 0$, $0 < y < 1$.

(c) $u(x, 0) = 1$, $u(x, 1) = 2$, $0 < x < 1$,
    $u(0, y) = 0$, $u(1, y) = 0$, $0 < y < 1$.
(d) $u(x, 0) = u(x, 1) = 0$, $0 < x < 1$,
    $u(0, y) = y$, $u(1, y) = 0$, $0 < y < 1$.
(e) $u(x, 0) = 1$, $u(x, 2) = -1$, $0 < x < 1$,
    $u(0, y) = y$, $u(1, y) = y$, $0 < y < 2$.

5. (a) Assume that $f(x + iy)$ and $g(x - iy)$ are twice differentiable functions of x and y in some region R of the (x, y)-plane common to both f and g. Verify that

$$u(x, y) = f(x + iy) + g(x - iy)$$

defines a solution u of the Laplace equation in R. (*Remark:* It can be shown that f and g will in fact turn out to be infinitely often differentiable if R is an **open** set.)
(b) Let $f(x + iy) = g(x + iy) = e^{x+iy}$. What is the function $u(x, y)$ defined in part (a)?

6. Describe all solutions of the Laplace equation $u_{xx} + u_{yy} = 0$ that are independent of the variable y and are defined for (x, y) satisfying $a < x < b$ for some $a < b$.

7. (a) The derivation of Equation 5.3 for the coefficients b_k in Equation 5.2 was based on the assumption that $u(x, q)$ was specified for $0 < x < p$. If instead we specify the values of $u_y(x, q)$ for $0 < x < p$ by $u_y(x, q) = f(x)$, show that Equation 5.3 is replaced by

$$b_k = \left(\cosh \frac{k\pi q}{p}\right)^{-1} \frac{2}{k\pi} \int_0^p f(x) \sin \frac{k\pi x}{p}\, dx.$$

(b) Solve the Laplace equation under the boundary conditions

$$u(x, 0) = 0, \qquad u_y(x, q) = 0, \qquad 0 < x < p,$$
$$u(0, y) = u(p, y) = 0, \qquad 0 < y < q.$$

8. The **Laplace equation in polar coordinates** (r, θ), where $x = r \cos \theta$, $y = r \sin \theta$, is of the form

$$r^2 \bar{u}_{rr} + r\bar{u}_r + \bar{u}_{\theta\theta} = 0;$$

here $\bar{u}(r, \theta) = u(r \cos \theta, r \sin \theta)$, where $u(x, y)$ satisfies the Laplace equation $u_{xx} + u_{yy} = 0$.
(a) Compute $\bar{u}_r$, $\bar{u}_{rr}$, and $\bar{u}_{\theta\theta}$ by applying the chain rule to $u(r \cos \theta, r \sin \theta)$. Then verify that $\bar{u}$ satisfies the polar form of the Laplace equation.
(b) Verify that the polar form of the Laplace equation has solutions of the form

$$\bar{u}_n(r, \theta) = r^n(a_n \cos n\theta + b_n \sin n\theta), \qquad n = 0, 1, 2, \ldots.$$

Define $\bar{u}(r, \theta) = \sum_{n=0}^{\infty} r^n(a_n \cos n\theta + b_n \sin n\theta)$, where

$$a_n = \frac{1}{R^n \pi} \int_{-\pi}^{\pi} F(\theta) \cos n\theta\, d\theta, \qquad b_n = \frac{1}{R^n \pi} \int_{-\pi}^{\pi} F(\theta) \sin n\theta\, d\theta.$$

Here $F(\theta) = f(R \cos \theta, R \sin \theta)$ is the boundary-value prescription of a function on the circular region $x^2 + y^2 \leq R^2$.

(c)* Show that the series defining $\bar{u}(r, \theta)$ converges uniformly if $0 \le r \le R'$ for any $R' < R$, and hence that $\bar{u}(r, \theta)$ is a solution of the Laplace equation in polar coordinates for $0 \le r < R$ and $-\pi \le \theta < \pi$.

6 STURM–LIOUVILLE EXPANSIONS

6A Orthogonal Functions

Many of the useful properties of Fourier expansions are shared by a large class of similar expansions in which the functions $\sin kx$ and $\cos kx$ are replaced by some sequence $\{\phi_k\}_{k=1, 2, 3 \ldots}$ of functions, all of which are mutually orthogonal. Recall that orthogonality of the trigonometric system amounts to having an integral of the product of two distinct functions equal to zero. The same is true for other systems of orthogonal functions, but since the precise integral that is to be computed varies from system to system, it is convenient to introduce a general notation for treating all systems at once. To the extent that the different systems have common features, the uniform notation allows us to display some of their similarities more clearly; furthermore, some formal geometric intuition can then be applied. To be specific, we define an **inner product** of real-valued functions $f(x)$ and $g(x)$, defined for $-\pi \le x \le \pi$, by

6.1
$$\langle f, g \rangle = \int_{-\pi}^{\pi} f(x)g(x) \, dx.$$

It follows by direct computation (see Section 2) that, if we set $\phi_1(x) = 1/\sqrt{2\pi}$, $\phi_{2n}(x) = (\cos nx)/\sqrt{\pi}$, and $\phi_{2n+1}(x) = (\sin nx)/\sqrt{\pi}$, then

$$\langle \phi_k, \phi_l \rangle = \begin{cases} 0, & k \ne l, \\ 1, & k = l. \end{cases}$$

Thus the sequence $\{\phi_k\}_{k=1, 2, \ldots}$ is not only **orthogonal** in the sense that $\langle \phi_k, \phi_l \rangle = 0$ when $k \ne l$, but is **orthonormal.** The "normal" part of the terminology comes from having $\langle \phi_k, \phi_k \rangle = 1$. This normalization comes, in the case of the trigonometric functions, from dividing by $\sqrt{\pi}$.

It turns out that many, although by no means all, consequences of defining the given inner product can be derived from four properties of a **general inner product.**

6.2 Positivity: $\langle f, f \rangle > 0$ except that $\langle 0, 0 \rangle = 0$.
Symmetry: $\langle f, g \rangle = \langle g, f \rangle$.
Additivity: $\langle f + g, h \rangle = \langle f, h \rangle + \langle g, h \rangle, \langle f, g + h \rangle = \langle f, g \rangle + \langle f, h \rangle$.
Homogeneity: $\langle cf, g \rangle = c\langle f, g \rangle, \langle f, cg \rangle = c\langle f, g \rangle$.

These properties can easily be verified for the particular inner product defined in Equation 6.1. For example, symmetry is just the statement that

$$\int_{-\pi}^{\pi} f(x)g(x)\,dx = \int_{-\pi}^{\pi} g(x)f(x)\,dx.$$

One reason for remarking on these four properties here is that they are shared by the dot product of coordinate vectors. Thus, if we define $\langle \mathbf{x}, \mathbf{y}\rangle = x_1 y_1 + x_2 y_2 + x_3 y_3$, where $\mathbf{x} = (x_1, x_2, x_3)$, $\mathbf{y} = (y_1, y_2, y_3)$, then it is also easy to check that this particular definition of $\langle \mathbf{x}, \mathbf{y}\rangle$ satisfies the relations 6.2 with $\mathbf{x}$ replacing f and $\mathbf{y}$ replacing g. With any definition of $\langle f, g\rangle$ that satisfies the relation 6.2, we can define a **norm** or **length** of a real-valued function f such that f^2 is integrable by

$$\|f\| = \langle f, f\rangle^{1/2}.$$

For our particular definition this amounts to

$$\|f\| = \left(\int_{-\pi}^{\pi} f^2(x)\,dx \right)^{1/2}.$$

More generally, we may consider functions $f(x)$ and $g(x)$ that are continuous for $a \le x \le b$. We denote the set of such functions by $C[a,b]$, and we define

$$\langle f, g\rangle = \int_a^b f(x)\,g(x)\,dx, \quad \|f\| = \left(\int_a^b f^2(x)\,dx \right)^{1/2}$$

for f and g in $C[a, b]$.

EXAMPLE 1   Let $f(x)$ and $g(x)$ be integrable functions. We could compute $\langle f + g, f - g\rangle$ directly from the definition (6.1) of $\langle,\rangle$. Alternatively, we can use the symmetry, additivity, and homogeneity properties:

$$\begin{aligned}
\langle f + g, f - g\rangle &= \langle f + g, f\rangle + \langle f + g, -g\rangle \\
&= \langle f, f\rangle + \langle g, f\rangle + \langle f, -g\rangle + \langle g, -g\rangle \\
&= \langle f, f\rangle + \langle f, g\rangle - \langle f, g\rangle - \langle g, g\rangle \\
&= \langle f, f\rangle - \langle g, g\rangle.
\end{aligned}$$

This much is true regardless of how f and g are chosen. If it should happen that $f^2(x) = g^2(x)$, for example $f(x) = x$, $g(x) = |x|$, then $\langle f, f\rangle = \langle g, g\rangle$; so we conclude that $\langle f + g, f - g\rangle = 0$. In other words, $f + g$ and $f - g$ are orthogonal. The conclusion is of course true also whenever $\langle f, f\rangle = \langle g, g\rangle$.

To see the importance of orthonormal sequences in general, we consider the following problem: Let $\langle f, g\rangle$ be an inner product, and let $\{\phi_k\}_{k=1, 2, \dots}$ be a sequence of elements, orthonormal with respect to the inner product. Using the norm defined by $\|f\| = \langle f, f\rangle^{1/2}$, we try to determine coefficients c_k, $k = 1, 2, \dots, N$, such that

$$\left\| g - \sum_{k=1}^{N} c_k \phi_k \right\|$$

is minimized for given g and N. The fact that the sequence ϕ_k is orthonormal makes the solution very simple; by adding and subtracting $\sum_{k=1}^{N} \langle g, \phi_k \rangle^2$ at the next-to-last step, we get

$$
\begin{aligned}
0 \le \left\| g - \sum_{k=1}^{N} c_k \phi_k \right\|^2 &= \left\langle g - \sum_{k=1}^{N} c_k \phi_k, \; g - \sum_{k=1}^{N} c_k \phi_k \right\rangle \\
&= \|g\|^2 - 2\sum_{k=1}^{N} c_k \langle g, \phi_k \rangle + \sum_{k=1}^{N} c_k^2 \\
&= \|g\| - \sum_{k=1}^{N} \langle g, \phi_k \rangle^2 \\
&\quad + \sum_{k=1}^{N} \left[\langle g, \phi_k \rangle^2 - 2c_k \langle g, \phi_k \rangle + c_k^2 \right] \\
&= \|g\|^2 - \sum_{k=1}^{N} \langle g, \phi_k \rangle^2 + \sum_{k=1}^{N} \left[\langle g, \phi_k \rangle - c_k \right]^2.
\end{aligned} \qquad (1)
$$

But the first two terms in the last expression are independent of the choice of the c_k's and the last sum is then minimized by taking $c_k = \langle g, \phi_k \rangle$. The numbers $\langle g, \phi_k \rangle$ are called the **Fourier coefficients** of g with respect to the orthonormal sequence $\{\phi_k\}$. Notice that the simplicity of the answer to the problem depends very much on the orthogonality of the ϕ_k's.

We can summarize what has just been proved as follows.

6.3   Theorem.   Let $\{\phi_k\}_{k=1, 2, \ldots}$ be an orthonormal sequence in a set with an inner product. Then, given an element g of the space, the distance

$$
\left\| g - \sum_{k=1}^{N} c_k \phi_k \right\|
$$

is minimized for $N = 1, 2, \ldots$ by taking c_k to be the Fourier coefficient $\langle g, \phi_k \rangle$.

The important thing about the conclusion of Theorem 6.3 is that the c_k's are uniquely determined, independently of N. In other words, if we wanted to improve the closeness of the approximation to g by increasing N, then Theorem 6.3 says that the c_k's already computed are to be left unchanged, and it is only necessary to compute additional coefficients $c_{N+1} = \langle g, \phi_{N+1} \rangle$, and so on.

As a by-product of the proof of Theorem 6.3, we have the following.

6.4   Bessel's Inequality.

$$
\|g\|^2 \ge \sum_{k=1}^{\infty} \langle g, \phi_k \rangle^2.
$$

Proof. The inequality

$$0 \le \|g\|^2 - \sum_{k=1}^{N} \langle g, \phi_k \rangle^2 + \sum_{k=1}^{N} [\langle g, \phi_k \rangle - c_k]^2$$

was established in the last step of equation (1). On taking $c_k = \langle g, \phi_k \rangle$, the inequality becomes

$$0 \le \|g\|^2 - \sum_{k=1}^{N} \langle g, \phi_k \rangle^2.$$

Bessel's inequality follows by letting N tend to infinity.

EXAMPLE 2 The approximation of g by a sum $\sum_{k=1}^{N} c_k \phi_k$ has been measured by a norm in Theorem 6.3. To see what this means for approximation by trigonometric polynomials, we use the inner product given by equation 6.1 on the space $C[-\pi, \pi]$ of continuous functions on $[-\pi, \pi]$. Given the orthonormal sequence $\phi_1(x) = 1/\sqrt{2\pi}$, $\phi_{2n}(x) = (\cos nx)/\sqrt{\pi}$, $\phi_{2n+1}(x) = (\sin nx)/\sqrt{\pi}$, we try to minimize, for given g in $C[-\pi, \pi]$, the norm

$$\left\| g - \sum_{k=1}^{2N+1} c_k \phi_k \right\|.$$

We have seen that this is done by taking $c_k = \langle g, \phi_k \rangle$. But, by the definition of the inner product,

$$\langle g, \phi_k \rangle = \begin{cases} \displaystyle\int_{-\pi}^{\pi} \frac{1}{\sqrt{2\pi}} g(t)\, dt, & k = 1, \\[2ex] \displaystyle\int_{-\pi}^{\pi} g(t) \frac{\cos nt}{\sqrt{\pi}}\, dt, & k = 2n, \\[2ex] \displaystyle\int_{-\pi}^{\pi} g(t) \frac{\sin nt}{\sqrt{\pi}}\, dt, & k = 2n + 1. \end{cases}$$

Hence the terms $c_k \phi_k$ become

$$\langle g, \phi_1 \rangle \phi_1(x) = \frac{1}{2\pi} \int_{-\pi}^{\pi} g(t)\, dt,$$

$$\langle g, \phi_{2n} \rangle \phi_{2n}(x) = \left[\frac{1}{\pi} \int_{-\pi}^{\pi} g(t) \cos nt\, dt \right] \cos nx,$$

$$\langle g, \phi_{2n+1} \rangle \phi_{2n+1}(x) = \left[\frac{1}{\pi} \int_{-\pi}^{\pi} g(t) \sin nt\, dt \right] \sin nx.$$

Then

$$\sum_{k=1}^{2N+1} c_k \phi_k(x) = \frac{a_0}{2} + \sum_{k=1}^{N} (a_k \cos kx + b_k \sin kx),$$

where a_k and b_k are the trigonometric Fourier coefficients as defined in Section 2. The square of the norm to be minimized takes the form

$$\int_{-\pi}^{\pi} \left[g(x) - \frac{a_0}{2} - \sum_{k=1}^{N} (a_k \cos kx + b_k \sin kx) \right]^2 dx,$$

and Theorem 6.3 says that the minimum will be attained for any fixed N by taking a_k and b_k to be the Fourier coefficients of g.

The minimization of an integral of the form

$$\int_{a}^{b} \left[g(x) - \sum_{k=1}^{N} c_k \phi_k(x) \right]^2 dx$$

is called a best **mean-square approximation** to g. In this sense we can say that the Fourier approximation provides the best mean-square approximation by a trigonometric polynomial.

EXERCISES

1. **(a)** Verify that

$$\langle f, g \rangle = \int_{a}^{b} f(x)g(x) \, dx$$

   defines an inner product on the set $C[a, b]$ of continuous functions defined on $[a, b]$; that is, verify Equations 6.2 of the text.
   (b) What condition is required of a continuous function w defined on $[a, b]$ in order that

$$\langle f, g \rangle = \int_{a}^{b} f(x)g(x)w(x) \, dx$$

   define an inner product?
2. Let $\{\phi_k\}$ be an orthonormal sequence of functions in $C[a, b]$, and let g be in $C[a, b]$. Show that if the real-valued function

$$\Delta(c_1, \ldots, c_N) = \int_{a}^{b} \left[g(x) - \sum_{k=1}^{N} c_k \phi_k(x) \right]^2 dx$$

   has a local minimum as a function of $(c_1, \ldots, c_N)$, then

$$c_k = \int_{a}^{b} g(x)\phi_k(x) \, dx.$$

   (*Hint*: Differentiate the formula for Δ under the integral sign. This is justified under the assumptions made here.)
3. Let $\{\phi_k\}$ be an orthonormal sequence of functions in $C[a, b]$ and let g be in $C[a, b]$. Use Bessel's inequality to show the following:
   (a) $\displaystyle\sum_{k=1}^{\infty} \left[\int_{a}^{b} g(x)\phi_k(x) \, dx \right]^2$ converges.

(b) If c_k is the kth Fourier coefficient of g with respect to $\{\phi_k\}$, then $\lim_{k\to\infty} c_k = 0$.

(c) Find an example of a function whose trigonometric Fourier series has coefficients a_k and b_k such that $\sum_{k=1}^{\infty} (a_k^2 + b_k^2)$ converges, but such that $\sum_{k=1}^{\infty} a_k$ or $\sum_{k=1}^{\infty} b_k$ does not converge.

4. Use the properties 6.2 of a general inner product, together with its relation to length, to prove the following relations. Can you give a geometric interpretation for (b) and (c)?

(a) $\|f - g\|^2 = \|f\|^2 + \|g\|^2 - 2\langle f, g\rangle$.

(b) $\|f + g\|^2 + \|f - g\|^2 = 2\|f\|^2 + 2\|g\|^2$.

(c) $\|f - g\|^2 = \|f - h\|^2 + \|h - g\|^2$ if and only if $\langle f - h, h - g\rangle = 0$.

5. The nth **Chebyshev polynomial** is defined by

$$T_n(x) = \cos(n \arccos x), \quad -1 \le x \le 1.$$

(a) Compute $T_n(x)$ for $n = 0, 1, 2, 3$.

(b) Show that $|T_n(x)| \le 1$ for $-1 \le x \le 1$.

(c) Verify that the relation

$$\langle f, g\rangle = \int_{-1}^{1} f(x)\, g(x)\, (1 - x^2)^{-1/2}\, dx$$

defines an inner product on $C[-1, 1]$ properties 6.2 of Section 6A. [Hint: Make the change of variable $x = \cos\theta$.]

(d) Prove the orthogonality relations

$$\int_{-1}^{1} T_k(x)\, T_l(x)\, (1 - x^2)^{-1/2}\, dx = \begin{cases} 0, & k \ne l, \\ \pi, & k = l \ne 0. \end{cases}$$

(*Hint:* Make the change of variable $x = \cos\theta$.)

(e) Show that $T_n(x)$ is always a polynomial by temporarily setting $\arccos x = \theta$ and expanding the right side of $\cos n\theta = \mathrm{Re}\,(\cos\theta + i\sin\theta)^n$ by the binomial theorem.

6B Sturm–Liouville Theory

It is easy to check that the functions $\cos kx$ and $\sin kx$ are solutions of the differential equation $y'' = -k^2 y$. The result can be stated by saying that $\cos kx$ and $\sin kx$ are **eigenfunctions** of the differential operator $d^2/(dx^2)$, corresponding to the **eigenvalue** $-k^2$. More generally, let L be a linear transformation defined on some vector space and having its range in the same space. We say that a *nonzero* function f is an **eigenfunction** of L, corresponding to the **eigenvalue** λ if $Lf = \lambda f$ for some number λ.

To see a connection between eigenfunctions and orthogonal sets of functions, we need one more definition. Let $\mathcal{F}$ be a set of functions with an inner product $\langle f, g\rangle$. We assume that for every pair f_1, f_2 of functions in $\mathcal{F}$ all linear combinations $c_1 f_1 + c_2 f_2$ are also in $\mathcal{F}$ for constant c_1, c_2. Let L be a linear operator from $\mathcal{F}$ to $\mathcal{F}$. Then L is **symmetric** with respect to the inner product if $\langle Lf, g\rangle = \langle f, Lg\rangle$ for all functions f and g for which Lf and Lg are defined as elements of $\mathcal{F}$. In the following examples we denote the set of continuous functions on $[a, b]$ by $C[a, b]$.

EXAMPLE 3   Let $\langle f, g \rangle$ be the inner product defined by equation 6.1. If we let $Lf = d^2f/dx^2$, then it is clear that Lf is in $C[-\pi, \pi]$ only for those f in $C[-\pi, \pi]$ that happen to have continuous second derivatives. Thus, for L to be symmetric, we must have $\langle Lf, g \rangle = \langle f, Lg \rangle$ for all twice continuously differentiable f and g on $[-\pi, \pi]$. Equivalently, we must have, because of the definition of the inner product,

$$\int_{-\pi}^{\pi} f''(x)g(x)\, dx = \int_{-\pi}^{\pi} f(x)g''(x)\, dx.$$

Integration by parts twice shows that

$$\int_{-\pi}^{\pi} f''(x)g(x)\, dx = f'(\pi)g(\pi) - f'(-\pi)g(-\pi) - f(\pi)g'(\pi)$$

$$+ f(-\pi)g'(-\pi) + \int_{-\pi}^{\pi} f(x)g''(x)\, dx.$$

Hence to make L symmetric we restrict its domain to some subspace of $C[-\pi, \pi]$ for which the nonintegrated terms will always add up to zero. This can be done in several ways. For example, we may restrict L to the subset consisting of those functions h in $C[-\pi, \pi]$ for which $h(\pi) = h(-\pi) = 0$, or to the subset for which $h'(\pi) = h'(-\pi) = 0$. With either of these restrictions, L becomes symmetric. Notice that a restriction of the required type is a boundary condition, in that it specifies the values of f at the endpoints π and $-\pi$ of its domain of definition.

The connection between orthogonal functions and symmetric operators is as follows.

6.5 Theorem.   Let L be a symmetric linear operator defined on a set of functions with an inner product. If f_1 and f_2 are eigenfunctions of L corresponding to distinct eigenvalues λ_1 and λ_2, then f_1 and f_2 are orthogonal.

Proof.   We assume that

$$Lf_1 = \lambda_1 f_1, \qquad Lf_2 = \lambda_2 f_2,$$

and prove that $\langle f_1, f_2 \rangle = 0$. We have

$$\langle Lf_1, f_2 \rangle = \langle \lambda_1 f_1, f_2 \rangle = \lambda_1 \langle f_1, f_2 \rangle$$

and
$$\langle f_1, Lf_2 \rangle = \langle f_1, \lambda_2 f_2 \rangle = \lambda_2 \langle f_1, f_2 \rangle.$$

Because L is symmetric, $\langle Lf_1, f_2 \rangle = \langle f_1, Lf_2 \rangle$, so $\lambda_1 \langle f_1, f_2 \rangle = \lambda_2 \langle f_1, f_2 \rangle$ or $(\lambda_1 - \lambda_2)\langle f_1, f_2 \rangle = 0$. Since λ_1 and λ_2 are not equal, we must have $\langle f_1, f_2 \rangle = 0$.

We consider the **Sturm–Liouville** differential operator, namely one of the form

$$Lf = (pf')' + qf, \tag{2}$$

where p is a continuously differentiable function and q is assumed only continuous. [The operator d^2/dx^2 is a special case if we set $p(x) = 1$, $q(x) = 0$.] We want to see what boundary conditions should be imposed on the domain of L in order to make L symmetric with respect to the inner product

$$\langle f, g \rangle = \int_a^b f(x)g(x)\, dx.$$

The following formula simplifies the problem.

6.6 Lagrange Formula.   If L is given by $Lf = (pf')' + qf$ on an interval $[a, b]$, then

$$\langle f_1, Lf_2 \rangle - \langle Lf_1, f_2 \rangle = [p(f_1 f_2' - f_1' f_2)]_a^b.$$

Proof. Starting with the definition of L, we rearrange $f_1(Lf_2) - (Lf_1)f_2$ as follows:

$$\begin{aligned}
f_1(Lf_2) - (Lf_1)f_2 &= f_1[(pf_2')' + qf_2] - f_2[(pf_1')' + qf_1] \\
&= f_1(pf_2')' - f_2(pf_1')' \\
&= f_1[pf_2'' + p'f_2'] - f_2[pf_1'' + p'f_1'] \\
&= p'[f_1 f_2' - f_1' f_2] + p[f_1 f_2'' - f_1'' f_2] \\
&= [p(f_1 f_2' - f_1' f_2)]'.
\end{aligned}$$

Integration of both sides from a to b gives the Lagrange formula.

Formula 6.6 shows that any condition on the coefficient p or on the space containing f_1 and f_2, which makes $[p(f_1 f_2' - f_1' f_2)]_a^b = 0$, will also make L symmetric.

EXAMPLE 4   The operator L, defined by $Lf(x) = (1 - x^2)f''(x) - 2xf'(x)$, has the form shown in equation (2) if we set $p(x) = (1 - x^2)$ and $q(x) \equiv 0$. We will consider L to be operating on twice continuously differentiable functions defined on $[-1, 1]$. To make L symmetric, we need to ensure that the right side of the Lagrange formula is always zero for $a = -1$, $b = 1$. But $p(x) = (1 - x^2)$, so $p(-1) = p(1) = 0$. Hence, L is symmetric on $C[-1, 1]$ without further restriction, and its domain consists of all twice continuously differentiable functions in $C[-1, 1]$.

The symmetric operator Lf defined in Example 4 is usually associated with the differential equation

$$(1 - x^2)y'' - 2xy' + n(n + 1)y = 0. \tag{3}$$

This is called the **Legendre equation** of index n, and it is satisfied by the nth

Legendre polynomial defined by

$$\textbf{6.7} \qquad P_n(x) = \frac{1}{2^n n!} \frac{d^n}{dx^n} (x^2 - 1)^n, \qquad n = 0, 1, 2, \ldots.$$

That P_n satisfies equation (3) can be verified by repeated differentiation. See Exercise 4. The significance of the fact that P^n satisfies the Legendre equation comes from writing the equation in the form $Ly = -n(n + 1)y$, where L is the symmetric operator $Ly = (1 - x^2)y'' - 2xy'$ on $C[-1, 1]$. Then P_n can be looked at as an eigenfunction of L, corresponding to the eigenvalue $-n(n + 1)$. Hence, by Theorem 6.5, the Legendre polynomials are orthogonal, that is,

$$\int_{-1}^{1} P_n(x)P_m(x) \, dx = 0, \qquad n \neq m.$$

Furthermore, a fairly complicated calculation (see Exercise 6) shows that

$$\int_{-1}^{1} P_n^2(x) \, dx = \frac{2}{2n + 1}, \qquad n = 0, 1, 2, \ldots.$$

Therefore, the normalized sequence $\{(\sqrt{2n + 1/2})P^n(x)\}$ $n = 0, 1, 2, \ldots$ is an orthonormal sequence in $C[-1, 1]$.

The Nth **Fourier–Legendre approximation** to a function g in $C[-1, 1]$ is the finite sum

$$\sum_{k=0}^{N} c_k P_k(x),$$

where

$$\textbf{6.8} \qquad c_k = \frac{2k + 1}{2} \int_{-1}^{1} g(x) P_k(x) \, dx.$$

Theorem 6.3 then implies that the best mean-square approximation to g by a linear combination of Legendre polynomials is given by the Fourier–Legendre approximation. In other words,

$$\int_{-1}^{1} \left[g(x) - \sum_{k=0}^{N} c_k P_k(x) \right]^2 dx$$

is minimized by computing c_k by equation (6.8).

EXERCISES

1. Find all eigenfunctions and corresponding eigenvalues of the differential operator d^2/dx^2 satisfying each of the following sets of boundary conditions. That is, solve $y'' = \lambda y$, subject to the following boundary conditions:

(a) $y(0) = y(\pi) = 0.$
(b) $y'(0) = y'(\pi).$
(c) $y(0) = y'(\pi) = 0.$

2. Find all eigenfunctions and corresponding eigenvalues of the differential operator

$$L = \frac{d^2}{dx^2} + 2\frac{d}{dx}$$

that satisfy each of the following sets of boundary conditions. That is, solve $L(y) = \lambda y$ subject to the following boundary conditions:
(a) $y(0) = y(\pi) = 0.$
(b) $y(0) = y(1) = 0.$
(c) $y'(0) = y'(\pi) = 0.$

3. Show that the differential operator d^2/dx^2 is symmetric with respect to the inner product

$$\langle f, g \rangle = \int_{-1}^{1} f(x)g(x)\, dx,$$

and with each of the following sets of boundary conditions.
(a) $f(-1) = f(1) = 0.$
(b) $f'(1) = f'(-1) = 0.$
(c) $c_1 f(-1) + d_1 f'(-1) = c_2 f(1) + d_2 f'(1) = 0$, where $c_i^2 + d_i^2 > 0.$

4. Verify that the Legendre polynomial P_n defined by Formula 6.7 satisfies the Legendre equation: $(1 - x^2)y'' - 2xy' + n(n + 1)y = 0.$ *Hint:* Let $u = (x^2 - 1)^n.$ Then $(x^2 - 1)u' = 2nxu.$ Differentiate both sides $(n + 1)$ times with respect to $x.$ Use the Leibnitz rule for the derivative of a product:

$$(fg)^{(n)} = \sum_{k=0}^{n} \binom{n}{k} f^{(k)} g^{(n-k)}.$$

5. Compute the Legendre polynomials P_0, P_1, and P_2.

6. (a) By using Formula 6.7 and repeated integration by parts, show that

$$\int_{-1}^{1} P_n^2(x)\, dx = \frac{(2n)!}{2^{2n}(n!)^2} \int_{-1}^{1} (1 - x^2)^n\, dx.$$

(b) Show that

$$\int_{-1}^{1} (1 - x^2)^n\, dx = 2 \int_{0}^{\pi/2} \sin^{2n+1} \theta\, d\theta.$$

(c) Show that

$$\int_{0}^{\pi/2} \sin^{2n+1} \theta\, d\theta = \frac{2 \cdot 4 \cdot 6 \cdot \cdots \cdot (2n)}{1 \cdot 3 \cdot 5 \cdot \cdots \cdot (2n + 1)}.$$

(d) Show that

$$\int_{-1}^{1} P_n^2(x)\, dx = \frac{2}{2n + 1}.$$

7. Prove that P_n, the nth Legendre polynomial, has n distinct roots in the interval $[-1, 1].$ (*Hint:* Use Formula 6.7 and Rolle's theorem.)

8. The three-dimensional Laplace equation in spherical coordinates (r, ϕ, θ) has the form

$$\frac{\partial}{\partial r}\left(r^2\frac{\partial u}{\partial r}\right) + \frac{1}{\sin\phi\,d\phi}\frac{\partial}{\partial \phi}\left(\sin\phi\frac{\partial u}{\partial\phi}\right) + \frac{1}{\sin^2\phi}\frac{\partial^2 u}{\partial\theta^2} = 0.$$

(a) Show that for solutions $u(r, \phi, \theta) = v(r, \phi)$ that are independent of θ, the equation has the form

$$r\frac{\partial^2}{\partial r^2}(rv) + \frac{1}{\sin\phi\,\partial\phi}\frac{\partial}{\partial\phi}\left(\sin\phi\frac{\partial v}{\partial\phi}\right) = 0.$$

(b) Show that the method of separation of variables applied to the equation of part (a) leads to the two ordinary differential equations

$$r^2 G'' + 2rG' = \lambda G,$$

$$\frac{1}{\sin\phi\,d\phi}\frac{d}{d\phi}\left(\sin\phi\frac{dH}{d\phi}\right) = -\lambda H.$$

(c) Show that the equation for $G(r)$ has solutions

$$r^{-(1/2)+\sqrt{\lambda+(1/4)}} \quad \text{and} \quad r^{-(1/2)-\sqrt{\lambda+(1/4)}}.$$

(*Hint:* Let $r = e^t$.)

(d) Show that the equation for H can be put in the form of the Legendre equation

$$(1 - x^2)H'' - 2xH' + \lambda H = 0.$$

(*Hint:* Let $x = \cos\phi$.)

(e) By setting $\lambda = n(n + 1)$, find a sequence of solutions to the partial differential equation of part (a).

9. Show that, in the case of the orthonormal sequence derived from the Legendre polynomials, the general expansion $\sum_{k=0}^{N}\langle g, \phi_k\rangle\phi_k$ reduces to $\sum_{k=0}^{N}c_k P_n(x)$, where c_k is given by equation (6.8) of the text.

10. Let $A(x)$, $B(x)$ and $C(x)$ be continuous on an interval, with $A(x) \neq 0$. Show that the expression

$$Ay'' + By' + Cy$$

can be written in the Sturm-Liouville form

$$(py')' + qy$$

after multiplication by M/A, where

$$M(x) = \exp\left(\int\frac{B(x)}{A(x)}\,dx\right).$$

What are p and q in terms of A, B and C?

9
Infinite Series

The study of numerical infinite series is an important branch of the theory of numerical approximation and can be regarded as the study of the limit behavior of numerical sequences. The techniques used for treating such limits are somewhat different from those usually encountered in calculus. However, calculus can be introduced as a powerful device for dealing with them through the study of infinite series of functions. The first two sections treat numerical series, and the remaining five sections show how to work with some special series of functions. Formally, an infinite series generalizes the idea of finite sums to infinite sums of the form

$$a_1 + a_2 + a_3 + \cdots,$$

where the three dots indicate that a term a_k is included for each of the infinitely many positive integers $k = 1, 2, 3, \ldots.$

1 EXAMPLES AND DEFINITIONS

1A Convergence and Divergence

The decimal expansion

$$\frac{1}{3} = 0.3333 \ldots$$

is really an example of an infinite series, because it can very naturally be written as an infinite sum in the form

$$\frac{1}{3} = \frac{3}{10} + \frac{3}{100} + \frac{3}{1000} + \frac{3}{10,000} + \cdots$$

$$= \frac{3}{10} + \frac{3}{10^2} + \frac{3}{10^3} + \frac{3}{10^4} + \cdots .$$

In both decimal and sum form, the dots at the end mean that the most obvious pattern is to be carried on indefinitely. To write infinite series more briefly, recall the Σ notation for finite sums, defined by

$$\sum_{k=m}^{n} a_k = a_m + a_{m+1} + \cdots + a_n, \qquad m \le n.$$

For example,

$$\sum_{k=1}^{n} k = 1 + 2 + 3 + \cdots + n.$$

A natural extension allows us to write a general *infinite* series in more compressed form like this:

$$a_1 + a_2 + a_3 + \cdots = \sum_{k=1}^{\infty} a_k.$$

Thus, for example,

$$\frac{3}{10} + \frac{3}{10^2} + \frac{3}{10^3} + \frac{3}{10^4} + \cdots = \sum_{k=1}^{\infty} \frac{3}{10^k}.$$

Assigning a natural numerical value to an infinite series can be done in many cases by finding the limit as $n \to \infty$ of the **nth partial sum** $s_n = \Sigma_{k=1}^{n} a_k$. Thus, by definition, the **sum** s of a series is

$$\sum_{k=1}^{\infty} a_k = \lim_{n \to \infty} \sum_{k=1}^{n} a_k = s,$$

if the limit exists. If the series has a sum in this sense, then the series is said to **converge** to s, and if the limit fails to exist, the series **diverges.**

EXAMPLE 1   Consider the **geometric series** with ratio r,

$$\sum_{k=0}^{\infty} r^k = 1 + r + r^2 + \cdots .$$

There is a simple formula for the partial sums if $r \ne 1$, given by

$$1 + r + r^2 + \cdots + r^n = \frac{1 - r^{n+1}}{1 - r} .$$

To verify the formula, multiply both sides by $1 - r$. If $|r| < 1$, then r^{n+1} gets smaller and in fact tends to 0 as $n \to \infty$. Thus the sum is

$$\sum_{k=0}^{\infty} r^k = \frac{1}{1-r}, \qquad \text{if } -1 < r < 1.$$

If $|r| > 1$, then r^{n+1} is unbounded as $n \to \infty$, so the sum fails to exist. If $r = -1$, the formula for s_n gives 0 if n is odd and 1 if n is even, so there is no limit then either. Finally, if $r = 1$ the formula for s_n is invalid, but in that case we see directly that $s_n = n + 1$, which tends to ∞ as $n \to \infty$.

We can get a good idea of convergence to a sum by looking at specific numerical examples. In each of the next three examples we can actually find a simple formula for the sequence of partial sums, something that is not possible for many commonly encountered infinite series.

EXAMPLE 2   Consider the geometric series with $r = \frac{1}{3}$. We have

$$\sum_{k=0}^{\infty} \frac{1}{3^k} = 1 + \frac{1}{3} + \frac{1}{3^2} + \frac{1}{3^3} + \cdots$$

$$= \lim_{n \to \infty} \sum_{k=0}^{n} \frac{1}{3^k} = \lim_{n \to \infty} \frac{1 - (1/3^{n+1})}{1 - (1/3)}$$

$$= \frac{1 - 0}{1 - (1/3)} = \frac{3}{2}.$$

EXAMPLE 3   Here is a particularly simple series.

$$\sum_{k=1}^{\infty} \left(\frac{1}{k} - \frac{1}{k+1} \right) = \lim_{n \to \infty} \sum_{k=1}^{n} \left(\frac{1}{k} - \frac{1}{k+1} \right)$$

$$= \lim_{n \to \infty} \left(1 - \frac{1}{2} \right) + \left(\frac{1}{2} - \frac{1}{3} \right) + \cdots + \left(\frac{1}{n} - \frac{1}{n+1} \right)$$

$$= \lim_{n \to \infty} 1 - \frac{1}{n+1} = 1 - \lim_{n \to \infty} \frac{1}{n+1} = 1 - 0 = 1.$$

This is an example of what is called a "telescoping" series, because the interior terms cancel in the partial sum. Because $(1/k) - (1/k+1) = 1/k(k+1)$, the series can also be written

$$\sum_{k=1}^{\infty} \frac{1}{k(k+1)} = 1.$$

EXAMPLE 4   The infinite series

$$\sum_{k=1}^{\infty} k = 1 + 2 + 3 + \cdots$$

is divergent. To see this, note that the nth partial sum is the sum of the first n integers:

$$s_n = 1 + 2 + \cdots + n = \frac{n(n + 1)}{2}.$$

Hence $\lim_{n \to \infty} s_n = \infty$.

EXAMPLE 5   The geometric series with $r = -2$ is

$$\sum_{k=0}^{\infty} (-2)^k = 1 - 2 + 2^2 - 2^3 + \cdots,$$

and the nth partial sum is, according to the formula in Example 1,

$$s_n = \frac{1 - (-2)^{n+1}}{1 + 2} = \frac{1 + (-1)^n 2^n}{3}.$$

As $n \to \infty$, the numerator oscillates between being large and positive and large and negative, so the series diverges.

1B Convergence of Sequences

For infinite series in general, we need a theory of convergence in which convergence can be demonstrated without actually identifying the sum of the series. The key property of real numbers that we assume is the following.

1.1. Nondecreasing Sequence Principle   Let $\{s_n\}$, $n = 1, 2, 3 \ldots$ be a nondecreasing sequence of real numbers, so that $s_n \leq s_{n+1}$ for $n = 1, 2, 3, \ldots$. Then either there is a finite upper bound b such that $s_n \leq b$ for all n, in which case

$$\text{(i)} \quad \lim_{n \to \infty} s_n = s, \qquad \text{for some number } s \leq b,$$

or else there is no finite upper bound, in which case

$$\text{(ii)} \quad \lim_{n \to \infty} s_n = +\infty.$$

This principle is a fundamental property of real numbers and can be proved from certain even more fundamental principles. We prefer to accept it here as a plausible geometric fact. Figure 1 illustrates both cases of the principle.

EXAMPLE 6   Every decimal expansion of the form

$$.b_1 b_2 b_3 \cdots,$$

where $0 \leq b_k \leq 9$, is really an infinite series with kth term $a_k = b_k 10^{-k}$. We observe that

$$\lim_{n \to \infty} s_n = s \leqslant b$$

(a)

$$\lim_{n \to \infty} s_n = \infty$$

(b)   **Figure 1**

$$s_n = \sum_{k=1}^{n} \frac{b_k}{10^k} \leq \sum_{k=1}^{n} \frac{9}{10^k}$$

$$= 9 \sum_{k=1}^{n} \frac{1}{10^k} = 9\left(-1 + \frac{1 - 10^{-n-1}}{1 - (1/10)}\right) = 1 - \frac{1}{10^n}.$$

Note that $s_{n+1} = s_n + b_{n+1}/10^{n+1} \geq s_n$, and that $s_n \leq b = 1$. Hence the nondecreasing sequence principle asserts that the series $\sum_{k=1}^{\infty} b_k 10^{-k}$, and hence the related decimal expansion, converges to some number $s \leq 1$. This is of course what we expect; the largest number we can represent in the decimal form $.b_1 b_2 b_3 \cdots$ is

$$1 = .9999 \cdots \cdot$$

EXAMPLE 7   The infinite series with kth term $a_k = k^{-1} 2^{-k}$ has nth partial sum

$$s_n = \sum_{k=1}^{n} \frac{1}{k 2^k} \leq \sum_{k=0}^{n} \frac{1}{2^k}$$

$$= 2(1 - 2^{-n-1}) \leq 2.$$

Since $s_{n+1} = s_n + 2^{-n-1}/(n + 1) > s_n$, and $s_n \leq b = 2$, the nondecreasing sequence principle shows that the series has a sum

$$s = \sum_{k=1}^{\infty} \frac{1}{k 2^k} \leq 2.$$

The equations

$$\lim_{n \to \infty} s_n = s \quad \text{and} \quad \lim_{n \to \infty} s_n - s = 0$$

are completely equivalent, and very often it is simpler to show that a sequence of numbers $s_n - s$ tends to 0 than to show that s_n tends to s. Here we record two important kinds of sequence that have zero for a limit.

$$\text{(a)} \quad \lim_{n \to \infty} \frac{1}{n^a} = 0, \qquad \text{if } a > 0.$$

1.2

$$\text{(b)} \quad \lim_{n \to \infty} \frac{1}{b^n} = 0, \qquad \text{if } |b| > 1.$$

The reason in each case is that the denominator, or its absolute value in (b), tends to infinity as $n \to \infty$.

1C Sums and Numerical Multiples

The general distributive and associative laws for finite sums can be written

$$c \sum_{k=1}^{n} a_k = \sum_{k=1}^{n} ca_k, \quad \text{and} \quad \sum_{k=1}^{n} a_k + \sum_{k=1}^{n} b_k = \sum_{k=1}^{n} (a_k + b_k).$$

Both laws extend to convergent infinite series as follows.

1.3 Theorem.   Suppose c is a fixed real number and that

$$\sum_{k=1}^{\infty} a_k \quad \text{and} \quad \sum_{k=1}^{\infty} b_k$$

are convergent series of real numbers. Then the series with kth terms ca_k and $a_k + b_k$, respectively, are convergent also, with

$$c \sum_{k=1}^{\infty} a_k = \sum_{k=1}^{\infty} ca_k, \quad \text{and} \quad \sum_{k=1}^{\infty} a_k + \sum_{k=1}^{\infty} b_k = \sum_{k=1}^{\infty} (a_k + b_k).$$

Proof.   Use of the distributive and associative laws for finite sums shows that the proof reduces to showing that if

$$s_n = \sum_{k=1}^{n} a_k, \quad \text{and} \quad t_n = \sum_{k=1}^{n} b_k,$$

then

$$c \lim_{n \to \infty} s_n = \lim_{n \to \infty} cs_n, \quad \text{and} \quad \lim_{n \to \infty} s_n + \lim_{n \to \infty} t_n = \lim_{n \to \infty} (s_n + t_n).$$

But these limit relations follow at once from the familiar and more general analogues for a continuous variable x:

$$c \lim_{x \to \infty} f(x) = \lim_{x \to \infty} cf(x), \quad \text{and} \quad \lim_{x \to \infty} f(x) + \lim_{x \to \infty} g(x) = \lim_{x \to \infty} f(x) + g(x).$$

EXAMPLE 8   We have, by Theorem 1.3,

(a) $3 \displaystyle\sum_{k=0}^{\infty} 2^{-k} = \sum_{k=0}^{\infty} 3 \cdot 2^{-k}.$

(b) $\displaystyle\sum_{k=0}^{\infty} 2^{-k} + \sum_{k=0}^{\infty} 3^{-k} = \sum_{k=0}^{\infty} (2^{-k} + 3^{-k}).$

(c) $2 \sum\limits_{k=0}^{\infty} 2^{-k} + 3 \sum\limits_{k=0}^{\infty} 3^{-k} = \sum\limits_{k=0}^{\infty} (2^{-k+1} + 3^{-k+1}).$

Here is a useful theorem.

1.4 Theorem.   Altering any finite number of terms in an infinite series has no effect on whether the series converges or diverges, although the sum of a convergent series may be changed by such an alteration.

Proof. Suppose all changes occur among the first M terms, replacing a_k by a'_k for $k = 1, 2, \ldots, M$. Thus, for $n > M$, the new partial sums s'_n differ from s_n by a fixed amount, $s'_n - s_n = d$, independent of n. Hence,

$$\lim_{n \to \infty} s'_n = d + \lim_{n \to \infty} s_n,$$

so the two series converge and diverge together.

EXAMPLE 9   We saw in Example 2 that

$$\sum_{k=0}^{\infty} \frac{1}{3^k} = 1 + \frac{1}{3} + \frac{1}{3^2} + \frac{1}{3^3} + \cdots = \frac{3}{2}.$$

Leaving out the first two terms, we get

$$\sum_{k=2}^{\infty} \frac{1}{3^k} = \frac{1}{3^2} + \frac{1}{3^3} + \cdots = \frac{3}{2} - 1 - \frac{1}{3} = \frac{1}{6}.$$

Increasing the first term by 2, we get

$$(1 + 2) + \sum_{k=1}^{\infty} \frac{1}{3^k} = 3 + \frac{1}{3} + \frac{1}{3^2} + \frac{1}{3^3} + \cdots = \frac{3}{2} + 2 = \frac{7}{2}.$$

EXERCISES

1. Using the definition of the Σ notation by

$$\sum_{k=m}^{n} a_k = a_m + a_{m+1} + \cdots + a_n,$$

write out and simplify the following:

(a) $\sum\limits_{k=1}^{5} \left(\frac{1}{k} - \frac{1}{k+1} \right).$

(b) $\sum\limits_{k=2}^{5} k - \sum\limits_{k=3}^{6} (k - 1).$

(c) $\sum\limits_{k=0}^{4} 2^{-k}.$

(d) $6 \sum\limits_{k=1}^{11} (-1)^k.$

2. Find a formula for the kth term, $k = 1, 2, 3, \ldots$, of an infinite series that is consistent with each of the following:

(a) $1 + \dfrac{1}{2} + \dfrac{1}{3} + \dfrac{1}{4} + \cdots$

(b) $\dfrac{1}{3} + \dfrac{1}{2 \cdot 3^2} + \dfrac{1}{3 \cdot 3^3} + \dfrac{1}{4 \cdot 3^4} + \cdots$

(c) $1 - \dfrac{1}{2} + \dfrac{1}{4} - \dfrac{1}{8} + \cdots$

(d) $\dfrac{1}{1 \cdot 3} + \dfrac{1}{2 \cdot 4} + \dfrac{1}{3 \cdot 5} + \dfrac{1}{4 \cdot 6} + \cdots$

(e) $\dfrac{4}{2 \cdot 3^2} + \dfrac{5}{3 \cdot 3^3} + \dfrac{6}{4 \cdot 3^4} + \cdots$

3. Verify that each of the following partial sum formulas is correct, and, if it exists, find the sum of the corresponding infinite series as $n \to \infty$.

(a) $\displaystyle\sum_{k=1}^{n} \frac{1}{2k(2k + 2)} = \frac{n}{4(n + 1)}.$

(b) $\displaystyle\sum_{k=0}^{n} \frac{3}{4^k} = 4 - 4^{-n}.$

(c) $\displaystyle\sum_{k=1}^{n} 2k = n(n + 1).$

(d) $\displaystyle\sum_{k=0}^{n} \left(\frac{2^k + 1}{2^k} - 1 \right) = 2 - 2^{-n}.$

(e) $\displaystyle\sum_{k=3}^{n} 2^{-k} = 2^{-2} - 2^{-n}.$

(f) $\displaystyle\sum_{k=4}^{n} (-1)^k = \begin{cases} 1, & n \text{ even,} \\ 0, & n \text{ odd.} \end{cases}$

4. The nondecreasing sequence principle 1.1 applies to some of the following sequences, but not to others. In those cases in which it does apply, draw whatever conclusion you can from the principle.

(a) $s_n = 1 + \dfrac{1}{n^2}.$

(b) $s_n = \dfrac{n}{n + 1}.$

(c) $s_n = \dfrac{1 + n^2}{n}.$

(d) $s_n = \dfrac{2^n + 3}{2^n + 2}.$

5. What is $\lim\limits_{n \to \infty} s_n$ for each sequence in Exercise 4 for which a limit exists?

6. What is the sum of each of the following geometric series?

(a) $\displaystyle\sum_{k=0}^{\infty} \left(\frac{1}{6} \right)^k.$

(b) $\displaystyle\sum_{k=0}^{\infty} \left(-\frac{3}{4} \right)^k.$

(c) $\displaystyle\sum_{k=1}^{\infty} \left(\frac{1}{1 + \pi} \right)^k.$

(d) $\displaystyle\sum_{k=0}^{\infty} \left(\frac{1}{1 + x^2} \right)^k.$

(e) $\displaystyle\sum_{k=0}^{\infty} e^{-2k}.$

(f) $\displaystyle\sum_{k=0}^{\infty} (0.01)^k.$

7. Write each of the following decimal expansions as an infinite geometric series, and find the sum.

(a) $0.123123123 \ldots$ **(b)** $0.10101010 \ldots$ **(c)** $0.888888 \ldots$

8. Write each of the following as a single infinite series.

(a) $\displaystyle\sum_{k=1}^{\infty} 2^{-k} + \sum_{k=1}^{\infty} 2^{-k+1}.$

(b) $2 \displaystyle\sum_{k=0}^{\infty} 2^{-k} - 4 \sum_{k=0}^{\infty} 2^{-k-1}.$

(c) $\displaystyle\sum_{k=0}^{\infty} \left(\tfrac{2}{3}\right)^k - \sum_{k=0}^{\infty} \left(\tfrac{2}{3}\right)^{k+1}.$

(d) $\displaystyle\sum_{k=1}^{\infty} \frac{k}{k^2 + 1} - \sum_{k=1}^{\infty} \frac{1}{k^2 + 1}.$

9. Find an infinite series with positive terms that converges to 3.

10. The symbol $k!$, called **k factorial**, is defined by

$$k! = \begin{cases} 1, & k = 0, \\ k(k - 1) \ldots 3 \cdot 2 \cdot 1, & k > 1. \end{cases}$$

Thus $1! = 1$, $2! = 2 \cdot 1 = 2$, $3! = 3 \cdot 2 \cdot 1 = 6$, and so on. Rewrite each of the following infinite series using factorial notation. Then write out the first three terms.

(a) $\displaystyle\sum_{k=1}^{\infty} \frac{3^{-k}}{1 \cdot 2 \ldots k}$.

(b) $\displaystyle\sum_{k=0}^{\infty} \frac{1}{1 \cdot 2 \cdot 4 \ldots (2k)}$.

(c) $\displaystyle\sum_{k=1}^{\infty} \frac{1 \cdot 3 \cdot 5 \ldots (2k-1)}{2 \cdot 4 \cdot 6 \ldots (2k)}$.

(d) $\displaystyle\sum_{k=1}^{\infty} \frac{1}{k(k+1)(k+2) \ldots (2k-1)(2k)}$.

(e) $\displaystyle\sum_{k=1}^{\infty} \frac{2^k}{2 \cdot 4 \cdot 6 \ldots (2k)}$.

(f) $\displaystyle\sum_{k=0}^{\infty} \frac{(-1)^k}{1 \cdot 2 \cdot 3 \ldots (2k+1)}$.

11. Write out s_n for $n = 1, 2,$ and 3. Also find $\displaystyle\lim_{n \to \infty} s_n$.

(a) $s_n = \left(1 + \dfrac{1}{n}\right)$.

(b) $s_n = \left(2 - \dfrac{1}{n}\right)$.

(c) $s_n = \dfrac{3n^2 - 1}{4n^2 + n}$.

(d) $s_n = \dfrac{2^n}{3^{n+1}}$.

(e) $s_n = \dfrac{3n + 4}{n}$.

(f) $s_n = \dfrac{n^{10} + 1}{n(n^9 + 1)}$.

12. Find a formula for the nth entry s_n in a sequence that starts in each of the following ways. Then find $\displaystyle\lim_{n \to \infty} s_n$.

(a) $\frac{1}{2}, \frac{2}{3}, \frac{3}{4}, \frac{4}{5} \ldots$

(b) $\frac{3}{2}, \frac{4}{3}, \frac{5}{4}, \frac{6}{4}, \ldots$

(c) $\frac{8}{1}, \frac{10}{2}, \frac{12}{3}, \frac{14}{4}, \ldots$

(d) $\frac{2}{3}, \frac{4}{9}, \frac{8}{27}, \frac{16}{81}, \ldots$

13. Given an infinite *sequence* of numbers s_n for $n = 1, 2, 3, \ldots$, there is always an infinite *series* $\sum_{k=1}^{\infty} a_k$ that has s_n for its nth partial sum.

(a) Show that if $a_1 = s_1$ and $a_k = s_k - s_{k-1}$ for $k = 2, 3, 4, \ldots$, then

$$\sum_{k=1}^{n} a_k = s_n \qquad \text{for } n = 1, 2, 3, \ldots.$$

(b) Find a simple formula for the kth term a_k of an infinite series that has $s_n = 1 + (1/n)$ for its nth partial sum. What is the sum of $\sum_{k=1}^{\infty} a_k$?

14. Verify each of the following equations:

(a) $\displaystyle\sum_{k=1}^{\infty} \frac{1}{10^k} = \frac{1}{9}$.

(b) $\displaystyle\sum_{k=2}^{\infty} \frac{1}{3^k} = \frac{1}{6}$.

(c) $\displaystyle\sum_{k=1}^{\infty} \left(-\frac{1}{4}\right)^k = -\frac{1}{5}$.

(d) $\displaystyle\sum_{k=3}^{\infty} \frac{1}{k(k+1)} = \frac{1}{3}$.

15. From the interval $[0, 1]$, the middle third $(\frac{1}{3}, \frac{2}{3})$ is deleted. Then the middle third is deleted from each of the two remaining intervals, then the middle third from each of the four remaining intervals, and so on.

(a) Find a formula for the amount c_n remaining after n steps.

(b) What is $\lim_{n \to \infty} c_n$?

16. Let $\sum_{k=1}^{\infty} a_k = s$ be a convergent series. Show that $\sum_{k=m}^{\infty} a_k$ is also convergent if $m > 1$. What is the sum?

17. Let $\sum_{n=1}^{\infty} a_k$ be a series of nonnegative terms. Show that the series either converges or else **diverges to infinity**, in the sense that

$$\lim_{n \to \infty} \sum_{k=1}^{n} a_k = +\infty.$$

18. Show that if $a_k \geq 0$ and $\Sigma_{k=1}^{\infty} a_k$ converges then so does $\Sigma_{k=1}^{\infty} a_{2k}$ and $\Sigma_{k=0}^{\infty} a_{2k+1}$. Is the conclusion true without the condition $a_k \geq 0$?

2 CONVERGENCE TESTS

There is no universal criterion, other than the definition, for deciding about the convergence or divergence of infinite series. However, the tests explained next have been found to be useful for large classes of series that have practical importance.

2A Series with Nonnegative Terms

Here is the simplest test of all to apply, but the only conclusion that we can draw from it is *divergence* of a series.

2.1   Term Test for Divergence.   If $\Sigma_{k=1}^{\infty} a_k$ converges, then $\lim_{k\to\infty} a_k = 0$. In other words, if a_k fails to tend to 0 as $k\to\infty$, then the series diverges.

Proof. Let $s_n = \Sigma_{k=1}^{n} a_k$. By assumption, $\lim_{n\to\infty} s_n = s$, for some finite number s. Hence $\lim_{n\to\infty} s_{n-1} = s$ also. It follows from $s_n - s_{n-1} = a_n$ that

$$\lim_{n\to\infty} a_n = \lim_{n\to\infty} (s_n - s_{n-1})$$

$$= \lim_{n\to\infty} s_n - \lim_{n\to\infty} s_{n-1}$$

$$= s - s = 0.$$

EXAMPLE 1   The series

$$\sum_{k=1}^{\infty} \frac{k}{k + 1} = \frac{1}{2} + \frac{2}{3} + \frac{3}{4} + \cdots$$

has kth term $a_k = k/(k + 1)$. Since

$$\lim_{k\to\infty} a_k = \lim_{k\to\infty} \frac{k}{k + 1}$$

$$= \lim_{k\to\infty} \frac{1}{1 + (1/k)} = 1,$$

the series fails to converge because a_n does not tend to 0.

Warning. It is *not* true, just because $\lim_{n\to\infty} a_n = 0$, that the series $\Sigma_{k=1}^{\infty} a_k$ converges. The **harmonic series** $\Sigma_{k=1}^{\infty}(1/k)$ discussed in Example 2, following, is a counterexample, because $\lim_{k\to 0} 1/k = 0$, but the series diverges.

There are close analogies between integrals over an infinite interval and infinite series. Thus the **improper integral** of $f(x)$ from a to ∞ is **convergent** if it has a finite value determined by

$$\int_a^\infty f(x)dx = \lim_{b \to \infty} \int_a^b f(x)\,dx;$$

otherwise, the integral is **divergent.** For example

$$\int_1^\infty e^{-px}\,dx = \lim_{b \to \infty} \int_1^b e^{-px}\,dx$$

$$= \lim_{b \to \infty} \frac{e^{-p} - e^{-pb}}{p} = \frac{e^{-p}}{p}, \quad \text{if } p > 0.$$

If $p < 0$, the computation shows that the improper integral is divergent to $+\infty$. (What about $p = 0$?) For another example, consider

$$\int_0^\infty \frac{dx}{1 + x^2} = \lim_{b \to \infty} \int_0^b \frac{dx}{1 + x^2}$$

$$= \lim_{b \to \infty} [\arctan b - \arctan 0] = \frac{\pi}{2}.$$

The obvious analogy with series is that to compute an improper integral you first compute "proper" integrals over finite intervals and then find their limit over intervals with length tending to ∞. The next theorem shows that there is also a useful connection between the convergence and divergence of particular infinite series and improper integrals.

2.2 Integral Test.   Let $\Sigma_{k=1}^\infty a_k$ be a series of positive terms, and suppose f is a decreasing function such that $f(k) = a_k$ for $k = 1, 2, 3 \ldots$. Then the series and improper integral,

$$\sum_{k=1}^\infty a_k \quad \text{and} \quad \int_1^\infty f(x)\,dx,$$

either both converge or both diverge.

Proof. Suppose first that the integral converges. Looking at Figure 2(a) shows that, by comparing areas, we have

$$\sum_{k=2}^N a_k \le \int_1^{N+1} f(x)\,dx \le \int_1^\infty f(x)\,dx < \infty.$$

The series converges because the partial sums are increasing (because $a_k > 0$) and bounded above by the value of the integral. Then the nondecreasing sequence principle (1.1) applies.

Figure 2(b) shows that

$$\int_1^N f(x)\,dx \le \sum_{k=1}^N a_k \le \sum_{k=1}^\infty a_k < \infty,$$

so again Principle 1.1 applies to show that the integral converges to a finite number as $N \to \infty$.

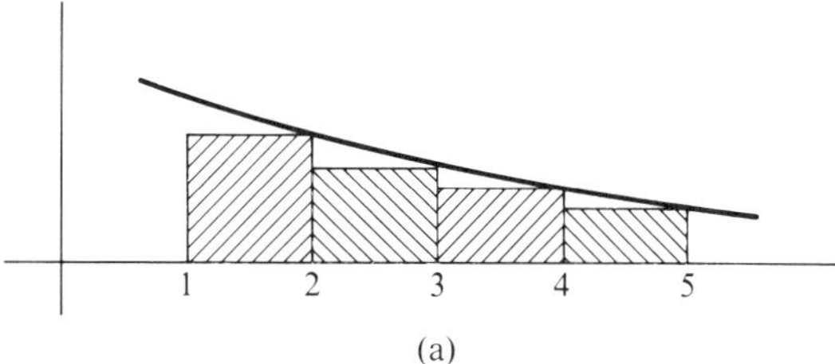

(a)

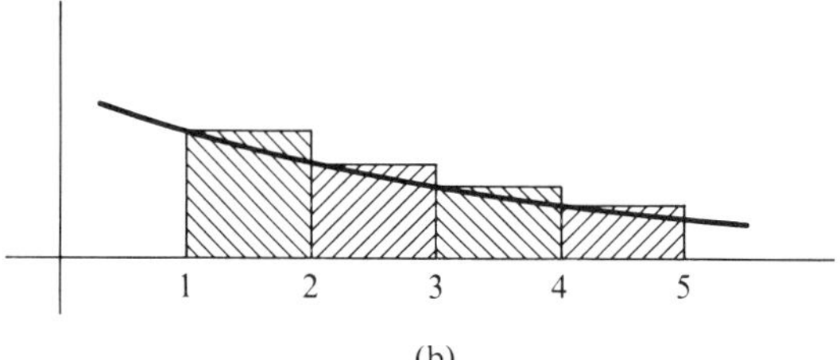

(b)   **Figure 2**

EXAMPLE 2 The **harmonic series** $\sum_{k=1}^{\infty} (1/k)$ diverges, because with $f(x) = 1/x$, we have $f(k) = 1/k$. But

$$\int_1^\infty \frac{1}{x}\, dx = \lim_{N\to\infty} \int_1^N \frac{1}{x}\, dx$$

$$= \lim_{N\to\infty} \log N = \infty.$$

Note, however, that the series diverges in spite of $\lim_{k\to\infty} (1/k) = 0$, as we pointed out in the warning about misapplication of the term test.

EXAMPLE 3 To decide about the **p-series** $\sum_{k=1}^{\infty} k^{-p}$ for $p > 0$, let $f(x) = 1/x^p$. We have, for $p \neq 1$,

$$\int_1^\infty \frac{1}{x^p}\, dx = \lim_{N\to\infty} \int_1^N \frac{1}{x^p}\, dx$$

$$= \lim_{N\to\infty} \left[\frac{1}{(1-p)x^{p-1}} \right]_1^N$$

$$= \lim_{N\to\infty} \frac{1}{1-p}\left[\frac{1}{N^{p-1}} - 1 \right] = \begin{cases} \dfrac{1}{p-1}, & p > 1, \\ +\infty, & 0 < p < 1. \end{cases}$$

Hence we have convergence for $p > 1$ and divergence for $p < 1$. The case $p = 1$ is the harmonic series, and when $p \leq 0$ the terms of the series fail to tend to zero, so the series diverges by the term test. For $p > 1$, the p-series defines a function $\zeta(p)$ called the **Riemann zeta function** and denoted by

$$\zeta(p) = \sum_{k=1}^{\infty} \frac{1}{k^p}.$$

For $p = 2$ and $p = 1.1$, the series

$$\zeta(2) = \sum_{k=1}^{\infty} \frac{1}{k^2} \quad \text{and} \quad \zeta(1.1) = \sum_{k=1}^{\infty} \frac{1}{k^{1.1}}$$

both converge. But for $p = \frac{1}{2}$ and $p = 0.9$, the series

$$\sum_{k=1}^{\infty} \frac{1}{\sqrt{k}} \quad \text{and} \quad \sum_{k=1}^{\infty} \frac{1}{k^{0.9}}$$

both diverge.

Application of the integral test usually depends on being able to compute some indefinite integral, but examples in which the computation is awkward can sometimes be handled by comparison with a related series.

2.3 Comparison Test.   Suppose $0 \le a_k \le b_k$.

(i). If $\sum_{k=1}^{\infty} b_k$ converges, then $\sum_{k=1}^{\infty} a_k$ converges.
(ii). If $\sum_{k=1}^{\infty} a_k$ diverges, then $\sum_{k=1}^{\infty} b_k$ diverges.

Proof. To prove (i), note that

$$s_n = \sum_{k=1}^{n} a_k \le \sum_{k=1}^{n} b_k \le \sum_{k=1}^{\infty} b_k = b.$$

Since $a_k \ge 0$, the partial sums s_n form a nondecreasing sequence, bounded by b. Then $\sum_{k=1}^{\infty} a_k$ converges by Principle 1.1(i). To prove (ii), note that

$$s_n = \sum_{k=1}^{n} a_k \le \sum_{k=1}^{n} b_k,$$

and that by Principle 1.1(ii), $\lim_{n \to \infty} s_n = \infty$. Hence $\sum_{k=1}^{\infty} b_k$ diverges also.

EXAMPLE 4   Consider the series $\sum_{k=2}^{\infty} \dfrac{1}{k^2 \log k}$. Since $\log k \ge 1$ for $k \ge 3$, we have

$$0 \le \frac{1}{k^2 \log k} \le \frac{1}{k^2}, \qquad \text{for } k \ge 3.$$

Since $\sum_{k=3}^{\infty} 1/k^2$ is a convergent p-series by the integral test, we know that $\sum_{k=3}^{\infty} 1/(k^2 \log k)$ converges by the comparison test. Hence the given series, with the additional term $1/(4 \log 2)$, converges also.

EXAMPLE 5   Consider the series

$$\sum_{k=1}^{\infty} \frac{1}{k^{1/2}(k+1)^{1/3}}.$$

Since, for $k \geq 1$,

$$\frac{1}{(k+1)^{5/6}} = \frac{1}{(k+1)^{1/2}(k+1)^{1/3}} \leq \frac{1}{k^{1/2}(k+1)^{1/3}},$$

the given series diverges by comparison with the divergent p-series for $p = \frac{5}{6}$.

2B   Absolute Convergence

If all terms of a series are nonpositive from some point on, we can simply multiply the series by (-1) and apply the tests of the previous subsection. For a series that has infinitely many terms a_k of both signs, we try if possible to show that the series with kth term $|a_k|$ converges. If

$$\sum_{k=1}^{\infty} |a_k|$$

converges we say that $\sum_{k=1}^{\infty} a_k$ is **absolutely convergent.** The terminology is justified because absolute convergence is in general *stronger* than ordinary convergence in the sense that every absolutely convergent series converges, but not conversely. The key theorem is this:

2.4 Theorem.   If $\sum_{k=1}^{\infty} |a_k|$ converges, then so does $\sum_{k=1}^{\infty} a_k$. In other words, absolute convergence implies convergence.

Proof. Since $0 \leq a_k + |a_k| \leq 2|a_k|$, the comparison test shows that $\sum_{k=1}^{\infty} (a_k + |a_k|)$ converges, because $2 \sum_{k=1}^{\infty} |a_k| = \sum_{k=1}^{\infty} 2|a_k|$ is assumed to converge. Hence

$$\sum_{k=1}^{\infty} a_k = \sum_{k=1}^{\infty} (a_k + |a_k|) - \sum_{k=1}^{\infty} |a_k|$$

converges also, because it is the difference of two convergent series.

EXAMPLE 6   The series

$$\sum_{k=1}^{\infty} (-1)^k \frac{1}{k^2} = -1 + \frac{1}{2^2} - \frac{1}{3^2} + \frac{1}{4^2} - \cdots$$

converges and even converges absolutely, because

$$\sum_{k=1}^{\infty} \left| (-1)^k \frac{1}{k^2} \right| = \sum_{k=1}^{\infty} \frac{1}{k^2}$$

is a convergent p-series for $p = 2$.

EXAMPLE 7   The series $\sum_{k=1}^{\infty} \dfrac{(-1)^k}{k\sqrt{k+1}}$ converges absolutely by the comparison test for series with positive terms. The reason is that

$$\left| \frac{(-1)^k}{k\sqrt{k+1}} \right| = \frac{1}{k\sqrt{k+1}} \le \frac{1}{k^{3/2}},$$

and $\sum_{k=1}^{\infty}(1/k^{3/2})$ is a convergent p-series with $p = \frac{3}{2}$.

For series with nonnegative terms, convergence is the same as absolute convergence, so any test that proves one proves the other also.

EXAMPLE 8   The geometric series $\sum_{k=0}^{\infty} r^k$ converges to $1/(1-r)$ for $|r| < 1$, by Example 1 of Section 1. We have

$$\sum_{k=0}^{\infty} \frac{1}{2^k} = 2, \qquad \text{with } r = \frac{1}{2},$$

and

$$\sum_{k=0}^{\infty} \frac{(-1)^k}{2^k} = \frac{2}{3}, \qquad \text{with } r = -1/2.$$

The series $\sum_{k=0}^{\infty} \dfrac{\pm 1}{2^k}$ converges absolutely, whatever choice of sign we make in each term because it converges when we always choose the plus sign. However, there is no general way to determine the sum of the series as there is for the two geometric series with all plus signs, or with alternating signs.

The following test can be applied to series with kth term $a_k \ne 0$ from some point on and deals directly with absolute convergence, or else with the case $a_k > 0$.

2.5 Ratio Test.   Let $\sum_{k=1}^{\infty} a_k$ be a series for which $\lim_{k\to\infty} |a_{k+1}|/|a_k|$ exists.

(i) If $\lim_{k\to\infty} \left| \dfrac{a_{k+1}}{a_k} \right| < 1$, the series converges absolutely.

(ii) If $\lim_{k\to\infty} \left| \dfrac{a_{k+1}}{a_k} \right| > 1$, or is infinite, the series fails to converge.

Note that if the limit of the ratio $|a_{k+1}/a_k|$ of successive terms fails to exist or if the limit is 1, no assertion is being made about convergence of the series. Note also that, if a series of positive terms fails to converge absolutely, then it fails to converge at all.

Proof.   We shall assume $a_k > 0$ since we are concerned only with terms of the form $|a_k|$ in the proof.

Case (i). Since the limit of a_{k+1}/a_k is less than 1, there is a number $r < 1$ such that $a_{k+1}/a_k \leq r$ for all sufficiently large values of k, say $k \geq N$. Thus $a_{k+1} \leq ra_k$ for $k = N, N + 1, \ldots$. Hence

$$a_{N+k} \leq ra_{N+k-1} \leq r^2 a_{N+k-2} \leq \cdots \leq r^k a_N.$$

Since $0 \leq r < 1$, the series

$$\sum_{k=0}^{\infty} r^k a_N = a_N \sum_{k=0}^{\infty} r^k$$

converges. Hence $\sum_{k=0}^{\infty} a_{N+k}$ converges also by the comparison test. Including the finitely many terms $a_1, \ldots, a_{N-1}$ shows that $\sum_{k=1}^{\infty} a_k$ converges also.

Case (ii). This time the limit of a_{k+1}/a_k is bigger than 1, so there is a number $r > 1$ for which $a_{k+1}/a_k \geq r$ for k sufficiently large, say $k \geq N$. Thus $a_{k+1} \geq ra_k$ for $k = N, N + 1, \ldots$. Hence

$$a_{N+k} \geq ra_{N+k-1} \geq \cdots \geq r^k a_N.$$

Since $r > 1$, r^k tends to ∞ a $k \to \infty$, and the series diverges by the term test.

EXAMPLE 9   The series $\sum_{k=1}^{\infty} k^2/2^k$ converges because, with $a_k = k^2/2^k$ and $a_{k+1} = (k + 1)^2/2^{k+1}$,

$$\lim_{k \to \infty} \frac{(k + 1)^2/2^{k+1}}{k^2/2^k} = \lim_{k \to \infty} \frac{(k + 1)^2 \, 2^k}{k^2 \, 2^{k+1}}$$

$$= \lim_{k \to \infty} \left(1 + \frac{1}{k} \right)^2 \cdot \frac{1}{2} = \frac{1}{2} < 1.$$

EXAMPLE 10   Consider the series $\sum_{k=0}^{\infty} x^k/k!$, where $k!$, **k factorial**, is $1 \cdot 2 \cdot 3 \ldots k$ if $k \geq 1$, and $0! = 1$. We have $a_k = x^k/k!$ and $a_{k+1} = x^{k+1}/(k + 1)!$

$$\lim_{k \to \infty} \left| \frac{a_{k+1}}{a_k} \right| = \lim_{k \to \infty} \left| \frac{x^{k+1}/(k + 1)!}{x^k/k!} \right|$$

$$= \lim_{k \to \infty} \frac{k! \, |x|}{(k + 1)!} = \lim_{k \to \infty} \frac{|x|}{k + 1} = 0 < 1.$$

Hence the series of terms depending on x converges by case (i) of Test 2.5 for all x.

2C Alternating Series

An **alternating series** is one in which the terms are alternately positive and negative. The **alternating harmonic series** is an example:

$$\sum_{k=1}^{\infty} (-1)^k \frac{1}{k} = 1 - \frac{1}{2} + \frac{1}{3} - \frac{1}{4} + \cdots .$$

Some of these series can be shown to converge by the following criterion.

2.6 Leibnitz Test. If $\sum_{k=1}^{\infty} a_k$ is an alternating series such that

$$\text{(i)} \quad |a_k| \geq |a_{k+1}| \quad \text{for } k = 1, 2, 3 \ldots ,$$

and

$$\text{(ii)} \quad \lim_{k \to \infty} a_k = 0,$$

then the series converges.

Proof. Suppose $a_1 = p_1$, $a_2 = -p_2$, $a_3 = p_3$, and so on, with $p_k \geq 0$. Then the partial sum $\sum_{k=1}^{2n} a_k$ can be written

$$s_{2n} = (p_1 - p_2) + (p_3 - p_4) + \cdots + (p_{2n-1} - p_{2n}),$$

where the sums $p_{2k-1} - p_{2k}$ are all nonnegative, because $p_k = |a_k| \geq |a_{k+1}| = p_{k+1}$. Hence s_{2n} is nondecreasing as n increases. For the same reason, grouping the terms differently shows that

$$p_1 \geq s_{2n} = p_1 - (p_2 - p_3) - \cdots - (p_{2n-2} - p_{2n-1}) - p_{2n}.$$

Hence the even-index partial sums are bounded above and nondecreasing, so by Principle 1.1, we have

$$\lim_{n \to \infty} s_{2n} = \lim_{n \to \infty} \sum_{k=1}^{2n} a_k = s,$$

for some number s. But $s_{2n+1} = s_{2n} + a_{2n+1}$, and since $\lim_{n \to \infty} a_{2n+1} = 0$ by (ii), we have

$$\lim_{n \to \infty} s_{2n+1} = \lim_{n \to \infty} \sum_{k=1}^{2n+1} a_k = s$$

also. Hence the given series converges to s.

EXAMPLE 11 The alternating harmonic series converges, because with $a_k = (-1)^{k+1}/k$, we have

$$\text{(i)} \quad \left| \frac{(-1)^{k+1}}{k} \right| > \left| \frac{(-1)^{k+2}}{k+1} \right| ,$$

and

$$\text{(ii)} \quad \lim_{k \to \infty} \frac{(-1)^{k+1}}{k} = 0.$$

Note that the alternating harmonic series fails to converge absolutely because $\sum_{k=1}^{\infty} 1/k$ is divergent.

EXAMPLE 12 The series

$$\sum_{k=2}^{\infty} \frac{(-1)^k}{\log k} = \frac{1}{\log 2} - \frac{1}{\log 3} + \frac{1}{\log 4} - \cdots$$

converges because (i) $1/\log k \geq 1/\log(k+1)$ and (ii) $\lim_{k \to \infty} 1/\log k = 0$. The Leibniz test implies convergence.

EXERCISES

1. Determine the convergence or divergence of each of the following infinite series by using the term test (for divergence) or the integral test (for convergence *or* divergence). Show carefully how the test you use applies in each case

(a) $\displaystyle\sum_{k=1}^{\infty} \frac{k^2}{k^2+1}.$ **(b)** $\displaystyle\sum_{k=1}^{\infty} \frac{k}{k^2+1}.$

(c) $\displaystyle\sum_{k=1}^{\infty} ke^{-k}.$ **(d)** $\displaystyle\sum_{k=1}^{\infty} \left(1+\frac{1}{k}\right)^k.$

(e) $\displaystyle\sum_{k=1}^{\infty} \frac{1}{k^2+1}.$ **(f)** $\displaystyle\sum_{k=1}^{\infty} \frac{1}{k(\log k)^2}.$

2. Determine the convergence or divergence of each of the following infinite series by using the comparison test or the ratio test. Show carefully how the test you use applies in each case.

(a) $\displaystyle\sum_{k=1}^{\infty} \frac{2^k}{3^{k+1}}.$ **(b)** $\displaystyle\sum_{k=1}^{\infty} \frac{1}{(k^2+k)^{3/2}}.$

(c) $\displaystyle\sum_{n=2}^{\infty} \frac{1}{n^2 \log n}.$ **(d)** $\displaystyle\sum_{n=1}^{\infty} \frac{1}{n 2^n}.$

(e) $\displaystyle\sum_{j=1}^{\infty} \frac{j}{j^3+1}.$ **(f)** $\displaystyle\sum_{j=1}^{\infty} \frac{j+2}{\sqrt{j+1}}.$

3. Determine whether each of the following series converges absolutely or not. For those alternating series that fail to converge absolutely, try to apply the Leibniz test for convergence.

(a) $\displaystyle\sum_{k=2}^{\infty} \frac{(-1)^k}{k^2 \log k}$ **(b)** $\displaystyle\sum_{k=1}^{\infty} \frac{(-1)^{k+1}}{k^2}.$

(c) $\displaystyle\sum_{j=1}^{\infty} \frac{(-1)^j}{\sqrt{j}}.$ **(d)** $\displaystyle\sum_{j=1}^{\infty} \frac{(-1)^j j}{j+1}.$

(e) $\displaystyle\sum_{m=0}^{\infty} \frac{(-1)^m}{m^2+1}.$ **(f)** $\displaystyle\sum_{m=0}^{\infty} \frac{(-1)^m m}{\sqrt{m^2+1}}.$

4. Determine the convergence, absolute convergence, or divergence of each of the following series.

(a) $\displaystyle\sum_{k=2}^{\infty} \frac{(-1)^k}{\log(1/k)}.$ **(b)** $\displaystyle\sum_{k=1}^{\infty} \frac{k}{k^3+1}.$

(c) $\displaystyle\sum_{k=1}^{\infty} \frac{(-2)^k}{k^2 + 1}$.

(d) $\displaystyle\sum_{k=1}^{\infty} \frac{k!}{(2k)!}$.

(e) $\displaystyle\sum_{k=1}^{\infty} \left(1 + \frac{1}{k}\right) 2^{-k}$.

(f) $\displaystyle\sum_{k=1}^{\infty} (-1)^k \frac{k!}{(k+1)!}$.

5. (a) Use L'Hopital's rule to show that $\lim_{x\to 0+} x^x = 1$, and hence that $\lim_{k\to\infty} (1/k)^{(1/k)} = 1$. (*Hint:* $x^x = e^{x \log x}$.)

(b) Show that $\sum_{k=1}^{\infty} \left(\dfrac{1}{k}\right)^{1/k}$ diverges.

(c) Show that $\sum_{k=1}^{\infty} k^{1/k}$ diverges.

6. Determine the convergence or divergence of each of the following series.

(a) $\displaystyle\sum_{k=1}^{\infty} \frac{k^2}{3^k}$.

(b) $\displaystyle\sum_{k=1}^{\infty} \frac{k^k}{k!}$.

(c) $\displaystyle\sum_{k=1}^{\infty} \frac{k!}{k^k}$.

(d) $\displaystyle\sum_{k=2}^{\infty} \frac{(-1)^k}{\sqrt{\log k}}$.

(e) $\displaystyle\sum_{k=1}^{\infty} \frac{\sin k}{k^2}$.

(f) $\displaystyle\sum_{n=1}^{\infty} \frac{1}{2n^2 - n}$.

7. For what values of x do the following series converge?

(a) $\displaystyle\sum_{k=1}^{\infty} \frac{1}{k} x^k$.

(b) $\displaystyle\sum_{k=1}^{\infty} \frac{1}{k^2} x^k$.

(c) $\displaystyle\sum_{k=1}^{\infty} \frac{\sin kx}{k^3}$.

(d) $\displaystyle\sum_{k=1}^{\infty} \frac{1}{2^k} (x - 1)^k$.

(e) $\displaystyle\sum_{j=1}^{\infty} \frac{1}{j} x^{2j}$.

(f) $\displaystyle\sum_{j=1}^{\infty} \frac{j}{j^2 + 1} x^j$.

(g) $\displaystyle\sum_{k=0}^{\infty} (\log x)^k$.

(h) $\displaystyle\sum_{k=0}^{\infty} \left(\frac{3}{1 + x^2}\right)^k$.

8. (a) Show that $\zeta(p)$, which is defined by the p-series as $\zeta(p) = \sum_{k=1}^{\infty} k^{-p}$, is decreasing as p increases, for $p > 1$.

(b) Show that $(1 - 2^{-p}) \zeta(p) = 1 + 1/3^p + 1/5^p + 1/7^p + \cdots$.

(c) Show that $(1 - 2^{1-p}) \zeta(p) = 1 - 1/2^p + 1/3^p - 1/4^p + \cdots$.

9. (a) Show that $\sum_{k=1}^{\infty} kx^k$ converges absolutely for $|x| < 1$.

(b) Show that $(1 - x) \sum_{k=1}^{\infty} kx^k = \sum_{k=1}^{\infty} x^k$, for $|x| < 1$.

(c) Show that $\sum_{k=1}^{\infty} x^k = x/(1 - x)$, for $|x| < 1$.

(d) Show that $\sum_{k=1}^{\infty} kx^k = x/(1 - x)^2$, for $|x| < 1$.

(e) Show that $\sum_{k=1}^{\infty} k2^{-k} = 2$.

10. Show that, if a is any real number and $|b| > 1$, then

$$\lim_{n\to\infty} \frac{n^a}{b^n} = 0,$$

by applying the ratio test to $\sum_{k=0}^{\infty} n^a b^{-n}$.

11. Show that

$$\sum_{k=2}^{\infty} \frac{1}{k (\log k)^a}$$

converges if $a > 1$ and diverges if $a < 1$.

3 TAYLOR SERIES

3A Taylor Expansions

A polynomial f of degree n is just a sum of numerical multiples of powers of x: $f(x) = c_0 + c_1 x + \cdots + c_n x^n$. To examine the behavior of f near a point $x = a$, it is useful to be able to write $f(x)$ as a polynomial in powers of $(x - a)$:

$$f(x) = a_0 + a_1(x - a) + a_2(x - a)^2 + \cdots + a_n(x - a)^n.$$

Here is the simple formula for the coefficients a_k in terms of derivatives $f^{(k)}(a)$ of f at a:

3.1
$$a_k = \frac{1}{k!} f^{(k)}(a),$$

where $k! = 1 \cdot 2 \cdot 3 \ldots k$ for $k \geq 1$ and $0! = 1$. Thus

$$a_0 = f(a), \qquad a_1 = \frac{1}{1!} f'(a), \qquad a_2 = \frac{1}{2!} f''(a), \qquad \text{etc.}$$

Formula 3.1 holds because k successive differentiations of $f(x)$ knock out the first k terms, starting with the constant a_0. All the remaining terms but

$$\frac{d^k}{dx^k} (a_k(x - a)^k) = 1 \cdot 2 \ldots (k - 1) \cdot k a_k$$

become zero when we set $x = a$. The result is $f^{(k)}(a) = k! \, a_k$; from this, Formula 3.1 follows upon division by $k!$.

EXAMPLE 1   Let $f(x) = 1 - x^2 + 2x^3$. To write f in terms of powers of $(x - a)$ with $a = 1$, we compute

$$f(1) = 2, \qquad f'(1) = 4, \qquad f''(1) = 10, \qquad f'''(1) = 12.$$

Hence

$$f(x) = 2 + \frac{4}{1!}(x - 1) + \frac{10}{2!}(x - 1)^2 + \frac{12}{3!}(x - 1)^3$$

$$= 2 + 4(x - 1) + 5(x - 1)^2 + 2(x - 1)^3.$$

The procedure described can be applied to any function with sufficiently many derivatives, the only problem being that in general things do not come out even at the end, and we have to supply a remainder term. What often makes the remainder usable is being able to find a simple estimate for its size; here is a particularly useful one:

3.2 Remainder Formula.   Suppose f has $n + 1$ continuous derivatives on an open interval containing $x = a$. Then, for all x in the interval,

$$f(x) = \sum_{k=0}^{n} \frac{1}{k!} f^{(k)}(a)(x - a)^k + R_n(x)$$

where

$$R_n(x) = \frac{1}{(n+1)!} f^{(n+1)}(c)(x-a)^{n+1}.$$

Here c is some number depending on x and situated between x and a.
Hence, if $\left| f^{(n+1)}(x) \right| \leq M$ on the interval, then

$$\left| R_n(x) \right| \leq \frac{M}{(n+1)!} \left| x - a \right|^{n+1}.$$

Proof. With x held fixed and $x \neq a$, define K by

$$f(x) = \sum_{k=0}^{n} \frac{1}{k!} f^{(k)}(a)(x-a)^k + K(x-a)^{n+1}.$$

With K determined in this way, define $g(t)$ by

$$g(t) = -f(x) + \sum_{k=0}^{n} \frac{1}{k!} f^{(k)}(t)(x-t)^k + K(x-t)^{n+1}.$$

Note that $g(a) = 0$ because of the way K is defined, and that $g(x) = 0$ no matter what value K has. Applying the product rule to the terms in the summation, we find that differentiation with respect to t gives

$$g'(t) = \sum_{k=0}^{n} \frac{1}{k!} f^{(k+1)}(t)(x-t)^k - \sum_{k=1}^{n} \frac{1}{(k-1)!} f^{(k)}(t)(x-t)^{k-1} - (n+1)K(x-t)^n.$$

But most of the terms cancel, so

$$g'(t) = \frac{1}{n!} f^{(n+1)}(t)(x-t)^n - (n+1)K(x-t)^n.$$

By Rolle's theorem, there is a number c between x and a such that $g'(c) = 0$, because $g(x) = g(a) = 0$. But the equation $g'(c) = 0$ allows us to solve for K to get what we wanted to show:

$$K = \frac{1}{(n+1)!} f^{(n+1)}(c).$$

The part of the Taylor formula exclusive of the remainder is called the **nth degree Taylor approximation** $T_n(x)$ to $f(x)$ about $x = a$. Note that, to increase the degree of the approximation, all we do is add another term without altering the previous terms. Without worrying about convergence, we can write the *infinite* **Taylor series**

3.3 $$\sum_{k=0}^{\infty} \frac{1}{k!} f^{(k)}(a)(x-a)^k =$$

$$f(a) + f'(a)(x-a) + \frac{1}{2!} f''(a)(x-a)^2 + \frac{1}{3!} f'''(a)(x-a)^3 + \cdots$$

and arrive at the nth degree Taylor approximation by stopping after $(n + 1)$ terms. Of course, the number of terms we can compute depends on having enough derivatives at $x = a$.

EXAMPLE 2   Let $f(x) = x^{1/2}$ for $x \geq 0$. The $f'(x) = (\frac{1}{2})x^{-1/2}$, $f''(x) = -(\frac{1}{4})x^{-3/2}$, $f'''(x) = (\frac{3}{8})x^{-5/2}$, and so on. We find, with $a = 1$ in Taylor's formula,

$$\sqrt{x} = 1 + \frac{1}{1!}\,(\tfrac{1}{2})\,(x - 1) + \frac{1}{2!}\,(-\tfrac{1}{4})\,(x - 1)^2 + R_2(x)$$

$$= 1 + \frac{1}{2}(x - 1) - \frac{1}{8}(x - 1)^2 + R_2(x),$$

where $R_2(x) = (1/3!)\,\tfrac{3}{8}\,c^{-5/2}(x - 1)^3$, and c is somewhere between x and 1. Note that the first two terms of the approximation give the function T_1 whose graph is the tangent line to the graph of $y = \sqrt{x}$ at $x = 1$. The first three terms describe a quadratic function T_2 that approximates $\sqrt{x}$ near $x = 1$. The graphs of both approximations are shown in Figure 3(a). Clearly, the approximation is better the closer we get to $x = 1$, because the factor $(x - 1)^3$ gets small as x, and hence c, get close to 1.

EXAMPLE 3   The Taylor approximations of $f(x) = e^x$ are particularly simple to compute, because $f^{(k)}(x) = e^x$ for $k = 1, 2, 3, \ldots$. Since $f^{(k)}(0) = e^0 = 1$, the formal infinite series 3.3, with $a = 0$, becomes

$$\sum_{k=0}^{\infty} \frac{1}{k!}\,x^k = 1 + \frac{1}{1!}\,x + \frac{1}{2!}\,x^2 + \frac{1}{3!}\,x^3 + \cdots .$$

The partial sums $T_1(x) = 1 + x$ and $T_2(x) = 1 + x + \frac{1}{2}x^2$ are compared with e^x graphically in Figure 3(b). The remainder after $n + 1$ terms is

$$R_n(x) = \frac{1}{(n + 1)!}\,e^c x^{n+1}.$$

We can then estimate that e^x differs from $\sum_{k=1}^{n} \frac{1}{k!}\,x^k$ by at most

$$e^x - \sum_{k=1}^{n} \frac{1}{k!}\,x^k = \frac{1}{(n + 1)!}\,e^c x^{n+1}$$

$$\leq \frac{e}{(n + 1)!}, \qquad \text{if } 0 \leq x \leq 1.$$

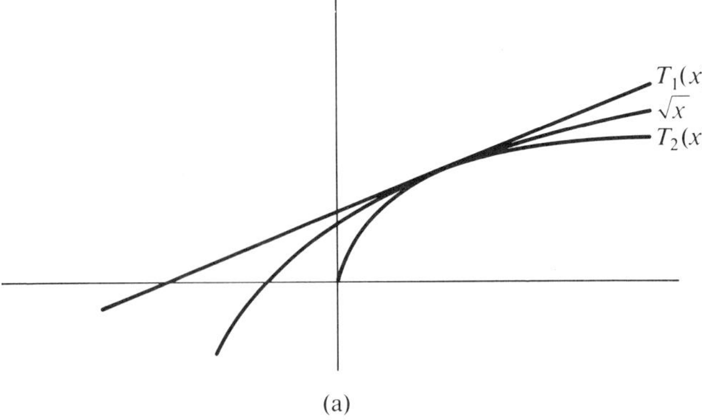

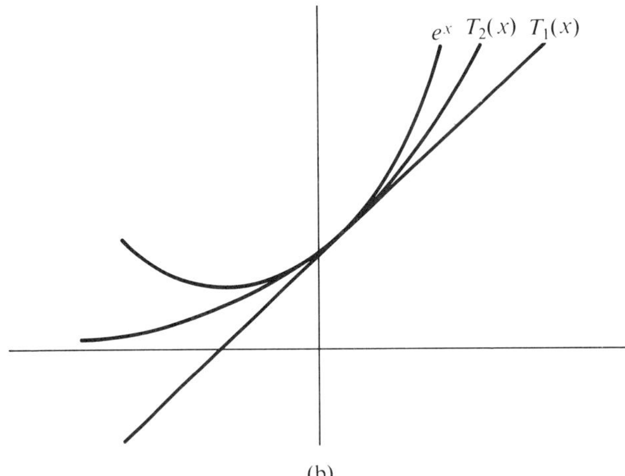

Figure 3

3B Convergence of Taylor Series

Taylor's formula in the form

$$f(x) - \sum_{k=0}^{n} \frac{1}{k!} f^{(k)}(a)(x - a)^k = R_n(x)$$

gives us a way of showing that in some cases

$$f(x) = \sum_{k=0}^{\infty} \frac{1}{k!} f^{(k)}(a)(x - a)^k,$$

with the series on the right converging for x in at least some subinterval of the domain of f. All we have to do is show that

$$\lim_{n \to \infty} R_n(x) = 0$$

for the values of x in question. This at once proves convergence of the series and shows that the sum at x is $f(x)$. Here is a list of examples for which equality holds.

3.4

(a) $e^x = \sum_{k=0}^{\infty} \frac{1}{k!} x^k = 1 + \frac{x}{1!} + \frac{x^2}{2!} + \frac{x^3}{3!} + \cdots, \quad -\infty < x < \infty.$

(b) $\cos x = \sum_{k=0}^{\infty} \frac{(-1)^k}{(2k)!} x^{2k} = 1 - \frac{x^2}{2!} + \frac{x^4}{4!} - \frac{x^6}{6!} + \cdots, \quad -\infty < x < \infty.$

(c) $\sin x = \sum_{k=0}^{\infty} \frac{(-1)^k}{(2k+1)!} x^{2k+1} = x - \frac{x^3}{3!} + \frac{x^5}{5!} - \frac{x^7}{7!} + \cdots, \quad -\infty < x < \infty.$

(d) $\ln(1+x) \sum_{k=1}^{\infty} \frac{(-1)^{k+1}}{k} x^k = x - \frac{x^2}{2} + \frac{x^3}{3} - \frac{x^4}{4} + \cdots, \quad -1 < x < 1.$

EXAMPLE 4   As we showed in Example 3, Taylor's formula for e^x with $a = 1$ is

$$e^x = 1 + \frac{x}{1!} + \frac{x^2}{2!} + \frac{x^3}{3!} + \cdots + \frac{x^n}{n!} + R_n(x),$$

where $R_n(x) = \frac{1}{(n+1)!} e^c x^{N+1}$ for some c between x and 0. To prove the infinite series expansion 3.4(a), what we have to do is show that the remainder satisfies

$$\lim_{n \to \infty} \frac{1}{(n+1)!} e^c x^{n+1} = 0$$

for all real numbers x. Pick a fixed value for x. Since $|c| \le |x|$, we see that $e^{|c|} < e^{|x|}$ for all relevant values of c. Hence, with x fixed, all we need to show is that

$$\lim_{n \to \infty} \frac{1}{(n+1)!} x^{n+1} = 0.$$

This can be proved directly (Exercise 14.) or else by a trick, as follows. In applying the ratio test for absolute convergence to the series

$$\sum_{k=0}^{\infty} \frac{x^k}{k!}$$

we found in Example 10 of Section 2 that the series converges (absolutely) for all x. But then by the term test, the terms of the series tend to 0. In other words, $\lim_{k \to \infty} x^k/k! = 0$. Now replace k by $(n+1)$ to get the limit in its originally stated form. Similar arguments apply to the other series listed previously, and these are left as exercises.

EXAMPLE 5   To find a series expansion for e^{-2x}, there is no need to start from scratch. Simply replace x by $-2x$ everywhere in 3.4(a):

$$e^{-2x} = \sum_{k=0}^{\infty} \frac{1}{k!}(-2)^k x^k.$$

Similarly, to find an expansion for $\log(1 - x)$ in powers of x, replace x by $-x$ in 3.4(d):

$$\log(1 - x) = \sum_{k=1}^{\infty} \frac{(-1)^{k+1}}{k}(-1)^k x^k, \qquad -1 < -x < 1,$$

$$= -\sum_{k=1}^{\infty} \frac{1}{k} x^k, \qquad -1 < x < 1.$$

EXERCISES

1. Write each of the following polynomials in the indicated form; that is, find the coefficients c_k by using Equation 3.1 or by some other method.
   (a) $1 + x + x^2 = c_0 + c_1(x - 2) + c_2(x - 2)^2$.
   (b) $2x - x^3 = c_0 + c_1(x - 1) + c_2(x - 1)^2 + c_3(x - 1)^3$.
   (c) $1 + x^2 = c_0 + c_1(x + 1) + c_2(x + 1)^2$.
   (d) $(1 - x) + (1 - x)^2 = c_0 + c_1 x + c_2 x^2$.

2. For each of the following functions, find the coefficient c_k in the indicated Taylor expansion. Compute the remainder as in Theorem 3.2.
   (a) $x^{1/3} = c_0 + c_1(x - 1) + c_2(x - 1)^2 + R_2(x)$.
   (b) $x/(1 + x^2) = c_0 + c_1 x + c_2 x^2 + R_2(x)$.
   (c) $1/x = c_0 + c_1(x + 1) + c_2(x + 1)^2 + R_2(x)$.
   (d) $e^{2x} = c_0 + c_1 x + c_2 x^2 + R_2(x)$.

3. Find infinite series expansions for each of the following functions about $x = 0$, by modifying the series given in 3.4.
   (a) e^{-x}.   (b) $\cos 2x$.
   (c) $\sin(x/2)$.   (d) $\log(1 - x^2)$.
   (e) e^{x^2}.   (f) xe^x.

4. Show that

   (a) $\cosh x = \dfrac{e^x + e^{-x}}{2} = \displaystyle\sum_{k=0}^{\infty} \frac{x^{2k}}{(2k)!}$.

   (b) $\sinh x = \dfrac{e^x - e^{-x}}{2} = \displaystyle\sum_{k=0}^{\infty} \frac{x^{2k+1}}{(2k + 1)!}$.

5. By using a Taylor expansion about $x = 0$, prove the binomial theorem:

$$(x + a)^n = \sum_{k=0}^{n} \binom{n}{k} a^{n-k} x^k,$$

where the binomial coefficients are given by

$$\binom{n}{k} = \frac{n!}{k!(n-k)!} = \frac{n(n-1)\ldots(n-k+1)}{k(k-1)\ldots 2\cdot 1}.$$

6. On the same coordinate axes, sketch the graphs of f and the Taylor approximation T_k. Estimate the remainders for all x.

(a) $f(x) = \cos x$ and $T_2(x) = 1 - \frac{1}{2}x^2$.

(b) $f(x) = \cos x$ and $T_4(x) = 1 - \frac{1}{2}x^2 + \frac{1}{24}x^4$.

(c) $f(x) = \sin x$ and $T_3(x) = x - \frac{1}{6}x^3$.

(d) $f(x) = \sin x$ and $T_5(x) = x - \frac{1}{6}x^3 + \frac{1}{120}x^5$.

7. (a) Show that the geometric series formula

$$\sum_{k=0}^{\infty} x^k = \frac{1}{1-x}, \qquad -1 < x < 1,$$

is the same as the Taylor expansion of $(1-x)^{-1}$ about $x = 0$.

(b) Derive from part (a) the expansion about $x = 1$

$$\frac{1}{x} = \sum_{k=0}^{\infty} (-1)^k (x-1)^k, \qquad 0 < x < 2.$$

(c) Use the identity

$$\frac{1}{1+x} = \frac{1}{2+(x-1)} = \frac{1}{2}\frac{1}{1+(x-1)/2}$$

to show that

$$(1+x)^{-1} = \sum_{k=0}^{\infty} (-1)^k 2^{-k-1}(x-1)^k.$$

8. (a) If $f(x) = \cos x$, show that the Taylor coefficients of f about $x = 0$ are

$$\frac{f^{(n)}(0)}{n!} = \begin{cases} 0, & n \text{ odd}, \\ \dfrac{(-1)^{n/2}}{n!}, & n \text{ even}. \end{cases}$$

(b) Using the method of Example 4, show that

$$\cos x = \sum_{k=0}^{\infty} \frac{(-1)^k}{(2k)!} x^{2k}, \qquad -\infty < x < \infty.$$

9. (a) If $f(x) = \sin x$, show that the Taylor coefficients of f about $x = 0$ are

$$\frac{f^{(n)}(0)}{n!} = \begin{cases} \dfrac{(-1)^{(n-1)/2}}{n!}, & n \text{ odd}, \\ 0, & n \text{ even}. \end{cases}$$

(b) Using the method of Example 4, show that

$$\sin x = \sum_{k=0}^{\infty} \frac{(-1)^k}{(2k+1)!} x^{2k+1}, \qquad -\infty < x < \infty.$$

10. Show that

$$\ln(1+x) = \sum_{k=1}^{\infty} \frac{(-1)^{k+1}}{k} x^k, \qquad \text{for } -\infty < x < \infty.$$

11. Show that

$$\ln 2 = \sum_{k=1}^{\infty} \frac{1}{k 2^k}$$

by letting $x = -1/2$ in Formula 3.4(d).

12. Show that

$$\left(\sum_{k=0}^{\infty} \frac{(-1)^k}{k!} \right) \left(\sum_{k=0}^{\infty} \frac{1}{k!} \right) = 1.$$

13. Show that

$$\sum_{k=0}^{\infty} \frac{(-1)^k \pi^k}{(2k)! \, 2^k} = 0.$$

14. Show that $\lim_{k\to\infty} x^k/k! = 0$ for $x > 0$ by choosing $k > l \geq 2x$ and writing

$$\frac{x^k}{k!} = \frac{x}{1} \cdot \frac{x}{2} \cdots \frac{x}{l} \cdot \frac{x}{l+1} \cdots \frac{x}{k}.$$

Then use the assumption $x/l \leq \frac{1}{2}$.

15. An even function f is a function such that $f(-x) = f(x)$ for all real x, and an odd function is a function such that $f(-x) = -f(x)$ for all real x.
   (a) Show that the odd-order derivatives $f^{(2k+1)}(0)$ of an even function are all zero. [For example, $f(x) = \cos x$.]
   (b) Show that the even-order derivatives $f^{(2k)}(0)$ of an odd function are all zero. [For example, $f(x) = \sin x$.]
   (c) What conclusions can you make about the Taylor expansions about $a = 0$ of even functions and odd functions?

4 UNIFORM CONVERGENCE

Let $f_k(\mathbf{x})$, $k = 1, 2, 3, \ldots$, be a sequence of real-valued functions defined for all $\mathbf{x}$ in some set S. Then for each $\mathbf{x}$, we consider the series $\sum_{k=1}^{\infty} f_k(\mathbf{x})$. If it converges for each $\mathbf{x}$ in S, we say that the series **converges pointwise** on S. Calling the limit $f(\mathbf{x})$ for each $\mathbf{x}$ in S, we write

$$f(\mathbf{x}) = \sum_{k=1}^{\infty} f_k(\mathbf{x})$$

$$= \lim_{N \to \infty} \sum_{k=1}^{N} f_k(\mathbf{x}).$$

Recall that this means that, for each $\mathbf{x}$ in S, there is a number $f(\mathbf{x})$ such that, given $\epsilon > 0$, there is an integer K sufficiently large that

$$\left| \sum_{k=1}^{N} f_k(\mathbf{x}) - f(\mathbf{x}) \right| < \epsilon,$$

whenever $N \geq K$.

EXAMPLE 1   The series $\sum_{k=0}^{\infty} x^k$ has as $(N + 1)$th partial sum the finite sum

$$\sum_{k=0}^{N} x_k = \begin{cases} \dfrac{1 - x^{N+1}}{1 - x}, & x \neq 1, \\ N + 1, & x = 1. \end{cases}$$

Then

$$\sum_{k=0}^{\infty} x^k = \lim_{N \to \infty} \sum_{k=0}^{N} x^k = \frac{1}{1 - x}, \qquad \text{for } -1 < x < 1.$$

For real values of x outside the interval $(-1, 1)$, the series fails to converge.

The trigonometric series $\sum_{k=1}^{\infty}(\sin kx)/k^2$ converges pointwise for all real x. The reason is that its terms can be compared with those of the convergent series $\sum_{k=1}^{\infty} 1/k^2$, by observing that

$$\left| \frac{\sin kx}{k^2} \right| \leq \frac{1}{k^2}, \qquad k = 1, 2, \ldots .$$

The result is that the given series even converges absolutely.

An infinite series $\sum_{k=1}^{\infty} f_k(\mathbf{x})$ that converges for each $\mathbf{x}$ in a set S to a number $f(\mathbf{x})$ defines a function f on S. However, in general, very little can be concluded about the properties of f from pointwise convergence alone. For this reason it is sometimes helpful to consider a stronger form of convergence on S. We say that $\sum_{k=1}^{\infty} f_k$ **converges uniformly** to a function f on a set S, if, given $\epsilon > 0$, there is an integer K such that for all $\mathbf{x}$ in S and for all $N \geq K$,

$$\left| \sum_{k=1}^{N} f_k(\mathbf{x}) - f(\mathbf{x}) \right| < \epsilon.$$

The definition just given should be compared carefully with that of point-wise convergence. Notice that uniform convergence implies pointwise convergence, but not conversely. Roughly speaking, uniform convergence of a series of functions defined on a set S means that the series converges with at least a certain minimum rate for all points in S. A pointwise convergent series may have points at which the convergence is arbitrarily slow. Figure 4 is a picture of uniform and nonuniform convergence to the same function f; $s_N(x)$ and $t_N(x)$ are Nth partial sums of two series.

To determine that a series converges uniformly, we have the following.

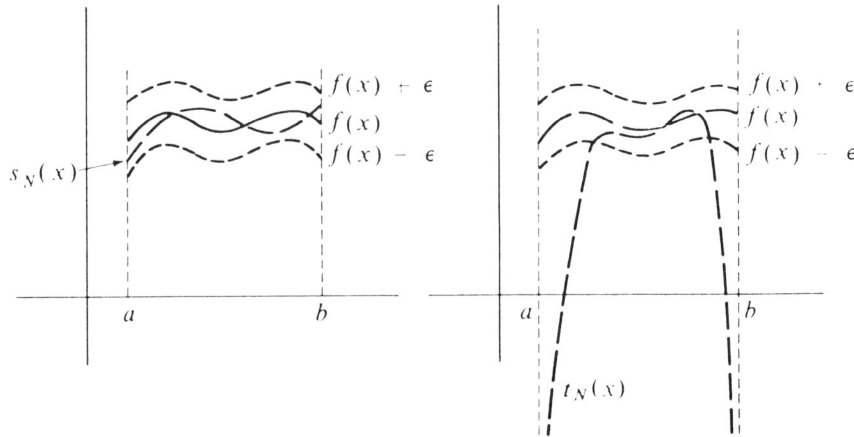

Figure 4

4.1 Weierstrass Test. Let $\Sigma_{k=1}^{\infty} f_k$ be a series of real-valued functions defined on a set S. If there is a constant series $\Sigma_{k=1}^{\infty} p_k$, such that

1. $|f_k(\mathbf{x})| \le p_k$ for all $\mathbf{x}$ in S and for $k = 1, 2, \ldots$,

2. $\displaystyle\sum_{k=1}^{\infty} p_k$ converges,

then $\Sigma_{k=1}^{\infty} f_k$ converges uniformly to a function f defined on S.

Proof. The comparison test for series shows that $\Sigma_{k=1}^{\infty} f_k(\mathbf{x})$ converges (even absolutely) for each $\mathbf{x}$ in S to a number that we will write $f(\mathbf{x})$. Hence we can write

$$f(\mathbf{x}) - \sum_{k=1}^{N} f_k(\mathbf{x}) = \sum_{k=1}^{\infty} f_k(\mathbf{x}) - \sum_{k=1}^{N} f_k(\mathbf{x})$$

$$= \sum_{k=N+1}^{\infty} f_k(\mathbf{x}).$$

It follows that

$$\left| f(\mathbf{x}) - \sum_{k=1}^{N} f_k(\mathbf{x}) \right| \le \sum_{k=N+1}^{\infty} |f_k(\mathbf{x})| \le \sum_{k=N+1}^{\infty} p_k.$$

Since $\Sigma_{k=1}^{\infty} p_k$ converges, we can, given $\epsilon > 0$, find a K such that $\Sigma_{k=N}^{\infty} p_k < \epsilon$ if $N > K$. This completes the proof, because the number K depends only on ϵ and not on $\mathbf{x}$.

EXAMPLE 2 The trigonometric series $\Sigma_{k=1}^{\infty}(\sin kx)/k^2$ converges uniformly for all real x, because

$$\left| \frac{\sin kx}{k^2} \right| \le \frac{1}{k^2},$$

and $\Sigma_{k=1}^{\infty} 1/k^2$ converges. However, the power series $\Sigma_{k=0}^{\infty} x^k$, while it converges pointwise for $-1 < x < 1$, fails to converge uniformly on $(-1, 1)$. See Exercise 6. The Weierstrass test can be applied on any closed subinterval $[-r, r]$ by observing that $|x^k| \le r^k$ for x on $[-r, r]$ and that $\Sigma_{k=0}^{\infty} r^k$ converges if $0 \le r < 1$. Hence the power series converges pointwise on $(-1, 1)$ and uniformly on $[-r, r]$ for any $r < 1$.

The next four theorems are about uniformly convergent series of functions. They all assert that certain limit operations can be interchanged with the summing of a series, provided that some series converges uniformly. If uniform convergence is replaced by pointwise convergence, then the resulting statements are false. See Exercise 9.

4.2 Theorem.   Let $f_1, f_2, f_3, \ldots$ be a sequence of functions defined on a set S in $\mathcal{R}^n$. Suppose $\mathbf{x}_0$ is a limit point of S, and suppose that the limit

$$\lim_{\mathbf{x} \to \mathbf{x}_0} f_k(\mathbf{x})$$

exists for $k = 1, 2, \ldots$. Then

$$\lim_{\mathbf{x} \to \mathbf{x}_0} \sum_{k=1}^{\infty} f_k(\mathbf{x}) = \sum_{k=1}^{\infty} \lim_{\mathbf{x} \to \mathbf{x}_0} f_k(\mathbf{x}),$$

provided the series of numbers on the right converges and the series on the left converges uniformly on S.

Proof. Let $\lim_{\mathbf{x} \to \mathbf{x}_0} f_k(\mathbf{x}) = a_k$. Then, adding and subtracting $\Sigma_{k=1}^{N} f_k(\mathbf{x})$ and $\Sigma_{k=1}^{N} a_k$, we get

$$\left| \sum_{k=1}^{\infty} f_k(\mathbf{x}) - \sum_{k=1}^{\infty} a_k \right| \le \left| \sum_{k=1}^{\infty} f_k(\mathbf{x}) - \sum_{k=1}^{N} f_k(\mathbf{x}) \right| \tag{1}$$

$$+ \left| \sum_{k=1}^{N} f_k(\mathbf{x}) - \sum_{k=1}^{N} a_k \right| + \left| \sum_{k=1}^{N} a_k - \sum_{k=1}^{\infty} a_k \right|.$$

Now let $\epsilon > 0$. Since $\Sigma_{k=1}^{\infty} f_k$ converges uniformly, we can choose K such that $N > K$ implies

$$\left| \sum_{k=1}^{\infty} f_k(\mathbf{x}) - \sum_{k=1}^{N} f_k(\mathbf{x}) \right| < \frac{\epsilon}{3}, \qquad \text{for all } \mathbf{x} \text{ in } S.$$

Then choose an $N > K$ such that

$$\left| \sum_{k=1}^{N} a_k - \sum_{k=1}^{\infty} a_k \right| < \frac{\epsilon}{3}.$$

Finally, pick $\delta > 0$ so that $|\mathbf{x} - \mathbf{x}_0| < \delta$ implies, by the relation

$$\lim_{\mathbf{x} \to \mathbf{x}_0} \sum_{k=1}^{N} f_k(\mathbf{x}) = \sum_{k=1}^{N} a_k,$$

that

$$\left| \sum_{k=1}^{N} f_k(\mathbf{x}) - \sum_{k=1}^{N} a_k \right| < \frac{\epsilon}{3}.$$

Then for $\mathbf{x}$ satisfying $|\mathbf{x} - \mathbf{x}_0| < \delta$, the left side of equation (1) is less than ϵ.

4.3 Corollary.   If $\Sigma_{k=1}^{\infty} f_k$ is a uniformly convergent series of continuous functions f_k defined on a set S in $\mathcal{R}_n$, then the function f defined by $f(\mathbf{x}) = \Sigma_{k=1}^{\infty} f_k(\mathbf{x})$ is continuous on S.

In the next two theorems we restrict ourselves to functions of one variable, although by treating one variable at a time, we can apply them to functions of several variables.

4.4 Theorem.   If the series $\Sigma_{k=1}^{\infty} f_k$ converges uniformly on the interval $[a, b]$, and the functions f_k are continuous on $[a, b]$, then

$$\sum_{k=1}^{\infty} \int_{a}^{b} f_k(x) \, dx = \int_{a}^{b} \left[\sum_{k=1}^{\infty} f_k(x) \right] dx.$$

Proof. By Theorem 4.3 the function $\Sigma_{k=1}^{\infty} f_k(x)$ is continuous on $[a, b]$ and so is integrable there. We have

$$\int_{a}^{b} \left[\sum_{k=1}^{\infty} f_k(x) \right] dx - \sum_{k=1}^{N} \int_{a}^{b} f_k(x) \, dx = \int_{a}^{b} \sum_{k=N+1}^{\infty} f_k(x) \, dx. \tag{2}$$

Let $\epsilon > 0$, and choose K so large that if $N > K$, then

$$\left| \sum_{k=N+1}^{\infty} f_k(x) \right| < \epsilon(b - a)^{-1}, \qquad \text{for all } x \text{ in } [a, b].$$

Then, using the fact that, for continuous g,

$$\left| \int_{a}^{b} g(x) \, dx \right| \leq (b - a) \max_{a \leq x \leq b} |g(x)|,$$

we have

$$\left| \int_{a}^{b} \sum_{k=N+1}^{\infty} f_k(x) \, dx \right| \leq (b - a) \cdot \epsilon \cdot (b - a)^{-1} = \epsilon, \qquad \text{for } N > K.$$

Thus the left side of equation (2) is less than ϵ in absolute value for $N > K$, which was to be shown.

The interchange of differentiation with the summing of a series requires somewhat more in the way of hypotheses than did the previous theorem on integration.

4.5 Theorem.   Let $f_1, f_2, f_3, \ldots$ be a sequence of continuously differentiable functions defined on an interval $[a, b]$. If $\Sigma_{k=1}^{\infty} f_k(x) = f(x)$ for all x in $[a, b]$ (pointwise convergence), and if $\Sigma_{k=1}^{\infty} df_k/dx$ converges uniformly on $[a, b]$, then f is continuously differentiable, and

$$\frac{d}{dx} \sum_{k=1}^{\infty} f_k(x) = \sum_{k=1}^{\infty} \frac{df_k}{dx}(x).$$

Proof. By the fundamental theorem of calculus,

$$\sum_{k=1}^{N} [f_k(x) - f_k(a)] = \sum_{k=1}^{N} \int_a^x f_k'(t)\, dt$$

$$= \int_a^x \left[\sum_{k=1}^{N} f_k'(t) \right] dt. \tag{3}$$

Letting N tend to infinity, we get $\Sigma_{k=1}^{\infty} f_k(x) = f(x)$; so

$$f(x) - f(a) = \int_a^x \left[\sum_{k=1}^{\infty} f_k'(t) \right] dt,$$

where we have used pointwise convergence on the left side of equation (3) and, on the right, have used uniform convergence, together with Theorem 4.4. Differentiation of both sides of the last equation gives

$$f'(x) = \sum_{k=1}^{\infty} f_k'(x),$$

which is the conclusion of the theorem.

EXAMPLE 3   Consider the trigonometric series

$$\sum_{k=1}^{\infty} \frac{\sin kx}{k^4}.$$

Clearly, the series converges for all real x. Furthermore, the series of derivatives of the terms of the given series is

$$\sum_{k=1}^{\infty} \frac{\cos kx}{k^3}.$$

This series converges uniformly for all x by the Weierstrass test, because

$$\left| \frac{\cos kx}{k^3} \right| \le \frac{1}{k^3},$$

and $\Sigma_1^{\infty} (1/k^3)$ converges. Hence, by Theorem 4.5,

$$\frac{d}{dx} \sum_{k=1}^{\infty} \frac{\sin kx}{k^4} = \sum_{k=1}^{\infty} \frac{\cos kx}{k^3}.$$

This same argument can be applied to give

$$\frac{d^2}{dx^2} \sum_{k=1}^{\infty} \frac{\sin kx}{k^4} = - \sum_{k=1}^{\infty} \frac{\sin kx}{k^2}.$$

EXERCISES

1. Show that the series $\sum_{k=0}^{\infty} x^k$ converges uniformly for $-d \le x \le d$ if $0 < d < 1$.

2. (a) Show that the trigonometric series $\sum_{k=1}^{\infty} (\cos kx / k^2)$ converges uniformly for all real x.
   (b) Prove that the series of part (a) defines a continuous function for all real x.

3. (a) Show that if a trigonometric series

$$\frac{a_0}{2} + \sum_{k=1}^{\infty} a_k \cos kx + b_k \sin kx$$

   converges uniformly on $[-\pi, \pi]$ then it converges uniformly for *all* real x.
   (b) Prove that the uniformly convergent series of part (a) is necessarily the Fourier series of the function it represents. (*Hint:* Use Theorem 4.4.)

4. (a) Show that if $|c_k| \le B$ for some fixed number B, then the series

$$u(x, t) = \sum_{k=1}^{\infty} c_k e^{-k^2 t} \sin kx$$

   is a solution of the differential equation

$$u_{xx} = u_t \qquad \text{for } t > 0 \quad \text{and} \quad x \text{ in } [0, \pi],$$

   satisfying $u(0, t) = u(\pi, t) = 0$. (*Hint:* Use Theorem 4.5.)
   (b) Show that, if $u(x, t)$ in part (a) is defined for $t = 0$ by a uniformly convergent series, then $u(x, t)$ is continuous on the set S in $\mathcal{R}^2$ defined by $0 \le t, 0 \le x \le \pi$.
   (c) Show that the function $u(x, t)$ is infinitely often differentiable with respect to both x and t, for $t > 0$.

5. Show that if a trigonometric series of the form shown in Exercise 3(a) satisfied $|a_n| \le A/n^2$, $|b_n| \le B/n^2$ for $n = 1, 2, 3, \ldots$, and some constants A and B, then the trigonometric series is a Fourier series.

6. By considering the partial sums of the power series $\sum_{k=0}^{\infty} x^k$ for $-1 < x < 1$, show that the series fails to converge uniformly on $(-1, 1)$.

7. Show that $\sum_{k=1}^{\infty} (-1)^k (1 - x)x^k$ converges uniformly on $[0, 1]$, but that $\sum_{k=1}^{\infty} (1 - x)x^k$ only converges pointwise on $[0, 1]$.

8. (a) Assume that the series $\sum_{k=1}^{\infty} k^2 a_k$ and $\sum_{k=1}^{\infty} k^2 b_k$ both converge absolutely. Show that

$$w(x, t) = \sum_{k=1}^{\infty} \sin kx (a_k \cos kat + b_k \sin kat)$$

   is a solution of the one-dimensional wave equation $a^2 w_{xx} = w_{tt}$. (*Hint:* Use the Weierstrass test and Theorem 4.5.)

(b) Show that the solution $w(x, t)$ of part (a) satisfies the boundary conditions $w(0, t) = w(\pi, t) = 0$ for $t \geq 0$ and an initial condition $w(x, 0) = h(x)$, where h is twice continuously differentiable.

9. Show that, with uniform convergence replaced by pointwise convergence, the statements of Theorems (a) 4.3, (b) 4.4, and (c) 4.5 become false.

5 POWER SERIES

A **power series** is an infinite series of the form

$$\sum_{k=0}^{\infty} a_k(x - a)^k = a_0 + a_1(x - a) + a_2(x - a)^2 + \cdots .$$

Such a series defines a function of x for all x for which the series converges. The Taylor series discussed in Section 4 are examples of power series associated with known functions such as e^x, $\cos x$, and $1/(1 - x)$. Our point of view here will be different in that we will start with the series, rather than with some other representation for the function, and then study the series directly as a function of x. This point of view is essential to the use of power series in solving ordinary differential equations.

5A Interval of Convergence

We speak of the series $\sum_{k=0} a_k(x - a)^k$ as being a power series "about" $x = a$, because the set of real numbers x for which such a series converges is always either an interval with its midpoint at $x = a$, or else is the whole real line. We will not prove this fact in general, but the examples will illustrate it. Figure 5(a) shows an interval of convergence; one-half its length is called the **radius of convergence**, denoted by R.

EXAMPLE 1   The power series

$$\sum_{k=0}^{\infty} (-1)^k \frac{(x - 1)^k}{2^k}$$

can be looked at as a geometric series of the form $\sum_{k=0}^{\infty} r^k$, with $r = -(x - 1)/2$. For a geometric series, we saw in Section 1 that convergence holds just exactly when $|r| = |-(x - 1)/2| < 1$, in other words, when $|x - 1| < 2$. Thus the interval of convergence is $-1 < x < 3$, with $x = 1$ as its midpoint. See Figure 5(b). The radius of convergence is $R = 2$.

EXAMPLE 2   The power series about $x = 0$ given by

$$\sum_{k=1}^{\infty} \frac{1}{k\,2^k} x^k$$

can be tested for absolute convergence by using the ratio test. The kth term is $a_k = 2^{-k} x^k / k$, so

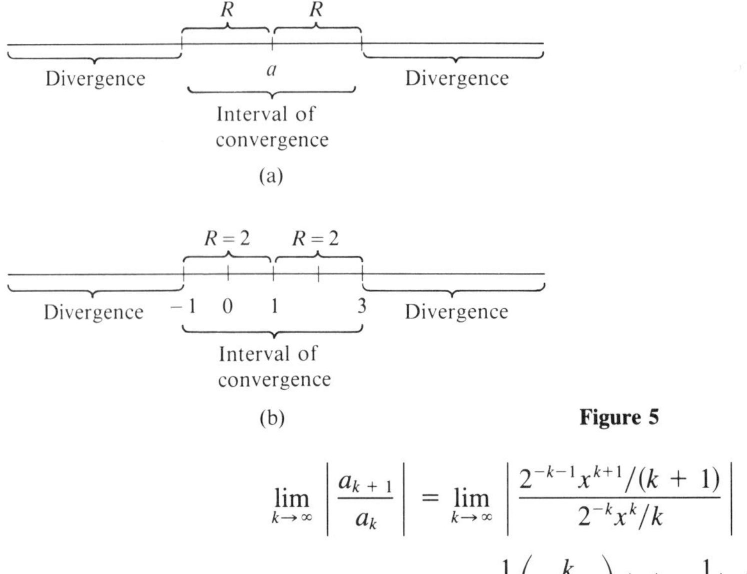

Figure 5

$$\lim_{k \to \infty} \left| \frac{a_{k+1}}{a_k} \right| = \lim_{k \to \infty} \left| \frac{2^{-k-1}x^{k+1}/(k+1)}{2^{-k}x^k/k} \right|$$

$$= \lim_{k \to \infty} \frac{1}{2}\left(\frac{k}{k+1} \right)|x| = \frac{1}{2}|x|.$$

By the ratio test, the series converges absolutely when $\frac{1}{2}|x| < 1$, that is, when $|x| < 2$, and diverges when $\frac{1}{2}|x| > 1$, that is, when $|x| > 2$. Thus the interval of convergence has radius $R = 2$ and is centered at $x = 0$. Because the ratio test gives no information when the limit, in this case $|x|/2$, is equal to 1, we have to check that case separately. The points satisfying $|x|/2 = 1$ are just the points $x = 2$ and $x = -2$. These points are the endpoints of the interval of convergence, and direct substitution into the series shows that at $x = 2$ we have $\sum_{k=1}^{\infty} 1/k$, which diverges; at $x = -2$ we have $\sum_{k=1}^{\infty} (-1)^k/k$, which is a convergent alternating series. Thus the precise interval of convergence is $-2 \le x < 2$.

EXAMPLE 3   The series

$$\sum_{k=0}^{\infty} \frac{1}{k!} x^k$$

is the Taylor expansion of the function e^x, and it was shown in Section 3 that the series converges, even absolutely, to e^x for all real values of x. Hence we see that the interval of convergence is $-\infty < x < \infty$, and it is customary to say that the radius of convergence is $R = \infty$.

EXAMPLE 4   The series

$$\sum_{k=0}^{\infty} \frac{1}{3^k(k+2)} x^{2k}$$

has kth term $a_k = x^{2k}/3^k(k+2)$. We apply the ratio test.

$$\lim_{k \to \infty} \left| \frac{a_{k+1}}{a_k} \right| = \lim_{k \to \infty} \left| \frac{x^{2k+2}/3^{k+1}(k+3)}{x^{2k}/3^k(k+2)} \right|$$

$$= \lim_{k \to \infty} |x|^2 \frac{k+2}{3(k+3)} = \frac{1}{3} x^2.$$

By the ratio test, we have convergence for $|x| < \sqrt{3}$ and divergence for $|x| > \sqrt{3}$. A separate check shows divergence for $x = \sqrt{3}$ and convergence for $x = -\sqrt{3}$.

5B Differentiation and Integration

One reason the power series representation of functions is so useful is that, in the interior of its interval of convergence, a power series can be differentiated and integrated term by term. To see what this means in practice, we first consider some examples. The term "interior" of an interval is meant specifically to exclude the endpoints of the interval.

EXAMPLE 5   The Taylor expansion

$$\frac{1}{1-x} = \sum_{k=0}^{\infty} x^k = 1 + x + x^2 + \cdots$$

is valid for $-1 < x < 1$. In the interior of the interval of convergence, that is for $-1 < x < 1$, we integrate both sides and include a constant of integration to get

$$-\ln(1-x) = c + \sum_{k=0}^{\infty} \frac{x^{k+1}}{k+1} = c + x + \frac{x^2}{2} + \frac{x^3}{3} + \cdots.$$

The constant c is determined by setting $x = 0$ in both sides. We get

$$-\log 1 = 0 = c,$$

so

$$-\ln(1-x) = \sum_{k=1}^{\infty} \frac{x^k}{k} = x + \frac{x^2}{2} + \frac{x^3}{3} + \cdots, \qquad -1 < x < 1.$$

Computing the successive derivatives of $-\log(1-x)$ at $x = 0$ shows that the preceding expansion is just the Taylor expansion of $-\log(1-x)$ about $x = 0$. It is important to notice that the function $\log(1-x)$ is defined for all $x < 1$, but that the series representation fails to converge when $x < -1$. The reason is that the radius of convergence is limited to $R = 1$ by the fact that $\log(1-x)$ is undefined at $x = 1$.

EXAMPLE 6   Consider the Taylor expansion

$$\sin x = \sum_{k=0}^{\infty} \frac{(-1)^k}{(2k+1)!} x^{2k+1} = x - \frac{x^3}{3!} + \frac{x^5}{5!} - \cdots,$$

valid for all real x. If we compute the derivative on both sides, we get

$$\cos x = \sum_{k=0}^{\infty} \frac{(-1)^k}{(2k+1)!} (2k+1)x^{2k}$$

$$= \sum_{k=0}^{\infty} \frac{(-1)^k}{(2k)!} x^{2k} = 1 - \frac{x^2}{2!} + \frac{x^4}{4!} - \cdots .$$

This is just the Taylor expansion of $\cos x$.

The theorem that justifies the preceding computations is as follows.

5.1 Theorem. A power series $\sum_{k=0}^{\infty} a_k(x-a)^k$ can be arbitrarily often differentiated or integrated term by term in the interior of any interval of convergence of the form $|x-a| < R$, where $R > 0$.

Proof. We prove first the part about integration, under the assumption that the series

$$f(x) = \sum_{k=0}^{\infty} a_k x^k$$

represents the function $f(x)$ in the interval $|x| < R$. The case of an expansion about a point other than $a = 0$ can then be handled by a simple change of variable (see Exercise 11). We will prove that if $-R < x_1 < R$ then

$$\int_0^{x_1} f(x)\, dx = \sum_{k=0}^{\infty} a_k \int_0^{x_1} x^k\, dx = \sum_{k=0}^{\infty} a_k \frac{x_1^{k+1}}{k+1}.$$

To use Theorem 4.4, we need to verify that the given power series converges uniformly on the interval between 0 and x_1. But if we choose s and r so that $R > s > r > x_1$, then the given series converges at $x = s$. Hence its terms tend to zero and so are bounded in absolute value by some number m: $|a_k s^k| \le m$. Then, for $x \le r < s$, we have

$$|a_k x^k| \le |a_k| r^k = |a_k| s^k \left(\frac{r}{s}\right)^k$$

$$\le m \left(\frac{r}{s}\right)^k .$$

The series with kth term $m(r/s)^k$ is a convergent geometric series, because $0 < r/s < 1$. This shows by Theorems 4.1 and 4.3 that the given series converges uniformly on $(-r, r)$ to a continuous function, which is necessarily $f(x)$ because the series is assumed to converge to $f(x)$ at each point x. Because r is any number such that $0 < r < R$, we can include any x in $(-R, R)$ in an interval of uniform convergence, so we can integrate term by term on any interval between 0 and x_1.

For the differentiation part of the theorem, we start with the same series for $f(x)$ and show that

$$f'(x) = \sum_{k=1}^{\infty} k a_k x^{k-1}.$$

To do this, we apply Theorem 4.5 by showing that the *differentiated* series converges uniformly on every interval $-r \leq x \leq r$, where $0 < r < R$. Choose a number c such that $r < c < R$. Since $\Sigma_{k=0}^{\infty} a_k c^k$ converges, there is a number b such that $\left| a_k \right| c^k \leq b$. Then for x in $[-r, r]$ we have

$$\left| k a_k x^{k-1} \right| \leq k \left| a_k \right| r^{k-1}$$

$$\leq k \left(\frac{b}{c^k} \right) r^{k-1} = k \left(\frac{b}{c} \right) \left(\frac{r}{c} \right)^{k-1}.$$

The series with kth term $k(b/c)(r/c)^{k-1}$ is geometric and convergent, because $0 < r/c < 1$. Hence the series with kth term $k a_k x^{k-1}$ converges uniformly on $[-r, r]$ by the Weierstrass test. This allows us to differentiate term by term on any interval $[-r, r]$ with $0 < r < R$. Finally, we can differentiate at any x, such that $-R < x < R$, simply by choosing r so that $\left| x \right| < r < R$.

Theorem 5.1 allows us to differentiate and integrate a power series repeatedly, because the result of performing one such operation on a power series convergent when $\left| x - a \right| < R$ is just another power series that is also convergent when $\left| x - a \right| < R$.

EXAMPLE 7   Starting with

$$\frac{1}{1 - x} = \sum_{k=0}^{\infty} x^k = 1 + x + x^2 + \cdots, \qquad \left| x \right| < 1,$$

we differentiate once to get

$$\frac{1}{(1 - x)^2} = \sum_{k=1}^{\infty} k x^{k-1} = 1 + 2x + 3x^2 + \cdots, \qquad \left| x \right| < 1.$$

Differentiating again gives

$$\frac{2}{(1 - x)^3} = \sum_{k=2}^{\infty} (k - 1)k x^{k-2} = 1 \cdot 2 + 2 \cdot 3x + 3 \cdot 4x^2 + \cdots, \qquad \left| x \right| < 1.$$

If we have a power series representation for a function f about a point $x = a$, then the power series is automatically the Taylor series for f about $x = a$. Hence we need not verify this in practice. Specifically, we have the following theorem.

5.2 Theorem.   If

$$f(x) = \sum_{k=0}^{\infty} a_k (x - a)^k$$

on some interval $\left| x - a \right| < R$, then the series is the Taylor series of f about $x = a$.

That is

$$a_n = \frac{1}{n!} f^{(n)}(a), \qquad n = 0, 1, 2 \cdots .$$

Proof. Differentiating n times in the expansion for f knocks out the first n terms, leaving

$$f^{(n)}(x) = \sum_{k=n}^{\infty} k(k-1) \cdots (k-n+1) a_k (x-a)^{k-n}$$

$$= n! \, a_n + (n+1)! \, a_{n+1}(x-a) + \cdots .$$

Now set $x = a$ on both sides. All terms on the right become zero except the first, so

$$f^n(a) = n! a_n,$$

which is what we wanted to show.

5C Finding Limits by Using Series

A convergent Taylor expansion

$$f(x) = \sum_{k=0}^{\infty} a_k (x-a)^k$$

about some point $x = a$ represents a differentiable and hence continuous function f. It follows that a limit of $f(x)$ as x approaches a can be computed by setting $x = a$ to get

$$\lim_{x \to a} f(x) = a_0.$$

This idea can be applied to calculate fairly complicated limits, as the following example shows.

EXAMPLE 8   To show that

$$\lim_{x \to 0} \frac{x - \sin x}{x^3} = \frac{1}{6}$$

we just observe that $x - \sin x = x - (x - \frac{1}{6} x^3 + \frac{1}{120} x^5 - \cdots)$. Hence $(x - \sin x)/x^3 = \frac{1}{6} - \frac{1}{120} x^5 + \cdots$. Taking the limit as $x \to 0$ amounts to setting $x = 0$ in the continuous function represented by the series. The resulting limit is $\frac{1}{6}$.

5D Products and Quotients

Power series can be multiplied and divided very much like polynomials. The product of two power series about the same point gives a third series called their **Cauchy product,** which is simply the series formed by collecting equal powers of x:

5.3 $(a_0 + a_1 x + a_2 x^2 + \cdots)(b_0 + b_1 x + b_2 x^2 + \cdots)$

$$= a_0 b_0 + (a_1 b_0 + a_0 b_1)x + (a_0 b_2 + a_1 b_1 + a_2 b_0)x^2 + \cdots .$$

The coefficient of x^k is $c_k = a_0 b_k + a_1 b_{k-1} + \cdots + a_{k-1} b_1 + a_k b_0$, and of course x may be replaced by $(x - a)$ in all three series. The relevant theorem states that the Cauchy product converges to the correct value in the interior of the common interval of convergence of the two factors.

EXAMPLE 9 Here are several simple examples.

(a) Recall that a polynomial in x is a (finite) power series:

$$(1 + x)(1 + x + x^2 + x^3 + \cdots) = 1 + 2x + 2x^2 + 2x^3 + \cdots ,$$

which is valid for $-1 < x < 1$.

(b) Multiplying together the powers series about $x = 0$ for $(x - 1)^{-1}$ and $\ln (1 - x)$, we get, after cancelling minus signs,

$$(1 + x + x^2 + x^3 + \cdots)(x + \tfrac{1}{2}x^2 + \tfrac{1}{3}x^3 + \cdots)$$

$$= x + (1 + \tfrac{1}{2})x^2 + (1 + \tfrac{1}{2} + \tfrac{1}{3})x^3 + \cdots = x + \tfrac{3}{2}x + \tfrac{11}{6}x^3 + \cdots ,$$

which is valid for $-1 < x < 1$.

(c) Multiplying together the power series for $1/x$ and $\ln x$ about $x = 1$, we get

$$(1 - (x - 1) + (x - 1)^2 - (x - 1)^3 + \cdots)$$

$$((x - 1) - \tfrac{1}{2}(x - 1)^2 + \tfrac{1}{3}(x - 1)^3 - \cdots)$$

$$= (x - 1) + (-1 - \tfrac{1}{2})(x - 1)^2 + (1 + \tfrac{1}{2} + \tfrac{1}{3})(x - 1)^3 + \cdots$$

$$= (x - 1) - \tfrac{3}{2}(x - 1)^2 + \tfrac{11}{6}(x - 1)^3 + \cdots .$$

Examples (b) and (c) illustrate the fact that it may be impossible to find a very simple formula for the kth coefficient in a power series. The best we can do in those two examples is to use a general expression of the form $\sum_{j=1}^{k} (1/j)$.

While division of power series is theoretically the reverse of multiplication, there are some things to watch out for in practice. Suppose we are looking for coefficients a_k that satisfy

$$\frac{c_0 + c_1 x + c_2 x^2 + \cdots}{b_0 + b_1 x + b_2 x^2 + \cdots} = a_0 + a_1 x + a_2 x^2 + \cdots .$$

We do need $b_0 \neq 0$, since $a_0 = c_0/b_0$. (However, if $b_0 = 0$ and $b_1 \neq 0$, we could factor an x from the denominator and then proceed.) Also, if the series in the denominator takes on the value zero for some x in its interval of convergence, that will limit the convergence interval of the quotient series. With those two observations in mind, we could then proceed to find the a_k by solution of the sequence of equations

$$a_0 b_0 = c_0, \qquad a_0 b_1 + a_1 b_0 = c_1, \qquad a_0 b_2 + a_1 b_1 + a_2 b_0 = c_2$$

for the desired values of the a_k. This process can be carried out as far as you like, because the solution of each equation depends only on the solution of the previous equations in the list. For this reason you can also truncate the series in the numerator and denominator and use polynomial long division to compute any preassigned number of terms.

EXAMPLE 10   To compute the first few terms in the expansion of $(1 + e^x)^{-1}$, note that $c_0 = 1$, $c_1 = c_2 = \cdots = 0$. Also $b_0 = 2$, $b_k = 1/k!$, $k = 1, 2, 3,$ $\ldots$. We then solve

$$2a_0 = 1, \qquad a_0 + 2a_1 = 0, \qquad \tfrac{1}{2}a_0 + a_1 + 2a_2 = 0,$$

$$\tfrac{1}{6}a_0 + \tfrac{1}{2}a_1 + a_2 + 2a_3 = 0, \ldots .$$

The result is $a_0 = \tfrac{1}{2}$, $a_1 = -\tfrac{1}{4}$, $a_2 = 0$, $a_3 = \tfrac{1}{48}$, $\ldots$. Hence

$$\frac{1}{1 + e^x} = \frac{1}{2} - \frac{1}{4}x + \frac{1}{48}x^3 + \cdots .$$

To then find the first three terms in the expansion of $\sin x/(1 + e^x)$ we compute

$$\left(x - \frac{1}{6}x^3 + \cdots\right)\left(\frac{1}{2} - \frac{1}{4}x + \frac{1}{48}x^3 + \cdots\right) = \frac{1}{2}x - \frac{1}{4}x^2 - \frac{1}{12}x^3 + \cdots .$$

EXERCISES

1. Using the ratio test, or by other means, find the interval of convergence of each of the following power series. In case the interval has finite endpoints, determine whether the series converges when x is equal to each of the endpoints, and sketch the interval.

   (a) $\displaystyle\sum_{k=1}^{\infty} \frac{1}{k}x^k.$

   (b) $\displaystyle\sum_{k=1}^{\infty} \frac{1}{k}(x - 2)^k.$

   (c) $\displaystyle\sum_{k=0}^{\infty} \frac{1}{(2k)!}x^{2k}.$

   (d) $\displaystyle\sum_{k=1}^{\infty} k^2(x + 1)^k.$

   (e) $\displaystyle\sum_{k=1}^{\infty} \frac{k}{k+1}(x + 2)^k.$

   (f) $\displaystyle\sum_{k=0}^{\infty} \frac{1}{\sqrt{k^2 + 1}}x^k.$

   (g) $\displaystyle\sum_{k=1}^{\infty} \frac{1}{\sqrt{k}}(x + 3)^{2k+1}.$

   (h) $\displaystyle\sum_{k=0}^{\infty} 2^k x^{2k}.$

2. Use the Taylor expansion $e^x = \sum_{k=0}^{\infty} \frac{1}{k!}x^k$ to derive Taylor expansions for each of the following functions about the point $a = 0$.

   (a) $e^{-x}.$

   (b) $(e^x + e^{-x})/2 = \cosh x.$

   (c) $(e^x - e^{-x})/2 = \sinh x.$

   (d) $xe^x.$

   (e) $e^{x^2}.$

   (f) $e^{5x}.$

3. Use the Taylor expansion $\log(1 - x) = -\sum_{k=1}^{\infty} \frac{1}{k} x^k$ to derive Taylor expansions, valid for $|x| < 1$, for each of the following functions.

   (a) $\log(1 + x)$.

   (b) $\frac{1}{2} \log\left(\frac{1 + x}{1 - x}\right)$.

   (c) $\log(1 + x^2)$.

   (d) $x^2 \log(1 - x^3)$.

4. Use the Taylor expansion $(1 - x)^{-1} = \sum_{k=0}^{\infty} x^k$ to derive Taylor expansions for each of the following.

   (a) $\frac{1}{1 + x^2}$, about $a = 0$.

   (b) $\frac{1}{1 - x^2}$, about $a = 0$.

   (c) $\frac{1}{x}$, about $a = 1$.

   (d) $\frac{x}{1 - x}$, about $a = 0$.

5. Use the relation $d(\arctan x)/dx = 1/(1 + x^2)$ and the result of Exercise 4(a) to derive the Taylor expansion of $\arctan x$ about $a = 0$.

6. Use the relation $d(1 + x^2)^{-1}/dx = -2x/(1 + x^2)^2$ and the result of Exercise 4(a) to derive the Taylor expansion of $(1 + x^2)^{-2}$ about $a = 0$.

7. Use the Taylor expansion $(1 - x)^{-1} = \sum_{k=0}^{\infty} x^k$ to show that

$$\frac{x(x + 1)}{(1 - x)^3} = \sum_{k=1}^{\infty} k^2 x^k, \qquad \text{for } |x| < 1.$$

   First find the expansion for $(1 - x)^{-3}$.

8. Use the Taylor expansions to find each of the following limits.

   (a) $\lim_{x \to 0} \frac{\log(1 + x^2)}{x^2}$.

   (b) $\lim_{x \to 0} \frac{\log(1 - x^5)}{x^5}$.

   (c) $\lim_{x \to 0} \frac{x + \log(1 - x)}{x^2}$.

   (d) $\lim_{x \to 0} \frac{\cos x - 1 + x^2/2}{x^4}$.

9. Show that $\sum_{k=0}^{\infty} x^{-k} = x/(x - 1)$ for $|x| > 1$.

10. (a) Show that, if $f(x) = (1 + x)^{\alpha}$, then $f^{(k)}(0) = (\alpha - 1) \cdots (\alpha - k + 1)$, so that the Taylor expansion of $(1 + x)^{\alpha}$ about 0 is

$$(1 + x)^{\alpha} = 1 + \sum_{k=1}^{\infty} \frac{(\alpha - 1) \cdots (\alpha - k + 1)}{k!} x^k.$$

   (b) Write out the expansion in part (a) for $\alpha = 3$, $\alpha = -3$, and $\alpha = \frac{1}{2}$.

11. (a) Derive the formula

$$\frac{d}{dx}\left(\sum_{k=0}^{\infty} a_k(x - a)^k\right) = \sum_{k=1}^{\infty} k a_k(x - a)^{k-1}, \qquad |x - a| < R,$$

   from the formula

$$\frac{d}{dx}\left(\sum_{k=0}^{\infty} a_k x^k\right) = \sum_{k=1}^{\infty} k a_k x^{k-1}, \qquad |x| < R.$$

   (b) Derive the formula

$$\int_0^{x_1} \sum_{k=0}^{\infty} a_k(x - a)^k \, dx = \sum_{k=0}^{\infty} a_k \frac{(x_1 - a)^{k+1}}{k + 1}, \qquad |x_1 - a| < R,$$

from the formula

$$\int_0^{x_1} \sum_{k=0}^{\infty} a_k x^k \, dx = \sum_{k=0}^{\infty} a_k \frac{x_1^{k+1}}{k+1}, \qquad |x_1| < R.$$

12. (a) Show that

$$\lim_{n \to \infty} n \, \log\left(1 + \frac{a}{n}\right) = a$$

by using the Taylor expansion for $\log(1 + x)$ about $x = 0$.
(b) Use part (a) to show that

$$\lim_{n \to \infty} \left(1 + \frac{a}{n}\right)^n = e^a.$$

13. Show that, if $f(x) = \sum_{k=0}^{\infty} c_k x^k$ converges in some interval, then

$$f''(x) + f(x) = \sum_{k=0}^{\infty} \left[c_k + (k+1)(k+2)c_{k+2}\right] x^k$$

in the interior of the same interval.
14. Show that, if two power series

$$\sum_{k=0}^{\infty} a_k (x - a)^k \quad \text{and} \quad \sum_{k=0}^{\infty} b_k (x - a)^k$$

converge to the same function on an interval containing $x = a$, then $a_k = b_k$ for $k = 0, 1, 2, 3 \ldots$. Use Theorem 5.2.
15. (a) Find the Taylor expansion of $f(x) = 1/(x + c)$ about $x = 0$ for $c \neq 0$.
(b) Find the Taylor expansion of $g(x) = 1/[(x + c)(x + d)]$ about $x = 0$, for $c \neq 0$, $d \neq 0$, and $c \neq d$, by expressing g as a sum of two fractions.

6 DIFFERENTIAL EQUATIONS

A differential equation of the form

$$y' = f(x, y),$$

if it has a solution $y = y(x)$ near $x = x_0$, determines the derivative $y'(x_0)$ by $y'(x_0) = f(x_0, y(x_0))$. If f is sufficiently differentiable, even higher derivatives are similarly determined at x_0.

EXAMPLE 1   The equation $y' = y^2$ has $y = (1 - x)^{-1}$ for a solution satisfying $y(0) = 1$. The power series expansion

$$y = \frac{1}{1 - x} = 1 + x + x^2 + x^3 + \cdots$$

is a Taylor series, so by the Taylor coefficient formulas we also have

$$y = y(0) + \frac{y'(0)}{1!}x + \frac{y''(0)}{2!}x^2 + \frac{y'''(0)}{3!}x^3 + \cdots .$$

Comparison of coefficients of x^k in the two expansions shows that $y^{(k)}(0)/k! = 1$, so $y^{(k)}(0) = k!$, $k = 0, 1, 2, \ldots$. Suppose, however, that we had no formula for the coefficients to begin with. (Most examples are of this sort.) Starting with the given differential equation, successive derivatives can be computed and then simplified by substitution from the earlier equations:

$$y' = y^2,$$

$$y'' = 2yy' = 2y^3,$$

$$y''' = 6y^2y' = 6y^4,$$

$$y^{(4)} = 24y^3y' = 24y^5.$$

The general pattern is evidently $y^{(k)} = k!y^{k+1}$. The formal Taylor expansion for a solution $y = y(x)$ about x_0 with $y(x_0) = y_0$ can then be written down:

$$y(x) = y_0 + y_0^2(x - x_0) + y_0^3(x - x_0)^2 + \cdots$$

$$= y_0(1 + y_0(x - x_0) + y_0^2(x - x_0)^2 + \cdots$$

$$= \frac{y_0}{1 - y_0(x - x_0)} .$$

Higher-order equations, for example

$$y'' = f(x, y, y'),$$

can be treated the same way.

EXAMPLE 2 Suppose we want successive derivatives $y^{(k)}(0)$ to a solution of

$$y'' = yy'$$

given that $y(0) = 1$ and $y'(0) = -1$. First compute from the given equation some formulas for higher derivatives. Then simplify by substituting the given values $y(0) = 1$, $y'(0) = -1$. We find

$$y'' = yy'; \qquad y''(0) = -1,$$

$$y''' = yy'' + (y')^2 = y^2y' + (y')^2; \qquad y'''(0) = 0,$$

$$y^{(4)} = 2y(y')^2 + y^2y'' + 2y'y'' = 4y(y')^2 + y^3y'; \; y^{(4)}(0) = 3.$$

The first five terms of the Taylor expansion of $y(x)$ about $x = 0$ then add up to

$$y(0) + \frac{y'(0)}{1!}x + \frac{y''(0)}{2!}x^2 + \frac{y'''(0)}{3!}x^3 + \frac{y^{(4)}(0)}{4!}x^4 = 1 - x - \frac{1}{2}x^2 + \frac{1}{8}x^4.$$

In this example there does not seem to be a simple pattern, but clearly we could compute as many terms as we had time and space for.

EXAMPLE 3 From the differential equation

$$y'' = xy' - y,$$

we compute, using substitution after differentiating,

$$y''' = xy'' = x(xy' - y)$$
$$= x^2 y' - xy,$$
$$y^{(4)} = x^2 y'' + xy' - y = x^2(xy' - y) + xy' - y$$
$$= (x^3 + x)y' - (x^2 + 1)y.$$

If we denote $y(0)$ by c_0 and $y'(0)$ by c_1, then

$$y''(0) = -c_0, \qquad y'''(0) = 0, \qquad y^{(4)}(0) = -c_0.$$

Thus the Taylor expansion of $y(x)$ about $x = 0$ has the form

$$y(x) = c_0 + c_1 x - \frac{c_0}{2!} x^2 + 0 - \frac{c_0}{4!} x^4 - \cdots$$

$$= c_0 \left(1 - \frac{1}{2} x^2 - \frac{1}{24} x^4 - \cdots \right) + c_1 x.$$

For comparison, note that if the original differential equation were replaced by

$$y'' = -y$$

then solutions would have the form

$$y(x) = c_0 \cos x + c_1 \sin x,$$

and that $y(0) = c_0$, $y'(0) = c_1$.

EXERCISES

1. Find the first three nonzero terms in the Taylor expansion about $x = 0$ of $y = y(x)$ if y satisfies each of the following relations.

   (a) $y' = y^2 + y$, $y(0) = 1$.

   [*Ans.* $y = 1 + 2x + 3x^2 + \cdots$]

   (b) $y' = y^2 + x$, $y(0) = -1$.

   [*Ans.* $y = -1 + x - \frac{1}{2} x^2 + \cdots$]

   (c) $y' = xy$, $y(0) = 2$.

   [*Ans.* $y = 2 + x^2 + \frac{1}{4} x^4 + \cdots$]

   (d) $y'' = xy$, $y(0) = 1$, $y'(0) = 0$.

   [*Ans.* $y = 1 + \frac{1}{6} x^3 + \frac{1}{180} x^6 + \cdots$]

2. Find the first four nonzero terms in the Taylor expansion of $y = y(x)$ about $x = 0$ if $y'' = yy'$ and $y(0) = y'(0) = 1$.

   [*Ans.* $y = 1 + x + \frac{1}{2} x^2 + \frac{1}{3} x^3 + \cdots$]

3. Find the first four nonzero terms in the Taylor expansion of $y = y(x)$ about $x = 1$ if $y''' = y$ and $y(1) = 2$, $y'(1) = 0$, $y''(1) = 1$.

$$[Ans.\ y = 2 + \frac{1}{2}(x - 1)^2 + \frac{1}{3}(x - 1)^3 + \frac{1}{120}(x - 1)^5 + \cdots]$$

4. Suppose $y'' = x^2 y$ while $y(0) = c_0$ and $y'(0) = c_1$. Show that $y = y(x)$ has the form

$$y = c_0\left(1 + \frac{1}{12}x^4 + \cdots\right) + c_1\left(x + \frac{1}{20}x^5 + \cdots\right).$$

5. Show that if $y''' = y^2 y'$ and $y(0) = y'(0) = y''(0) = 1$ then

$$y = 1 + x + \frac{1}{2}x^2 + \frac{1}{6}x^3 + \frac{1}{8}x^4 + \frac{3}{40}x^5 + \cdots.$$

7 SERIES SOLUTIONS

The solutions of many of the differential equations we have studied can be represented in terms of their Taylor expansions. For example, polynomials, the elementary transcendental functions $\cos x$, $\sin x$, e^x, and linear combinations of all these have Taylor expansions that are valid for all x. In fact, there is a large and important class of differential equations that has solutions representable by power series. Even if the solution so represented is not a combination of elementary functions at all, the infinite series expansion may serve to define a new function nearly as important as some of the more familiar ones. Furthermore, the partial sums of a series expansion often give useful approximations to the true solution.

Recall that a Taylor expansion of a function f is an infinite power series of the form

$$\sum_{k=0}^{\infty} \frac{f^{(k)}(x_0)}{k!}(x - x_0)^k = f(x_0) + \frac{1}{1!}f'(x_0)(x - x_0) + \frac{1}{2!}f''(x_0)(x - x_0)^2 + \cdots.$$

Such series always converge absolutely in an interval $x_0 - r < x < x_0 + r$ that is symmetric about x_0; for functions f that are called **analytic,** the series actually converges to $f(x)$ for all x in that interval. Furthermore, Taylor expansions about the same point x_0 can be conveniently added and multiplied, while any Taylor expansion can be differentiated or integrated term by term to produce the Taylor expansion of the derivative or integral of the expanded function within the interval of convergence.

EXAMPLE 1   The Taylor expansions

$$e^x = \sum_{k=0}^{\infty} \frac{x^k}{k!}, \qquad -\infty < x < \infty,$$

$$\cos x = \sum_{k=0}^{\infty} \frac{(-1)^k x^{2k}}{(2k)!}, \qquad -\infty < x < \infty,$$

$$\sin x = \sum_{k=0}^{\infty} \frac{(-1)^k x^{2k+1}}{(2k+1)!}, \qquad -\infty < x < \infty,$$

$$\frac{1}{1-x} = \sum_{k=0}^{\infty} x^k, \qquad -1 < x < 1,$$

$$\log x = \sum_{k=1}^{\infty} \frac{(-1)^{k+1}(x-1)^k}{k}, \qquad 0 < x < 2,$$

can all be computed directly from the general Taylor formula and can be shown to converge to the value of the function on the left for each x in the indicated interval. Furthermore, we can compute the derivative of a function such as $\cos 2x$ as follows:

$$\frac{d}{dx}\cos 2x = \frac{d}{dx} \sum_{k=0}^{\infty} \frac{(-1)^k (2x)^{2k}}{(2k)!}$$

$$= \sum_{k=0}^{\infty} \frac{d}{dx} \frac{(-1)^k (2x)^{2k}}{(2k)!}$$

$$= \sum_{k=1}^{\infty} \frac{(-1)^k (2k)(2x)^{2k-1}(2)}{(2k)!}$$

$$= 2 \sum_{k=1}^{\infty} \frac{(-1)^k (2x)^{2k-1}}{(2k-1)!}$$

$$= -2 \sum_{k=0}^{\infty} \frac{(-1)^k (2x)^{2k+1}}{(2k+1)!} = -2 \sin 2x.$$

In the last step we simply replaced k by $k+1$ throughout in order to make the expansion look more like the expansion in the preceding examples.

Many of the most important examples of series solutions of differential equations are expressible as power series about the point $x_1 = 0$. Since Taylor expansions about zero are also a little easier to work with, most of our examples will be of that kind. The next example shows how to solve a familiar differential equation using series.

EXAMPLE 2   To solve $y'' + y = 0$, we try to find a solution of the form

$$y(x) = \sum_{k=0}^{\infty} c_k x^k.$$

This form of the expansion is particularly appropriate if we want to solve an initial-value problem with $y(0)$ and $y'(0)$ specified, because then $c_0 = y(0)$ and $c_1 = y'(0)$, if there is a Taylor expansion for the solution about $x = 0$. Proceeding under that assumption for the moment, we compute

$$y'(x) = \sum_{k=1}^{\infty} kc_k x^{k-1},$$

$$y''(x) = \sum_{k=2}^{\infty} (k-1)kc_k x^{k-2}$$

$$= \sum_{k=0}^{\infty} (k+1)(k+2)c_{k+2} x^k.$$

We have shifted the index of summation by 2 in the expression for $y''(x)$ in order to make it more convenient to add to the one for $y(x)$. We find that for $y(x)$ to represent a solution we must have

$$y''(x) + y(x) = \sum_{k=0}^{\infty} c_k x^k + \sum_{k=0}^{\infty} (k+1)(k+2)c_{k+2} x^k$$

$$= \sum_{k=0}^{\infty} [c_k + (k+1)(k+2)c_{k+2}]x^k = 0.$$

Next we use the fundamental fact that, for a Taylor expansion to be identically zero, all the coefficients must be zero. Hence

$$c_{k+2} = -\frac{c_k}{(k+1)(k+2)}, \qquad k = 0, 1, 2, \ldots .$$

Recalling that we can specify a particular solution by determining the numbers $c_0 = y(0)$, $c_1 = y'(0)$, it is natural to compute recursively

$$c_2 = -\frac{c_0}{1\cdot 2} = -\frac{c_0}{2!}, \qquad\qquad c_3 = -\frac{c_1}{2\cdot 3} = -\frac{c_1}{3!},$$

$$c_4 = -\frac{c_2}{3\cdot 4} = \frac{c_0}{4!}, \qquad\qquad c_5 = -\frac{c_3}{4\cdot 5} = \frac{c_1}{5!},$$

$$c_6 = -\frac{c_4}{5\cdot 6} = -\frac{c_0}{6!}, \qquad\qquad c_7 = -\frac{c_5}{6\cdot 7} = -\frac{c_1}{7!},$$

$$\vdots \qquad\qquad\qquad\qquad \vdots$$

$$c_{2k} = -\frac{c_{2k-2}}{(2k-1)(2k)} = (-1)^k \frac{c_0}{(2k)!}, \quad c_{2k+1} = -\frac{c_{2k-1}}{2k(2k+1)} = (-1)^k \frac{c_1}{(2k+1)!}.$$

If we take $y(0) = c_0$ and $y'(0) = c_1 = 0$, then only the first column contains nonzero entries, and we get the solution

$$y_0(x) = c_0 - \frac{c_0}{2!} x^2 + \frac{c_0}{4!} x^4 - \cdots + (-1)^k \frac{c_0}{(2k)!} x^{2k} + \cdots$$

$$= c_0 \cos x.$$

On the other hand, the choice $y(0) = c_0 = 0$ and $y'(0) = c_1$ makes the entries in the first column all zero, so we get another solution from the second column:

$$y_1(x) = c_1 - \frac{c_1}{3!}x^3 + \frac{c_1}{5!}x^5 - \cdots + (-1)\frac{c_1}{(2k+1)}x^{2k+1} + \cdots$$

$$= c_1 \sin x.$$

Thus the general solution is

$$y(x) = c_0 \cos x + c_1 \sin x$$

as usual.

EXAMPLE 3   Solutions of Airy's equation $y'' + xy = 0$ are not obtainable in terms of elementary functions, so we try

$$y(x) = \sum_{k=0}^{\infty} c_k x^k,$$

$$y''(x) = \sum_{k=2}^{\infty} (k-1)k c_k x^{k-2}.$$

Substitution into the differential equation gives

$$y''(x) + xy(x) = \sum_{k=2}^{\infty} (k-1)k c_k x^{k-2} + \sum_{k=0}^{\infty} c_k x^{k+1}$$

$$= \sum_{k=0}^{\infty} (k+1)(k+2) c_{k+2} x^k + \sum_{k=1}^{\infty} c_{k-1} x^k = 0,$$

where this time we shifted the index in the first summation up by 2 and in the second summation down by 1 in order to make the powers of x agree. Thus we get a single term, the constant $2c_2$, in the first summation that does not correspond to a term in the second summation. We can write then

$$2c_2 + \sum_{k=1}^{\infty} [(k+1)(k+2) c_{k+2} + c_{k-1}]x^k = 0;$$

setting all coefficients equal to zero gives

$$c_2 = 0,$$

$$c_{k+2} = -\frac{c_{k-1}}{(k+1)(k+2)}.$$

We find as a result that the terms are determined in sequences with indexes differing by 3, and that

$$0 = c_2 = c_5 = c_8 = \cdots = c_{3k+2} = \cdots.$$

However, if c_0 or c_1 is not zero, it is worth computing as follows:

$$c_3 = -\frac{c_0}{2 \cdot 3},$$

$$c_6 = -\frac{c_3}{5 \cdot 6} = \frac{c_0}{2 \cdot 3 \cdot 5 \cdot 6},$$

$$c_9 = -\frac{c_6}{8 \cdot 9} = -\frac{c_0}{2 \cdot 3 \cdot 5 \cdot 6 \cdot 8 \cdot 9},$$

$$\vdots$$

$$c_{3k} = -\frac{c_{3k-3}}{(3k-1)3k} = \frac{(-1)^k c_0}{2 \cdot 3 \cdot 5 \cdot 6 \ldots (3k-1)3k},$$

$$c_4 = -\frac{c_1}{3 \cdot 4},$$

$$c_7 = -\frac{c_4}{6 \cdot 7} = \frac{c_1}{3 \cdot 4 \cdot 6 \cdot 7},$$

$$c_{10} = -\frac{c_7}{9 \cdot 10} = -\frac{c_1}{3 \cdot 4 \cdot 6 \cdot 7 \cdot 9 \cdot 10},$$

$$\vdots$$

$$c_{3k+1} = -\frac{c_{3k-2}}{3k(3k+1)} = \frac{(-1)^k c_1}{3 \cdot 4 \cdot 6 \cdot 7 \ldots 3k(3k+1)}.$$

The solution determined by $y(0) = c_0 = 1$, $y'(0) = c_1 = 0$ is

$$y_0(x) = 1 + \sum_{k=1}^{\infty} \frac{(-1)^k x^{3k}}{2 \cdot 3 \cdot 5 \cdot 6 \ldots (3k-1)3k},$$

and the solution determined by $y(0) = c_0 = 0$, $y'(0) = c_1 = 1$ is

$$y_1(x) = x + \sum_{k=1}^{\infty} \frac{(-1)^k x^{3k+1}}{3 \cdot 4 \cdot 6 \cdot 7 \ldots 3k(3k+1)}.$$

The power series for y_0 and y_1 both converge for all x because the denominators of the kth terms each contain increasing integer factors, $2k$ in number. Hence each series has terms dominated by those of an everywhere convergent series, for example

$$\left| \frac{(-1)^k x^{3k}}{2 \cdot 3 \cdot 5 \cdot 6 \ldots (3k-1)3k} \right| \le \frac{|x|^{3k}}{(2k)!}.$$

In fact, estimates of this kind can be used to test the accuracy of stopping with a specified number of terms in a Taylor expansion. In this example, to get an estimate when $|x| \le 1$, we estimate the tail of the factorial series by a geometric series with ratio $1/4n^2$:

$$(2n)! \sum_{k=n}^{\infty} \frac{1}{(2k)!} = 1 + \frac{1}{(2n+1)(2n+2)} + \frac{1}{(2n+1)\ldots(2n+4)} + \cdots$$

$$\le 1 + \frac{1}{4n^2} + \frac{1}{(4n^2)^2} + \cdots$$

$$= \frac{4n^2}{4n^2 - 1}.$$

Thus the error in stopping after $n - 1$ terms on the interval $-1 \le x \le 1$ is at most

$$\frac{4n^2}{4n^2 - 1} \cdot \frac{1}{(2n)!}.$$

For $n = 5$, that is, keeping terms of degree 4, the error is at most $3 \cdot 10^{-7}$.

Since the two solutions y_0 and y_1 are linearly independent, the general solution of $y'' + xy = 0$ has the form

$$y(x) = c_0 y_0(x) + c_1 y_1(x).$$

The solutions of $y'' + cy = 0$ have infinitely many zero values if $c > 0$ and at most one such when $c < 0$. Analogously it can be shown that a solution of the Airy equation has infinitely many positive zeros, but at most one negative zero.

EXAMPLE 4   The Legendre equation of index m is

$$(1 - x^2)y'' - 2xy' + m(m+1)y = 0,$$

and it is discussed in Chapter 8. There a formula is given for polynomial solutions in case m is a nonnegative integer. To find other solutions in the form of power series, we let

$$y(x) = \sum_{k=0}^{\infty} c_k x^k, \qquad y'(x) = \sum_{k=1}^{\infty} kc_k x^{k-1}, \qquad y''(x) = \sum_{k=2}^{\infty} (k-1)kc_k x^{k-2},$$

so that the c_k are determined by

$$(1 - x^2)\sum_{k=2}^{\infty} (k-1)kc_k x^{k-2} - 2x \sum_{k=1}^{\infty} kc_k x^{k-1} + m(m+1)\sum_{k=0}^{\infty} c_k x^k = 0.$$

Shifting the index by 2 in the first sum allows us to write

$$[2c_2 + m(m + 1)c_0] + [2 \cdot 3c_3 - 2c_1 + m(m + 1)c_1]x$$

$$+ \sum_{k=2}^{\infty} [(k + 1)(k + 2)c_{k+2} - ((k - 1)k + 2k - m(m + 1))c_k]x^k = 0.$$

Setting the coefficient of each power of x equal to zero gives

$$c_2 = -\frac{m(m + 1)}{2} c_0, \qquad c_3 = -\frac{(m - 1)(m + 2)}{2 \cdot 3} c_1,$$

$$c_{k+2} = \frac{(k + 1 + m)(k - m)}{(k + 1)(k + 2)} c_k, \qquad k \geq 2.$$

Since the recurrence relation contains a shift by 2, it is natural to split the coefficients into those of even and those of odd index:

$$c_4 = \frac{(3 + m)(2 - m)}{3 \cdot 4} c_2 = \frac{(m - 2)m(m + 3)(m + 1)}{4!} c_0,$$

$$c_6 = \frac{(5 + m)(4 - m)}{5 \cdot 6} c_4 = -\frac{(m - 4)(m - 2)m(m + 5)(m + 3)(m + 1)}{6!} c_0,$$

$$\vdots$$

and

$$c_5 = \frac{(4 + m)(3 - m)}{4 \cdot 5} c_3 = \frac{(m - 3)(m - 1)(m + 4)(m + 2)}{5!} c_1,$$

$$c_7 = \frac{(6 + m)(5 - m)}{6 \cdot 7} c_5 = -\frac{(m - 5)(m - 3)(m - 1)(m + 6)(m + 4)(m + 2)}{7!} c_1.$$

$$\vdots$$

Clearly, if $m = 2l$ is a positive even integer, then all even coefficients are zero beyond $2l$. Thus the series expansion with even powers reduces to an even polynomial in that case. Similarly, if $m = 2l + 1$ is a positive odd integer, the series expansion with odd powers reduces to an odd polynomial. For example, when $m = 4$ and $c_0 = c_1 = 1$, we get the solutions

$$P_4(x) = 1 - \frac{4 \cdot 5}{2!} x^2 + \frac{2 \cdot 4 \cdot 7 \cdot 5}{4!} x^4,$$

$$Q_4(x) = x - \frac{3 \cdot 6}{3!} x^3 + \frac{1 \cdot 3 \cdot 8 \cdot 6}{5!} x^5 - \frac{-1 \cdot 1 \cdot 3 \cdot 10 \cdot 8 \cdot 6}{7!} x^7 + \cdots.$$

Using the ratio test for convergence (see Exercise 7) shows that the infinite series solution converges for $-1 < x < 1$. These two solutions form the basis for the collection of all solutions of the homogeneous Legendre equation of index 4 on the interval $-1 < x < 1$. In general, the Legendre equation of integer index m evidently has independent solutions of which one is a polynomial and the other is not. The importance of the polynomial solutions P_m is that they are orthogonal eigenfunctions, with corresponding eigenvalues $-m(m + 1)$, of the differential operator L, where

$$L(y) = (1 - x^2)y'' - 2xy'.$$

EXERCISES

1. Use the method of power series to derive the general solution

$$y = c_0 \cosh x + c_1 \sinh x$$

to the differential equation $y'' - y = 0$.

2. (a) Show that, if $y = f(x)$ is a solution of $y'' + xy = 0$, then $y = f(-x)$ is a solution of

$$y'' - xy = 0.$$

   (b) Use the result of part (a) together with the result of Example 3 of the text to find a power series expansion for the general solution of $y'' - xy = 0$.

3. (a) Apply the method of power series to solve the first-order differential equation

$$y' + 2xy = 0.$$

   (b) Solve the differential equation in part (a) by finding an exponential multiplier and then integrating.

   (c) Do the results of parts (a) and (b) agree for all x?

4. (a) Apply the power series method to find the general solution of the differential equation

$$y'' - xy' = 0$$

in the form $y(x) = c_0 y_0(x) + c_1 y_1(x)$.

   (b) Solve the differential equation in part (a) by solving the equivalent system

$$y' = u,$$

$$u' = xu.$$

   (c) Do the results of parts (a) and (b) agree for all x?

5. (a) Apply the power series method to find the general solution of the differential equation

$$y'' + xy' + y = 0$$

in the form $y(x) = c_0 y_0(x) + c_1 y_1(x)$.

   (b) Show that a special case of the general solution found in part (a) is the solution $y(x) = e^{-(x^2/2)}$.

6. (a) Apply the power series method to find a solution of the differential equation

$$y'' + xy' + y = x.$$

(b) Combine the result of part (a) with that of Exercise 5 to write the general solution of
$y'' + xy' + y = x.$

7. Apply the ratio test for series convergence to the series solution found for the Legendre equation in Example 4 of the text. Show that the series converges for $-1 < x < 1$. (*Hint:* Split into even and odd parts; show that each converges separately.)

8. The **Bessel equation** of index n is

$$y'' + \frac{1}{x} y' + \left(1 - \frac{n^2}{x^2}\right) y = 0.$$

(The case $n = \frac{1}{2}$ is treated in Example 4 of Chapter 6, Section 1.)

(a) Show that when $n = 0$ a solution of the form $\sum_{k=0}^{\infty} c_k x^k$ satisfies $(k + 2)^2 c_{k+2} = -c_k$.

(b) Show that, if we choose $c_0 = 1$, $c_1 = 0$ in part (a), we get the solution

$$J_0(x) = \sum_{k=0}^{\infty} (-1)^k 2^{-2k} (k!)^{-2} x^{2k},$$

called a Bessel function of order 0.

(c) Show that $J_1(x) = -J_0'(x)$ defines a solution of the Bessel equation of index 1.

9. A Bessel function of integer order n is defined by

$$J_n(x) = \left(\frac{x}{2}\right)^n \sum_{k=0}^{\infty} (-1)^k \frac{x^{2k}}{2^{2k} k! (n + k)!}.$$

(a) Show that J_n satisfies the Bessel equation of index n given in Exercise 8.

(b) Show that, if y_n is a solution of the Bessel equation of index n, and $u_n(x) = \sqrt{x}\, y_n(x)$, then u_n satisfies

$$u'' + \left(1 - \frac{4n^2 - 1}{4x^2}\right) u = 0.$$

Answers and Hints

Chapter 1

Section 1 (pp. 6–8)

1. **(a)** Yes. **(b)** Yes. **(c)** Yes. **(e)** Yes. **(g)** Yes. **(k)** Yes.
2. **(a)** Solution $x(t) = 0$.
3. **(a)** $y = 7e^{3x}$. **(c)** $x = -2 + \ln t$. **(e)** $y = -2 \cos x$.
4. **(a)** $y' - 3y = 0$. **(c)** $x' = 1/t$. **(e)** $y' = -y \tan x$. 5. **(a)** $r = -2$.
6. **(a)** $P(0) = ce^{k0} = c$. **(c)** $P(t_0 + \ln R/k) = Rce^{kt_0} = RP(t_0)$.
7. **(a)** $Q(0) = ce^{-k0} = c$. **(c)** $Q(t_0 - \ln r/k) = rce^{-kt_0} = rQ(t_0)$.
8. **(c)** $\displaystyle\lim_{n \to \infty} c_0[1 - e^{-nat}]/(1 - e^{-at}) = c_0/(1 - e^{-at})$.
9. **(c)** Show $s = \dfrac{1}{2} g \left(\dfrac{v}{g}\right)^2$.

Section 2 (pp. 12–13)

1. **(a)** Slopes change only when x changes.
   (c) All slopes are the same.
   (e) Slopes are positive in first and third quadrants.
2. **(a)** Isoclines are ellipses.
   (c) Isoclines are lines of slope 1.
   (e) Isoclines are parabolas.
3. **(b)** Let $y = 0$ if $x \le x_0$.
4. **(a)** Isoclines are of the form $y = cx$.
5. All slopes are at least 1.

```
6.  PLOTTER < name >
    SET WINDOW -2,2,-2,2
    PLOT AXES 2
    DEF F(X,Y)=EXP(-X*X-Y*Y)
    FOR X=-2 TO 2 STEP .25
    FOR Y=-2 TO 2 STEP .25
    LET D= SQR(1+F(X,Y)^2)
    PLOT X-.1/D,Y-(.1/D)*F(X,Y);
    PLOT X+.1/D,Y+(.1/D)*F(X,Y)
    NEXT Y
    NEXT X
    END
```

Section 3A (p. 15)

1. (a) $1 + \dfrac{1}{2}x^2 - \dfrac{1}{3}x^3$.   (c) $1 - \dfrac{1}{2}\ln(1 - x^2)$.   (e) $1 + 2x - \sin x$.

2. $388,199$ ft at 155.3 secs.

3. 567.45 ft at 17.6 secs.

4. (a) 4983.9 ft at 1 sec.
   (b) 401.3 ft per sec.

5. $a = -5$; 40 ft.

6. (b) At the midpoint by $\dfrac{5}{384} RL^4$.

Section 3B, C (pp. 21–22)

1. (a) $y = \left[\dfrac{1}{3}(2x^3 - 13)\right]^{1/2}$.   (c) $y = (3e^{-x} - 2)^{1/3}$.
   (e) $y = \arcsin(1 - \cos v)$.

2. (a) No solution if $2x^3 < 13$.

3. (b) $u|u|^n = (-1)^n u^{n+1}$ if $u < 0$; otherwise $u|u|^n = u^{n+1}$.

4. $u = 10\,e^{-t/2} + 10$.

5. (a) $y = ke^x - 2x - 5$.   (c) $y = \dfrac{x}{2}\left[\sqrt{3}\tan\left(\dfrac{\sqrt{3}}{2}\ln x + c\right) + 1\right]$.

7. Show that $P(x,y)/Q(x,y) = P(1, y/x)/Q(1, y/x)$.

8. $f(x) = \cosh(x + c)$.

Section 4 (pp. 25–26)

1. (a) $xy + x^2y^2 = 2$.   (c) $xy + xe^y = 1$.   (e) Not exact.

2. (a) $k = 1$.   (d) All real numbers k; $\ln|x + ky| + x = c$.

Section 5 (pp. 34–35)

1. (a) e^{2x}.   (c) x^2 (e) $e^{-(1/2)e^{-2x}}$.

3. $y = x - 1 + ce^{-x}$, $0 \le x < 1$; $y = x^2/3 + c/x$, $1 \le x$.

4. $s = 10e^{-(1/50)t}$.   **5.** $s = \dfrac{40}{13}(1 - e^{-2t})$.

8. Maximum value $4 - e^{-20}$; $\lim\limits_{t \to \infty} u(t) = 1$.

Section 6 (pp. 43–46)

1. $P_0 = 200/3$; $k = \dfrac{1}{10} \ln \dfrac{3}{2}$; $P(-10) = \dfrac{400}{9}$.

2. $D \ln 3/\ln 2$.   **3.** $P'' = kP' = k(kP) = k^2P$; use induction.

4. Estimated 1890 population 72,250,000.

5. **(a)** $P(t) = \left[\dfrac{ke^{kqt} P_\infty}{1 + ke^{kqt}} \right]^{1/q}$.   **6.** $a = k$, $b = k/P_\infty$.

8. $T(1) = 0.847$ grams; $T(60) = 4.89 \times 10^{-5}$ grams; $L(2) = 0.283$ grams.

9. 16,100 years.   **10.** 5.75 per min.   **11.** 272 years.

12. **(a)** 0.756.   **(c)** $u(10,000) = 0.757$; $T(10,000) = 0.580$.

13. Show that $T'(0) > 0$ and that $T'(t) = 0$ for a unique $t > 0$ if $mU_0 > nT_0$.

14. $L(t) = 1 - (1 - r)\left[\dfrac{k_3 e^{-k_1 t} - k_1 e^{-k_3 t}}{k_3 - k_1} \right] - r\left[\dfrac{k_2 e^{-k_1 t} - k_1 e^{-k_2 t}}{k_2 - k_1} \right]$.

15. **(a)** $v(t) = v_0 + v_e \ln \dfrac{m(0)}{m(t)}$.

16. **(a)** $v(t) = v_0 + v_e \ln \left(\dfrac{m_1 + m_0}{m_1 + m_0 - at} \right) - gt$.

17. **(d)** $s = v_0 \sqrt{\dfrac{m_1}{a}} \arctan \dfrac{\sqrt{a}}{m_1} t$.

18. $v(t) = \dfrac{v_e a}{k}\left[1 - \left(\dfrac{m_1 + m_0 - at}{m_1 + m_0} \right)^{k/a} \right]$.

Section 7 (pp. 52–54)

1. **(a)** $y = e^x$;

x_k	.1	.2	.3	.4	.5
y_k	1.1	1.21	1.331	1.4641	1.61051
$y(x_k)$	1.1057	1.2214	1.34986	1.49182	1.64872

(b) $y = e^x - x - 1$;

x_k	.1	.2	.3	.4	.5
y_k	0	.01	.031	.0641	.11051
$y(x_k)$	.00517	.0214	.04985	.09182	.148721

2. **(a)** $y = e^x$;

x_k	.1	.2	.3	.4	.5
y_k	1.105	1.221	1.34923	1.4909	1.64745
$y(x_k)$	1.105	1.2214	1.34986	1.49182	1.64872

(b) $y = e^x - x - 1$;

x_k	.1	.2	.3	.4	.5
y_k	.005	.02103	.04923	.09090	.14745
$y(x_k)$	.00517	.0214	.04985	.09182	.148721

5.
```
PLOTTER < name >
SET WINDOW −4,4,−4,4
PLOT AXES2
DEF F(X,Y)=X−Y
FOR X=−4 TO 4 STEP .5
FOR Y=−4 TO 4 STEP .5
LET D= SQR(1+F(X,Y)^2)
PLOT X−.2/D,Y−(.2/D)*F(X,Y);
PLOT X+.2/D,Y+(.2/D)*F(X,Y)
NEXT Y
NEXT X
LET X=−1
LET Y=−1
LET H=.1
PLOT X,Y
FOR X=−1 TO 4 STEP H
LET P=Y+H*F(X,Y)
LET Y=Y+.5*H*(F(X,Y)+F(X+H,P))
PLOT X+H,Y;
NEXT X
END
```

6.
```
DEF F(X,Y)= X*Y
LET X=0
LET Y=1
LET H=.1
FOR K=1 TO 20
LET M1=H*F(X,Y)
LET M2=H*F(X+H/2,Y+M1/2)
LET M3=H*F(X+H/2,Y+M2/2)
LET M4=H*F(X+H,Y+M3)
LET Y=Y+(M1+2*M2+2*M3+M4)/6
LET X=X+H
PRINT X,Y,EXP(X^2/2)
NEXT K
END
```

12. **(b)**
$$V = \frac{aVe}{k}\left(1 - \left(\frac{m_0 + m_1 - at}{m_0 + m_1}\right)^{k/a}\right) + \frac{g}{a-k}(m_0 + m_1 - at)$$
$$- \frac{g}{a-k}(m_0 + m_1)\left(\frac{m_0 + m_1 - at}{m_0 + m_1}\right)^{k/a}.$$

Chapter 2

Section 1 (pp. 62–64)

1. **(a)** $-e^{-2x}$.   **(b)** $2e^x$.   **(c)** $27e^{3x}$ **(d)** $\cos x - 2\sin x$.   **(e)** $-2\sin x$.

2. **(a)** $c_1 e^{-3x} + c_2 e^{2x}$ **(c)** $c_1 e^{-x} + c_2 x e^{-x}$.   **(e)** $c_1 + c_2 e^x$.   **(g)** $c_1 e^{x/2} + c_2 e^x$.

3. **(a)** $\dfrac{2}{5}e^{-3x} + \dfrac{8}{5}e^{2x}$.   **(c)** $e^{-x} + 3xe^{-x}$.   **(e)** $-\dfrac{e}{1-e} + \dfrac{e^x}{1-e}$.   **(g)** 0.

4. **(a)** Roots: -1, double; $y'' + 2y' + y = 0$.
    (c) Roots: 0, double; $y'' = 0$.

5. **(a)** $(D+1)^2$.   **(c)** $2(D - 1/\sqrt{2})(D + 1/\sqrt{2})$.   **(e)** DD.

6. **(a)** $c_1 + c_2 e^{3x}$.   **(c)** $-1 + c_1 e^x + c_2 e^{-x}$.

7. **(b)** $c_1 x + c_2/x$.   **(b)** $c_2 + c_1 \ln x$.

9. **(b)** Use the conclusion of Exercise 8(a).

10. First find the general solution of the second order equation.

Section 2 (pp. 71–72)

1. **(a)** $e^{i\pi/2}$.   **(c)** $e^{-i\pi/4}$.

2. **(a)** $c_1 = c_2 = 1/2$.   **(c)** $c_1 = -\pi i/2$, $c_2 = \pi i/2$.

3. **(b)** $2\pi/|\beta|$.   **4.** Note that $\sin x$ is odd and $\cos x$ is even.

6. **(a)** $1 + c_1 e^{ix} + c_2 e^{-ix}$.   **(c)** $c_1 e^{i\sqrt{2}x} + c_2 e^{-i\sqrt{2}x}$.

8. **(a)** $c_1 \cos \sqrt{2}\, x + c_2 \sin \sqrt{2}\, x$.   **(c)** $c_1 e^x \cos x + c_2 e^x \sin x$.
    (e) $c_1 e^{x/2} + c_2 e^{-x}$.

9. **(a)** $\dfrac{1}{\sqrt{2}} \sin \sqrt{2}\, x$.   **(c)** 0.   **(e)** $\dfrac{4}{3} e^{x/2} - \dfrac{4}{3} e^{-x}$.

10. $\theta = \pi/3$, $r = 2$.   **12.** Represent f and g in terms of their real and imaginary parts.

13. **(a)** $c_1 e^x + c_2 e^{-x/2} \cos \dfrac{\sqrt{3}}{2} x + c_3 e^{-x/2} \sin \dfrac{\sqrt{3}}{2} x$.

    (c) $c_1 + c_2 e^{\sqrt{2}x} + c_2 e^{-\sqrt{2}x}$.

14. **(a)** $y'' + 4y = 0$.   **(c)** $y'' - 2y' + 5y = 0$.
    (e) $y^{(4)} - 4y''' + 2y'' - 32y' + 64y = 0$.

Section 3 (pp. 81–84)

1. Use the fact that D is linear.

2. **(a)** Linear.   **(c)** Not linear.

3. **(a)** $c_1 \cos x + c_2 \sin x + x$.   **(c)** $c_1 e^{x/\sqrt{2}} + c_2 e^{-x/\sqrt{2}} + e^x$.

4. **(a)** $y'' - 3y' + 2y = 0$.   **(d)** $y'' - 2y' + y = 0$.

5. **(a)** $c_1 e^{-x} + c_2 e^x + \dfrac{1}{3} e^{2x}$.   **(c)** $c_1 e^{-x} + c_2 x e^{-x} + \dfrac{1}{4} e^x$.

    (e) $c_1 e^{-x} + c_2 e^x + \dfrac{1}{2} x e^x - x$.   **(g)** $c_1 \cos x + c_2 \sin x + \dfrac{1}{2} x \sin x$.

    (i) $c_1 \cos x + c_2 \sin x + \dfrac{1}{4} x \cos x + \dfrac{1}{4} x^2 \sin x$.

6. **(a)** $-\dfrac{1}{3} e^{-x} + \dfrac{1}{3} e^{2x}$.   **(c)** $-\dfrac{1}{4} e^{-x} + \dfrac{1}{2} x e^{-x} + \dfrac{1}{4} e^x$.

(e) $-\dfrac{3}{4}e^{-x} + \dfrac{3}{4}e^{x} + \dfrac{1}{2}xe^{x} - x.$ **(g)** $\sin x + \dfrac{1}{2}x\sin x.$

(i) $\dfrac{3}{4}\sin x + \dfrac{1}{4}x\cos x + \dfrac{1}{4}x^{2}\sin x.$

7.   **(a)** $\dfrac{1}{2}(e^{(x-t)} - e^{-(x-t)}).$ **(b)** $(x - t)e^{-(x-t)}.$

    (f) $\dfrac{1}{2\sqrt{2}}(e^{\sqrt{2}(x-t)} - e^{-\sqrt{2}(x-t)}.$ **(g)** $\sin(x - t).$ **(h)** $(x - t).$

8.   **(a)** $-x + \dfrac{1}{2}e^{x} - \dfrac{1}{2}e^{-x}.$ **(b)** $x - \sin x$ **(d)** $0.$

9.   **(a)** $-\dfrac{1}{4} + \dfrac{1}{8}(e^{2x} + e^{-2x}),\ 0 \le x;\ 0,\ x < 0.$

    (b) $0,\ 0 \le x;\ -\dfrac{1}{4} + \dfrac{1}{8}(e^{2x} + e^{-2x}),\ x < 0.$

    (e) $0,\ x < 1;\ \dfrac{1}{4}(1 - x) + \dfrac{1}{16}(e^{2(x-1)} - e^{-2(x-1)}),\ 1 \le x.$

    (f) $-\dfrac{1}{8}(2x^{2} + 1) + \dfrac{1}{16}(e^{2x} + e^{-2x}).$

10.   **(a)** $-\dfrac{1}{2}e^{x} + c_{1}e^{2x} + c_{2}e^{-x}.$

14.   **(b)** $\dfrac{3}{2}\sin t - \dfrac{1}{2}t\cos t.$

Section 4A, B (pp. 89–92)

1.   **(a)** 3.52 secs.   **(b)** 113.49 ft/sec.   **(c)** 155.28 ft.

3.   **(c)** $\dfrac{mg}{k}t + \left(\dfrac{V_0 mk - m^2 g}{k^2}\right)(1 - e^{-kt/m}).$ **4.**   **(b)** Yes.

5.   **(a)** $k = .556.$ **(b)** 1173.9 ft.   **6.**   **(b)** $v_\infty = \sqrt{mg/k}.$

7. $A = (L^{3/2}/g)\,\dot\theta(0);$ period $= 2\pi\sqrt{L/g}.$ **8.**   3.26 ft.

9.   **(b)** $k < 2m\sqrt{gl}.$

Section 4C (pp. 99–101)

1.   **(a)** Critically damped.   **(c)** Harmonic.   **(e)** Underdamped

2.   **(a)** $c_1 e^{-t} + c_2 t e^{-t}.$ **(c)** $c_1\cos pt + c_2\sin pt.$

    (e) $e^{-at/2}\left(c_1\cos a\sqrt{\dfrac{3}{2}}\,t + c_2\sin a\sqrt{\dfrac{3}{2}}\,t\right).$

3.   **(a)** $\dfrac{1}{2}\sin 2t;\ \ddot x + 4x = 0.$ **(c)** $-te^{-2t};\ \ddot x + 4\dot x + 4 = 0.$

(e) $-\dfrac{5}{2}e^{-2t} + \dfrac{3}{2}e^{-4t}$; $\ddot{x} + 6\dot{x} + 8x = 0$.

4. (a) $-(14/85)\cos 3t + (12/85)\sin 3t$; about 5 sec.

5. (a) Critical relation is $R^2 = 4L/C$.

Section 5 (pp. 106–108)

1. (b) LET T0=0
    ```
    DEF F(T,Y,Z)=-16*SIN(Y)
    LET Y=0
    LET Z=.5
    LET H=.01
    FOR K=1 TO 1000
       LET P=Y+H*Z
       LET Q=Z+H*F(T,Y,Z)
       LET S=Y
       LET Y=Y+.5*H*(Z+Q)
       LET Z=Z+.5*H*(F(T,S,Z)+F(T+H,P,Q))
       LET T=T0+K*H
       PRINT T,Y,Z
    NEXT K
    END
    ```

Chapter 3

Section 1 (pp. 116–117)

3. (a) $4s/(s^2 + 4)^2$.   (b) $(s + 2)/(s^2 + 1)$.   (c) $(-s^2 + 2s + 2)/s^3$.
    (d) $(s\cos a - \sin a)/(s^2 + 1)$.   (e) $(s - 1)/(s - 3)^2$.   (f) $2s/(s^2 - 1)$.

4. (a) $(e^t - e^{-t})/2$.   (b) $(e^{2t} + e^{-2t})/2$.   (c) $\sin 2t$.   (d) $t\sin 2t$.
    (e) $(\tfrac{1}{3})e^{2t}\sin 3t$.   (f) $t - te^t$.

5. (a) $3e^t - t - 1$.   (b) $(e^{-2t} + 1)/2$.   (c) $(-\tfrac{3}{13})e^{-3t} + (\tfrac{3}{13})\cos 2t + (\tfrac{2}{13})\sin 2t$
    (d) $(-\tfrac{5}{2})\cos t + (\tfrac{1}{2})\sin t + (\tfrac{1}{2})e^{-t} + 1$.   (e) $\tfrac{456}{111}e^{t/2} - \tfrac{12}{111}\cos 3t - \tfrac{2}{111}\sin 3t - 4$.

6. (d) $(\tfrac{1}{2})(t - a)^2 H_a(t) + 1$.

9. *Hint.* Break $[0,\infty)$ at kp, $k = 1, 2, 3, \cdots$.

Section 2 (pp. 121–122)

1. (a) $e^{-t} + t - 1$.   (b) $(t^5 + 10t^3)/30$.   (c) $t^2/2$.

2. (a) $e^{-t} + t - 1$.   (b) $e^{2t} - e^t$.   (c) $(t - 2)^2/2$.

3. (a) $\int_0^t ue^{-3u}\,du = (-\tfrac{1}{3})te^{-3t} - (\tfrac{1}{9})e^{-3t} + \tfrac{1}{9}$.
    (b) $\tfrac{1}{2}\int_0^{t-2}\sin 2u\,du = (1 - \cos 2(t - 2))/4$.
    (c) $e^{-t}\sin t$.   (d) $(1/\sqrt{2})\sin\sqrt{2}t$.
    (e) $1 - H_1(t) = 1$, $0 \le t \le 1$, 0 otherwise.   (f) $\cos\sqrt{10}t$.

4. (a) $(\tfrac{7}{10})e^t + (\tfrac{13}{10})e^{-t} - (\tfrac{1}{5})\sin 2t - 1$.   (b) $(\sqrt{2}/4)\sin\sqrt{2}t + t/2$.
    (c) $5 - 3e^{-t} - te^{-t} - t + t^2/2$.

5. *Hint.* Transform the equation, using Formulas 3 and 4 in Table II.
    Ans. $y(t) = (ae^{at} - be^{bt})/\sqrt{3}$, $a = -\tfrac{1}{2} + \tfrac{1}{2}\sqrt{5}$, $b = -\tfrac{1}{2} - \tfrac{1}{2}\sqrt{5}$.

Section 3 (pp. 128–129)

1. **(a)**

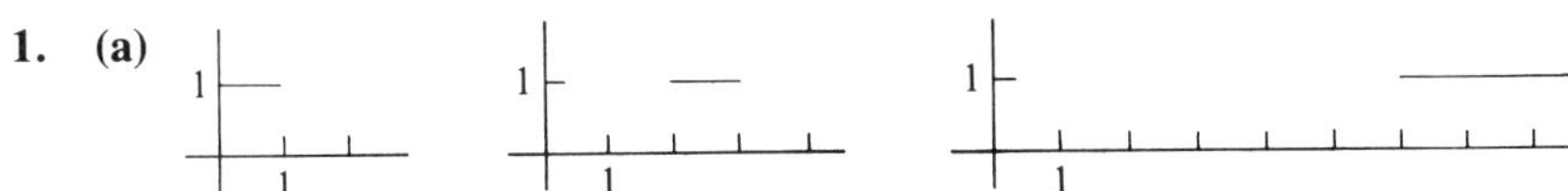

4. **(a)** 0, $0 \le t \le t_0$; $(t - t_0)/h$, $t_0 \le t \le t_0 th$; 1, $t_0 + h \le t$.   **(c)** $\lim\limits_{h \to 0+} y(t) = H_{t_0}(t)$.

5. $y(t) = (t - t_0) H_{t_0}(t)$, $t_0 > 0$.

6. **(a)** $y_\pi = \sin(t - \pi) H_\pi(t)$; $y_{2\pi} = \sin t\, H_{2\pi}(t)$.
   (c) Note that $y_{2\pi}(t) = -y_\pi(t) = \sin t$ for $2\pi < t$.

7. **(a)** $\sin t$, $0 \le t < 1$; $\sin t + \sin(t - 1)$, $1 \le t$.
   (c) 0, $0 \le t < 2$; $e^{t-2} - e^{2-t}$, $2 \le t$.

8. 0, $0 \le t < 1$; $(t - 1)e^{1-t}$, $1 \le t$.

11. **(a)** $\dfrac{1}{s}(e^{-s} - e^{-2s})$.   **(c)** $\dfrac{1}{s}(1 + e^{-s} - 2e^{-2s})$.

   (e) $\dfrac{1}{s^2}(1 - 2(s + 1)e^{-s} + (2s + 1)e^{-2s})$.

Chapter 4

Section 1 (pp. 138–139)

1. **(a)** $x = c_1 e^t - 1$, $y = c_2 e^t$; $x = 2e^t - 1$, $y = 2e^t$.
   (c) $x = c_1 e^t$, $y = c_2 e^{t/2}$, $z = c_3 e^{t/3}$; $x = 0$, $y = e^{t/2}$, $z = -e^{t/3}$.

2. **(i)** $F(x, y) = (x + 1, y)$; $v = \sqrt{(x + 1)^2 + y^2}$.
   (ii) $F(x, y, z) = (x, y/2, z/3)$; $v = \sqrt{x^2 + y^2/4 + z^2/9}$.

3. **(a)**

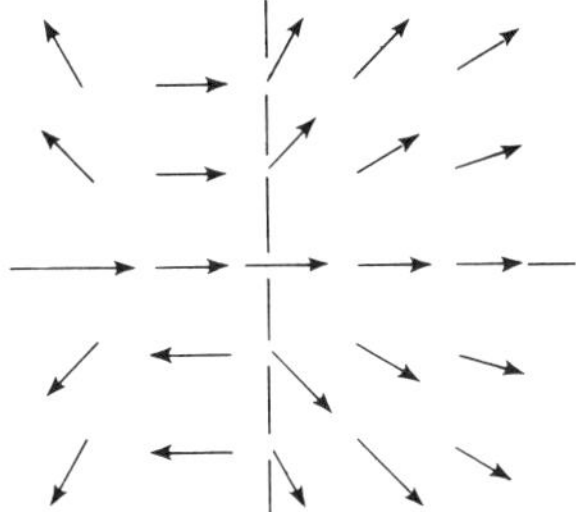

4. **(a)**
```
PLOTTER < name >
CLEAR
PRINT "FIELD LIMITS"
INPUT X1,X2,Y1,Y2
PRINT "SPACING";
INPUT D
SET WINDOW X1,X2,Y1,Y2
PLOT AXES2
DEF F(X,Y)=.1*(X+Y)
DEF G(X,Y)=.1*(X-Y)
```

```
FOR X=X1 TO X2 STEP D
FOR Y=Y1 TO Y2 STEP D
CALL POINT
PLOT X,Y;
PLOT X+F(X,Y),Y+G(X,Y)
NEXT Y
NEXT X
SUB POINT
FOR T= 0 TO 6 STEP .5
PLOT X+.03*D*COS(T),Y+.03*D*SIN(T);
NEXT T
PLOT
END SUB
END
```

6. (a) $\dot{y} = x,\ y(0) = 1$
$\dot{x} = -x - y,\ x(0) = 1.$

7. (a) $y(t) = e^{-t/2}\left(\cos\dfrac{\sqrt{3}}{2}\,t + \sqrt{3}\,\sin\dfrac{\sqrt{3}}{2}\,t\right).$

8. (a) $\dot{x} = y$ (c) $\dot{x} = y,\ \dot{y} = z,\ \dot{z} = -xy + z - t.$
$\dot{y} = -x^2 - y^2 + e^t.$

9. (a) $y = -x - 1 + ce^x.$ (c) $y^2 - x^2 = c.$

Section 2 (pp.144–147)

1. (a) Nonlinear.   (c) Linear.   (e) Linear.

2. (a) $x = c_1 e^{2t} + c_2 e^{-2t},\ y = -\dfrac{1}{2}c_1 e^{2t} - c_2 e^{-2t}.$

(c) $x = c_1 e^{(1 + \sqrt{2})t} + c_2 e^{(1-\sqrt{2})t} - 2t + 4,$

$y = \dfrac{1}{\sqrt{2}}c_1 e^{(1+\sqrt{2})t} - \dfrac{1}{\sqrt{2}}c_2 e^{(1-\sqrt{2})t} + t - 3.$

3. Eliminate x.

4. (a) $\dfrac{dx}{dt} = \dfrac{1}{2}x + \dfrac{1}{2}t$ (c) $\dfrac{dx}{dt} = x - 3y + t$

$\dfrac{dy}{dt} = -\dfrac{1}{2}x + \dfrac{1}{2}t.$ $\dfrac{dy}{dt} = -3x + y - t.$

5. (a) $x = 3e^{t/2} - t - 2$ (c) $x = -e^{4t} + \dfrac{1}{4}e^{-2t} - \dfrac{1}{6}t - \dfrac{1}{4}$

$y = -3e^{t/2} + \dfrac{1}{2}t^2 + t + 2.$ $y = e^{4t} - \dfrac{3}{4}e^{-2t} + \dfrac{1}{2}t - \dfrac{1}{4}.$

6. (a) $x = t + 1 - e^t$ (c) $x = 2 - 2e^{-t} + \dfrac{1}{2}t^2 - t$

$y = t.$ $y = 2e^{-t} - \dfrac{1}{2}t^2 + t - 2.$

7. (*a*) $\dfrac{dx}{dt} = u,\ \dfrac{dy}{dt} = v,\ \dfrac{dy}{dt} = x - y - v,\ \dfrac{dv}{dt} = x - y - u.$

8. (*a*) No solutions.   (**c**) $x = t,\ y = \dfrac{1}{2}t^2 - t + c.$

11. $x = c_1 e^{\sqrt{2}\,t} + c_2 e^{-\sqrt{2}\,t},\ u = \dfrac{c_1}{\sqrt{2}} e^{\sqrt{2}\,t} - \dfrac{c_2}{\sqrt{2}} e^{-\sqrt{2}\,t} - \dfrac{c_3}{\sqrt{2}} \sin \sqrt{2}\,t + \dfrac{c_4}{\sqrt{2}} \cos \sqrt{2}\,t,$

$y = c_3 \cos \sqrt{2}\,t + c_4 \sin \sqrt{2}\,t,$

$v = \dfrac{c_1}{\sqrt{2}} e^{\sqrt{2}\,t} - \dfrac{c_2}{\sqrt{2}} e^{-\sqrt{2}\,t} + \dfrac{c_3}{\sqrt{2}} \sin \sqrt{2}\,t - \dfrac{c_4}{\sqrt{2}} \cos \sqrt{2}\,t.$

Section 3A, B (pp. 151–152)

1. (**b**) $y = 25 e^{-t/50} - 15 e^{-3t/50},\ z = \dfrac{25}{2} e^{-t/50} + \dfrac{15}{2} e^{-3t/50}.$

2. (**b**) $c_1 = (2x_0 + 3y_0 - 100)/5,\ c_2 = (-6x_0 + 6y_0 + 50)/5.$

3. (**a**) $x = \dfrac{1}{2} \cos t + \dfrac{1}{2} \sin t - \dfrac{1}{2} \cos \sqrt{3}\,t + \dfrac{\sqrt{3}}{2} \sin \sqrt{3}\,t,$

$y = \dfrac{1}{2} \cos t + \dfrac{1}{2} \sin t + \dfrac{1}{2} \cos \sqrt{3}\,t - \dfrac{\sqrt{3}}{2} \sin \sqrt{3}\,t.$

4. (**b**) $t = 0.1.$

Section 3C (pp. 155–156)

2. 11,077 m/sec; 11,077 m/sec; 11,086 m/sec.

4. $\ddot{x}_k = \displaystyle\sum_{j=1}^{3} \dfrac{Gm_j}{r_{jk}^3}(x_j - x_k),\ k = 1, 2, 3;$ similarly for $\ddot{y}_k$ and $\ddot{z}_k.$

5. $\ddot{\mathbf{x}}_1 = \displaystyle\sum_{j=2}^{4} \dfrac{Gm_j}{r_{j1}^3}(\mathbf{x}_j - \mathbf{x}_1);$ similarly for $\ddot{\mathbf{x}}_2,\ \ddot{\mathbf{x}}_3$ and $\ddot{\mathbf{x}}_4.$

7. $\ddot{\mathbf{x}} = \dfrac{-G(m_1 + m_2)\mathbf{x}}{|\mathbf{x}|^{p+1}}.$

8. Show that $x_2(t) - x_1(t) = e^{-t},\ 0 \le t.$

9. (**c**) $f(P) = \dfrac{P^a}{e^{bP}} = k\dfrac{e^{cH}}{H^d}.$ Show that $f(P)$ increases to a unique maximum and then decreases. How many P-values correspond to each H-value? Similarly, consider

$g(H) = \dfrac{H^d}{e^{cH}}.$

(**d**) Start with the differential equation found in part (b).

Section 3D (pp. 158–159)

1. (**a**) $\dot{y} = z,\ \dot{z} = -y.$ (**c**) $\dot{y} = z,\ \dot{z} = ty + t.$

2. (**a**) $y = 1 + \cos t,\ z = -\sin t.$ (**c**) $y = 0,\ z = 0.$

4. PLOTTER ⟨name⟩
    CLEAR
    LET S=.5
    READ Y1,Y2,Z1,Z2
    DATA −4,4,−4,4
    SET WINDOW Y1,Y2,Z1,Z2
    PLOT AXES2
    DEF F(Y,Z)=Z
    DEF G(Y,Z)=1−Y
    FOR Y=Y1 TO Y2 STEP S
    FOR Z=Z1 TO Z2 STEP S
    CALL POINT
    LET D=.01+2*SQR(F(Y,Z)^2+G(Y,Z)^2)
    PLOT Y,Z;
    PLOT Y+F(Y,Z)/D,Z+G(Y,Z)/D
    NEXT Z
    NEXT Y
    SUB POINT
    FOR T=0 TO 6 STEP .5
    PLOT Y+.03*S*COS(T),Z+.03*S*SIN(T);
    NEXT T
    PLOT
    END SUB
    END

Section 3E (pp. 163–164)

1. (a) $I_1 = 4, I_2 = 2, I_3 = 6.$
    (c) $I_1 = 22, I_2 = 11, I_3 = 33.$
2. (a) $\dot{I}_1 + 200I_1 + 150I_2 = 1100,$
    $3I_1 + 5I_2 = 22, I_3 = I_1 + I_2.$

    (c) $I_1 = 4 - e^{-110t}, I_2 = 2 + \dfrac{3}{5}e^{-110t}, I_3 = 6 - \dfrac{2}{5}e^{-110t}.$

3. (c) $I_1 = \dfrac{1225}{16}e^{-20t} - \dfrac{2253}{16}e^{-100t} + \dfrac{269}{4}$

    $I_2 = -\dfrac{735}{8}e^{-20t} + \dfrac{751}{8}e^{-100t}$

    $I_3 = -\dfrac{245}{16}e^{-20t} - \dfrac{751}{16}e^{-100t} + \dfrac{269}{4}.$

4. (a) $\dot{I}_1 + 200I_1 + 150I_2 = 10E$
    $3\dot{I}_1 + 5\dot{I}_2 + 250I_2 = 0$
    $I_1 + I_2 = I_3.$

Section 4 (pp. 170–172)

1. *(Euler)*   *(Modified Euler)*
    DEF F(T,X,Y)=Y   DEF F(T,X,Y)=Y

```
DEF G(T,X,Y)=X              DEF G(T,X,Y)=X
LET T0=0                    LET T0=0
LET X=1                     LET X=1
LET Y=2                     LET Y=2
LET H=.1                    LET H=.1
FOR K=1 T0 5                FOR K=1 TO 5
LET X=X+H*F(T,X,Y)          LET P=X+H*F(T,X,Y)
LET Y=Y+H*G(T,X,Y)          LET Q=Y+H*G(T,X,Y)
LET T=T0+K*H                LET W=X
PRINT T,X,Y                 LET X=X+.5*H*(F(T,X,Y)+F(T+H,P,Q))
NEXT K                      LET Y=Y+.5*H*(G(T,W,Y)+G(T+H,P,Q))
END                         LET T=T0+K*H
                            PRINT T,X,Y
                            NEXT K
                            END
```

4. (c)
```
DEF F(T,X,Y)=-X/(X*X+Y*Y)^1.5
DEF G(T,X,Y)=-Y/(X*X+Y*Y)^1.5
LET T0=0
LET U=0
LET V = .7
LET X=1
LET Y=0
LET H=.0
FOR K=1 TO 4000
LET R=U+H*F(T,X,Y)
LET S=V+H*G(T,X,Y)
LET P=X+H*U
LET Q=Y+H*V
LET W=U
LET Z=V
LET U=U+.5*H*(F(T,X,Y)+F(T+H,P,Q))
LET V=V+.5*H*(G(T,X,Y)+G(T+H,P,Q))
LET X=X+.5*H*(W+R)
LET Y=Y+.5*H*(Z+S)
LET T=T0+K*H
PRINT T,X,Y
NEXT K
END
```

8.
```
PLOTTER ⟨name⟩
SET WINDOW -1.5,1.5,-1.5,1.5
! GRAVITATIONAL CONSTANT G DEPENDS ON UNITS:
!   SUN'S MASS=1,MEAN RADIUS OF EARTH ORBIT=1, 58.01 DAYS=1 MAKE G=1;
!   SUN'S MASS=1,MEAN RADIUS OF EARTH ORBIT=1,YEAR=1 MAKE G=39.634;
!   EARTH'S MASS=1,EARTH'S RADIUS=1,HOUR=1 MAKE G=20.0125.
LET G=1
DIM X(12),F(12),Q(12)
SUB FIELD
```

```
            LET D12=((X(1)−X(3))^2+(X(2)−X(4))^2)^(−1.5)
            LET D13=((X(1)−X(5))^2+(X(2)−X(6))^2)^(−1.5)
            LET D23=((X(3)−X(5))^2+(X(4)−X(6))^2)^(−1.5)
            LET F(1)=X(7)
            LET F(2)=X(8)
            LET F(3)=X(9)
            LET F(4)=X(10)
            LET F(5)=X(11)
            LET F(6)=X(12)
            LET F(7)=G*M2*(X(3)−X(1))*D12+G*M3*(X(5)−X(1))*D13
            LET F(8)=G*M2*(X(4)−X(2))*D12+G*M3*(X(6)−X(2))*D13
            LET F(9)=G*M1*(X(1)−X(3))*D12+G*M3*(X(5)−X(3))*D23
            LET F(10)=G*M1*(X(2)−X(4))*D12+G*M3*(X(6)−X(4))*D23
            LET F(11)=G*M1*(X(1)−X(5))*D13+G*M2*(X(3)−X(5))*D23
            LET F(12)=G*M1*(X(2)−X(6))*D13+G*M2*(X(4)−X(6))*D23
END SUB
DATA .0005,100
READ H,M
DATA 0,20
READ T0,T1
DATA 1,.01,.001
READ M1,M2,M3
DATA 0,0,.9,0,1,0
DATA 0,0,0,0,.−1,.5
      MAT READ X                          ! POSITIONS, VELOCITIES
      LET X(1)=(−M2*X(3)−M3*X(5))/M1      !PUTS CENTER OF MASS AT ORIGIN
      LET X(2)=(−M2*X(4)−M3*X(6))/M1
      LET X(7)=(−M2*X(9)−M3*X(11))/M1     !MAKES TOTAL MOMENTUM ZERO
      LET X(8)=(−M2*X(10)−M3*X(12))/M1
      LET T=T0
      PLOT X(1),X(2)
      PLOT X(3),X(4)
      PLOT X(5),X(6)
      LET L=INT(ABS((T1−T0)/H/M))
      FOR K=0 TO L
          FOR J=1 TO M
              MAT Q=X
              CALL FIELD
              MAT X=X+H*F
              LET T=T+H
              CALL FIELD
              MAT X=.5*(X+Q+H*F)
          NEXT J
          PLOT X(1),X(2)
          PLOT X(3),X(4)
          PLOT X(5),X(6)
      NEXT K
      END
```

Section 5A (pp. 180–181)

1. **(a)** $\lambda = \pm 1$; saddle, unstable. **(c)** $\lambda = 2 \pm i$; unstable spiral.
   (e) $\lambda = \pm\sqrt{2}i$; stable center. **(g)** $\lambda = 0, 1$; degenerate.

2. **(a)** $\lambda = \pm 1$; saddle, unstable. **(c)** $\lambda = (-1 \pm \sqrt{3}i)/2$;
   asymptotically stable spiral.

3. **(a)** Stable center, $k = 0$; asymptotically stable spiral, $0 < k < 2$; asymptotically
   stable star, $k = 2$; asymptotically stable node, $2 < k$.

Section 5B (pp. 185–187)

1. **(a)** $(0, 0)$, saddle, unstable.
   (c) $(0, 0)$, stable center; $(-1, 0)$ saddle, unstable.
   (e) $(0, 0)$, saddle; $(1, 1)$ asymptotically stable node; $(-1, -1)$, asymptotically stable
   node.

2. **(a)** $((2k + 1)\pi, 1)$, saddles; $(2k\pi, 1)$, stable centers (Note the solutions
   $(y - 1)^2 = 2(a - \cos x), 0 < a < 1$.)
   (b) Only $(0, 0)$ is nondegenerate.

3. **(b)** $y^2 = \dfrac{2g}{l}\left(\cos x + \dfrac{lc}{2g}\right)$, $l|c|/2g < 1$, has two real solutions for y when
   $\cos x + lc/2g > 0$.
   (c) $\cos x + lc/2g > 0$ for all x if $c > 2g/l$, so there are two real square roots for
   all x.
   (d) $y^2 = \dfrac{2g}{l}(\cos x + 1)$ if $c = 2g/l$ so there are always two real square roots with

   graphs coinciding when x is an odd multiple of π.

4. **(d)** $r^2 = -2/(\alpha t + c_1)$, $\theta = -t + c_2$. If $\alpha < 0$, $dr/dt < 0$ and $\lim\limits_{t\to\infty} r^2 = 0$. If
   $\alpha > 0$, $dr/dt > 0$ and $\lim\limits_{t\to 2/\alpha r_0^2} r^2 = \infty$.

7. $(0, 0)$ is a saddle.

8. Equilibrium at $(A, B/A)$ can be an unstable spiral, an unstable node or a saddle, depending on A and B.

Chapter 5

Section 1 A–E (pp. 198–201)

1. **(a)** $\begin{pmatrix} 1 & -2 \\ 1 & 1 \end{pmatrix}$. **(c)** $\begin{pmatrix} 4 & 14 \\ 1 & 7 \end{pmatrix}$. **(e)** $\begin{pmatrix} 7 \\ 6 \\ 4 \end{pmatrix}$.

2. **(a)** $\begin{pmatrix} 1 & -1 \\ 1 & 1 \end{pmatrix}\begin{pmatrix} x \\ y \end{pmatrix} = \begin{pmatrix} 1 \\ 2 \end{pmatrix}$ **(c)** $\begin{pmatrix} 1 & 1 \\ 0 & 1 \end{pmatrix}\begin{pmatrix} x \\ y \end{pmatrix} = \begin{pmatrix} 1 \\ 1 \end{pmatrix}$.

   (e) $\begin{pmatrix} 1 & 1 & 0 \\ 0 & 1 & 1 \\ 1 & 0 & -1 \end{pmatrix}\begin{pmatrix} x \\ y \\ z \end{pmatrix} = \begin{pmatrix} 0 \\ 0 \\ 1 \end{pmatrix}$.

3. **(a)** $\begin{pmatrix} \dot{x} \\ \dot{y} \end{pmatrix} = \begin{pmatrix} 2 & 3 \\ 2 & -4 \end{pmatrix} \begin{pmatrix} x \\ y \end{pmatrix} + \begin{pmatrix} 0 \\ 0 \end{pmatrix}$.   **(c)** $\begin{pmatrix} \dot{x} \\ \dot{y} \\ \dot{z} \end{pmatrix} = \begin{pmatrix} 1 & 1 & -1 \\ 1 & -1 & 1 \\ 1 & t & 0 \end{pmatrix} \begin{pmatrix} x \\ y \\ z \end{pmatrix} + \begin{pmatrix} 0 \\ 0 \\ 0 \end{pmatrix}$.

4. **(a)** $\begin{pmatrix} \dot{x} \\ \dot{y} \end{pmatrix} = \begin{pmatrix} 0 & 2 \\ -\frac{1}{3} & 0 \end{pmatrix} \begin{pmatrix} x \\ y \end{pmatrix} + \begin{pmatrix} 1 \\ \frac{2}{3} \end{pmatrix}$.   **(c)** $\begin{pmatrix} \dot{x} \\ \dot{y} \end{pmatrix} = \begin{pmatrix} 0 & 1 \\ 0 & -1 \end{pmatrix} \begin{pmatrix} x \\ y \end{pmatrix} + \begin{pmatrix} 0 \\ 0 \end{pmatrix}$.

(e) $\begin{pmatrix} \dot{x} \\ \dot{y} \\ \dot{z} \end{pmatrix} = \begin{pmatrix} 0 & 2 & 0 \\ 0 & 0 & -1 \\ -1 & 0 & 0 \end{pmatrix} \begin{pmatrix} x \\ y \\ z \end{pmatrix} + \begin{pmatrix} 1 \\ 0 \\ 0 \end{pmatrix}$.

5. **(a)** $e^t \mathbf{e}_1 + e^t \mathbf{e}_2 + t^2 \mathbf{e}_3$.   **(c)** $(t + 1)\mathbf{e}_1 + (t - 1)\mathbf{e}_2 + (t + 2)\mathbf{e}_3$.
(e) $\cos t\, \mathbf{e}_1 + \sin t\, \mathbf{e}_2$.

6. **(a)** $\begin{pmatrix} e^t & (t+1)e^t \\ 0 & e^t \end{pmatrix}$.   **(c)** $\begin{pmatrix} e^t & 0 & 0 \\ 0 & 2e^{2t} & 0 \\ 0 & 0 & 3e^{3t} \end{pmatrix}$.

9. **(a)** $\begin{pmatrix} 2 \\ 5 \\ 8 \end{pmatrix}$.   **(c)** (39).   **(e)** $\begin{pmatrix} 1 & -1 & 1 \\ -1 & 1 & -1 \\ -1 & 1 & -1 \end{pmatrix}$.

11. **(a)** $A^2 = \begin{pmatrix} 2 & 3 \\ 0 & 1 \end{pmatrix}$, $A^3 = \begin{pmatrix} 4 & 5 \\ 0 & 1 \end{pmatrix}$.

(c) $A^2 = \begin{pmatrix} 4 & 0 \\ 3 & 1 \end{pmatrix}$, $A^3 = \begin{pmatrix} 8 & 0 \\ 7 & 1 \end{pmatrix}$.

12. **(c)** $\begin{pmatrix} a & b \\ c & d \end{pmatrix}^2 = \begin{pmatrix} a^2 + bc & b(a + d) \\ c(a + d) & d^2 + bc \end{pmatrix} = \begin{pmatrix} 1 & 0 \\ 0 & 1 \end{pmatrix}$ implies what?

13. **(a)** Coefficients are -2 and 1.   **(b)** No. Why?

16. **(a)** $\begin{pmatrix} 2 & -1 \\ -1 & 1 \end{pmatrix}$.   **(b)** No inverse.   **(d)** $\begin{pmatrix} -3 & -\frac{5}{3} \\ 4 & \frac{7}{3} \end{pmatrix}$.

17. **(a)** $x = 5/11$, $y = -1/11$.   **(b)** $x = -2$, $y = 6$.

18. **(c)** Multiply the equation $AB = I$ by C on the left.

Section 1F (pp. 208–209)

1. **(a)** $AB = \begin{pmatrix} -4 & 7 \\ 2 & 0 \end{pmatrix}$, $BA = \begin{pmatrix} 3 & 1 \\ -7 & -7 \end{pmatrix}$; $\det AB = \det BA = -14$.

2. $\det(2B) = 16$.

3. Use induction.

4. Depends on the number n, where A is n-by-n.

7. Show that $\begin{pmatrix} A & 0 \\ 0 & I \end{pmatrix} \begin{pmatrix} I & 0 \\ 0 & B \end{pmatrix} = \begin{pmatrix} A & 0 \\ 0 & B \end{pmatrix}$.

8. **(a)** 32.   **(b)** 12.   **9.** **(a)** -12.

10. **(a)** $\begin{pmatrix} 1 & 2 & -1 \\ 4 & 5 & 3 \\ 1 & 6 & 7 \end{pmatrix}$.   **(b)** -52.

11. Use Theorem 1.9.

13. **(a)** $\begin{pmatrix} 1 & 0 & 0 \\ -13 & 1 & -5 \\ 2 & 0 & 1 \end{pmatrix}$ **(b)** No inverse.

(d) $\begin{pmatrix} \frac{1}{t} & 0 & 0 \\ 0 & \frac{1}{2} & 0 \\ 0 & 0 & 1 \end{pmatrix}$, $t \neq 0$.   **(i)** $\begin{pmatrix} 1 & 0 & 0 \\ 0 & e^{-t} & -te^{-t} \\ 0 & 0 & e^{-t} \end{pmatrix}$.

15. $\det C = \sum_{j=1}^{n} (-1)^{j+i}(a_{ij} + b_{ij}) \det C_{ij}$. But $C_{ij} = A_{ij} = B_{ij}$ for $j = 1, 2, \ldots, n$. Now break the sum into two parts.

16. **(a)** If the ith row of A is multiplied by r, then the new matrix A_r satisfies

$$\det A_r = \sum_{j=1}^{n} (-1)^{i+j} r a_{ij} \det A_{ij}.$$

Now factor out r.

Section 2 (pp. 217–220)

1. $\lambda_1 = 7$ has among its eigenvectors $\begin{pmatrix} 2 \\ 1 \end{pmatrix}$ and $\begin{pmatrix} -4 \\ -2 \end{pmatrix}$.

$\lambda_2 = -5$ has among its eigenvectors $\begin{pmatrix} -2 \\ 1 \end{pmatrix}$.

2. **(a)** $3, -1$.   **(c)** $0, 4$.

3. **(a)** $\lambda_1 = 3$, $\begin{pmatrix} 2 \\ 1 \end{pmatrix}$; $\lambda_2 = -1$, $\begin{pmatrix} -2 \\ 1 \end{pmatrix}$.

(c) $\lambda_1 = 0$, $\begin{pmatrix} -2 \\ 1 \end{pmatrix}$; $\lambda_2 = 4$, $\begin{pmatrix} 2 \\ 1 \end{pmatrix}$.

5. **(d)** $d_1 e^{2t} \begin{pmatrix} \cos t \\ \sin t \end{pmatrix} + d_2 e^{2t} \begin{pmatrix} \sin t \\ \cos t \end{pmatrix}$.

6. **(a)** $x = 2e^{t} - e^{-t}$, $y = 4e^{t} + e^{-t}$.   **(c)** $x = e^{5t}$, $y = e^{5t}$.

7. **(a)** $c_1 e^{t} \begin{pmatrix} 1 \\ 4 \\ -1 \end{pmatrix} + c_2 e^{3t} \begin{pmatrix} 1 \\ 2 \\ 1 \end{pmatrix} + c_3 e^{-2t} \begin{pmatrix} 1 \\ -1 \\ -1 \end{pmatrix}$.

(b) Find a constant solution $\mathbf{x}$, for which $d\mathbf{x}/dt = 0$, and add it to the homogeneous solution found in part (a).

(c) Set $t = 0$ in the solution found in part (b) and set the result equal to $\mathbf{x}(0)$. Then solve for c_1, c_2 and c_3.

8. **(a)** By the elimination method
$x = c_2 \cos t + c_3 \sin t$, $y = -c_2 \sin t + c_3 \cos t$, $z = c_1 e^{-t}$.

(b) Eigenvalues associated with these solutions are $-1, i, -i$.

9. **(a)** $x = c_1 e^{-t} + c_2 t e^{-t}$, $y = c_2 e^{-t}$.

(b) Eigenvectors are $\begin{pmatrix} c \\ 0 \end{pmatrix}$, $c \neq 0$, and these are inadequate for representing the most general solution.

12. **(a)** Note that $U\mathbf{u}_k = \mathbf{e}_k$, the kth column of the identity matrix I. Then
$$A\mathbf{u}_k = UDU^{-1}\mathbf{u}_k = UD\mathbf{e}_k = U\lambda_k \mathbf{e}_k.$$

(b) Use part (a); $\begin{pmatrix} -3 & 4 \\ -6 & 7 \end{pmatrix}$.

13. **(a)** Solutions oscillate and tend to 0 as t tends to infinity.

(c) Solutions oscillate and become unbounded as t tends to infinity.

Section 3A, B (pp. 226–228)

1. **(a)** $\begin{pmatrix} e^{-t} & 0 \\ 0 & e^{t} \end{pmatrix}$.   **(c)** $\begin{pmatrix} e^{it} & 0 \\ 0 & e^{-it} \end{pmatrix}$.

5. **(a)** $x = (c_1 + \tfrac{1}{4}c_2)e^{5t} + c_2 t e^{5t},\ y = c_1 e^{5t} + c_2 t e^{5t}$.

9. **(a)** $\begin{pmatrix} 3e^{2t} - 2e^{3t} & -e^{2t} + e^{3t} \\ 6e^{2t} - 6e^{3t} & -2e^{2t} + 3e^{3t} \end{pmatrix}$.

(b) $\begin{pmatrix} e^{t} & 0 & 0 \\ 0 & \tfrac{1}{2}e^{2t} + \tfrac{1}{2}e^{t} & -\tfrac{1}{2}e^{2t} + \tfrac{1}{2}e^{t} \\ 0 & -\tfrac{1}{2}e^{2t} + \tfrac{1}{2}e^{t} & \tfrac{1}{2}e^{2t} + \tfrac{1}{2}e^{t} \end{pmatrix}$.

Section 3C (p. 231)

1. **(a)** $c_1 \begin{pmatrix} 1 \\ 0 \end{pmatrix} + c_2 \begin{pmatrix} 0 \\ e^{t} \end{pmatrix}$.   **(c)** $c_1 \begin{pmatrix} e^{t} \\ 0 \\ 0 \end{pmatrix} + c_2 \begin{pmatrix} 0 \\ e^{2t} \\ 0 \end{pmatrix} + c_3 \begin{pmatrix} 0 \\ 0 \\ e^{3t} \end{pmatrix}$.

2. **(a)** $\begin{pmatrix} 0 & 0 \\ 0 & 1 \end{pmatrix}$.   **(c)** $\begin{pmatrix} 1 & 0 & 0 \\ 0 & 2 & 0 \\ 0 & 0 & 3 \end{pmatrix}$.

3. **(a)** Compute the derivative of the matrix; this should equal the matrix A such that the given matrix is e^{tA}.

Section 4 (pp. 237–239)

1. **(a)** $\mathbf{x}_p(t) = \begin{pmatrix} -\tfrac{1}{3}e^{2t} - 2 \\ -\tfrac{4}{3}e^{2t} - 3 \end{pmatrix}$.   **(b)** $\mathbf{x}_p(t) = \begin{pmatrix} 1 - \tfrac{1}{4}e^{t} - te^{t} \\ -\tfrac{1}{4}e^{t} \end{pmatrix}$.

2. **(a)** $\mathbf{x} = \begin{pmatrix} -4e^{t} + \tfrac{16}{3}e^{-t} - \tfrac{1}{3}e^{2t} - 2 \\ -2e^{t} + \tfrac{16}{3}e^{-t} - \tfrac{4}{3}e^{2t} - 3 \end{pmatrix}$

(c) $\mathbf{x} = \begin{pmatrix} \tfrac{5}{4}e^{5t} - \tfrac{9}{4}e^{-t} + 1 \\ \tfrac{5}{4}e^{5t} - \tfrac{1}{4}e^{-t} \end{pmatrix}$.

5. Solutions to this problem will differ from the corresponding solutions to Exercise 1 at most by the addition of a solution to the homogeneous equation.

6. **(a)** $\mathbf{x}_p = \begin{pmatrix} \tfrac{1}{6}t^{4} \\ \tfrac{7}{6}t^{3} \end{pmatrix}$.

7. **(b)** Let $X(t) = \begin{pmatrix} e^{t} & 1 \\ 2e^{2t} & e^{t} \end{pmatrix}$. Then $A(t) = X'(t)X^{-1}(t)$.

Section 5 (pp. 245–246)

1. **(a)** $\begin{pmatrix} 6e^{3t} - 5e^{2t} & 3e^{2t} - 3e^{3t} \\ 10e^{3t} - 10e^{2t} & 6e^{2t} - 5e^{3t} \end{pmatrix}$.

(c) $\begin{pmatrix} e^{t} & te^{t} & e^{2t} - e^{t} \\ 0 & e^{t} & 0 \\ 0 & 0 & e^{2t} \end{pmatrix}$.

(e)
$$\begin{pmatrix} \frac{1}{2} + \frac{1}{2}e^t & \frac{1}{2}e^{2t} - \frac{1}{2} & \frac{1}{2}e^{2t} - \frac{1}{2} & 0 \\ e^t - \frac{1}{2} - \frac{1}{2}e^{2t} & e^t + \frac{1}{2} - \frac{1}{2}e^{2t} & \frac{1}{2} - \frac{1}{2}e^{2t} & 0 \\ e^{2t} - e^t & e^{2t} - e^t & e^{2t} & 0 \\ (t+1)e^t - \frac{1}{2} - \frac{1}{2}e^{2t} & te^t + \frac{1}{2} - \frac{1}{2}e^{2t} & \frac{1}{2} - \frac{1}{2}e^{2t} & e^t \end{pmatrix}.$$

2. Exponential matrix is
$$\begin{pmatrix} (1-t)e^{3t} & 0 & te^{3t} \\ -te^{3t} & e^{3t} & te^{3t} \\ -te^{3t} & 0 & (t+1)e^{3t} \end{pmatrix}.$$

3. Exponential matrix is
$$\begin{pmatrix} e^{2t} & -\frac{1}{2}e^t\cos t + \frac{1-t}{2}e^{2t} & te^{2t} & -\frac{1}{2}e^t\sin t + \frac{t}{2}e^{2t} \\ 0 & e^t\cos t & 0 & e^t\sin t \\ 0 & \frac{1}{2}e^t\sin t + \frac{1}{2}e^t\cos t - \frac{1}{2}e^{2t} & e^{2t} & \frac{1}{2}e^t\sin t - \frac{1}{2}e^t\cos t + \frac{1}{2}e^{2t} \\ 0 & -e^t\sin t & 0 & e^t\cos t \end{pmatrix}.$$

4. Replace t by $-t$ in e^{tA}.

5. **(a)** $\begin{pmatrix} 21e^{3t} - 19e^{2t} \\ 35e^{3t} - 38e^{2t} \end{pmatrix}.$ **(c)** $\begin{pmatrix} te^t \\ e^t \\ 0 \end{pmatrix}.$

6. **(a)** $\begin{pmatrix} e^{\alpha t} & \frac{1}{\alpha - \beta}(e^{\alpha t} - e^{\beta t}) \\ 0 & e^{\beta t} \end{pmatrix}.$ **(b)** $\begin{pmatrix} e^t & te^t \\ 0 & e^t \end{pmatrix}.$

7. $e^{tI} = \sum\limits_{k=0}^{\infty} \dfrac{t^k I^k}{k!}.$

9. **(a)** $A^2 = 5A + 6I.$ **(c)** $A^3 = 4A^2 - 5A + 2I.$
  (e) $A^4 = 4A^3 - 6A^2 + 2A.$

Section 6A, B (pp. 252–253)

1. **(a)** $\begin{pmatrix} 3 & -2 \\ 1 & -3 \end{pmatrix}\begin{pmatrix} x \\ y \end{pmatrix} = \begin{pmatrix} 1 \\ 2 \end{pmatrix}.$ **(c)** $\begin{pmatrix} 1 & 1 & 0 \\ 0 & 1 & -1 \\ 1 & 0 & 1 \end{pmatrix}\begin{pmatrix} x \\ y \\ z \end{pmatrix} = \begin{pmatrix} 1 \\ 1 \\ 0 \end{pmatrix}.$

2. **(a)** $x + 2y = 1$ **(c)** $x + z = 0$
  $\ 3x + y = 0.$ $\ y = 1$
  $x + y = 0.$

3. **(a)** $x = -\frac{1}{7},\ y = -\frac{5}{7}.$ **(c)** $x = -t,\ y = t + 1,\ z = t,\ -\infty < t < t.$
  (d) No solutions.

4. **(a)** $x = -\frac{1}{5},\ y = \frac{3}{5}.$ **(c)** $x = -1,\ y = z = 1.$ **(d)** No solutions.

5. **(a)** $x = \frac{24}{15},\ y = \frac{13}{15},\ z = \frac{-7}{15}.$

6. **(a)** $\begin{pmatrix} 1 & 0 & -1 \\ 0 & 1 & -1 \end{pmatrix}.$ **(c)** $\begin{pmatrix} 1 & 0 & 0 \\ 0 & 0 & 1 \\ 0 & 1 & 0 \end{pmatrix}.$

7. **(a)** $x = y = z = t,\ -\infty < t < \infty.$ **(c)** $x = y = z = 0.$

8. (a) $\mathbf{e}_1 = 2\begin{pmatrix} 2 \\ 3 \end{pmatrix} - 3\begin{pmatrix} 1 \\ 2 \end{pmatrix}$, $\mathbf{e}_2 = 2\begin{pmatrix} 1 \\ 2 \end{pmatrix} - \begin{pmatrix} 2 \\ 3 \end{pmatrix}$.

(c) $(5, 0, 1, 2) = (1, 2, 1, 0) + 2(2, -1, 0, 1)$.

Section 6C, D (pp. 256–257)

1. (a) $\begin{pmatrix} 1 & 0 & 0 \\ -13 & 1 & -5 \\ 2 & 0 & 1 \end{pmatrix}$. (b) No inverse. (c) $\begin{pmatrix} 0 & 1 & 0 \\ \frac{7}{4} & -\frac{11}{2} & 2 \\ -\frac{3}{4} & \frac{5}{2} & -1 \end{pmatrix}$.

2. $\begin{pmatrix} \frac{1}{2} & \frac{11}{2} & -6 \\ -\frac{1}{4} & -\frac{17}{4} & 5 \end{pmatrix}$.

3. (a) $\begin{pmatrix} 1 & 0 & 0 \\ 0 & \frac{1}{2} & 0 \\ 0 & 0 & 1 \end{pmatrix}$. (c) $\begin{pmatrix} 1 & -1 & 0 \\ 0 & -1 & 1 \\ 0 & 0 & 1 \end{pmatrix}$.

5. (a) Independent. (b) Not independent. (c) Independent.
(d) Not independent. (e) Independent. (f) Independent.

6. (a) $\begin{pmatrix} 2 \\ 6 \\ 0 \end{pmatrix} = -4\begin{pmatrix} 1 \\ -1 \\ 2 \end{pmatrix} + 0\begin{pmatrix} 2 \\ 1 \\ 2 \end{pmatrix} + 2\begin{pmatrix} 3 \\ 1 \\ 4 \end{pmatrix}$.

(c) $\begin{pmatrix} 3 \\ 4 \end{pmatrix} = \frac{5}{3}\begin{pmatrix} 1 \\ 2 \end{pmatrix} + \frac{2}{3}\begin{pmatrix} 2 \\ 1 \end{pmatrix}$.

7. Show that there is a vector $\mathbf{V} \neq 0$ such that the dot products $\mathbf{V} \cdot \mathbf{X}_1$ and $\mathbf{V} \cdot \mathbf{X}_2$ are both zero. Then show that $\mathbf{V}$ can't be a linear combination of $\mathbf{X}_1$ and $\mathbf{X}_2$.

Chapter 6

Section 1 (pp. 263–265)

1. (a) Independent. (c) Independent. (e) Dependent.

2. (a) xe^{2x}. (c) x. (e) $x^2 + 1$.

3. (a) $c_1 \int e^{-x^2/2}\, dx + c_2$.

4. (a) $-1 + \frac{1}{2}x \ln\left|\dfrac{1+x}{1-x}\right|$. **5.** (a) $\displaystyle\sum_{k=0}^{\infty} \frac{(-1)^k x^{2k}}{(2k-1)k!}$.

6. $c_1 x + c_2 x^2 + x^2 \ln x$. **10.** (a) $c_1 x + c_2/x$. (c) $\dfrac{c_1}{x} + \dfrac{c_2}{x} \ln x$.

12. (b) $y'' + 2 \cot xy' + (\cot^2 x + \csc^2 x)y = 0$.

Section 2 (pp. 271–272)

1. (a) $c_1 e^{2x} + c_2 xe^{2x} + e^x$. (c) $c_1 x^3 + c_2 x + \frac{1}{3}x^4$.

2. (a) $c_1 e^{-2x} + c_2 e^x + \frac{1}{4}e^{2x}$.
(c) $c_1 \cos x + c_2 \sin x + x \sin x + \cos x \ln|\cos x|$.
(e) $c_1 x + c_2 x^2 + \frac{1}{2}$.

3. (a) $\frac{1}{2}x^2 e^x$. (c) $(e^{-2x} + e^{-x}) \ln(1 + e^x)$.

(e) $\sin x \displaystyle\int \frac{e^x}{\cos x - \sin x}\,dx - e^x \int \frac{\sin x}{\cos x - \sin x}\,dx.$

4. **(a)** $e^{(t-x)} - e^{2(t-x)} = G(x, t).$ **(c)** $t(\ln x - \ln t) = G(x, t).$

5. **(a)** $-3e^{-2x} + 4e^{-x} + y_p(x)$, where $y_p(x) = 0$ if $x < 1$
and $y_p(x) = \frac{1}{2} + \frac{1}{2}e^{2(1-x)} - e^{(1-x)}$ if $1 \le x.$
(c) $x + \ln x.$

Section 3 (pp. 280–282)

1. **(a)** Unique solution.   **(c)** Infinitely many solutions.

3. **(a)** $G(x, t) = -\dfrac{1}{\sqrt{2}}(\cos \sqrt{2}\,x \sin \sqrt{2}\,x)$ if $0 \le t \le x,$

$\qquad G(x, t) = -\dfrac{1}{\sqrt{2}}(\sin \sqrt{2}\,x \cos \sqrt{2}\,t)$ if $x \le t \le \pi/(2\sqrt{2}).$

(c) $G(x, t) = -\dfrac{2}{\sqrt{3}}e^{-1/2\,(x-t)} \cos \dfrac{\sqrt{3}}{2}x \sin \dfrac{\sqrt{3}}{2}t$ if $0 \le t \le x,$

$\qquad G(x, t) = -\dfrac{2}{\sqrt{3}}e^{-1/2(x-t)} \sin \dfrac{\sqrt{3}}{2}x \cos \dfrac{\sqrt{3}}{2}t$ if $x \le t \le \pi/\sqrt{3}.$

4. **(a)** $\dfrac{x}{2} - \dfrac{\pi}{4\sqrt{2}} \sin \sqrt{2}\,x.$ **(c)** $\sin x - e^{1/2\,(\pi/\sqrt{3}-x)} \sin \dfrac{\pi}{\sqrt{3}} \sin \dfrac{\sqrt{3}}{2}x.$

5. **(a)** $\dfrac{x}{2} + \cos \sqrt{2}\,x - \dfrac{\pi}{4\sqrt{2}} \sin \sqrt{2}\,x.$

(c) $\sin x - e^{1/2\,x} \cos \dfrac{\sqrt{3}}{2}x - \sin \dfrac{\pi}{\sqrt{3}} e^{1/2\,(\pi/\sqrt{3}-x)} \sin \dfrac{\sqrt{3}}{2}x.$

6. $\alpha \cos \sqrt{2}\,x + c \sin \sqrt{2}\,x$ if $\alpha = \beta$, otherwise no solution. Choose $c = \dfrac{1}{\sqrt{2}}$ to make
$y'(0) = 1.$

7. $y_k = c_1 \cos \left(\dfrac{k\pi x}{b-a}\right) + c_2 \sin \left(\dfrac{k\pi x}{b-a}\right).$

8. **(c)** Pairs a, b such that $b - a = k\pi/\beta$, where $\beta = \sqrt{4mh - k^2}/(2m).$

10. **(a)** $\dfrac{RL^3}{24}x - \dfrac{RL}{12}x^3 + \dfrac{R}{24}x^4.$

(c) $\dfrac{RL^2}{4}x^2 - \dfrac{RL}{6}x^3 + \dfrac{R}{24}x^4.$

Section 4 (pp. 286–287)

1. **(a)** $z_3 = \dfrac{1-e}{1+e}$; y_3 agrees with the given solution.

(b) $G(x, t) = \dfrac{1}{2}(e^{(x-t)} - e^{-(x-t)}).$

2. **(a)** The denominator $y_1(1) - y_2(1)$ will be zero for every choice of $y_1(x)$ and $y_2(x).$
(b) $y(x) = 1 + c \sin x$, c constant.

3.
```
! SOLVES BOUNDARY-VALUE PROBLEM Y''=F(T,Y,Y') BY SHOOTING
! WITH Y(X0)=Y0 AND Y(X1)=Y1. PLOTS AFTER M STEPS, N TIMES.
! ZA AND ZB ARE THE FIRST TWO TRIAL VALUES FOR Y'(X0);
! TRY TO CHOOSE ZA,ZB SO THAT THE CORRESPONDING VALUES
! Y1A AND Y1B OF Y(X1) STRADDLE THE TARGET VALUE Y1.
! FOR MORE EFFICIENCY REFINE YOUR ESTIMATES FOR ZA,ZB.
! NOTE:SOMETIMES IT'S BETTER TO SHOOT FROM RIGHT TO LEFT.
DEF F(X,Y,Z)=-.05*X*Z-SIN(Y)
READ X0,X1  ! ENDPOINTS OF INTERVAL
DATA 0,3
READ Y0,Y1   ! VALUES OF Y(X0) AND Y(X1)
DATA 1,-1
READ ZA,ZB  ! TRIAL Y'(X0) VALUES
DATA -1,1
READ M,N
DATA 10,50
     ! 1ST SHOT: Y(X0)=Y0, Y'(X0)=ZA
LET D=1 ! D COUNTS THE SHOTS
LET X=X0
LET Y=Y0
LET Z=ZA
LET H=(X1-X0)/M/N
CALL SHOT
SUB SHOT      ! MODIFIED EULER METHOD
    FOR K=1 TO N
        FOR J=1 TO M
            LET P=Y+H*Z
            LET Q=Z+H*F(X,Y,Z)
            LET S=Y
            LET Y=Y+.5*H*(Z+Q)
            LET Z=Z+.5*H*(F(X,S,Z)+F(X+H,P,Q))
            LET X=X+H
        NEXT J
PRINT X,Y
   NEXT K
PRINT "END OF SHOT NO.";D
PRINT "WITH Y'(X0)=";Z
END SUB
LET Y1A=Y
LET D=2        ! Y(X0)=Y0, Y'(X0)=ZB
LET X=X0
LET Y=Y0
LET Z=ZB
CALL SHOT
LET Y1B=Y
DO
PRINT "SHOOT AGAIN";
INPUT R$
```

```
IF R$="YES" OR R$="yes" THEN
   LET D=D+1   ! Y(X0)=Y0, Y'(X0) INTERP.
   LET X=X0
   LET Y=Y0
   LET Z=ZB+(Y1−Y1B)*(ZA−ZB)/(Y1A−Y1B)
   LET ZL=Z ! LATEST Y'(X0)
 CALL SHOT
       LET Y1A=Y1B
       LET Y1B=Y
       LET ZA=ZB
       LET ZB=ZL
 END IF
 LOOP WHILE R$="YES" OR R$="yes"
 END
```

4. **(b)** Show that the homogeneous equation has arctan x and 1 for independent solutions; use these to form a Green's function. Note also that $\tan(\theta_1 - \theta_2) = (\tan \theta_1 - \tan \theta_2)/(1 + \tan \theta_1 \tan \theta_2)$.

Chapter 7

Section 1 (pp. 296–297)

1. **(a)** $x(t) = 1 + \int_0^t (ux(u) + u^2)\, du.$

2. **(a)** $x_1(t) = 1 + \frac{1}{2}t^2 + \frac{1}{3}t^3$, $x_2(t) = 1 + \frac{1}{2}t^2 + \frac{1}{3}t^3 + \frac{1}{6}t^4 + \frac{1}{12}t^5.$

3. **(b)** $x_1(t) = 1 + t^2$, $y_1(t) = 2 + \frac{1}{2}t^2.$

4. **(a)** $\dot{y} = x$, $y(0) = 0$ **(c)** $\begin{pmatrix} y_1 \\ x_1 \end{pmatrix} = \begin{pmatrix} t \\ 1 \end{pmatrix}$, $\begin{pmatrix} y_2 \\ x_2 \end{pmatrix} = \begin{pmatrix} t \\ 1 - \frac{1}{3}t^3 \end{pmatrix}.$

$\dot{x} = -ty$, $x(0) = 1.$

5. **(a)** $L = 2.$ **(c)** $L = 2.$

Section 2 (p. 302)

1. **(b)** $y = e^{x^2/2}.$ **(c)** $y = xe^{-x}.$

2. **(c)** Consider the equivalent system

$$\dot{y} = x, \; y(0) = 0$$
$$\dot{x} = -2x - y, \; x(0) = 1.$$

3. **(c)** Solutions that are never zero have the form $x = (\frac{1}{4}t^2 + c)^2$, $c > 0$; if $c \le 0$ solutions can be pieced together using this formula together with segments of the t-axis.

Section 3 (pp. 309–310)

1. **(a)** Independent.   **(d)** No conclusion possible. Why?

2. **(a), (c).**

3. **(a)** Independent.   **(c)** Independent.

4. (a), (c).

5. (a).

6. (a) If A has two equal columns, then $\det A = 0$.

   (b) $y'' + \dfrac{2}{1 + \tan t} y' - \dfrac{1 - \tan t}{1 + \tan t} y = 0.$

7. (a) $2e^{6t}$.

Chapter 8

Section 1 (pp. 314–316)

2. (a) $u(x, y) = G(x) + H(y)$, G and H differentiable.

   (b) $u(x, y) = xy + G(x) + H(y)$.

3. $u(x, y) = f(y)e^{-x} + g(y)xe^{-x} + y$, f and g twice differentiable.

5. (a) Elliptic.   (c) Hyperbolic.

Section 2 (p. 321)

1. (a) $\displaystyle\sum_{k=1}^{\infty} \frac{2(-1)^{k+1}}{k} \sin kx.$ (c) $\dfrac{\pi^2}{3} + \displaystyle\sum_{k=1}^{\infty} \frac{4(-1)^k}{k^2} \cos kx.$

   (e) $\dfrac{1}{2} + \displaystyle\sum_{k=1}^{\infty} \frac{1 + (-1)^{k+1}}{\pi k} \sin kx.$

4. (b) $\cos^3 x = \frac{1}{2}\cos x + \frac{1}{2}\cos x \cos 2x.$

Section 3 (pp. 328–329)

1. $\displaystyle\sum_{k=1}^{\infty} \frac{4(-1)^k}{k\pi} \sin \frac{k\pi}{2} x.$

2. $\dfrac{5}{2} - \dfrac{1}{\pi}\left(\sin 2\pi x + \dfrac{1}{2}\sin 4\pi x + \dfrac{1}{3}\sin 6\pi x + \cdots\right).$

3. (a) $\displaystyle\sum_{k=1}^{\infty} \frac{2[1 - (-1)^k]}{\pi k} \sin kx.$

   (c) $\dfrac{8}{\pi} \displaystyle\sum_{k=1}^{\infty} \frac{k}{(4k^2 - 1)} \sin 2kx.$

   (e) $-\displaystyle\sum_{k=1}^{\infty} \left(\frac{2(-1)^k}{k\pi} + \frac{9 \sin (2k\pi/3)}{k^2 \pi^2}\right) \sin \left(\frac{k\pi}{3}\right)x.$

4. (a) $1.$ (c) $\dfrac{2}{\pi} - \dfrac{4}{\pi} \displaystyle\sum_{k=1}^{\infty} \frac{1}{4k^2 - 1} \cos 2kx.$

   (e) $\dfrac{1}{6} + \displaystyle\sum_{k=1}^{\infty} \frac{6}{k^2 \pi^2} \left(\frac{k\pi}{3} \sin \frac{k\pi}{3} + \cos \frac{k\pi}{3} - 1\right) \cos \frac{k\pi}{3} x.$

5. (c) Complete expansion agrees with (b). Why?

8. If $f(x)$ is even, or odd,

$$\int_p^p f(x)\,dx = \int_{-p}^0 f(x)\,dx + \int_0^p f(x)\,dx$$

$$= \int_0^p f(-x)\,dx + \int_0^p f(x)\,dx.$$

Section 4A, B (pp. 334–335)

1. (b) $-\dfrac{8}{\pi}\displaystyle\sum_{k=1}^{\infty}\dfrac{1}{k^3}e^{-k^2a^2t}\sin kx.$

2. (a) $v(x) = x - 1.$ **(c)** $v(x) = x + 1.$

3. (a) $\left(1 + \dfrac{2}{p}x\right) + \left(\dfrac{2}{p} - \dfrac{8}{\pi}\right)\sin\dfrac{\pi x}{p} + \displaystyle\sum_{k=2}^{\infty}\dfrac{2}{k\pi}(3(-1)^k - 1)e^{-k^2a^2\pi^2t/p^2}\sin\dfrac{k\pi x}{p}.$

(c) $x + \displaystyle\sum_{k=1}^{\infty}\dfrac{2}{k\pi}e^{-4k^2a^2\pi^2t}\sin k\pi x.$

5. (a) $(c_1e^{r_1x} + c_2e^{r_2x})e^{-\lambda^2t},\ r_1, r_2 = \dfrac{1}{2}(-1 \pm \sqrt{1 - 4\lambda^2}).$

(c) $c_1x^{-\lambda^2}e^{-\lambda^2t/2}.$

7. (a) $v(x) = x^2 + 1.$

(c) $v(x) = \dfrac{1}{6}x^3 + \dfrac{5}{6}x + 1.$

Section 4C (pp. 341–342)

1. (a) $\cos(at)\sin x.$ **(c)** $\dfrac{2}{\pi a}\displaystyle\sum_{k=1}^{\infty}\left[\cos\dfrac{k\pi}{2} - (-1)^k\right]\sin(akt)\sin kx.$

2. (b) $v(x) = \dfrac{g}{2a^2}(px - x^2).$

5. Form products of solutions of the equations $a^2X'' + \lambda^2X = 0,\ T'' + kT' + \lambda^2T = 0.$

Section 5 (pp. 347–349)

3. 0.

4. (a) $\displaystyle\sum_{k=0}^{\infty}\dfrac{4(\sin h(2k + 1)\pi)^{-1}}{(2k + 1)\pi}\sin(2k + 1)\pi[\sin h(2k + 1)\pi(1 - y)$

$$+ \sin h(2k + 1)\pi y].$$

(c) $\displaystyle\sum_{k=0}^{\infty}\dfrac{4(\sin hk\pi)^{-1}}{k\pi}\sin k\pi x[\sin hk\pi(1 - y) + 2\sin hk\pi y].$

(e) Add four simple solutions.

6. Solve $u_{xx} = 0.$

Section 6A (pp. 353–354)

1. (b) $w(x) \le 0$ cannot hold on any subinterval of $a \le x \le b.$

5. (a) $T_0 = 1,\ T_1(x) = x,\ T_2(x) = 2x^2 - 1,\ T_3(x) = 4x^3 - 3x.$

Section 6B (pp. 357–359)

1. **(a)** $y_k = c \sin kx$, $k = 1, 2, 3, \cdots$.   **(b)** $y_k = c \cos kx$, $k = 1, 2, \cdots$.
   (c) $y_k = c \sin (k + \frac{1}{2})x$, $k = 0, 1, 2, \cdots$.

2. **(a)** $y_k = ce^{-x} \sin kx$, $k = 1, 2, 3, \cdots$.
   (c) $y_k = ce^{-x}[\sin kx + k \cos kx]$, $k = 1, 2, 3, \cdots$.

Chapter 9

Section 1 (pp. 366–369)

1. **(a)** $\dfrac{5}{6}$.   **(c)** $\dfrac{31}{16}$.   2. **(a)** $\dfrac{1}{k}$.   **(c)** $\left(-\dfrac{1}{2}\right)^{k-1}$.   **(e)** $\dfrac{k + 3}{(k + 1)2^k}$.

3. **(a)** $\dfrac{1}{4}$.   **(c)** Diverges to ∞.   **(e)** $\dfrac{1}{4}$.

4. **(a)** Convergent.   **(c)** Diverges to ∞.   5. **(a)** 1.

6. **(a)** $\dfrac{6}{5}$.   **(c)** $\dfrac{1}{\pi}$.   **(e)** $\dfrac{e^2}{e^2 - 1}$.

7. **(a)** $\dfrac{123}{999}$.   **(c)** $\dfrac{8}{9}$.   8. **(a)** $\displaystyle\sum_{k=1}^{\infty} \dfrac{3}{2^k}$.   **(c)** $\displaystyle\sum_{k=0}^{\infty} \dfrac{2^k}{3^{k+1}}$.

10. **(a)** $\displaystyle\sum_{k=1}^{\infty} \dfrac{3^{-k}}{k!}$.   **(c)** $\displaystyle\sum_{k=1}^{\infty} \dfrac{(2k)!}{2^{2k}(k!)^2}$ **(e)** $\displaystyle\sum_{k=1}^{\infty} \dfrac{1}{k!}$.

11. **(a)** 1.   **(c)** $\dfrac{3}{4}$.   **(e)** 3.   12. **(a)** 1.   **(c)** 2.

14. **(a)** $\displaystyle\sum_{k=1}^{\infty} \dfrac{1}{10^k} = -1 + \sum_{k=0}^{\infty} \left(\dfrac{1}{10}\right)^k = -1 + \dfrac{1}{1 - 1/10} = \dfrac{1}{9}$.

15. **(a)** $c_n = \left(\dfrac{2}{3}\right)^n$.

Section 2 (pp. 377–378)

1. **(a)** Diverges by term test.   **(c)** Converges by integral test.
   (e) Converges by integral test.

2. **(a)** Converges by ratio test.   **(c)** Converges by comparison.
   (e) Converges by comparison.

3. **(a)** Converges absolutely.   **(c)** Converges, but not absolutely.
   (e) Converges absolutely.

4. **(a)** Converges.   **(c)** Diverges.   **(e)** Converges absolutely.

6. **(a)** Converges.   **(c)** Converges.   **(e)** Converges.

7. **(a)** $-1 \le x < 1$.   **(c)** All real x.   **(e)** $-1 < x < 1$.   **(g)** $e^{-1} < x < e$.

9. **(a)** Use the ratio test.   **(c)** Use a geometric series.

11. Use the integral test.

Section 3 (pp. 384–386)

1. **(a)** $7 + 5(x - 2) + (x - 2)^2$.   **(c)** $2 - 2(x + 1) + (x + 1)^2$.

2. **(a)** $R_3(x) = \dfrac{1}{81 c^{8/3}} (x - 1)^3$, c between x and 1.

   (c) $R_3(x) = -\dfrac{1}{c^4} (x + 1)^3$, c between x and 1.

3. **(a)** $\displaystyle\sum_{k=0}^{\infty} \frac{(-1)^k}{k!} x^k$.   **(c)** $\displaystyle\sum_{k=0}^{\infty} \frac{(-1)^k}{(2k + 1)! \, 2^{2k+1}} x^{2k+1}$.

4. **(a)** $e^x + e^{-x} = \displaystyle\sum_{k=0}^{\infty} \frac{1 + (-1)^k}{k!} x^k = 2 \sum_{j=0}^{\infty} \frac{1}{(2j)!} x^{2j}$.

5. Note that the kth derivative of $(x + a)^n$ is $\dfrac{n!}{(n - k)!} (x + a)^{n-k}$ for $0 \le k \le n$, and is

   zero for $n < k$.

6. **(a)** $|R_2(x)| \le \dfrac{1}{6} |x|^3$.   **(c)** $|R_3(x)| \le \dfrac{1}{24} |x|^4$.

11. $\ln 2 = -\ln \left(\dfrac{1}{2} \right) = -\ln \left(1 - \dfrac{1}{2} \right)$.

13. Set $x = \pi/2$ in an expansion of $\cos x$.

Section 4 (pp. 392–393)

1. Apply the Weierstrass test using $\displaystyle\sum_{k=0}^{\infty} d^k$.

2. Apply the Weierstrass test using $\displaystyle\sum_{k=1}^{\infty} k^{-2}$.

5. Use Theorem 4.1 and Exercise 3(b).

6. Show that $\displaystyle\sum_{k=0}^{\infty} x^k - \sum_{k=0}^{n} x^k = \frac{x^{n+1}}{1 - x}$.

7. Show that the first series converges to $(x^2 - x)/(1 + x)$.

Section 5 (pp. 404–405)

1. **(a)** Converges for $-1 \le x < 1$.   **(c)** Converges for all x.
   (e) Converges for $-3 < x < -1$.   **(g)** Converges for $-4 \le x < -2$.

2. **(a)** $\displaystyle\sum_{k=0}^{\infty} \frac{(-1)^k}{k!} x^k$.   **(c)** $\displaystyle\sum_{k=0}^{\infty} \frac{x^{2k+1}}{(2k + 1)!}$.   **(e)** $\displaystyle\sum_{k=0}^{\infty} \frac{x^{2k}}{k!}$.

3. **(a)** $\displaystyle\sum_{k=1}^{\infty} \frac{(-1)^{k+1}}{k} x^k$.   **(c)** $\displaystyle\sum_{k=1}^{\infty} \frac{(-1)^{k+1}}{k} x^{2k}$.

4. **(a)** $\displaystyle\sum_{k=0}^{\infty} (-1)^k x^{2k}$.   **(c)** $\displaystyle\sum_{k=0}^{\infty} (-1)^k (x - 1)^k$.

6. $\displaystyle\sum_{k=0}^{\infty} (-1)^{k+1} (k + 1) x^{2k}$.   8. **(a)** 1.   **(c)** $-\dfrac{1}{2}$.

9. Replace x by $1/x$ in the series $\displaystyle\sum_{k=0}^{\infty} x^k$.

Section 6 answers are in the text.

Section 7 (pp. 412–413)

1. Let $y = \displaystyle\sum_{k=0}^{\infty} c_k x^k$.   **2.** If $y = f(-x)$, then $y'' = f''(-x)$.

3. **(a)** $c_0 \displaystyle\sum_{k=0}^{\infty} \frac{(-1)^k}{k!} x^{2k}$.   **(b)** ce^{-x^2}.

4. **(a)** $c_0 + c_1 \displaystyle\sum_{k=0}^{\infty} \frac{1}{(2k+1)k!\, 2^k} x^{2k+1}$.   **(b)** $c_0 + c_1 \displaystyle\int e^{x^2/2} dx$.

5. **(a)** $y_0(x) = \displaystyle\sum_{k=0}^{\infty} \frac{x^{2k}}{2^k\, k!}$, $y_1(x) = \displaystyle\sum_{k=0}^{\infty} \frac{(-1)^k x^{2k+1}}{(2k+1)(2k-1)\cdots 5\cdot 3}$.

   (b) $y_0 = e^{x^2/2}$.

6. **(a)** $y_p(x) = \dfrac{1}{2} x$ is a solution.

   (b) Use the solutions found in Exercise 5.

Index